Devonian Events and Correlations

The Geological Society of London
Books Editorial Committee

Chief Editor

Bob Pankhurst (UK)

Society Books Editors

John Gregory (UK)
Jim Griffiths (UK)
John Howe (UK)
Phil Leat (UK)
Nick Robins (UK)
Jonathan Turner (UK)

Society Books Advisors

Mike Brown (USA)
Eric Buffetaut (France)
Reto Gieré (Germany)
Jon Gluyas (UK)
Doug Stead (Canada)
Randell Stephenson (The Netherlands)

IUGS/GSL publishing agreement

This volume is published under an agreement between the International Union of Geological Sciences and the Geological Society of London and arises from the International Commission on Stratigraphy's Subcommission on Devonian Stratigraphy.

GSL is the publisher of choice for books related to IUGS activities, and the IUGS receives a royalty for all books published under this agreement.

Books published under this agreement are subject to the Society's standard rigorous proposal and manuscript review procedures.

It is recommended that reference to all or part of this book should be made in one of the following ways:

Becker, R. T. & Kirchgasser, W. T. (eds) 2007. *Devonian Events and Correlations*. Geological Society, London, Special Publications, **278**.

Brice, D., Legrand-Blain, M. & Nicollin, J. P. (2007) Brachiopod faunal changes across the Devonian–Carboniferous boundary in NW Sahara (Morocco, Algeria). *In*: Becker, R. T. & Kirchgasser, W. T. (eds) *Devonian Events and Correlations*. Geological Society, London, Special Publications, **278**, 261–271.

GEOLOGICAL SOCIETY SPECIAL PUBLICATION NO. 278

Devonian Events and Correlations

EDITED BY

R. T. BECKER
Westfälische Wilhelms-Universität, Germany

AND

W. T. KIRCHGASSER
SUNY Potsdam, USA

2007
Published by
The Geological Society
London

THE GEOLOGICAL SOCIETY

The Geological Society of London (GSL) was founded in 1807. It is the oldest national geological society in the world and the largest in Europe. It was incorporated under Royal Charter in 1825 and is Registered Charity 210161.

The Society is the UK national learned and professional society for geology with a worldwide Fellowship (FGS) of over 9000. The Society has the power to confer Chartered status on suitably qualified Fellows, and about 2000 of the Fellowship carry the title (CGeol). Chartered Geologists may also obtain the equivalent European title, European Geologist (EurGeol). One fifth of the Society's fellowship resides outside the UK. To find out more about the Society, log on to www.geolsoc.org.uk.

The Geological Society Publishing House (Bath, UK) produces the Society's international journals and books, and acts as European distributor for selected publications of the American Association of Petroleum Geologists (AAPG), the Indonesian Petroleum Association (IPA), the Geological Society of America (GSA), the Society for Sedimentary Geology (SEPM) and the Geologists' Association (GA). Joint marketing agreements ensure that GSL Fellows may purchase these societies' publications at a discount. Hard copies of books can be purchased via the Society's online bookshop at www.geolsoc.org.uk. Electronic versions of key Society book series are available via it Lyell Collection at www.lyellcollection.org.

To find out about joining the Society and benefiting from substantial discounts on publications of GSL and other societies worldwide, consult www.geolsoc.org.uk, or contact the Fellowship Department at: The Geological Society, Burlington House, Piccadilly, London W1J 0BG: Tel. +44 (0)20 7434 9944; Fax +44 (0)20 7439 8975; E-mail: enquiries@geolsoc.org.uk.

For information about the Society's meetings, consult *Events* on www.geolsoc.org.uk. To find out more about the Society's Corporate Affiliates Scheme, write to enquiries@geolsoc.org.uk.

Published by The Geological Society from:
The Geological Society Publishing House, Unit 7, Brassmill Enterprise Centre, Brassmill Lane, Bath BA1 3JN, UK

(*Orders*: Tel. +44 (0)1225 445046, Fax +44 (0)1225 442836)
Online bookshop: www.geolsoc.org.uk/bookshop

The publishers make no representation, express or implied, with regard to the accuracy of the information contained in this book and cannot accept any legal responsibility for any errors or omissions that may be made.

© The Geological Society of London 2007. All rights reserved. No reproduction, copy or transmission of this publication may be made without written permission. No paragraph of this publication may be reproduced, copied or transmitted save with the provisions of the Copyright Licensing Agency, 90 Tottenham Court Road, London W1P 9HE. Users registered with the Copyright Clearance Center, 27 Congress Street, Salem, MA 01970, USA: the item-fee code for this publication is 0305-8719/07/$15.00.

British Library Cataloguing in Publication Data

A catalogue record for this book is available from the British Library.

ISBN 978-1-86239-222-9

Typeset by Techset, Salisbury, UK

Printed by Cromwell Press, Trowbridge, UK.

Distributors

North America
For trade and institutional orders:
The Geological Society, c/o AIDC, 82 Winter Sport Lane, Williston, VT 05495, USA
Orders: Tel. +1 800-972-9892
            Fax +1 802-864-7626
            E-mail: gsl.orders@aidcvt.com

For individual and corporate orders:
AAPG Bookstore, PO Box 979, Tulsa, OK 74101-0979, USA
Orders: Tel. +1 918-584-2555
            Fax +1 918-560-2652
            E-mail: bookstore@aapg.org
            Website http://bookstore.aapg.org

India
Affiliated East-West Press Private Ltd, Marketing Division, G-1/16 Ansari Road, Darya Ganj, New Delhi 110 002, India
Orders: Tel. +91 11 2327-9113/2326-4180
            Fax +91 11 2326-0538

Contents

BECKER, R. T. & KIRCHGASSER, W. T. Devonian events and correlations—a tribute to the lifetime achievements of Michael Robert House (1930–2002) — 1

JANSEN, U., LAZREQ, N., PLODOWSKI, G., SCHEMM-GREGORY, M., SCHINDLER, E. & WEDDIGE, K. Neritic-pelagic correlation in the Lower and basal Middle Devonian of the Dra Valley (Southern Anti-Atlas, Moroccan Pre-Sahara) — 9

VER STRAETEN, C. A. Basinwide stratigraphic synthesis and sequence stratigraphy, upper Pragian, Emsian and Eifelian stages (Lower to Middle Devonian), Appalachian Basin — 39

DESANTIS, M. K., BRETT, C. E. & VER STRAETEN, C. A. Persistent depositional sequences and bioevents in the Eifelian (early Middle Devonian) of eastern Laurentia: North American evidence of the Kačák Events? — 83

BARTHOLOMEW, A. J. & BRETT, C. E. Correlation of Middle Devonian Hamilton Group-equivalent strata in east-central North America: implications for eustasy, tectonics, and faunal provinciality — 105

MARSHALL, J. E. A., ASTIN, T. R., BROWN, J. F., MARK-KURIK, E. & LAZAUSKIENE, J. Recognizing the Kačák Event in the Devonian terrestrial environment and its implications for understanding land–sea interactions — 133

EBBIGHAUSEN, V., BECKER, R. T., BOCKWINKEL, J. & ABOUSSALAM, Z. S. Givetian (Middle Devonian) brachiopod–goniatite–correlation in the Dra Valley (Anti-Atlas, Morocco) and Bergisch Gladbach-Paffrath Syncline (Rhenish Massif, Germany) — 157

HARTKOPF-FRÖDER, C., KLOPPISCH, M., MANN, U., NEUMANN-MAHLKAU, P., SCHAEFER, R. G. & WILKES, H. The end-Frasnian mass extinction in the Eifel Mountains, Germany: new insights from organic matter composition and preservation — 173

RIQUIER, L., AVERBUCH, O., TRIBOVILLARD, N., EL ALBANI, A., LAZREQ, N. & CHAKIRI, S. Environmental changes at the Frasnian–Famennian boundary in Central Morocco (Northern Gondwana): integrated rock-magnetic and geochemical studies — 197

BLIECK, A., CLEMENT, G., BLOM, H., LELIEVRE, H., LUKSEVICS, E., STREEL, M., THOREZ, J. & YOUNG, G. C. The biostratigraphical and palaeogeographical framework of the earliest diversification of tetrapods (Late Devonian) — 219

KAISER, S. I., BECKER, R. T. & EL HASSANI. A. Middle to Late Famennian successions at Ain Jemaa (Moroccan Meseta)—implications for regional correlation, event stratigraphy and synsedimentary tectonics of NW Gondwana — 237

BRICE, D., LEGRAND-BLAIN, M. & NICOLLIN, J.-P. Brachiopod faunal changes across the Devonian–Carboniferous boundary in NW Sahara (Morocco, Algeria) — 261

Index — 273

Devonian events and correlations—a tribute to the lifetime achievements of Michael Robert House (1930–2002)

R. T. BECKER[1] & W. T. KIRCHGASSER[2]

[1]*Westfälische Wilhelms-Universität, Geologisch-Paläontologisches Institut, Corrensstr. 24, D-48149 Münster, Germany (e-mail: rbecker@uni-muenster.de)*

[2]*Potsdam, New York State, USA (e-mail: kirchgwt@potsdam.edu)*

The Devonian was a peculiar time in the Phanerozoic evolution of the Earth. Most continents, including the large Gondwana and Laurussia cratons, formed a Pangea-type assembly around the tropical Prototethys and an increasingly hot, global, greenhouse climate prevailed, with a complete lack of major ice sheets, even in polar areas. There was gradual and increasing flooding of the continents, creating huge epicontinental seas that have no modern analogues. Under these conditions the plants finally conquered the land, with the innovation of deep roots in the Emsian, the appearance of seed precursors and trees in the Givetian, and the spread of vegetation into dry uplands in the late Famennian. In the marine realm, the largest-known Phanerozoic tropical reef belts surrounded craton margins and tropical islands. It was the time of the sudden radiation of early ammonoids, of the earliest episodic blooms of calcareous-shelled, pelagic zooplankton (tentaculitoids), the rise to dominance of fishes, mostly of armoured forms and with giants reaching 10 m in length, but also including the first sharks, and the appearance of earliest tetrapods in marginal settings. However, the tropical and subtropical areas reaching up to 45° latitude were hardly a paradise. A combination of climatic, plate tectonic/magmatic and still poorly understood palaeoceanographic factors caused the recurrent sudden perturbation of stable ecological conditions by short-term global events of variable magnitude (e.g. House 1985), including two of the biggest mass extinctions that the Earth's biosphere has experienced—the Upper Kellwasser Event at the Frasnian–Famennian boundary and the Hangenberg Events at the close of the Devonian.

The global and regional correlation of Devonian rocks relies strongly on the available fossil record, on the facies distribution of marker fossils, on the recognition of eustatic sea-level change in sequence stratigraphy, and, as a relatively new development, on the application of stable isotope geochemistry and magnetic susceptibility. The Prototethys seaway, fortunately, allowed a free exchange of pelagic faunas through all oceans and seas. In the warm-water areas, there was only moderate endemism in conodonts, ammonoids, pelagic ostracodes and tentaculitoids, which are the main biostratigraphic marker groups. Low diversity benthic assemblages of the outer shelf also include many cosmopolitan groups but they differ strongly from fossil associations of the nearshore, neritic and photic zones. In the latter areas, including reef and biostrome complexes, the evolution and ecologically controlled distribution of brachiopods, stromatoporoids, rugose and tabulate corals, and trilobites provides refined regional time frameworks but the faunas are mostly characterized by endemism that prevents simple correlation across oceans and seaways. Palynomorphs that were washed into the seas enable the correlation of marine, fluvial, limnic and terrestrial deposits.

The highly variable palaeoecology and taxonomic composition of the terrestrial, neritic and pelagic faunal assemblages are the fundamental problems of cross-facies correlations within the Devonian. Apart from overlapping facies and palaeobiogeographic ranges of some key taxa, event and physical stratigraphy are the major tools in reconstructing global changes in the Devonian world at high level of time resolution. The International Subcommission on Devonian Stratigraphy (SDS) was the first International Union of Geological Sciences (IUGS) Subcommission to complete the formal designation of all its series and stage subdivisions, but all ratified GSSPs (Global Stratotype Sections and Points), as the norm, were defined in the pelagic facies realm. Their recognition on all continents and in all other facies belts is still a major task to be resolved. As a first step to reach some progress in this wide, international and multidisciplinary scientific field, the Devonian Subcommission and the Institut Scientifique University Mohammed V, Rabat, especially with a major organizational input by its Titular Member and Vice-Chairman, Prof. Ahmed El Hassani, organized in March 2004 an international symposium (El Hassani 2004*a*) on 'Devonian neritic–pelagic correlation and events', followed by an excursion with the same topic to the continually inaccessible Dra Valley of SW Morocco (El Hassani 2004*b*).

From: BECKER, R. T. & KIRCHGASSER, W. T. (eds) *Devonian Events and Correlations.*
Geological Society, London, Special Publications, 278, 1–8. DOI: 10.1144/SP278.1
0305-8719/07/$15 © The Geological Society of London 2007.

The Rabat meeting and this Volume honour the spirit and continue the major scientific interest of one of the most prominent Devonian stratigraphers of the last five decades, Prof. Michael Robert House (Figs 1–3), who passed away on 6 August 2002. Consequently, the proceedings of the Rabat meeting and this volume are dedicated to his memorial and in honour of his outstanding lifetime achievements in Devonian stratigraphy. Michael, as most Devonian workers knew him, not only promoted significantly and at an early stage the international approach of stratigraphic correlation, and recognized the complex sequence of global events and extinctions in the Emsian to Famennian, he also had a strong focus on Morocco and would have loved to take part in the Rabat meeting and the Dra Valley excursion, an area that he visited briefly during one of the excursions organized by the famous Henry Hollard.

Michael R. House was born on 27 August 1930 in Blandford Forum, Dorset, and his family soon moved to the Weymouth area where he developed an early interest in Jurassic fossils, including ammonites, that he could collect close to his home. His geography teacher at school introduced him to the Geological Survey Memoir on the area of the Dorset coast by the famous W. J. Arkell, which broadened his geological interests. This wide approach, from palaeontology to sedimentology, oil geology, cartography and tectonics, and his sense of the broad context of all observations, characterized his subsequent work and allowed him to remain innovative and inspiring to his students and colleagues. After army service in Reading, Michael went up to Cambridge in October 1951 and was soon introduced to W. J. Arkell himself who provided the best support and encouragement that an undergraduate student could get. Both W. B. R. King, then head of the department, and Arkell did not accept the logical idea that Michael should work on Jurassic ammonites, but, to the fortune of the Devonian community, suggested instead that he explore a much less-known scientific field at that time, the Devonian of SW England, whose stratigraphy and ammonoid faunas had been badly neglected since the beginning of the 20th century. Michael was offered a position as Lecturer at Durham University before finishing his final examination. Although his first two publications (House 1955, 1956a) dealt with the Mesozoic of his Dorset home area, his Devonian PhD studies soon led to discoveries that allowed him to publish on Devonian goniatites from North Cornwall (House 1956b). The love for Devonian ammonoids never left him and made him eventually one of the very few world-leading authorities on this fossil group (House 1970, 1971a, 1979, 1981a, b, 1988a, b, 1989c, 1993a, b; Becker & House 2000a). They proved to be a key for global correlations and understanding the causes of rapid and severe evolutionary change. Early significant contributions included the discovery of *Wocklumeria* faunas in North Cornwall (House & Selwood 1957), Upper Devonian ammonoids in the Dartmoor area (House 1959), and palaeopathological observations (House 1960a), now referred to as 'Housean pits', in Lower and Middle Devonian goniatites.

For two years (1958 and 1959), as a Commonwealth Fund Fellow, Michael crossed the Atlantic, at that time still by ship, to work at Harvard, Cornell and the Smithsonian Institution (Washington) on North American Devonian ammonoids. At Cornell, he was well received and supported by John W. Wells and his 'Friends of the Devonian'. Despite the long research history in New York State and adjacent states, he soon doubled the number of regionally known genera (House 1962), introduced detailed morphometry to the taxonomy of specific groups (House 1965b), straightened the myth of alleged Frasnian clymenids (House 1960b), extended his studies to Canada (House & Pedder 1963) and Nevada (House 1965a), and based on his extensive knowledge of European faunas, was able to correlate faunas across the two continents (House 1962, 1967, 1976). The fact that he was given an international research perspective early in his scientific career influenced significantly his subsequent work. He considered the

Fig. 1. Michael Robert House (1930–2002) in his Hull office.

Fig. 2. Michael R. House during field work in NW Australia (Summer 1990), overlooking the famous marginal slope deposits of the Frasnian reef at Windjana Gorge; with W. T. Kirchgasser, P. E. Playford and G. Klapper in the middle ground (from left to right).

whole global Devonian world in terms of biogeography, plate tectonics, palaeoecology and climate after his return to England (Friend & House 1964; House 1964a, 1971b, 1973a, b, 1974, 1975a, b), where he was appointed Lecturer in Palaeontology in Oxford (1963) and later offered the Chair of Geology at Hull (1967). Studies on North American goniatites, however, continued throughout his life (House 1965a, 1978, 1981c; Kirchgasser & House 1981; House & Blodgett 1982; House et al. 1986; House & Kirchgasser 1993) and, with the almost finished 'Opus' on the Frasnian of New York, even until his untimely death (House & Kirchgasser in press).

Fig. 3. A typical field photo of Michael R. House (from October 1997), well camouflaged by field gear, in one of his favourite areas of North Devon, showing the Upper Famennian 'Grand Slump' at the top of the Baggy Sandstone.

At Durham, Oxford (partly as Dean of St. Peter's College) and Hull, Michael continued his work on the Devonian of Devon and Cornwall, which lead to a series of papers by himself or with co-authors (House & Butcher 1962, 1973; House 1963, 1964b, 1981d; House & Selwood 1964; Hendricks et al. 1971; Gauss & House 1972; House et al. 1978a, b). His enthusiasm, energy, outstanding teaching and administrative abilities earned him the highest reputation. He was active in the Yorkshire Geological Society (President from 1972–1974), the Geological Society of London (council member from 1966–1968, Chair of the stratigraphy committee 1989–1992), the Palaeontological Association (President from 1972–1974), the Systematics Association (President from 1978–1981), the Hull Geological Society, the Geologists Association (Vice-President from 1970–1973), the Association for the Advancement of Science (President of Section C in 1977), and the Palaeontographical Society (President from 1989–1994). He was a founding member of the Ussher Society in 1962 and donated many significant papers on Devonian stratigraphy to its journal, which he sometimes questioned since these papers became not as widely known as he wished. His output of papers was so high that almost nobody realized that important parts of his PhD thesis were never published. As if he unconsciously felt that something was wrong, two important manuscripts on Givetian and Frasnian goniatites from SW England were eventually finished and submitted (House 2003a, b) before he learnt of his illness. The Devonian community also may not have recognized that Michael continued to write about his Dorset home area, that he was also heavily involved in the mapping of Malta and Gozo (lacking any Palaeozoic rocks), and that he was interested in topics such as growth banding in bivalves and the post-mortal drift of Nautilus.

The Devonian remained his first interest, which led him to organize an international Devonian Symposium in England in 1978, resulting in the now-famous volume on 'The Devonian System' (House et al. 1979). It provided the only up-to-date and comprehensive review of Devonian biostratigraphy and facies at its time. From the beginning (1975) of activities concerning the subdivision of the Devonian into formally defined series and stages, Michael joined the SDS as a most active Corresponding and Titular Member, Secretary and, from 1992–1996, as Chairman. Over the same period he was deeply involved with the task of the International Working Group on the Devonian–Carboniferous boundary (e.g. House & Sevastopulo 1984; Price & House 1984). His research and leadership were crucial to define the Middle/Upper Devonian series boundary (House 1982; Klapper et al. 1987; House et al. 2000a), the base of the Famennian (e.g. Becker et al. 1988; Becker & House 1990; Klapper et al. 1993; House et al. 2000b), the base of the Givetian (Walliser et al. 1995), and the base of the Emsian (Yolkin et al. 1998, 2000). His major aim was to conclude the definition of all Devonian time units whilst he was chairing the SDS. He provided chronostratigraphic summaries (House 1989a; House & Gradstein, 2005) and the correlation of British successions with the newly established series and stages (House & Dineley 1985; Bluck et al. 1989; Marshall & House 2000; House 2001).

Michael's scientific authority and international reputation were recognized by receiving a range of honorary medals from many geological societies: the William Bolitho Gold Medal of the Royal Geological Society of Cornwall (1970), the Neville George Medal of the Geological Society of Glasgow (1984), the Sorby Medal of the Yorkshire Geological Society (1985) and, outstandingly, the Murchison Medal of the Geological Society (1991). The international search for boundaries and stratotypes required global investigations, correlation, communication and a lot of travel—something that Michael most appreciated, especially the joint field work with friends (Fig. 2) and the international symposia that over time formed a circle of closely befriended specialists. His interests in global Devonian ammonoid faunas led him to start research projects in the Carnic Alps (House & Price 1980, with most data still unpublished), in the Montagne Noire (House & Price 1985; House et al. 1985; Becker & House 1990, 1994a), in Germany (House & Ziegler 1977), in the Tafilalt and Maider of southern Morocco (e.g. Becker et al. 1988; Becker & House 1991, 1994b; Becker et al. 2002), in the Moroccan Meseta (Becker & House 2000b), in the Timan of polar Russia (Becker et al. 2000; House et al. 2000c; Menner et al. 2000), and in the isolated Kimberleys of Western Australia (Fig. 2), where co-operation with Phil Playford (Perth) enabled the collection of huge faunas (Becker et al. 1991, 1993; Becker & House 1997, 2007 in prep.) that are still only partly published. Apart from these regions, Michael has visited most other important Devonian basins, at least those with a significant goniatite record, leaving out only politically too difficult regions, such as Central Kazakhstan, Novaya Zemlya and southern Algeria. Although being himself a perfect committee leader, his distaste for the interference of politics on science was most seriously confirmed when the Thatcher Government decided to re-organize the geological sciences of England. This resulted in the closure of his Institute at Hull, where he was Head of Department, Dean of Science (1976–1978) and Pro-Vice-Chancellor (1980–1983). The forced move to the University of Southampton in 1988 brought him closer to his Dorset home area but left some bitterness concerning the British science system.

Two scientific ideas and important developments additionally shaped Michael's research from the 1980s on. The recognition of major extinction events, possibly or partly caused by impact events, but more usually linked with the global punctuation of sedimentary sequences; and the impact of orbital cycles on sedimentation and faunas. Together with O. H. Walliser from Göttingen, Michael was at the forefront to recognize a complex succession of Devonian extinction events (House 1983, 1985) and recognized that ammonoids, as a type of 'bioseismometer', were especially sensitive to global environmental change. Several studies dealt with the two mass extinctions known as the Kellwasser (Becker *et al.* 1989, 1991; Becker & House 1990, 1994*a*; House 1997) and Hangenberg Events (Price & House 1984; House 1992, 1996*b*) but he gave more emphasis to the complete event set (House 1987*a*, *b*, 1988*b*, 1989*b*, *c*, 1993*a*, 1996*a*, *c*, 1998, 2002; Racki & House 2002) with many shared similarities between events that require a uniform causation model and that contradicted the simple impact scenarios of other contemporaneous authors. The links between biodiversity and eustatic sea-level change were recognized to be strong and he was the first to propose a Devonian global eustatic chart (House 1983). Since this was published in the Proceedings of the Ussher Society, the subsequent Johnson *et al.* (1985) sea-level curve became much more well-known. The interest in global sea-level change and, therefore, in sequence stratigraphy, fuelled his field work as much as his enthusiasm for goniatites and led to a series of publications (e.g. Becker *et al.* 1993; House & Kirchgasser 1993; Becker & House 1997; House & Ziegler 1997; House *et al.* 2000*c*). The interest in cyclic stratigraphy arose first from his Jurassic observations (House 1986) and he realized the enormous potential for a much more refined, extremely precise future dating method. His attempts to interpret Devonian successions in terms of Milankovitch cyclicity (House 1991, 1995) have to be seen as pioneering work that left a wide-open field for future research and with the highest potential for correlations at a scale much beyond our current time resolution.

After retiring from Southampton University (Fig. 3) to a new home in Weymouth on the Dorset coast, Michael's scientific productivity never slowed down. He returned to writing on Dorset Geology, but was actively continuing the publication of the wealth of ammonoid material from Morocco, New York (House & Kirchgasser in press), and the Canning Basin (Becker & House in press), as well as revising his planned chapter on Devonian Goniatitida for the Treatise on Invertebrate Paleontology. It is remarkable that his enormous achievements in Devonian palaeontology and stratigraphy and huge amounts of other academic work did not prevent him from being truly a family man, raising two children, Sue and Jim. Of course, this would have been impossible without the permanent and highest support of his wife Felicity, better known as Flick to those students and friends who had the pleasure to stay at one of the homes of the House family.

The Volume presented here includes significant contributions of the Rabat meeting as well as some papers invited by the editors because they fit the aim of the current book, which is to continue the work and ideas of M. R. House and broaden its approach. Contributions cover pelagic, neritic, and terrestrial environments (**Blieck *et al.***; **Marshall *et al.***) and all of the Devonian System. They are presented roughly in ascending stratigraphical order. There are case studies on rare neritic fossils found in pelagic settings, and *vice versa* (**Ebbighausen *et al.***), on specific short-term global events (**Marshall *et al.***), including the main mass extinction periods near the Frasnian-Famennian (**Riquier *et al.***; **Hartkopf-Fröder *et al.***) and Devonian–Carboniferous boundary (**Brice *et al.***), on geochemistry, magnetostratigraphy and on the Eovariscan tectonic influence on the correlation of regional sequences (**Kaiser *et al.***). Since the meeting took place in Morocco, it is logical that several papers deal with Moroccan successions (**Jansen *et al.***; **Kaiser *et al.***; **Brice *et al.***), but all contributions aim at global correlation or at the regional application of established global time definitions. Three papers cover the wide Lower and Middle Devonian of one of the most classical Devonian areas—eastern North America (**DeSantis *et al.*;** **Bartholomew & Brett**; **Ver Straeten *et al.***)—where incredible stratigraphic progress has been enabled by an holistic approach combining lithostratigraphy, bentonite stratigraphy, biostratigraphy and sequence stratigraphy. Other investigated areas lie in Germany, France, the north of Brittany, and Algeria. Each of the papers offers important progress in its region, facies or part of the stratigraphic column and can be taken as an example of how research will have to continue. Devonian stratigraphy has progressed to such an extent in recent time that there is no further excuse to look at any basin only in terms of its regional geological context. It is the aim of the international community of Devonian stratigraphers to reconstruct the whole Devonian world, its ecosystems and sedimentary history at the finest available global timescale and, herewith, we offer several steps towards this goal.

References

BECKER, R. T. & HOUSE, M. R. 1990. The Montagne Noire goniatite record around the Frasnian/Famennian boundary. *Document submitted to the International Subcommission on Devonian Stratigraphy, Frankfurt 1990*, 1–21.

BECKER, R. T. & HOUSE, M. R. 1991. Eifelian to Givetian goniatites at Bou Tchrafine and Jebel Amelane (Anti-Atlas, Morocco). In: WALLISER, O. H. (ed.) *Moroccan Field Meeting of the Subcommission on Devonian Stratigraphy*. IUGS, Göttingen, 59–73.

BECKER, R. T. & HOUSE, M. R. 1994a. Kellwasser Events and goniatite successions in the Montagne Noire with comments on possible causations. *Courier Forschungsinstitut Senckenberg*, **169**, 45–77.

BECKER, R. T. & HOUSE, M. R. 1994b. International Devonian goniatite zonation, Emsian to Givetian, with new records from Morocco. *Courier Forschungsinstitut Senckenberg*, **169**, 79–132.

BECKER, R. T. & HOUSE, M. R. 1997. Sea-level changes in the Upper Devonian of the Canning Basin. *Courier Forschungsinstitut Senckenberg*, **199**, 129–146.

BECKER, R. T. & HOUSE, M. R. 2000a. Devonian ammonoid zones and their correlation with established stage boundaries. *Courier Forschungsinstitut Senckenberg*, **220**, 113–151.

BECKER, R. T. & HOUSE, M. R. 2000b. Sedimentary and faunal succession of the allochthonous Upper Devonian at Gara d'Mrirt (Eastern Moroccan Meseta). *Notes et Mémoires du Service Géologique*, **399**, 49–56.

BECKER, R. T. & HOUSE, M. R. 2007. Devonian ammonoid biostratigraphy of the Canning Basin. *Geological Survey of Western Australia, Bulletin*, in press.

BECKER, R. T., FEIST, R., HOUSE, M. R. & KLAPPER, G. 1989. Frasnian–Famennian extinction events in the Devonian at Coumiac, southern France. *Compte Rendu, Academie des Sciences, Paris Série II*, **309**, 259–266.

BECKER, R. T., HOUSE, M. R. & ASHOURI, A.-R. 1988. Potential stratotype section for the Frasnian/Famennian boundary at El Atrous, Tafilalt, Morocco. *Document submitted to the International Subcommission on Devonian Stratigraphy, Rennes, August 1988*, 1–9.

BECKER, R. T., HOUSE, M. R., KIRCHGASSER, W. T. & PLAYFORD, P. E. 1991. Sedimentary and faunal changes across the Frasnian/Famennian boundary in the Canning Basin of Western Australia. *Historical Biology*, **5**, 183–196.

BECKER, R. T., HOUSE, M. R. & KIRCHGASSER, W. T. 1993. Devonian goniatite biostratigraphy and timing of facies movements in the Frasnian of the Canning Basin. In: HAILWOOD, E. A. & KIDD, R. B. (eds) *High Resolution Stratigraphy*. Geological Society, Special Publications, **70**, 293–321.

BECKER, R. T., HOUSE, M. R., MENNER, V. V. & OVNATANOVA, N. S. 2000. Revision of ammonoid biostratigraphy in the Frasnian (Upper Devonian) of the Southern Timan. *Acta Geologica Polonica*, **50**, 67–97.

BECKER, R. T., HOUSE, M. R., BOCKWINKEL, J., EBBIGHAUSEN, V. & ABOUSSALAM, Z. S. 2002. Famennian Ammonoid Zones of the Eastern Anti-Atlas (southern Morocco). *Münster'sche Forschungen zur Geologie und Paläontologie*, **93**, 159–205.

BLUCK, B. J., HAUGHTON, P. D. H., HOUSE, M. R., SELWOOD, E. B. & TURNBRIDGE, I. P. 1989. Devonian of England, Wales and Scotland. *Canadian Society of Petroleum Geologists, Memoir*, **14**(I), 305–324.

EL HASSANI, A. (ed.) 2004a. Devonian neritic–pelagic correlation and events. *International Meeting on Stratigraphy, Rabat, March 1–10, 2004, Oral Presentations and Posters*. 1–87, Institut Scientifique, University Mohammed V-Agdal, Rabat, Morocco.

EL HASSANI, A. (ed.) 2004b. Devonian neritic–pelagic correlation and events in the Dra Valley (Western Anti-Atlas, Morocco). *Documents de l'Institut Scientifique*, **19**, 1–100.

FRIEND, P. F. & HOUSE, M. R. 1964. The Devonian Period. In: HARLAND, W. B., SMITH, A. G. & WILCOCK, B. (eds) *The Phanerozoic Time Scale*. Geological Society of London, Memoirs, **10**, 233–236.

GAUSS, G. A. & HOUSE, M. R. 1972. The Devonian successions in the Padstow area, North Cornwall. *Journal of the Geological Society*, **128**, 151–172.

HENDRICKS, E. M. L., RHODES, F. H. T. & HOUSE, M. R. 1971. Evidence bearing on the stratigraphical successions in South Cornwall. *Proceedings of the Ussher Society*, **2**, 270–275.

HOUSE, M. R. 1955. New records from the Red Nodule Beds near Weymouth. *Proceedings of the Dorset Natural History and Archaeology Society*, **75**, 134–135.

HOUSE, M. R. 1956a. The structure of the northern margin of the Poxwell pericline. *Proceedings of the Dorset Natural History and Archaeology Society*, **76**, 129–135.

HOUSE, M. R. 1956b. Devonian goniatites from North Cornwall. *Geological Magazine*, **93**, 257–262.

HOUSE, M. R. 1959. Upper Devonian ammonoids from north-west Dartmoor, Devonshire. *Proceedings of the Geologists Association*, **70**, 315–321.

HOUSE, M. R. 1960a. Abnormal growth in some Devonian goniatites. *Palaeontology*, **3**, 129–136.

HOUSE, M. R. 1960b. *Acanthoclymenia*, the supposed earliest Devonian clymenid is a *Manticoceras*. *Palaeontology*, **3**, 471–476.

HOUSE, M. R. 1962. Observations on the ammonoid succession of the North American Devonian. *Journal of Paleontology*, **36**, 247–284.

HOUSE, M. R. 1963. Devonian ammonoid successions and facies in Devon and Cornwall. *Quarterly Journal of the Geological Society*, **119**, 1–27.

HOUSE, M. R. 1964a. Devonian northern-hemisphere ammonoid distribution and marine links. In: NAIRN, A. E. M. (ed.) *Problems in Palaeoclimatology*. Interscience Publications, London, 262–269, 299–301.

HOUSE, M. R. 1964b. A new goniatite locality at Babbacombe and its problems. *Proceedings of the Ussher Society*, **1**, 124–125.

HOUSE, M. R. 1965a. Devonian goniatites from Nevada. *Neues Jahrbuch für Geologie und Paläontologie, Abhandlungen*, **122**, 337–342.

HOUSE, M. R. 1965b. A study in the Tornoceratidae: the succession of *Tornoceras* and related genera in the North American Devonian. *Philosophical Transactions of the Royal Society, B*, **250**, 79–130.

HOUSE, M. R. 1967. Devonian ammonoid zonation and correlations between North America and Europe. In: OSWALD, D. H. (ed.) *International Symposium on the Devonian System, Calgary 1967*, 2, 1061–1068.

HOUSE, M. R. 1970. On the origin of the clymenid ammonoids. *Palaeontology*, **13**, 664–676.

HOUSE, M. R. 1971a. The goniatite wrinkle-layer. *Smithsonian Contributions to Palaeobiology*, **3**, 23–332.

HOUSE, M. R. 1971b. Devonian faunal distributions. In: MIDDLEMISS, F. A., RAWSON, P. F. & NEWALL, G. F. (eds) *Faunal Provinces in Space and Time*. Seel House Press, Liverpool, 77–94.

HOUSE, M. R. 1973a. An analysis of Devonian goniatite distributions. *Special Papers in Palaeontology*, **12**, 305–317.

HOUSE, M. R. 1973b. Devonian goniatites. *In*: HALLAM, E. (ed.) *Atlas of Palaeobiogeography*. Elsevier, Amsterdam, 97–104.

HOUSE, M. R. 1974. The Devonian Period. *Encyclopaedia Brittanica Macropaedia*, **5**, 671–679.

HOUSE, M. R. 1975a. Facies and time in Devonian tropical areas. *Proceedings of the Yorkshire Geological Society*, **40**, 233–288.

HOUSE, M. R. 1975b. Faunas and time in the marine Devonian. *Proceedings of the Yorkshire Geological Society*, **40**, 459–490.

HOUSE, M. R. 1976. Criteria for international correlation using Devonian ammonoids. *25th International Geological Congress*, **1**, 272–273.

HOUSE, M. R. 1978. Devonian ammonoids from the Appalachians and their international correlation and zonation. *Special Papers in Palaeontology*, **21**, 1–70.

HOUSE, M. R. 1979. Biostratigraphy of the early Ammonoidea. *Special Papers in Palaeontology*, **23**, 263–280.

HOUSE, M. R. 1981a. On the origin, classification and evolution of the early Ammonoidea. *In*: HOUSE, M. R. & SENIOR, J. R. (eds) *The Ammonoidea: The Evolution, Classification, Mode of Life and Geological Usefulness of a Major Fossil Group*. Systematics Association, Special Volume, **18**, 3–36.

HOUSE, M. R. 1981b. Early ammonoids in space and time. *In*: HOUSE, M. R. & SENIOR, J. R. (eds) *The Ammonoidea: The Evolution, Classification, Mode of Life and Geological Usefulness of a Major Fossil Group*. Systematics Association, Special Volume, **18**, 359–367.

HOUSE, M. R. 1981c. Lower and Middle Devonian goniatite biostratigraphy. *In*: OLIVER, W. A. & KLAPPER, G. (eds) *Devonian Biostratigraphy of New York, Subcommission on Devonian Stratigraphy*, Washington, 33–38.

HOUSE, M. R. 1981d. Mudrocks of the Devonian. *Quarterly Journal of Engineering Geologists*, **14**, 253–256.

HOUSE, M. R. 1982. The Middle/Upper Devonian Series boundary and decisions of the International Geological Congress. *Courier Forschungsinstitut Senckenberg*, **55**, 449–462.

HOUSE, M. R. 1983. Devonian eustatic events. *Proceedings of the Ussher Society*, **5**, 396–405.

HOUSE, M. R. 1985. Correlation of mid-Palaeozoic ammonoid evolutionary events with global sedimentary perturbations. *Nature*, **313**, 17–22.

HOUSE, M. R. 1986. Are Jurassic sedimentary microrhythms due to orbital forcing? *Proceedings of the Ussher Society*, **6**, 299–311.

HOUSE, M. R. 1987a. Geological rhythms, cycles and other revolutions. *Geological Magazine*, **124**, 273–276.

HOUSE, M. R. 1987b. Devonian bioevents. *Episodes*, **10**, 138–139.

HOUSE, M. R. 1988a. Major features of Cephalopod evolution. *In*: WIEDMANN, J. & KULLMANN, J. (eds) *Cephalopods, Present and Past*, 2nd International Cephalopod Symposium, J. Schweizerbart Verlag, 1–16.

HOUSE, M. R. 1988b. Extinction and Survival in the Cephalopoda. Systematics Association, Special Volume, **34**, 139–154.

HOUSE, M. R. 1989a. International definition of Devonian System boundaries. *Proceedings of the Ussher Society*, **7**, 41–46.

HOUSE, M. R. 1989b. Analysis of mid-Palaeozoic extinctions. *Bulletin de la Societé Belge de Géologie*, **98**(2), 99–107.

HOUSE, M. R. 1989c. Ammonoid extinction events. *Philosophical Transactions, Royal Society of London*, **B325**, 307–326.

HOUSE, M. R. 1991. Devonian sedimentary microrhythms and a Givetian time scale. *Proceedings of the Ussher Society*, **7**, 392–395.

HOUSE, M. R. 1992. Earliest Carboniferous goniatite recovery after the Hangenberg Event. *Annales de la Societé Géologique de Belgique*, **115**, 559–579.

HOUSE, M. R. 1993a. Fluctuations in Ammonoid Evolution and Possible Environmental Controls. Systematics Association, Special Volume, **47**, 13–34.

HOUSE, M. R. (ed.) 1993b. *The Ammonoidea: Environment, Ecology and Evolutionary Change*. Systematics Association, Special Volume, **47**, 1–354.

HOUSE, M. R. 1995. *Devonian Precessional and Other Signatures for Establishing a Givetian Timescale*. Geological Society, Special Publications, **85**, 37–49.

HOUSE, M. R. 1996a. Juvenile Goniatite Survival Strategies following Devonian Extinction Events. Geological Society, Special Publications, **102**, 163–185.

HOUSE, M. R. 1996b. An *Eocanites* fauna from Chile and its palaeogeographic implications. *Annales de la Societé Géologique de Belgique*, **117**, 95–105.

HOUSE, M. R. 1996c. The Middle Devonian Kacak Event. *Proceedings of the Ussher Society*, **9**, 79–84.

HOUSE, M. R. 1997. The late Devonian mass extinction: the Frasnian/Famennian crisis. *Historical Biology*, **12**, 291–301.

HOUSE, M. R. 1998. Evolution and environmental controls: Palaeozoic Black Deaths. *In*: BINTLIFF, J. (ed.) *Structure and Contingency. Evolutionary Processes of Life and Human Society*. Leicester University Press, Leicester, 14–30.

HOUSE, M. R. 2001. Chronostratigraphic framework for the Devonian of the Old Red Sandstone. *In*: FRIEND, P. F. & WILLIAMS, B. P. J. (eds) *New Perspectives on the Old Red Sandstone*. Geological Society, Special Publications, **180**, 23–27.

HOUSE, M. R. 2002. Strength, timing, setting and cause of Mid-Palaeozoic extinctions. *Palaeogeography, Palaeoclimatology, Palaeoecology*, **181**, 5–25.

HOUSE, M. R. 2003a. Devonian (Givetian) goniatites from Wolborough, Barton and Lummaton. *Geosciences in South-West England*, **10**, 281–292.

HOUSE, M. R. 2003b. Devonian (Frasnian) goniatites from Waterside Cove and Staverton, South Devon. *Geosciences in South-West England*, **10**, 267–280.

HOUSE, M. R. & BLODGETT, R. B. 1982. The Devonian goniatite genera *Pinacites* and *Foordites* from Alaska. *Canadian Journal of Earth Sciences*, **19**, 1873–1874.

HOUSE, M. R. & BUTCHER, N. E. 1962. Excavations in the Devonian and Carboniferous rocks of the Chudleigh area, South Devon. *Proceedings of the Ussher Society*, **1**, 28–29.

HOUSE, M. R. & BUTCHER, N. E. 1973. Excavations in the Upper Devonian and Carboniferous rocks near Chudleigh, South Devon. *Transactions of the Royal Geological Society of Cornwall*, **20**, 199–220.

HOUSE, M. R. & DINELEY, D. L. 1985. Devonian Series Boundaries in Great Britain. *Courier Forschungsinstitut Senckenberg*, **75**, 301–310.

HOUSE, M. R. & GRADSTEIN, F. 2005. The Devonian Period. *In*: GRADSTEIN, F. & OGG, J. (eds) *A Geological Time Scale*. Cambridge University Press, Cambridge.

HOUSE, M. R. & KIRCHGASSER, W. T. 1993. Devonian goniatite biostratigraphy and timing of facies movements in the Frasnian of Eastern North America. *In*: HAILWOOD, E. A. & KIDD, R. B. (eds) *High Resolution Stratigraphy*. Geological Society, Special Publications, **70**, 267–292.

HOUSE, M. R. & KIRCHGASSER, W. T. (in press). Late Devonian goniatites (Cephalopoda, Ammonoidea) from New York State. *Bulletins of American Paleontology*.

HOUSE, M. R. & PEDDER, A. E. H. 1963. Devonian goniatites and stratigraphical correlations in Western Canada. *Palaeontology*, **6**, 491–539.

HOUSE, M. R. & PRICE, J. D. 1980. Devonian ammonoid faunas of the Carnic Alps. *In*: SCHÖNLAUB, H. (ed.) *Guidebook, Second European Conodont Symposium*, Vienna, 14–15.

HOUSE, M. R. & PRICE, J. D. 1985. New late Devonian genera and species of tornoceratid goniatites. *Palaeontology*, **28**, 159–188.

HOUSE, M. R. & SELWOOD, E. B. 1957. Discovery of ammonoids of the Upper Devonian Wocklumeria Zone in North Cornwall. *Nature*, **179**, 832.

HOUSE, M. R. & SELWOOD, E. B. 1964. Palaeozoic Palaeontology in Devon and Cornwall. *In*: HOSKING, K. F. G. & SHRIMPTON, G. J. (eds) *Present views on Some Aspects of the Geology of Cornwall and Devon*. Royal Geological Society of Cornwall, 45–86.

HOUSE, M. R. & SEVASTOPULO, G. D. 1984. The Devonian–Carboniferous boundary in the British Isles; a summary. *Courier Forschungsinstitut Senckenberg*, **67**, 15–22.

HOUSE, M. R. & ZIEGLER, W. 1977. The goniatite and conodont sequences in the early Upper Devonian at Adorf, Germany. *Geologica et Palaeontologica*, **11**, 69–108.

HOUSE, M. R. & ZIEGLER, W. (eds) 1997. On sea-level fluctuations in the Devonian. *Courier Forschungsinstitut Senckenberg*, **199**, 1–146.

HOUSE, M. R., MOURAVIEFF, N. & BEESE, A. P. 1978a. North Cornwall. *In*: SCRUTTON, C. T. (ed.) *A field guide to selected areas of the Devonian of South-West England, International Symposium on the Devonian System, P.A.D.S. 78*, The Palaeontological Association, 57–68.

HOUSE, M. R., RICHARDSON, R. B., CHALONER, W. G., ALLEN, J. R. L., HOLLAND, C. H. & WESTOLL, T. S. 1978b. *A Correlation of the Devonian Rocks of the British Isles*. Geological Society of London, Special Report, **8**, 1–110.

HOUSE, M. R., SCRUTTON, C. T. & BASSETT, M. G. (eds) 1979. The Devonian System. *Special Papers in Palaeontology*, **23**, 1–353.

HOUSE, M. R., KIRCHGASSER, W. T., PRICE, J. D. & WADE, G. 1985. Goniatites from the Frasnian (Upper Devonian) and related strata of the Montagne Noire. *Hercynica*, **1**, 1–19.

HOUSE, M. R., GORDON, M. & HLAVIN, W. J. 1986. Late Devonian ammonoids from Ohio and adjacent states. *Journal of Paleontology*, **60**, 126–144.

HOUSE, M. R., FEIST, R. & KORN, D. 2000a. The Middle/Upper Devonian GSSP at Puech de la Suque, Southern France. *Courier Forschungsinstitut Senckenberg*, **225**, 49–58.

HOUSE, M. R., BECKER, R. T., FEIST, R., FLAJS, G., GIRARD, C. & KLAPPER, G. 2000b. The Frasnian/Famennian boundary GSSP at Coumiac, southern France. *Courier Forschungsinstitut Senckenberg*, **225**, 59–75.

HOUSE, M. R., MENNER, V. V., BECKER, R. T., KLAPPER, G., OVNATANOVA, N. S. & KUŹMIN, A. V. 2000c. Reef episodes, anoxia and sea-level changes in the Frasnian of the southern Timan (NE Russian platform). *In*: INSALACO, E., SKELTON, P. W. & PALMER, T. J. (eds) *Carbonate Platform Systems: Components and Interactions*. Geological Society, Special Publications, **178**, 147–176.

JOHNSON, J. G., KLAPPER, G. & SANDBERG, C. A. 1985. Devonian eustatic fluctuations in Euramerica. *Geological Society of America, Bulletin*, **96**, 567–587.

KIRCHGASSER, W. T. & HOUSE, M. R. 1981. Upper Devonian goniatite biostratigraphy. *In*: OLIVER, W. A. & KLAPPER, G. (eds) *Devonian Biostratigraphy of New York, Subcommission on Devonian Stratigraphy*, Washington, 39–45.

KLAPPER, G., FEIST, R. & HOUSE, M. R. 1987. Decision on the boundary stratotype for the Middle/Upper Devonian Series boundary. *Episodes*, **10**, 97–101.

KLAPPER, G., FEIST, R., BECKER, R. T. & HOUSE, M. R. 1993. Definition of the Frasnian/Famennian Stage Boundary. *Episodes*, **16**, 433–440.

MARSHALL, J. E. A. & HOUSE, M. R. 2000. Devonian stage boundaries in England, Wales and Scotland. *Courier Forschungsinstitut Senckenberg*, **225**, 83–90.

MENNER, V. V., SHUVALOVA, G. A., OBUKOVSKAYA, T. G., OVNATANOVA, N. S., HOUSE, M. R. & BECKER, R. T. 2000. The correlation of the regional stages and the Frasnian spore, conodont, and ammonoid assemblages of the Timan–Pechora province with standard zones. *Ichthyolith Issues, Special Publication*, **6**, 77–79.

PRICE, J. D. & HOUSE, M. R. 1984. Ammonoids near the Devonian–Carboniferous boundary. *Courier Forschungsinstitut Senckenberg*, **67**, 15–22.

RACKI, G. & HOUSE, M. R. (eds) 2002. Late Devonian biotic crisis: ecological, depositional and geochemical records. *Palaeography, Palaeoclimatology, Palaeoecology*, **118**, 1–374.

WALLISER, O. H., BULTYNCK, P., WEDDIGE, K., BECKER, R. T. & HOUSE, M. R. 1995. Definition of the Eifelian–Givetian Stage Boundary. *Episodes*, **18**(3), 107–115.

YOLKIN, E. A., KIM, A. I., WEDDIGE, K., TALENT, J. A. & HOUSE, M. R. 1998. Definition of the Pragian/Emsian Stage Boundary. *Episodes*, **20**, 235–240.

YOLKIN, E. A., KIM, A. I., WEDDIGE, K., TALENT, J. A. & HOUSE, M. R. 2000. The basal Emsian GSSP in Zinzil'ban Gorge, Uzbekistan. *Courier Forschungsinstitut Senckenberg*, **225**, 17–25.

Neritic–pelagic correlation in the Lower and basal Middle Devonian of the Dra Valley (Southern Anti-Atlas, Moroccan Pre-Sahara)

U. JANSEN[1], N. LAZREQ[2], G. PLODOWSKI[1], M. SCHEMM-GREGORY[1], E. SCHINDLER[1] & K. WEDDIGE[1]

[1]*Forschungsinstitut Senckenberg, Senckenberganlage 25, D-60325 Frankfurt am Main, Germany (e-mail: Ulrich.Jansen@senckenberg.de)*

[2]*Université Cadi Ayadd, Faculté des Sciences Semlalia, Département de Géologie, BP 2390, Marrakech, Morocco*

Abstract: Marine Lower Devonian successions are widely exposed in the Dra Valley (Southern Anti-Atlas, Moroccan Pre-Sahara). Resulting from new studies, especially on brachiopods, conodonts, and dacryoconarid tentaculitids, the chronostratigraphic assignments of the Lower Devonian formations are revised. Thanks to lateral and vertical facies variations, it is possible to correlate pelagic and neritic successions and corresponding biostratigraphies. Pelagic conodont, dacryoconarid and goniatite faunas allow correlations and dating in the sense of the Bohemian and global chronostratigraphies, whereas units of the traditional Rhenish subdivision can be identified by means of neritic brachiopods.

Devonian strata are almost continuously exposed over 400 km in a SW–NE direction on strike in the Dra Valley along the southeastern flank of the Anti-Atlas Anticlinorium (Fig. 1). The Lower Devonian to Eifelian strata bear a complex variety of marine facies and faunas described by Hollard in a number of pioneer stratigraphic works (Hollard 1963*a–c*, 1965, 1967, 1977, 1978, 1981*a, b*). Palaeontological research is still ongoing. Previous faunal studies focused on trilobites (e.g. Alberti & Hollard 1963; Hollard 1963*b*; Morzadec 1988, 2001; Schraut 1998*a, b*, 2000), brachiopods (e.g. Drot 1964, 1971, 1975; Jansen 1999, 2001; Aït Malek *et al.* 2000), ammonoids (Hollard 1960, 1963*c*), conodonts (Bultynck & Hollard 1980), and ostracodes (G. Becker *et al.* 2003, 2004 *in* Jansen, G. Becker *et al.* 2004*a*, p. 24). Bultynck & Walliser (2000) summarized stratigraphic assignments in their comprehensive survey of the Devonian in the Moroccan Anti-Atlas.

The Lower Devonian to Eifelian successions in the Dra Valley offer outstanding possibilities for correlation mainly due to the fact that neritic and pelagic units follow one above the other in four main cycles called 'riches' (Choubert 1956). A 'rich' generally consists of a thin (few metres) basal limestone, 100 to 150 m of silty and sandy shales in the middle part and a 50 to 120 m thick sandstone unit at the top (Hollard 1981*a*). In essence, this succession is repeated four times and, accordingly, the whole complex may be subdivided into four formations, 'Rich 1' to 'Rich 4' (Hollard 1963*a*, 1967, 1981*a*). Single units may be traced and correlated over tens of kilometres (Fig. 2).

However, significant lateral facies changes are present within these cycles, and new names have been proposed for special regional facies developments (Hollard 1978, 1981*b*; R. T. Becker, Jansen *et al.* 2004). Essentially, the facies in the studied area changes from the SW, i.e. from Aouinet Torkoz to about Foum el Hassane, to the NE, i.e. from about Foum el Hassane to Foum Zguid (Figs 1 & 2). The calcareous and shaly parts of the 'riches' correspond to the 'Hercynian' resp. (hemi-)pelagic facies and the sandy parts to the 'Rhenish' resp. neritic facies. The neritic sandstones contain diverse and abundant brachiopod faunas allowing precise assignments to the regional Rhenish subdivision as defined in the Rheinisches Schiefergebirge (Germany) and in the Ardennes (Belgium) (Jansen 1999, 2000, 2001). On the other hand, conodonts, dacryoconarid tentaculitids and goniatites are present in the intercalated limestones which can be attributed to global pelagic biozones, the classical Bohemian subdivision, and the global chronostratigraphy (Bultynck & Hollard 1980; Lazreq & Ouanaimi 1998).

Unfortunately, the use of different subdivisions has caused much confusion. There is a Pragian in the original sense of Bohemia and a shorter Pragian in the GSSP sense according to the decisions of the international *Subcommission on Devonian Stratigraphy* (SDS). Similarly, there is an Emsian in the traditional German (Rhenish) sense and an Emsian in the Belgian (Ardennan) sense slightly differing in the age of their bases; both concepts are used in the Rhenish facies of Central and Western Europe and North Africa.

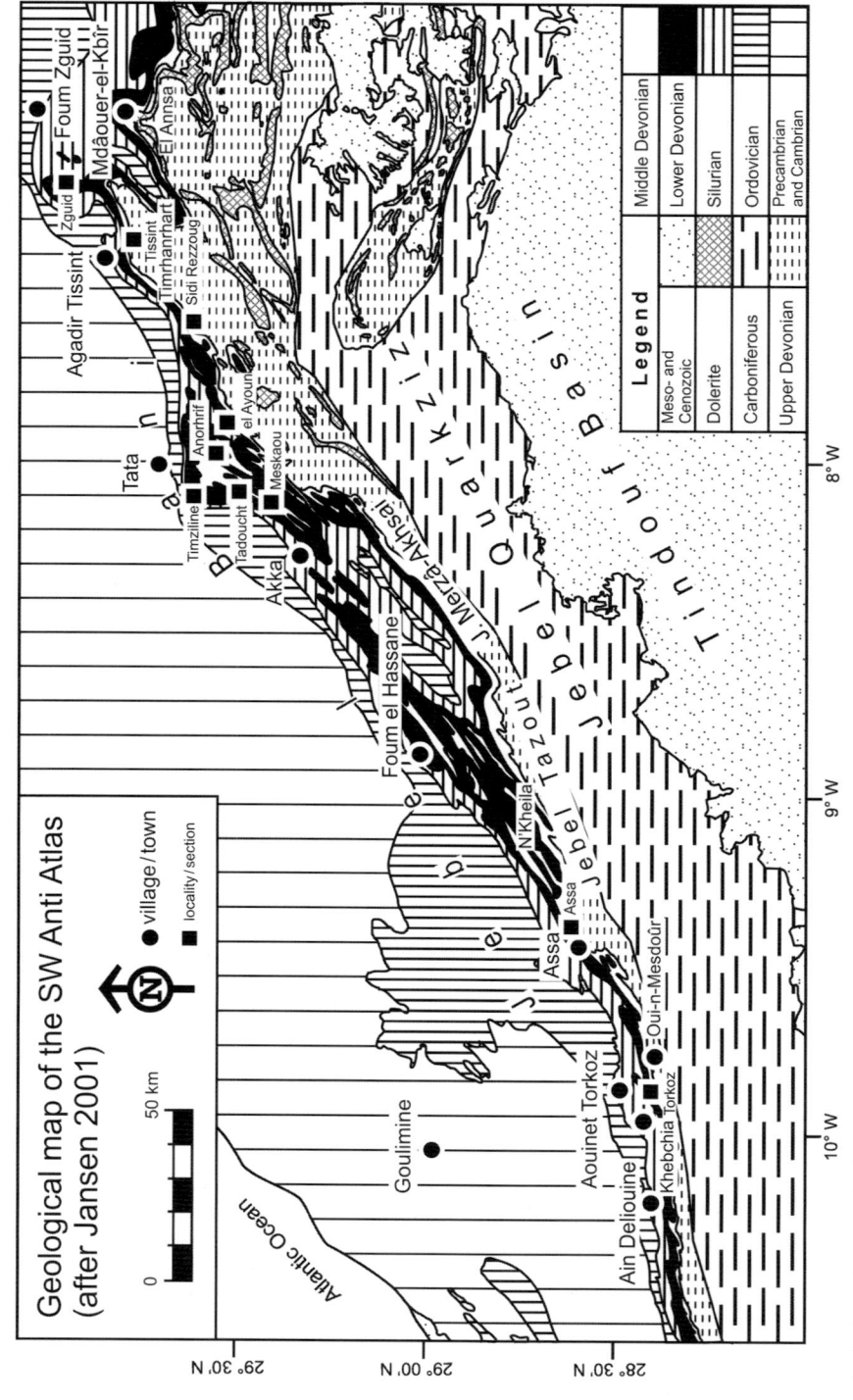

Fig. 1. Geological map of the Southern Anti-Atlas, after Jansen (2001), based on Hollard & Jacquemont (1956). The location of the main sections is indicated.

Finally, there is a third Emsian in the GSSP sense as defined in the Zinzilban section in Uzbekistan (Yolkin et al. 1997). Therefore, it is always necessary to explain in which sense one of these chronostratigraphic units is meant. The Rhenish stages Gedinnian and Siegenian are still used, because they are regionally important. Unfortunately, some authors use the terms Lochkovian and Pragian just as synonyms of Gedinnian and Siegenian—and by doing so increase the confusion.

The Lower Devonian strata in the Ardenno-Rhenish Mountains are not continuously marine. Fortunately, the resulting gaps in the documentation of marine faunas, which have proved to be substantial, could at least partly be closed by correlations with sections in the Armorican Massif (NW France), the Cantabrian Mountains (North Spain), the Celtiberian Chains (NE Spain), and in North Africa where the marine documentation is more continuous (e.g. Plusquellec 1980; Morzadec et al. 1981; Carls 1987; Gourvennec 1989; García-Alcalde et al. 1990; Carls et al. 1993; García-Alcalde 1996; Jansen 2001). As a result, the biostratigraphical content of the regional stages Gedinnian, Siegenian and Emsian is now much better known. The regional chronostratigraphic boundaries Lower/Upper Gedinnian, Gedinnian/Siegenian and Lower/Middle Siegenian could be redefined in marine successions in W and SW Europe (Carls 1987). When discussing the stratigraphic position of the Rhenish faunas, this subdivision is followed herein. Thanks to detailed sampling in many sections and palaeontological studies, evolutionary lineages of brachiopods could be reconstructed (orthids, strophomenids: Renouf 1972; Carls 1974; chonetids: Racheboeuf 1981; Carls 1985; spiriferids: Gourvennec 1989; Carls et al. 1993; Jansen 2001; Schemm-Gregory & Jansen 2004, 2005; athyridids: Alvarez 1990).

Based on brachiopod comparisons Jansen (2000, 2001) was able to show that the Lower Devonian 'riches' in the Dra Valley are generally older with regard to the Rhenish stages than formerly considered, e.g. by Hollard (1963a, 1981a).

The Forschungsinstitut Senckenberg (Frankfurt am Main) and Moroccan Institutions (Université Cadi Ayyad, Marrakech; Institut Scientifique, Rabat) cooperate in a successful research project focusing on the Devonian of the Anti-Atlas. The project benefited from its integration into the UNESCO-IGCP Project 421 'North Gondwanan mid-Palaeozoic bioevent/biogeography patterns in relation to crustal dynamics' from 1997 to 2003. The most recent activities are embedded in the follow-up IGCP Project 499 'Devonian land–sea interaction: evolution of ecosystems and climate' (DEVEC), to which this article is a contribution.

New results of these investigations have recently been presented during a joint meeting of SDS and IGCP 499 in Rabat (Morocco) and a subsequent field trip (Jansen, Plodowski et al. 2004a, b; Jansen, G. Becker et al. 2004a, b; R. T. Becker, Jansen et al. 2004).

The present work summarizes these results based on brachiopods, conodonts, and dacryoconarid tentaculitids from the Lower and basal Middle Devonian in the Dra Valley, drawing special attention to interfacial correlations. Stratigraphic data from other faunal groups are taken from the literature, especially information on the trilobites and goniatites. All mentioned localities and sections are described in detail in Jansen (2001), Jansen, Plodowski et al. (2004b), and Jansen, G. Becker et al. (2004a, b). The Rhenish stratigraphic terminology used below follows Weddige (1996c) and Weddige et al. (2002). The global Pragian defined by GSSPs is informally subdivided into lower, middle and upper Pragian following the three conodont zones in Carls & Weddige (1996). The position of litho-, bio- and chronostratigraphic units is often given in terms of the Devonian Correlation Table (DCT) edited by K. Weddige as a series in the past 10 years (Weddige 1996a, 1998a, b, 2000a, b, 2001, 2003, 2004, 2005a, b); it is indicated by values on a scale from 0 to 20 cm (for introduction, see Weddige 2000a) for both the Lower Devonian (di) and the Middle Devonian (dm). The material referred to is stored in the Senckenberg Museum, Frankfurt am Main ('SMF' numbers).

Lmhaïfid Formation

Distribution, thickness and facies

The Lmhaïfid Formation (Hollard 1981a, b; = 'Série de Passage' sensu Hollard 1963a) consists of up to more than 600 m of sandy shales, sandy limestones and limestones which are further subdivided into five subunits (Hollard 1963a, b, 1981b). The facies is predominantly of the 'Hercynian' resp. 'Bohemian' type. The formation is known to us only from the section el Ayoun S of Tata in the NE Dra Valley. Because of the considerable variation in facies and strong diachrony from SW to NE, the concept of the formation apparently needs revision; a division into separate formations has to be considered.

Conodont faunas C 1 to C 4

Conodonts of the Lmhaïfid Formation (Fig. 3) were collected from the Section el Ayoun by N. Lazreq and K. Weddige and determined with the kind assistance of Mike Murphy (Davis, California) during a stay at Senckenberg. The section and sample numbers are illustrated in Lazreq &

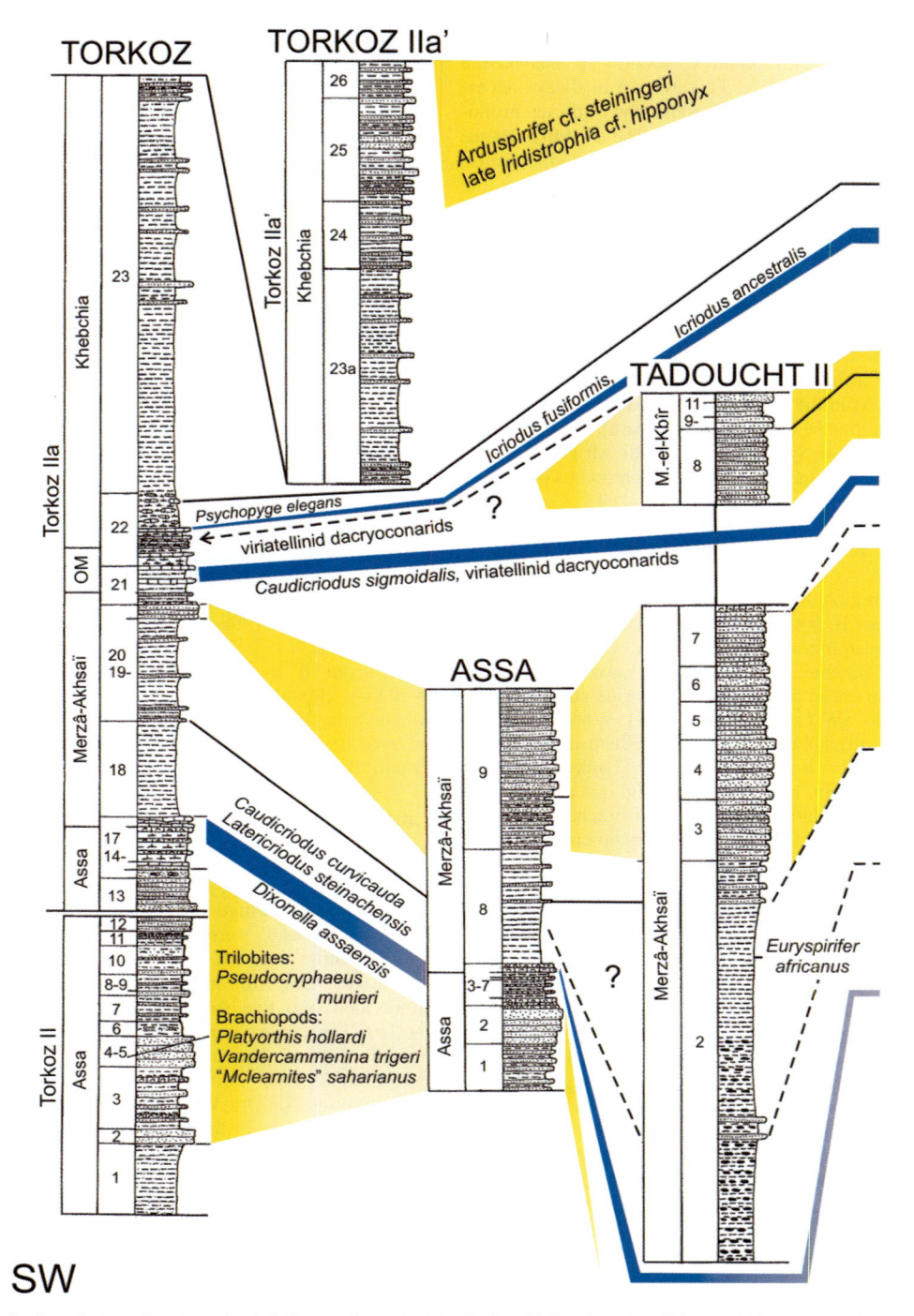

Fig. 2. Correlation of sections (mainly Lower Devonian) in the Dra Valley based on litho- and biostratigraphical criteria (after Jansen 2001, R. T. Becker, Jansen *et al.* 2004). For location of the sections, see Figure 1. OM, Oui-n'-Mesdoûr; O-el-M, Oued-el-Mdâouer; M.-el-Kbîr, Mdâouer-el-Kbîr.

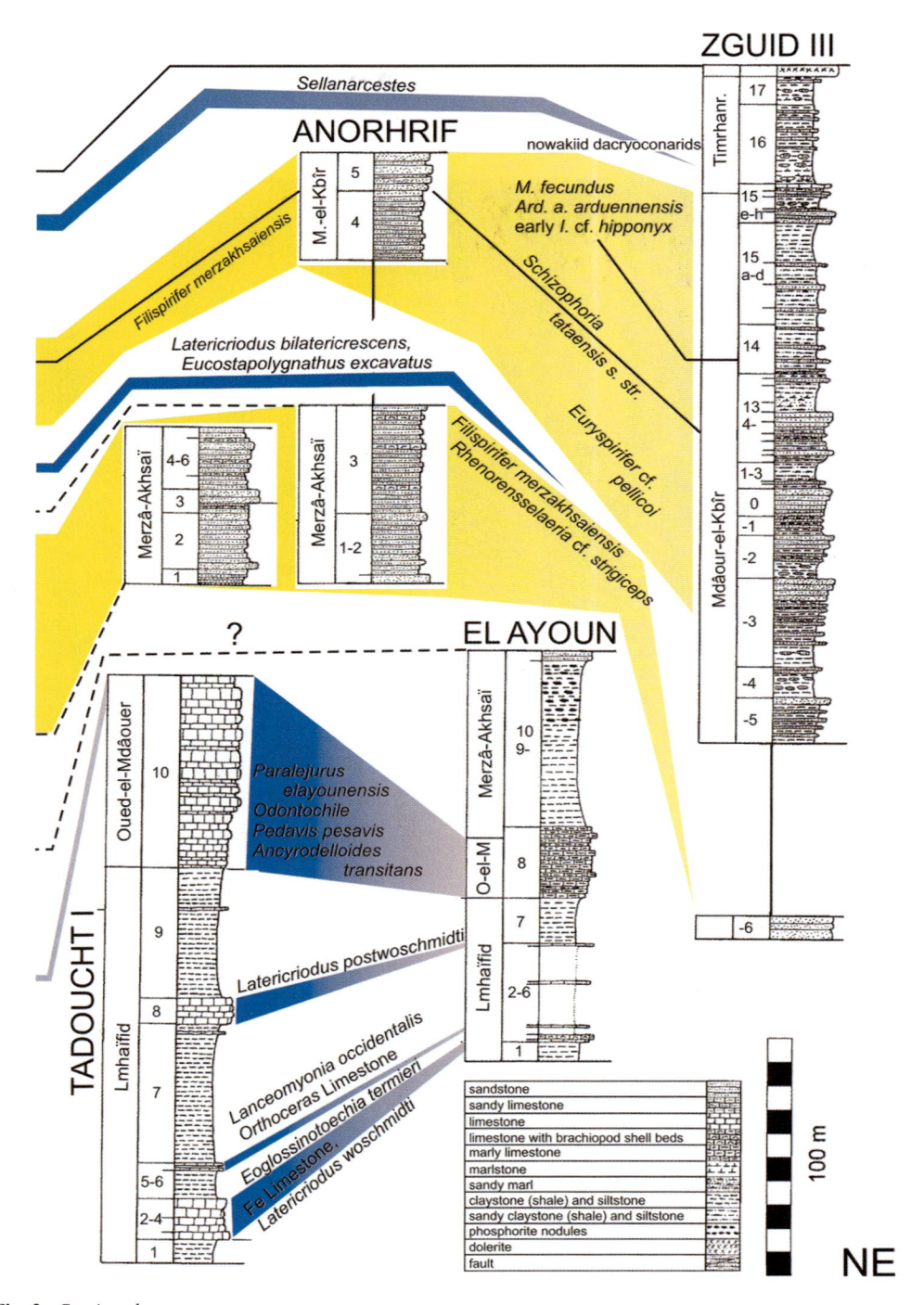

Fig. 2. *Continued.*

Ouanaimi (1998, pp. 224–227, text-fig. 2). Herein (cf. Fig. 5, Table 1), Lazreq's and Ouanaimi's samples CH1, CH5, L1 represent conodont fauna C 1, sample L2 is C 2, L5 is C 3, L8 is C 4, L15 is C 5, and L16 is C 6.

The basal limestones with the conodont fauna C 1 are poor in conodonts, but single fragments of ?'*Ozarkodina*' *eosteinhornensis* (Walliser 1964) [*Remark*: The genus assignment is still under revision, e.g. it was assigned to a 'New Genus W' by

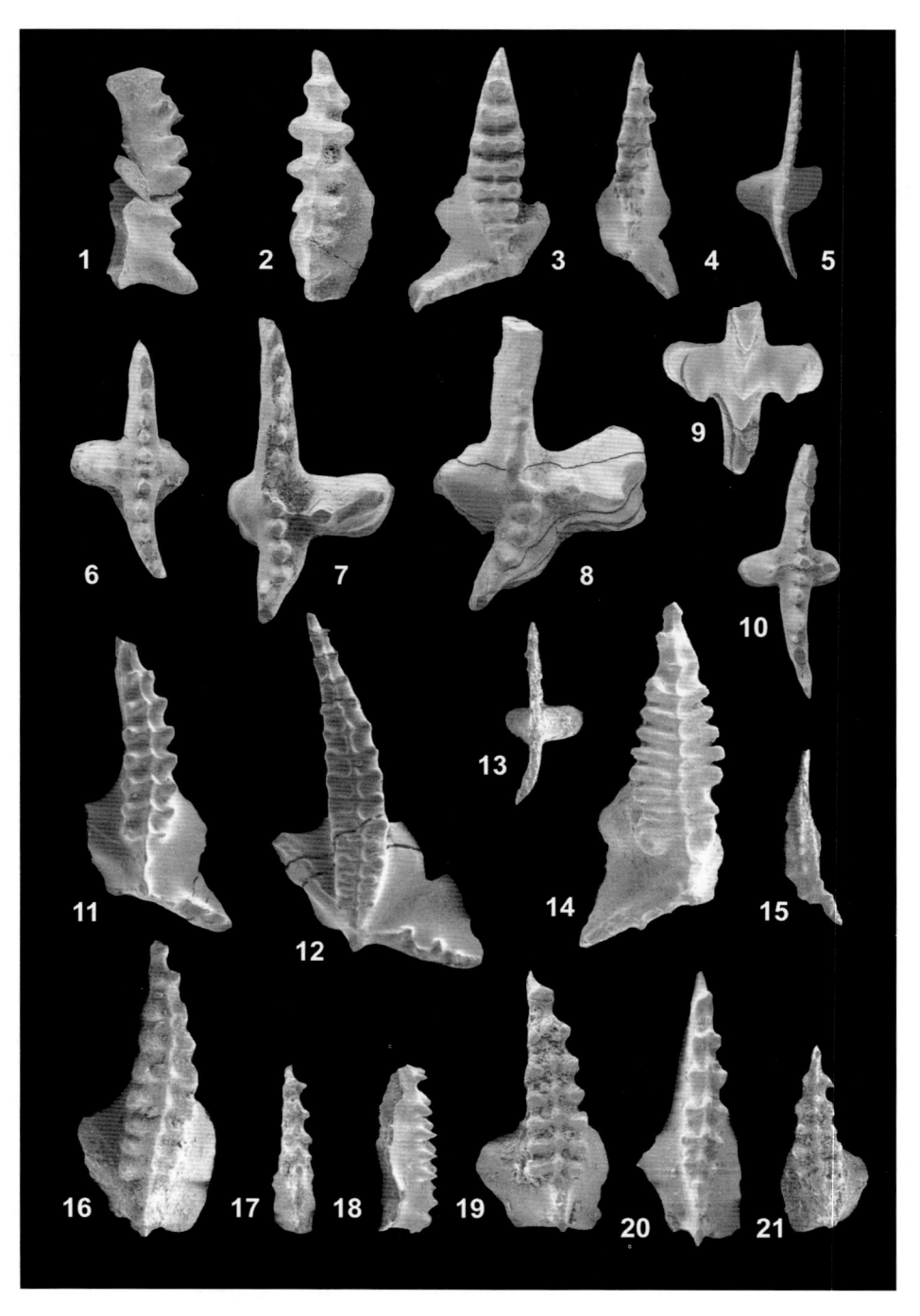

Murphy et al. (2004). According to Peter Carls (pers. comm. 2006), the true '*eosteinhornensis*' does not occur during the Lochkovian, and thus the present Moroccan specimens identified as '*eosteinhornensis*' may represent forms of the '*remscheidensis* group'] and ?*Caudicriodus angustoides* (Carls & Gandl 1969) have been found. They questionably suggest a chronostratigraphic position within the Lower Lochkovian, about DCT interval from di 0.0 to 2.6 cm, i.e. *woschmidti-postwoschmidti-*, and *eurekaensis* zones (see Carls & Weddige 1996) resp. *hesperius-, eurekaensis-* and *sexidentata* zones *sensu* Murphy (2000).

The following conodont fauna C 2 with *Caudicriodus angustoides* (Carls & Gandl 1969), *Latericriodus postwoschmidti* (Mashkova 1968), *Lanea omoalpha* Murphy & Valenzuela-Ríos 1999, and ?'*Ozarkodina*' *eosteinhornensis* (Walliser 1964) documents the *omoalpha* Zone *sensu* Murphy (2000, DCT interval di 2.7 to 3.1 cm) which after Murphy & Valenzuela-Ríos (1999, p. 322, text-fig. 1) is rather similar with the range of the zone index.

The *omoalpha* Zone is also represented by the conodont fauna C 3 because of the occurrence of *Ancyrodelloides asymmetricus* (Bischoff & Sannemann 1958), *A. carlsi* (Boersma 1974), ?'*Ozarkodina*' *steinhornensis* (Ziegler 1956), *Lanea omoalpha* (Murphy & Valenzuela-Ríos 1999), and *Ozarkodina repetitor* (Carls & Gandl 1969). Since Murphy & Valenzuela-Ríos (1999, p. 322, table 1, text-fig. 1) regard *A. asymmetricus* as a possible variant of *A. transitans* (Bischoff & Sannemann 1958), the entry of which is later than that of *A. omoalpha*, the fauna C 3 may document a high level within the *omoalpha* Zone (DCT interval di 2.9 to 3.1 cm).

The succeeding conodont fauna C 4 contains *Ancyrodelloides transitans* (Bischoff & Sannemann 1958) represented by its early morphotype which is characterized by an open basal cavity. Therefore, C 4 represents the lower half of the *A. transitans* range *sensu* Murphy & Valenzuela-Ríos (1999, p. 322, text-fig. 1) or the upper *omoalpha* Zone *sensu* Murphy (2000, DCT interval di 2.9 to 3.1 cm).

Conodont faunas C 2 to C 4 as a whole distinctly indicate the *omoalpha* Zone resp. the lower part of the (traditional) *delta* Zone, i.e. the lower Middle Lochkovian [*Remark*: Valenzuela-Ríos & Murphy (1997, p. 133, fig. 3) and Murphy & Valenzuela-Ríos (1999, p. 322, text-fig. 1) indicate the range of the conodont genus *Ancyrodelloides* as 'Middle Lochkovian'. It starts with the *omoalpha* Zone and ends with the base of the *pandora* beta Zone of the global conodont zonation, and it is rather coincident with the *delta* Zone of the Cordilleran conodont zonation.] according to Murphy & Valenzuela-Ríos (1999) and Murphy (2000). Thus, the Lmhaïfid Formation of the el Ayoun section corresponds well with the '*Orthoceras*' Limestone horizons in the Achguig area of the Tafilalt region (SE Morocco) which are approximately of the same age (Bultynck & Walliser 2000, Unit A).

Fig. 3. All magnifications of specimens about ×60. **1–8**, conodonts from the Lmhaïfid Formation (Lochkovian). **1–2**, *Caudicriodus angustoides* (Carls & Gandl 1969). **1**, SMF 69627, lateral view—Section el Ayoun (Tata), bed CH5. **2**, SMF 69625, upper view—Section el Ayoun (Tata), bed L2. **3**, *Caudicriodus postwoschmidti* (Mashkova 1968); SMF 69624 upper view—Section el Ayoun (Tata), bed L2. **4**, *Icriodus eolatericrescens* (Mashkova 1968); SMF 69652 upper view—Section el Ayoun (Tata), bed L2. **5**, '*Ozarkordina*' *steinhornensis* (Ziegler 1956); SMF 69650 upper view—Section el Ayoun (Tata), bed L5. **6**, *Lanea omoalpha* Murphy & Valenzuela-Rios 1999; SMF 69631 upper view—Section el Ayoun (Tata), bed L5. **7**, *Ancyrodelloides asymmetricus* (Bischoff & Sannemann 1958); SMF 69629 upper view—Section el Ayoun (Tata), bed L5. **8**, *Ancyrodelloides carlsi* (Boersma 1974); SMF 69638 upper view—Section el Ayoun (Tata), bed L5. **9–10**, conodonts from the Oued-el-Mdâouer Formation (Lochkovian to Pragian). **9**, *Ancyrodelloides transitans* (Bischoff & Sannemann 1958); SMF 69635 upper view—Section el Ayoun (Tata), bed L15. **10**, *Ancyrodelloides transitans* (Bischoff & Sannemann 1958); SMF 69634 upper view—Section el Ayoun (Tata), bed L15. **11–15**, Conodonts from the Oui-n-Mesdoûr Formation (Zlichovian to Lower Dalejian resp. Lower to lower Upper Emsian). **11**, *Latericriodus beckmanni sinuatus* (Klapper, Ziegler & Mashkova 1978); SMF 69961 upper view—Section Torkoz IIa (Dra Valley), bed 21 g. **12**, *Latericriodus multicostatus* (Carls & Gandl 1969); SMF 69962 upper view—Section Torkoz IIa (Dra Valley), bed 21 g. **13**, '*Ozarkordina*' *steinhornensis miae* (Bultynck 1971); SMF 69963 upper view—Section Torkoz IIa (Dra Valley), bed 21e. **14**, *Caudicriodus sigmoidalis* (Carls & Gandl 1969); SMF 69964 upper view—Section Torkoz IIa (Dra Valley), bed 21d. **15**, *Polygnathus inversus* Klapper & Johnson 1975 (juvenile form); SMF 69965 upper view—Section Torkoz IIa (Dra Valley), bed 21e. **11–21**, Conodonts from the Khebchia Formation (Dalejian resp. Upper Emsian). **16**, *Icriodus corniger ancestralis* Weddige 1977; SMF 69966 upper view—Section Torkoz IIa (Dra Valley), bed 22c. **17**, *Icriodus culicellus* Bultynck 1976; SMF 69967 upper view—Section Torkoz IIa (Dra Valley), bed 22d. **18**, *Icriodus culicellus* Bultynck 1976; SMF 69968 lateral view—Section Torkoz IIa (Dra Valley), bed 22e. **19**, *Icriodus corniger rectirostratus* Bultynck 1970; SMF 69969 upper view—Section Torkoz IIb (Dra Valley), 54.4 m. **20**, *Icriodus corniger leptus* Weddige 1977; SMF 69970 upper view—Section Torkoz IIa (Dra Valley), bed 22 g. **21**, *Icriodus* sp., aff. *I. werneri* Weddige 1977; SMF 69971 upper view—Section Torkoz IIa (Dra Valley), bed 22d.

Hollard (1977) reported '*I. woschmidti woschmidti* Ziegler 1960' from limestones in the lower Lmhaïfid Formation in the SW Dra Valley at Aïn-Deliouine; if this determination is correct (if it is actually not *Icriodus postwoschmidti*), then the formation would have an early *woschmidti-postwoschmidti* age there, corresponding to the DCT interval di 0.0 to 0.85 cm (see Carls 1996a). According to Hollard (1977), a limestone layer near the top of the Lmhaïfid Formation in the SW Dra Valley has yielded *Pedavis pesavis* (Bischoff & Sannemann 1958), the range of which would indicate an age within the interval upper Middle Lochkovian to Upper Lochkovian or lowermost Pragian (DCT interval about di 3.5 to 6.55 cm).

Brachiopod faunas B 1 to B 2

In the lower brachiopod limestones of the el Ayoun section, i.e. in the level of conodont fauna C 1 mentioned above, *Eoglossinotoechia termieri* (Drot 1964) has been found in association with other rhynchonellids and atrypids (brachiopod fauna B 1: Table 1 and Fig. 5); *E. termieri* has been reported from strata attributed to the Lower Gedinnian [Remark: Carls (1987) has proposed a formal definition of Lower and Upper Gedinnian] in Morocco (Drot 1964; Hollard 1977). Higher in the section, corresponding to the fauna C 2, *Lanceomyonia occidentalis* (Drot 1964) occurs in masses (brachiopod fauna B 2: Table 1 and Fig. 5). According to Carls (1987), *L. occidentalis* occurs in the upper part of the Lower Gedinnian (Carls 1996d, DCT interval di <2.7 to 3.5 cm) in the Celtiberian Chains and the Sierra de Guadarrama (Spain). A faunal element showing a typical Gedinnian aspect from the SW Dra Valley is *Howellella* ex gr. *mercurii* (Gosselet 1880) as described by Drot (1964) from the 'zone à *Acastella tiro*'.

Trilobites and graptolites

Hollard (1963b) described a succession of typical Gedinnian *Acastella* species from the Lmhaïfid Formation at Aïn-Deliouine in the SW Dra Valley including, e.g. *A. heberti* (Gosselet 1888), *A. tiro* (Rud. & E. Richter 1954) and *A. tanzidensis* Hollard 1963 (Carls 1996b, DCT interval di 0.0 to 1.8 cm, di 2.7 to 3.5 cm, and di 4.5 to 6.0 cm). Apparently, the Lmhaïfid Formation here ends later than at the el Ayoun section, confirmed by the occurrence of Pragian fossils in the basal Assa Formation following above (see below). Hollard (1977) reported *Monograptus uniformis* Přibyl 1940, the zonal index fossil of the first graptolite zone in the Lochkovian.

Correlation and age

Due to the conodont data, the Lmhaïfid Formation as it was identified in the el Ayoun section ranges from the Lower to the lower Middle Lochkovian. Brachiopods from the same section in the NE Dra Valley indicate lower to upper parts of the Lower Gedinnian.

According to Hollard (1977) the SW succession of the Lmhaïfid Formation at Aïn-Deliouine begins above the *Scyphocrinites* Beds near the Silurian–Devonian boundary which can be recognized mainly by graptolites and conodonts. Trilobites point to the Lower Gedinnian and lower part of the Upper Gedinnian in terms of the Rhenish subdivision. The top of the formation is younger and could reach there even into the lowermost Pragian. Accordingly, Pragian index fossils have been reported from the basal limestone of the overlying Assa Formation.

Oued-el-Mdâouer Formation

Distribution, thickness and facies

In the NE Dra Valley, crinoidal limestones yielding corals and bryozoans and marly limestones of the Oued-el-Mdâouer Formation overlie the Lmhaïfid Formation. The thickness of this formation reaches about 30 to 80 m near Tata (Jansen 2001). The facies is of 'Hercynian' resp. 'Bohemian' type.

Conodont faunas C 5 to C 6

Although the Oued-el-Mdâouer Formation in the el Ayoun section is dominated by limestones conodonts are restricted to the samples L15 and L16 (see sample numbers in Lazreq & Ouanaimi 1998, pp. 224–227, text-fig. 2).

In sample L15, the conodont fauna C 5 occurs with *Icriodus eolatericrescens* (Mashkova 1968) and, as a distinct guide fossil, *Ancyrodelloides transitans* (Bischoff & Sannemann 1958) represented by a morph which possesses a basal cavity constricted in large individuals; thus it is obviously more advanced than individuals with open basal cavities from fauna C 4 of the Lmhaïfid Formation (see above). According to Murphy & Valenzuela-Ríos (1999, text-fig. 1) this late *transitans* morph, however, suggests an age still within the Middle Lochkovian, i.e. still within the (traditional) *delta* Zone resp. below the *pesavis* Zone (see Carls & Weddige 1996, DCT interval about di 3.1 to 3.8 cm).

On the contrary, sample L16 resp. conodont fauna C 6 from the overlying bed yielded *Caudicriodus curvicauda* (Carls & Gandl 1969), according to Lazreq & Ouanaimi (1998, p. 225, text-fig. 2). Due to Bultynck (2003, fig. 1), this would indicate an age not earlier than the upper *kindlei* Zone of the middle Pragian resp. DCT height di 8.4 cm. As a consequence, the whole *pesavis* Zone, the Lochkovian/Pragian boundary and even the lower Pragian would be represented by the interval between samples L15 and L16. Such a long time interval, however, is rather contrary to the limited thickness of the interval in the section, and therefore the C 6 datation should be affirmed by further samples. Moreover, a correlation of the upper Oued-el-Mdâouer Formation with the Assa Formation towards the SW because of a Pragian datation should be affirmed, too. Towards the NE, in any case, the el Ayoun limestone sequence of both the upper Lmhaïfid Formation and the lower Oued-el-Mdâouer Formation corresponds obviously well with the 'Orthoceras' Limestone in the Achguig region with regard to carbonate facies development and presumably to the datation mainly as *delta* Zone (Bultynck & Walliser 2000, Unit A). It is not clear whether the uppermost limestone beds may already represent the basal Merzâ-Akhsaï Formation.

Trilobites

Uppermost parts of the limestone succession in the el Ayoun section have yielded a Pragian trilobite fauna including the genera *Odontochile* and *Paralejurus* (see Schraut 2000). It is, however, not clear whether these beds already correlate with the basal Merzâ-Akhsaï Formation.

Correlation and age

According to conodont faunas, the Oued-el-Mdâouer Formation in the el Ayoun section starts in the *delta* Zone resp. the Middle Lochkovian and reaches into the Pragian. The formation represents an equivalent of upper parts of the Lmhaïfid Formation and at least partly of the Assa Formation in the SW Dra Valley. Upper beds of the Oued-el-Mdâouer Formation may correlate with the limestone beds of the uppermost Assa Formation; uppermost limestone beds in the section el Ayoun with Pragian trilobites possibly represent already the basal Merzâ-Akhsaï Formation. Towards the NE, the Oued-el-Mdâouer Formation is most probably an equivalent of the 'Orthoceras' Limestone in the Achguig region.

Assa Formation ('Rich 1')

Distribution, thickness and facies

The Assa Formation (Hollard 1963*a*, content modified by Hollard 1981*a*) or 'Rich 1' is typically developed as a 'rich' including basal limestones, middle shales and upper sandstones in the SW Dra Valley, between Aïn Deliouine in the SW and Foum el Hassane in the NE. In the type area near the town of Assa, the formation reaches a thickness of approximately 300 m (Hollard 1981*a*).

Conodont fauna C 7

In the limestones just below the top of the Assa Formation in the Assa I section (samples 3 and 5, Unit 5; Fig. 5, Table 1: conodont fauna C 7), questionably determinable fragments of ?*Caudicriodus angustoides castilianus* (Carls 1969) were found but particularly specimens of *Latericriodus steinachensis* (Al-Rawi 1977). The latter is a distinct index for the uppermost Lochkovian and lower to middle Pragian (Schönlaub *in* Chlupáč *et al.* 1985 pp. 24–27; Murphy 1989; Weddige 1987; Slavik 2004; Slavik & Hladil 2004, fig. 3). Because no overlap of *L. steinachensis* and *Eoctenopolygnathus pireneae* (Boersma 1974) is known world-wide up to now, it can be concluded that the middle Pragian resp. the Assa Formation is older than the *pireneae* Zone and ranges within the interval of the *sulcatus* and *kindlei* zones (see Carls & Weddige 1996, DCT interval di ∼6.0 to 9.0 cm).

Brachiopod faunas B 3 and B 4

Hollard (1977) reported *Eucharitina oehlerti* (Bayle 1878) already from the basal limestones of the Assa Formation; the genus supports a latest Lochkovian to Pragian age (Brice *et al.* 2000). *Platyorthis* sp. described by Drot (1975) from the same level represents an ancestral form still close to the Gedinnian *P. monnieri* (Rouault 1851).

The brachiopod fauna B 3 (Table 1 and Fig. 5) from the upper sandstones of the Assa Formation in the sections Assa I and Torkoz II includes a number of stratigraphically significant brachiopod taxa (Fig. 4) allowing stratigraphic alignments in the sense of the regional Rhenish subdivision as improved and modified by Carls (1987) and Jansen (2001): *Iridistrophia* sp. X *sensu* Jansen (2001) represents a primitive evolutionary stage because of its short dental lamellae; it is certainly more primitive than *I. maior* (Fuchs 1915) of the Rhenish Lower Emsian and more evolved than the Gedinnian *I. euzona* (Fuchs 1919).

Ctenochonetes ex gr. *aremoricensis* Racheboeuf 1976 is an advanced taxon which is similar to congeneric forms from the Siegenian of Western Europe (comp. Racheboeuf 1981). *Mclearnites saharianus* Jansen 2001 is more primitive than *M. cherguiensis* Jansen 2001 from the overlying Merzâ-Akhsaï Formation which supports a Middle Siegenian age (see below). *Platyorthis hollardi* Jansen 1999 is phylogenetically intermediate between the Gedinnian *P. monnieri* (Rouault 1851) and the Middle Siegenian *P. circularis taunica* (Fuchs 1915), being more similar to *P. c. taunica* than to *P. monnieri*. Very similar forms of *Platyorthis* occur in upper parts of the Nogueras Formation of the Celtiberian Chains, i.e. the d2c beta (material studied by U. Jansen, see also

Jansen 2001), representing upper parts of the Lower Siegenian according to Carls (1987). Accordingly, in the sense of the DCT (Carls 1996d), the upper Assa Formation falls into the interval di 8.4 to 9.1 cm.

Rhenoschizophoria torkozensis (Jansen 2001) is much more developed than *R. runegatensis* (Renouf 1972) from the Gedinnian of the Armorican Massif in NW France and slightly less developed than

R. provulvaria (Maurer 1886) from the Middle Siegenian of the Rheinisches Schiefergebirge in Germany.

Filispirifer cf. *merzakhsaiensis* Jansen 2001 is a large, plicate spiriferid of 'Siegenian aspect', with strong apical callosities and lacking free crural plates. Forms determined as *Vandercammenia trigeri* (de Verneuil 1850) from the Rhenish Middle Siegenian (Dahmer 1940, material re-studied by U. Jansen) are more advanced than the specimens from the upper Assa Formation as indicated by a trend towards a greater shell-size and more plications. The latter are comparable to Lower Siegenian forms from the Néhou Formation in Brittany described by Gourvennec (1989) and contemperaneous ones from the upper Nogueras Formation in the Celtiberian Chains (Carls 1996c), whereas forms from the Middle or Upper Siegenian Faou Formation in Brittany (Gourvennec 1989, Pl. 7, text-figs 3–5) and the top of the Santa Cruz Formation (coll. Carls) are again representing a higher phylogenetic development.

The genus *Dixonella* is transitional between *Hysterolites* and *Euryspirifer*, the first being regarded as the ancestor of the latter. *Euryspirifer* appears first in the Middle Siegenian (Carls 1987). *Dixonella assaensis* Jansen 2001 from the shales near the top and above the terminal sandstones of the Assa Formation (Assa I section, Unit 4b, brachiopod fauna B 4: Table 1 and Fig. 5) is similar to 'Acrospiriferinae n. gen. N, n. sp. N' *sensu* Carls & Heddebaut (1980) from high in the Nogueras Formation in the Celtiberian Chains, unit d2c beta 8 (Carls & Heddebaut 1980, p. 22; Carls 1987, p. 101, text-fig. 7) corresponding to DCT di 8.9 to 9.1 cm (Carls 1996d). *Dixonella rouaulti* (Gourvennec 1988) from the upper Céneré Formation of the Armorican Massif may be more primitive and somewhat older. In the same unit, again *Ctenochonetes* ex gr. *aremoricensis* Racheboeuf 1976 and *Oligoptycherhynchus* cf. *daleidensis* (C. F. Roemer 1844) occur.

In the lowermost Siegenian of the Armorican Massif and the Celtiberian Chains, species of *Platyorthis*, *Schizophoria*, *Mclearnites*, *Ctenochonetes* and *Vandercammenina* occur, which are more primitive than their congeneric relatives from the upper Assa Formation (Carls 1987 and observations by U. Jansen). Finally, the absence of *Euryspirifer* and *Arduspirifer* in spite of suitable facies suggests that the Assa Formation is older than the Middle Siegenian. From the total brachiopod fauna, it is concluded that the upper Assa Formation has a rather late Early Siegenian age.

Tentaculitids and trilobites

Nowakia acuaria (Rh. Richter 1854) has been described from the Assa Formation (Lardeux 1969, p. 96) having its entry in the basal limestones (Hollard 1977, pp. 191). Therefore, this level indicates already the Pragian.

Schraut (2000) described several trilobite taxa of Rhenish aspect from the upper Assa Formation which he considered to be of late Pragian age. However, *Pseudocryphaeus munieri* (Oehlert 1877) is known from the d2c alpha and beta in the Celtiberian Chains (Gandl 1972, pp. 100–101), which is well consistent with the brachiopod-based correlation. In general, Siegenian trilobites are still little known and require revision.

Fig. 4. 1–7, Brachiopods from the upper Merzâ-Akhsaï Formation ('Rich 2' sandstones, Middle to Upper Siegenian). 1–2, *Rhenorensselaeria* cf. *strigiceps* (Roemer 1843). **1**, Internal mould of conjoined specimen; SMF 65040—Section Anorhrif I. **2** Internal mould of conjoined specimen and latex replica; SMF 65044—Section Anorhrif I. **2a**, Internal mould, posterior view. **2b**, latex replica of apical region. **3**, '*Mclearnites*' *cherguiensis* Jansen 2001; Internal mould of ventral valve; holotype SMF 59295—Section Anorhrif II, Unit 3. **4–5**, *Tropidoleptus carinatus* ssp. A *sensu* Jansen 2001. **4**, Internal mould of dorsal valve; SMF 59058—Loc. Meskaou I. **5**, External mould of dorsal valve; SMF 59046—Loc. Meskaou I. **6**, *Hysterolites* sp.: Internal mould of dorsal valve; SMF 65029—Loc. Tiguisselt. **7**, *Arduspirifer maroccanicus* Jansen 2001: Internal mould of ventral valve; holotype SMF 65064/24—Loc. Anorhrif III. **8–20**, Brachiopods from the upper Assa Formation (Rich 1 sandstones, upper Lower Siegenian, lower or middle Pragian). **8–9**, *Platyorthis hollardi* Jansen 1999. **8**, Internal mould of dorsal valve; SMF 59208—Section Assa I, Unit 2. **9**, Internal mould of ventral valve; SMF 59203—Section Assa I, Unit 2. **10–11**, *Rhenoschizophoria torkozensis* (Jansen 1999). **10**, Internal mould of dorsal valve; SMF 59228—Section Torkoz II, Unit 9. **10a**, posterior view; **10b**, dorsal view. **11**, Internal mould of ventral valve; SMF 59234a—Section Torkoz II, Unit 6. **12–14**, *Ctenochonetes* ex gr. *aremoricensis* Racheboeuf 1976. **12**, External mould of ventral valve; SMF 59021—Section Assa I. **13**, Internal mould of dorsal valve; SMF 59231—Section Torkoz II, Unit 6. **14**, Internal mould of ventral valve; SMF 59321—Section Assa I. **15**, '*Mclearnites*' *saharianus* Jansen 2001: Internal mould of ventral valve; holotype, SMF 59275—Section Timziline, Unit 1. **16**, *Iridistrophia* sp. X *sensu* Jansen 2001: Internal mould of ventral valve; SMF 65833—Loc. Assa II. **17–18**, *Dixonella assaensis* Jansen 2001. **17**, Internal mould of conjoined specimen; SMF 59338—Section Assa I, Unit 4b. **17a**, dorsal view. **17b**, ventral view. **18**, Internal mould of ventral valve; SMF 65241—Section Assa I, Unit 4b. **19**, *Vandercammenina trigeri* (de Verneuil 1850): Internal mould of ventral valve; SMF 59181—Section Assa I, Unit 3. **20**, *Filispirifer* cf. *merzakhsaiensis* Jansen 2001: Internal mould of ventral valve; SMF 65022—Section Assa I, Unit 3.

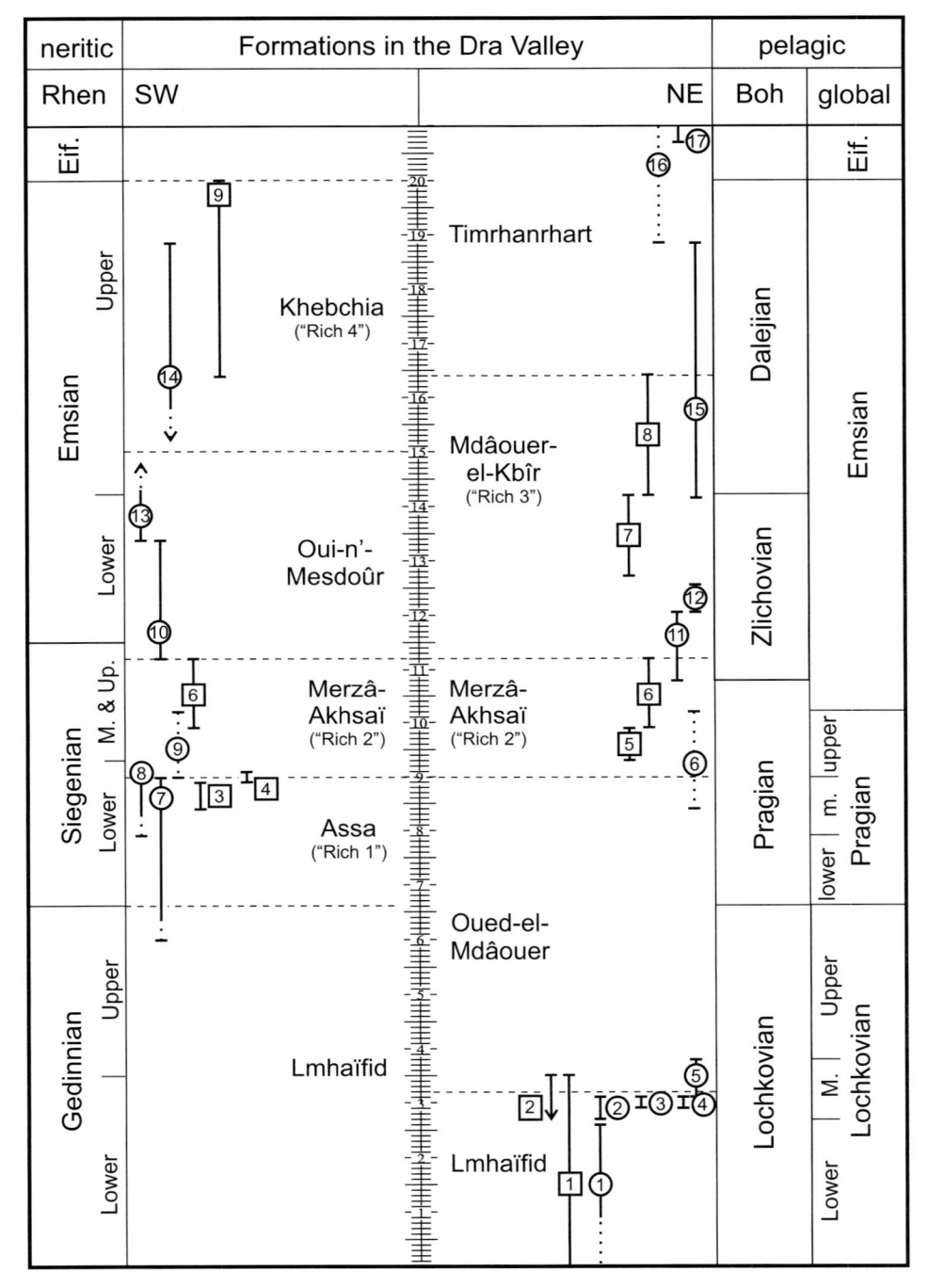

Fig. 5. Ranges of Lower Devonian brachiopod and conodont faunas of the Dra Valley. Numbers of brachiopod (in square frames) and conodont faunas (in circles) are positioned with respect to their lithostratigraphic occurrences within the sequence of Dra Valley formations. The vertical range line, adjacent to each number, demonstrates the uncertainty interval of the biostratigraphic dating with respect to the DCT time ruler (=median cm scale). 'Rhen', Rhenish chronostratigraphy; 'Boh', Bohemian chronostratigraphy; 'global', global chronostratigraphy with stage boundaries defined by GSSPs. For detailed lithostratigraphy, see R. T. Becker, Jansen *et al.* (2004).

Correlation and age

Due to the age of the underlying Lmhaïfid Formation, the Assa Formation probably starts in the uppermost Lochkovian. However, the occurrence of *Nowakia acuaria* in the basal Assa Formation—as mentioned above—pleads already for a Pragian age; a diachronous onset of the formation has also to be taken into consideration. The Assa Formation has no age equivalents documented by marine faunas in the Lower Devonian of the Rhenish Mountains. Phylogenetically, the brachiopod fauna of the upper Assa Formation contains forms which are intermediate between the classical Gedinnian and Middle Siegenian faunas. In the Celtiberian Chains, the Cantabrian Mountains and the Armorican Massif, similar faunas occur in units regarded as upper Lower Siegenian by Carls (1987). In terms of the DCT (Carls 1996d), a level from 8.4 to 9.1 cm is most probable. The icriodid conodont fauna from the upper Assa Formation shows an early to middle Pragian age in the GSSP sense, older than the *pireneae* Zone resp. lower than 9.0 cm in the DCT.

Merzâ-Akhsaï Formation ('Rich 2')

Distribution, thickness and facies

The Merzâ-Akhsaï Formation or 'Rich 2' (Hollard 1963a, 1981a) is typically developed between Aouinet Torkoz in the SW and Tata in the NE. It is a typical 'rich' with 10 to 20 m basal limestone beds, followed by approximately 100 m shales, siltstones and single sandstone beds and about 50 m of 'Rich 2' sandstone beds at the top.

Conodont faunas C 8 to C 9

The basal limestones of the Merzâ-Akhsaï Formation have yielded single conodont elements in a rather poor preservation.

In the section Torkoz IIa, only one conodont fragment could be determined as *Latericriodus* sp., aff. *L. beckmanni* (Ziegler 1956).

In the section Assa I, however, in its 2.7 m basal limestone beds (Unit 7) conodont fragments are more diverse, at 0.3–0.35 m (Fig. 5, Table 1: conodont fauna C 8) with individuals of *Latericriodus steinachensis* (Al-Rawi 1977), ?*Caudicriodus angustoides castilianus* (Carls & Gandl 1969), ?*C. curvicauda* (Carls & Gandl 1969), and at 0.75 m (Fig. 5, Table 1: conodont fauna C 9) with ?*L. claudiae* (Klapper 1980), *Latericriodus* sp., aff. *L. beckmanni* (Ziegler 1956), and *Wurmiella excavata* (Branson & Mehl 1933). The association of *L. steinachensis* and ?*C. curvicauda* in C 8 points to a middle Pragian age of about the *kindlei* Zone in the sense of the Czech regional stages (Schönlaub in Chlupáč et al. 1985, pp. 24–27; Weddige 1987; Slavik 2004), i.e. DCT interval di ~7.9 to 9.0 cm. Because of the icriodid form aff. *L. beckmanni*, Fauna C 9, however, this possibly represents a later Pragian age above the *L. steinachensis* range, e.g. the *pireneae* Zone resp. the DCT interval di 9.0 to 10.2 cm.

Brachiopod faunas B 5 and B 6

The Merzâ-Akhsaï Formation contains diverse brachiopod faunas (see Figures 4 and 6). *Euryspirifer africanus* Jansen 2001 from middle shales (brachiopod fauna B 5: Table 1 and Fig. 5) of the formation is an early representative of the genus that is well comparable with forms from the Middle and Upper Siegenian of the Rheinisches Schiefergebirge and the Ardennes (Godefroid 1994; Siegenian forms determined as *dunensis;=Euryspirifer* sp. 3 *sensu* Jansen 2001). In the same level, an ancestral *Arduspirifer* sp. occurs.

The brachiopod fauna B 6 (Table 1 and Fig. 5) from the upper sandstones includes a number of biostratigraphically indicative taxa. *Arduspirifer maroccanicus* Jansen 2001 from the upper sandstones is more primitive than the Lower Emsian *A. arduennensis prolatestriatus* Mittmeyer 1973. *Hysterolites* sp. is a typical Siegenian form. *Filizpirifer merzakhsaiensis* Jansen 2001 (='*Acrospirifer*' in older literature) shows close similarities to congeneric Middle or Upper Siegenian forms of the upper Santa-Cruz Formation in the Celtiberian Chains (Carls & Valenzuela-Ríos 1998) and the Montguyon Formation in the Armorican Massif (Gourvennec 1989).

The large strophomenid '*Mclearnites*' *cherguiensis* Jansen 2001 which occurs in many sections of the Dra Valley is similar to *Boucotstrophia herculea* (Drevermann 1904) from the Middle Siegenian of the Rheinisches Schiefergebirge. The differences between the two genera, mainly in the shell profile, are regarded as rather ecomorphotypic than phylogenetic, so that a comparable evolutionary level is assumed (Jansen 2001). *Iridistrophia anorhrifensis* Jansen 2001 is especially closely related to a congeneric form occurring in the Augustenthal Beds of the Rhenish Middle Siegenian (Dahmer 1932); it is also close to the Lower Emsian *I. maior* (Fuchs 1915) (see discussion in Jansen 2001).

Tropidoleptus carinatus ssp. A *sensu* Jansen (2001) is very similar to Upper Siegenian to Lower Emsian *T. c. rhenanus* Frech 1897 from the Rheinisches Schiefergebirge. *Proschizophoria* sp. F *sensu* Jansen (2001) is—as indicated by its large shell and diffuse ventral muscle-bounding ridges—an advanced form within the genus and closely related to the Rhenish *P. personata* (Zeiler 1857) from the Middle to Upper Siegenian.

Table 1. Succession of studied Dra Valley brachiopod and conodont faunas. For numbers of faunas and DCT interval values compare with Figure 5 and text

Brachiopod faunas		Conodont faunas	
		C 17 *Eucostapolygnathus costatus costatus* ?*Linguipolygnathus cooperi cooperi* ?*Linguipolygnathus linguiformis pinguis* ?*Polygnathus benderi* ?*Polygnathus trigonicus*	dm 0.7–2.3
B 9 *Arduspirifer* cf. *steiningeri* *Alatiformia* sp. *Iridistrophia* cf. *hipponyx* late morph	di 16.4–≥20.0	C 16 *Eucostapolygnathus costatus* ssp. C 15 *Caudicriodus* ?*culicellus* *Icriodus fusiformis* *Icriodus corniger rectirostratus*	di ?18.85–dm ?3.4 di 14.15–18.85
B 8 *Arduspirifer* ard. *arduennensis* *Euryspirifer* cf. *robustiformis* *Iridistrophia* cf. *hipponyx* early morph *Platyorthis circularis* cf. *circularis* *Pachyschizophoria tataensis* late morph	di 14.2–16.4	C 14 *Caudicriodus culicellus culicellus* *Iciodus corniger ancestralis* *Iciodus corniger leptus* *Icriodus corniger rectirostratus* *Icriodus fusiformis* *Icriodus* sp., aff. *I. werneri*	di ~14.15–18.85
B 7 *Arduspirifer* ard. cf. *arduennensis* *Euryspirifer* cf. *pellicoi* *Iridistrophia* cf. *maior* *Pachyschizophoria tataensis* early morph	di 12.7–14.2	C 13 '*Ozarkordina*' *steinhornensis miae* ?*Eucostapolygnathus inversus* *Latericriodus bilatericrescens*	di 13.35–~14.2
		C 12 '*Ozarkordina*' *steinhornensis miae* *Latericriodus beckmanni sinuatus* *Latericriodus multicostatus* *Lat. bilatericrescens bilatericrescens* *Latericriodus celtibericus*	di 12.05–2.55
		C 11 *Caudicriodus sigmoidalis* *Eucostapolygnathus excavatus* *Latericriodus celtibericus*	di 11.2–12.05
		C 10 *Caudicriodus sigmoidalis* *Caudicriodus* cf. *ultimus* *Latericriodus beckmanni* *Latericriodus bilatericrescens*	di 11.2–13.35
B 6 '*Mclearnites*' *cherguiensis* *Iridistrophia anorhifensis* *Tropidoleptus carinatus* ssp. A *Proschizophoria* sp. F *Hysterolites* sp. *Arduspirifer maroccanicus* *Rhenorensselaeria* cf. *strigiceps* B 5 *Euryspirifer africanus*	di 9.1–11.5	C 9 ?*Latericriodus claudiae* *Latericriodus* sp., aff. *L. beckmanni* *Wurmiella excavata*	di ?9.0–?10.2

	C 8 ?*Caudicriodus angustoides castilianus*	di ~7.9–9.0
	?*Caudicriodus curvicauda*	
	Latericriodus steinachensis	
	C 7 *Latericriodus steinachensis*	di ~6.0–9.0
	?*Caudicriodus angustoides castilianus*	
B 4 *Dixonella assaensis* di 8.4–9.1		
Ctenochonetes ex gr. *aremoricensis*		
Oligoptycherhynchus cf. *daleidensis*		
B 3 *Iridistrophia* sp. X		
Ctenochonetes ex gr. *aremoricensis*		
Mclearnites saharianus		
Platyorthis hollardi		
Rhenoschizophoria torkozensis		
Vandercammenina trigeri		
	C 6 ?*Caudicriodus curvicauda*	di ?8.4–?10.2
	C 5 *Ancyrodelloides transitans* late morph	di 3.1–3.8
	Latericriodus eolatericrescens	
	C 4 *Ancyrodelloides transitans* early morph	di 2.9–3.1
	C 3 *Ancyrodelloides asymmetricus*	di 2.9–3.1
	?'*Ozarkordina*' *steinhornensis*	
	Lanea omoalpha	
	Ozarkodina repetitor	
B 2 *Lanceomyonia occidentalis* di 2.9–3.5	C 2 *Caudicriodus angustoides*	di 2.7–3.1
	?'*Ozarkordina*' *eosteinhornensis*	
	Lanea omoalpha	
	Latericriodus postwoschmidti	
B 1 *Eoglossinotoechia termieri* di 0.0–2.6	C 1 *Caudicriodus angustoides*	di <0.0–2.6
	?'*Ozarkordina*' *eosteinhornensis*	

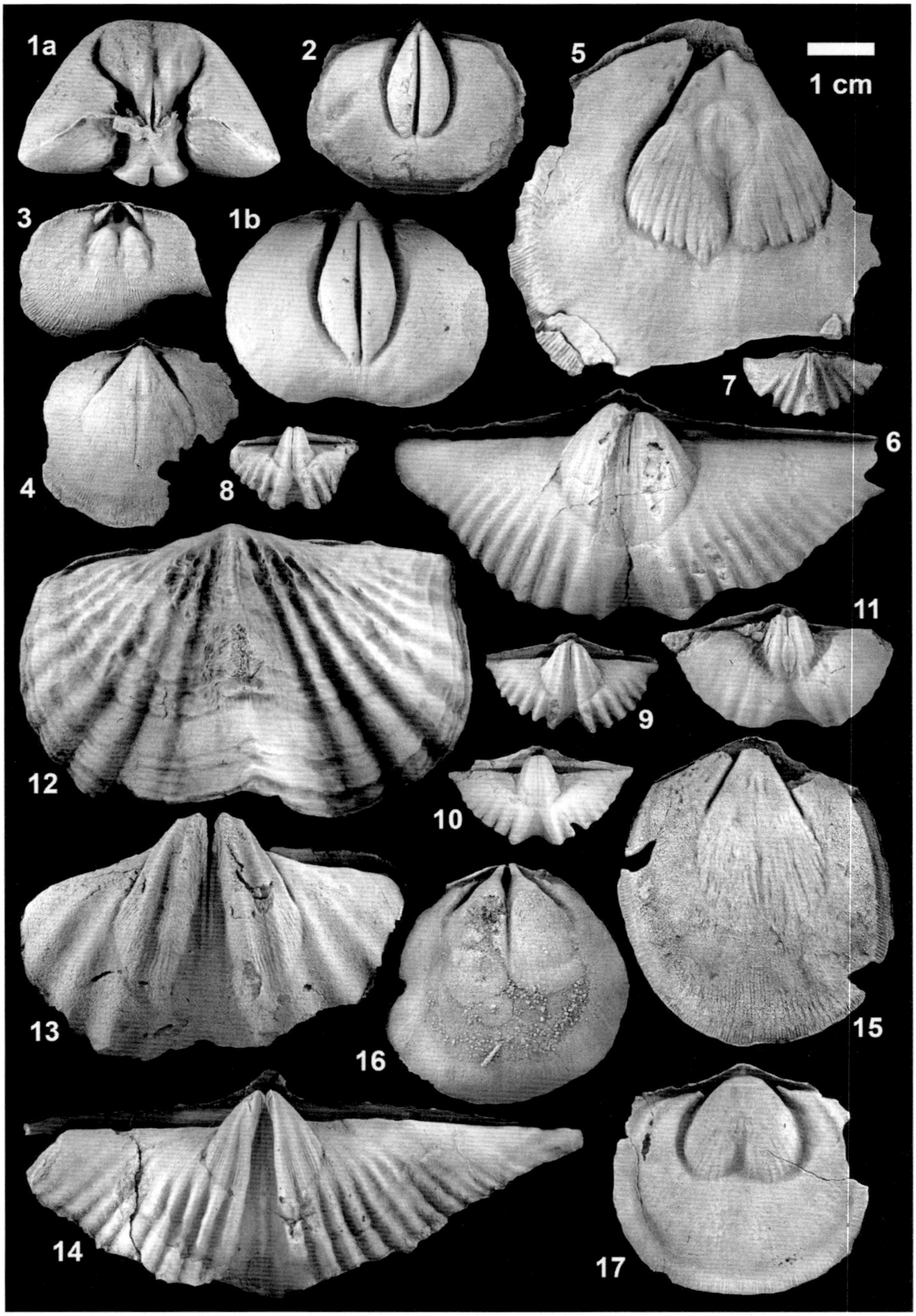
1a
2
5
1 cm
3
1b
7
6
4
8
12
11
9
10
13
15
16
14
17

The genus *Proschizophoria* has generally been considered to have gone extinct before the classical Siegenian/Emsian boundary in the German sense (Carls 1974); however, at least one specimen has recently been found in the Lower Emsian of the Celtiberian Chains (Mariposas Formation, top of the d4a beta; Carls, pers. comm. 2005), but this seems to be a very rare exception.

Finally, the occurrence of the terebratulid *Rhenorensselaeria* cf. *strigiceps* (C. F. Roemer 1844) is a strong argument in favour of the Middle to Upper Siegenian, because *R. strigiceps* is confined to this interval in the Rheinisches Schiefergebirge (Mittmeyer 1982). From the brachiopod fauna it is concluded that middle to high parts of the Merzâ-Akhsaï Formation have a Middle to Late Siegenian age corresponding to the DCT interval di 9.3 to 11.5 cm (Carls 1996c).

Trilobites

The basal limestone of the Merzâ-Akhsaï Formation at section Assa I contains specimens of the genus *Odontochile* indicating the Pragian stage. Schraut (2000) considered the trilobite faunas from the middle and upper Merzâ-Akhsaï Formation as 'uppermost Pragian' in the GSSP sense.

Correlation and age

Basal limestone beds of the Merzâ-Akhsaï Formation point to a middle to upper Pragian age (in the GSSP sense) according to the icriodid conodonts, whereas middle and upper parts of the formation can be dated as Middle to Late Siegenian in the sense of the Rhenish subdivision. Since the overlying formations start within the *excavatus* Zone resp. within the regional Zlichovian stage, it is concluded that the Merzâ-Akhsaï Formation ranges into the upper parts of the Pragian (in the sense of the Czech regional stages) resp. into the lowermost Emsian (in the GSSP sense). In comparison with the Achguig region, there is a coeval equivalent recognizable in the 'Pragian' Limestone (cf. Bultynck & Walliser 2000).

Mdâouer-el-Kbîr Formation ('Rich 3')

Distribution, thickness and facies

The Mdâouer-el-Kbîr Formation (Hollard 1978; =most of the former 'Rich d'El Annsar' or 'Rich 3' sensu Hollard, 1963a, 1965, 1967) is typically developed as a 'rich' in the NE Dra Valley between Akka and Foum Zguid. Maximal thickness of the formation (295 m) is reached near Foum Zguid at its type locality (Hollard 1978). The formation was studied mainly in the Foum Zguid section.

Conodont faunas C 11, C 12 and C 15

In the section Sidi Rezzoug I, which could not be investigated in detail, two limestone beds at the base of the Mdâouer-el-Kbîr Formation ('Rich 3') have yielded distinct conodont faunas. In the lowermost part (Fig. 5, Table 1: conodont fauna C 11), *Eucostapolygnathus excavatus* (Carls & Gandl

---

Fig. 6. 1–11, Brachiopods from the upper Mdâouer-el-Kbîr Formation ('Rich 3' sandstones, upper Lower Emsian to lower Upper Emsian). 1–2, *Pachyschizophoria tataensis* (Jansen 2001). 1, Internal mould of conjoined specimen; holotype SMF 59138a—Section Anorhrif II, Unit 5. 1a, posterior view 1b, ventral view, boundary interval Lower/Upper Emsian. 2, Internal mould of dorsal valve; SMF 65528—Section Foum Zguid, Unit 9, boundary interval Lower/Upper Emsian. 3–4, *Platyorthis circularis* ssp. 3, Internal mould of dorsal valve; SMF 66094—Section Foum Zguid, Unit 14c-d, basal Upper Emsian. 4, Internal mould of ventral valve; SMF 66093—Section Foum Zguid, Unit 14c-d, basal Upper Emsian. 5, *Iridistrophia* cf. *hipponyx* (Schnur 1851): Internal mould of ventral valve; SMF 65835—Section Foum Zguid, Unit 14, basal Upper Emsian. 6, *Euryspirifer* cf. *pellicoi* (de Verneuil & d'Archiac 1845): Internal mould of ventral valve, SMF 65706—Section Tissint I, boundary interval Lower/Upper Emsian. 7–10, *Arduspirifer arduennensis* cf. *arduennensis* (Schnur 1853). 7, Internal mould of dorsal valve; SMF 66102—Section Foum Zguid, Unit 5, boundary interval Lower/Upper Emsian, more probably upper Lower Emsian. 8, Internal mould of ventral valve; SMF 65583—Section Foum Zguid, top Unit 12, boundary interval Lower/Upper Emsian. 9, Internal mould of ventral valve; SMF 66101—Section Foum Zguid, Unit 5, upper Lower Emsian. 10, Internal mould of ventral valve; SMF 65587—Section Foum Zguid, top Unit 12, boundary interval Lower/Upper Emsian. 11, *Arduspirifer arduennensis arduennensis* (Schnur 1853). Internal mould of ventral valve; SMF 66105—Section Foum Zguid, Unit 14a–b, basal Upper Emsian. 12–17, Brachiopods from the Merzâ-Akhsaï Formation ('Rich 2' sandstones, Middle to Upper Siegenian). 12–13, *Filispirifer merzakhsaiensis* Jansen 2001. 12, Internal mould of dorsal valve; holotype SMF 59103a—Section Tadoucht II. 13, Internal mould of ventral valve; SMF 59121—Loc. Meskaou I. 14, *Euryspirifer africanus* Jansen 2001: Internal mould of ventral valve; holotype SMF 59061—Section Tadoucht II, Unit 3. 15, *Iridistrophia anorhrifensis* Jansen 2001: Internal mould of ventral valve; holotype SMF 59837—Section Anorhrif II. 16–17, *Proschizophoria* sp. F *sensu* Jansen (2001). 16, Internal mould of ventral valve; SMF 59393—Section Torkoz IIa, base of Unit 21a. 17, Internal mould of dorsal valve; SMF 59288—Section Torkoz IIa, base of Unit 21a.

1969), *Caudicriodus sigmoidalis* (Carls & Gandl 1969), and *Latericriodus celtibericus* (Carls & Gandl 1969) indicate the *excavatus* Zone—most probably its lower part, because this conodont association defines 'conodont step 17' *sensu* Carls (1996a, DCT interval di 11.2 to 12.05 cm). In black limestones a little above (Fig. 5, Table 1: conodont fauna C 12), *L. multicostatus* (Carls & Gandl 1969), *L. bilatericrescens bilatericrescens* (Ziegler 1956), *L. celtibericus*, and ?'*Ozarkodina*' *steinhornensis miae* (Bultynck 1971) are present; this association represents the 'conodont step 18' *sensu* Carls (1996a, DCT interval di 12.05 to 12.55 cm). The base of the Mdâouer-el-Kbîr Formation thus appears to be distinctly correlative with the lower Zlichovian stage in Bohemia.

Bultynck & Hollard (1980, pp. 26, 33, samples 27-1 and 27-2) identified *Caudicriodus* ?*culicellus* (Bultynck 1976), *Icriodus fusiformis* Carls & Gandl 1969, and *I. corniger rectirostratus* Bultynck 1970 from shell beds intercalated in the upper sandstones of the formation (Fig. 5, Table 1: conodont fauna C 15). The fauna correlates with Upper Emsian faunas from the 'Grauwacke de Hierges' in the Ardenne Mountains (Bultynck & Hollard 1980, p. 26, Faune Vb) attributed to the *laticostatus* to *serotinus* zones (after Bultynck *et al.* 2000, cf. Weddige & Requadt 1985) resp. DCT interval of 14.15 to 18.85 cm (Bultynck *et al.* 2004). Moreover, Becker, Bockwinkel *et al.* (2004) suggested that such icriodid faunas are of high correlative value in the basal Upper Emsian below the onset of polygnathid faunas of the *serotinus* Zone.

Tentaculitids

Bultynck & Hollard (1980, p. 22) reported *Viriatellina pseudogeinitziana* Bouček 1964 and *V. hercynica* Bouček 1964 from the base of the Mdâouer-el-Kbîr Formation and noted the absence of *Nowakia acuaria* (Rh. Richter 1854). They attributed these limestones to the Lower Emsian. Lardeux (1969, pp. 106–108, text-fig. 77) described *N. praecursor* Bouček 1964 from the basal limestone of the Mdâouer-el-Kbîr Formation. The species characterizes the middle Zlichovian in Bohemia (e.g. Chlupáč 1982, p. 362; Lukeš & Chlupáč 1998) resp. DCT interval di 10.75 to 12.4 cm (Alberti 1996); it is also present in the Lower Emsian Hunsrück-Schiefer of the Rheinisches Schiefergebirge (Alberti 1982). This is in good agreement with the datation by conodont faunas C 11 and C 12 from the basal limestones.

Brachiopod faunas B 7 and B 8

Brachiopods from sandstones of the upper Mdâouer-el-Kbîr Formation (see Fig. 6) are of outstanding biostratigraphic potential. Most informative in this respect is the Foum Zguid section in the NE Dra Valley (see Fig. 1 and Jansen, G. Becker *et al.* 2004a) bearing highly diverse and abundant faunas.

In previous works (Drot 1964; Hollard 1978; Bultynck & Hollard 1980), all specimens of *Arduspirifer* from the Mdâouer-el-Kbîr Formation have been determined as *A. arduennensis arduennensis* (Schnur 1853). According to the range of this subspecies in the Ardenno-Rhenish Mountains, the productive beds as a whole were attributed to the Upper Emsian. However, more recent studies (Jansen 2001) have shown that several taxa are present. Some specimens from the Foum Zguid section, units 5 and 12 (Fig. 6, specimens 7–10), which are provisionally determined as *A. a.* cf. *arduennensis* herein, are also rather similar to *A. adradensis* García-Alcalde 2004 from the upper Lower Emsian La Pedrosa Formation in the Province of Leon, North Spain (see García-Alcalde 2004). Starting with Unit 14 of the section Foum Zguid, some specimens (e.g. Fig. 6, specimen 11) are hardly distinguishable from Upper Emsian *A. a. arduennensis* as known from the Rheinisches Schiefergebirge.

Euryspirifer cf. *pellicoi* (de Verneuil & d'Archiac 1845) occurs very abundantly in the section; it is typical of the interval around the Lower/Upper Emsian boundary (Truyóls-Massoni & García-Alcalde 1994). Specimens of *Euryspirifer* from units 9 to 14 are very similar to *E. robustiformis* Mittmeyer 1972 from the basal Upper Emsian of the Rheinisches Schiefergebirge (Emsquarzit and Wiltz formations).

Representatives of *Iridistrophia* from the Foum Zguid section, Unit 7, are close to *I. maior* (Fuchs 1915) known from the Rhenish Lower Emsian whereas forms from Unit 14 resemble early morphs of *I.* cf. *hipponyx* (Schnur 1851) *sensu* Jansen (2001) from the Lahnstein substage (lower Upper Emsian) in the Rhein/Mosel/Lahn area of the Rheinisches Schiefergebirge (Weddige 1996c, DCT interval di 14.2 to 16.35 cm).

Stropheodontids of the *gigas* group ('*Boucotstrophia*' pro parte) range from the Middle Siegenian to the lower Upper Emsian in the Rheinisches Schiefergebirge (Lahnstein substage). In the sandstones of the upper Mdâouer-el-Kbîr Formation, the group is represented by '*Stropheodonta*' *jahnkei* (Aït Malek *et al.* 2000).

Platyorthis circularis ssp. from units 14c–d resembles *Pl. c. circularis* (Sowerby 1842) from the Stadtfeld Formation (upper Lower Emsian) of the Eifel region, but seems to be even more similar to *Pl. c. transfuga* (Walther 1903) from the lower Upper Emsian Hohenrhein Formation of the Middle Rhein/Mosel/Lahn region; however, both taxa still need to be revised. The latter occur especially in units 14c–d. *Pachyschizophoria tataensis* (Jansen 2001) from Unit 9 of the Foum Zguid section represents a less advanced evolutionary level than *Pachyschizophoria* sp. C *sensu* Jansen (2001) from the lower Upper Emsian (Emsquarzit Formation) of the Rheinisches Schiefergebirge. A very similar form is '*Schizophoria vulvaria*' *sensu* Mélou (1981) (non von Schlotheim 1820) from the Marettes Formation of the Armorican Massif (Mélou 1981, Pl. 16 text-figs 5–10). The accompanying tentaculitids *Nowakia* cf. *praecursor* Bouček 1964 and *N.* cf. *zlichovensis* Bouček 1964 indicate a Zlichovian age (Morzadec *et al.* 1981). Closer to the Rhenish species is a late morph of the same species from Unit 14 which already supports a lower Upper Emsian position.

It is concluded from the brachiopod faunas B 7 and B 8 (Table 1 and Fig. 5) that the upper Mdâouer-el-Kbîr Formation reaches from the upper Lower Emsian to the lower Upper Emsian in the Rhenish sense, resp. from about the Vallendar to the Lahnstein substages (DCT interval di 12.7 to 16.4 cm). The Lower/Upper Emsian boundary may be drawn in the Foum Zguid section a little below Unit 14. According to the observed changes, the brachiopod fauna B 7 from units 2 to 13 shows a less advanced phylogenetic development than the fauna B 8 from units 14 and 15.

The conodont fauna C 15 (Table 1 and Fig. 5) from shell beds high in the sandstones (see above) attributed to the *laticostatus* to *serotinus* zones correlates well with Upper Emsian faunas in the Rheinisches Schiefergebirge, where accompanying brachiopod faunas with *Arduspirifer arduennensis arduennensis* (Schnur 1853) allow an assignment to the lower and middle Upper Emsian (Weddige & Requadt 1985).

Current studies show that the upper boundary of the Mdâouer-el-Kbîr Formation is slightly older in the SW, e.g. in the section Oufrane near Tata (Ebbighausen *et al.* 2004).

Goniatites

The report of the primitive goniatite *Erbenoceras advolvens* (Erben 1960) in the basal limestones of the Mdâouer-el-Kbîr Formation (see Hollard 1963c, 1978; Bultynck & Hollard 1980, p. 22) indicates a Zlichovian age (R. T. Becker 1996b, DCT interval di 11.3 to 12.1 cm).

In Unit 14b of the Foum Zguid section, two goniatite fragments have been found which have been determined by O. H. Walliser as *Mimagoniatites fecundus* (Barrande 1865) indicating an interval from the upper Zlichovian to the lower Dalejian in the sense of the Bohemian subdivision (Chlupáč & Turek 1983; Chlupáč 1998b, DCT interval di 13.5 to 16.1 cm). Hollard (1978, Table 2) reported ('niveau 2') *Latanarcestes* cf. *noeggerathi* (von Buch 1832) from the upper Mdâouer-el-Kbîr Formation and *Latanarcestes* spp. (R. T. Becker 1996a, b, DCT interval di 15.1 to 16.4 cm). Chlupáč & Turek (1983, p. 136) attributed this level to the lower Dalejian. The Mdâouer-el-Kbîr Formation is overlain by trilobite limestones and the *Sellanarcestes* Limestone of the Timrhanrhart Formation which are clearly within the Upper Emsian (R. T. Becker 1996a, b, DCT interval di 16.4 to 17.5 cm; approx. *serotinus* Zone).

Correlation and age

The abrupt change from sandstones of the Merzâ-Akhsaï Formation to basal limestones of the Mdâouer-el-Kbîr Formation resp. the Oui-n'-Mesdoûr Formation (see below) has been referred to the Basal Zlíchov Event (*sensu* Chlupáč & Kukal 1988) which is visible in many sections of Western and Central Europe (e.g. García-Alcalde 1997). However, more research is required to prove the synchronicity of this event. The basal limestone of the Mdâouer-el-Kbîr Formation starts within the *excavatus* Zone and contains dacryoconarid tentaculitids, conodonts, and goniatites of the Zlichovian stage in Bohemia resp. the middle Lower Emsian (in the GSSP sense). Hollard (1978) assumed a break in sedimentation or a very condensed interval in the lowermost Emsian (in the Rhenish sense). It is to be stressed that it is not yet clear how long is the time span between the Middle/Late Siegenian brachiopod faunas below and the first anetoceratids above, so that our observations do not necessarily suggest that the first anetoceratids occur low in the Lower Emsian in the German sense as appears to be consistent with the probable Ulmen age of the Hunsrück-Schiefer. On the other hand, observations in the Tafilalt region by Klug (2001) suggest that anetoceratids appear earlier than thought before.

The basal limestones correlate well with Klug's Unit A, yielding a similar conodont fauna associated with the mimosphinctid goniatite *Chebbites* Klug 2001 and the Unit B 1 with *Erbenoceras advolvens* above. With respect to the Achguig area, the lower parts of the basal limestone well correspond with the '*Jovellania*' Limestone, because that limestone is also assigned to the *excavatus* Zone; above, shales with very diverse early goniatites (Klug 2001) and the *Anetoceras* Limestone follow (Bultynck & Walliser 2000).

The upper sandstones of the Mdâouer-el-Kbîr Formation contain diverse brachiopod faunas showing an evolutionary development from the late Early Emsian to the early Late Emsian (both in the Rhenish sense). The boundary between the Lower and the Upper Emsian (in the Rhenish sense) can be approximately recognized by brachiopods and roughly corresponds to the Zlichovian/Dalejian boundary in Bohemia according to the goniatites. The top of the formation is obviously positioned within the Upper Emsian resp. Dalejian demonstrated by all fossil groups. According to conodonts and particularly to ammonoids, a position from the *laticostatus* to *serotinus* zones is most probable.

Oui-n'-Mesdoûr Formation

Distribution, thickness and facies

The Oui-n'-Mesdoûr Formation (Hollard 1978) is developed in Hercynian facies; it consists of limestones and marly limestones reaching a thickness of up to 45 m. It is confined to the SW Dra Valley. Its type locality is situated near Aouinet Torkoz. The formation has been subdivided into Akhal Tergoua and Black Marl members (R. T. Becker, Jansen *et al.* 2004).

Conodont faunas C 10 and C 13

The limestone beds become more distinct towards the top of the bed 21b in the section Torkoz IIa. Although the conodont elements belong to the shallow water facies, are fragmentary, and stratigraphically less significant, the Oui-n'-Mesdoûr Formation in its basal 0.3 m (with the sample horizons 21b-1 to 21b-4; Fig. 5, Table 1: conodont fauna C 10) include particularly *Caudicriodus sigmoidalis* (Carls & Gandl 1969), but also *Latericriodus beckmanni* (Ziegler 1956), *L. bilatericrescens* (Ziegler 1956) and *C.* cf. *ultimus* Weddige 1985. This conodont fauna can be assigned more or less to the *excavatus* Zone of the Lower Emsian (see Carls & Weddige 1996, DCT interval di 11.2 to 13.35 cm). In the Barrandian area, this conodont zone indicates the Zlíchov Formation resp. the 'Zlichovian' as a Czech regional stage (see Chlupáč 1998a, DCT interval di 10.5 to 13.8 cm).

In the beds 21b-5 to 21b-7 (Fig. 5, Table 1: conodont fauna C 13), which lie 6.5 to 8.2 m above the base of the formation, a recognizably younger Emsian conodont fauna occurs because of ?*Eucostapolygnathus inversus* (Klapper & Johnson 1975), ?'*Ozarkodina*' *steinhornensis miae* (Bultynck 1971), *Latericriodus bilatericrescens* (Ziegler 1956) and *L. beckmanni sinuatus* (Klapper *et al.* 1978). This assemblage pleads for the *nothoperbonus* Zone (see Carls & Weddige 1996, DCT interval di 13.35 to 14.2 cm). The *inversus* Zone, however, cannot be excluded, because of a questionably determinable ?*E. inversus* fragment; that would indicate already the stratigraphic transition to the Dalejian as a Barrandian regional stage (see Chlupáč 1998a, DCT ≥ di 14.2 cm).

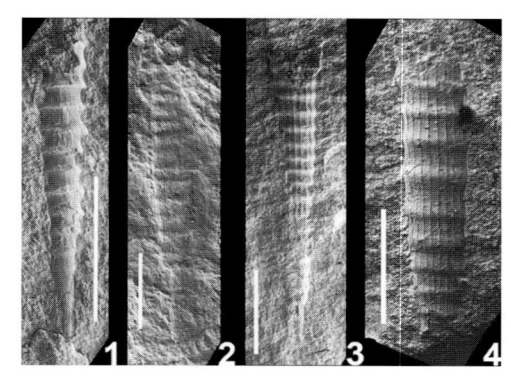

Fig. 7. Selected tentaculitids. **1**, *Nowakia* (*Dmitriella*) *sulcata sulcata* (F. A. Roemer 1843). Partly preserved shell; SMF 37798, Timrhanrhart Formation, section Foum Zguid 16-k-1. Scale 1 mm. **2**, *Viriatellina* cf. *pseudogeinitziana* Bouček 1964. Poorly preserved specimen; SMF 37799, Oui-n'-Mesdoûr Formation, section Torkoz IIa, middle part of unit 21. Scale 1 mm. **3**, *Viriatellina* cf. *hercynica* Bouček 1964. Partly preserved shell; SMF 37800, lower Khebchia Formation, section Torkoz IIa 22b$_3$. Scale 1 mm. **4**, *Viriatellina* cf. *hercynica* Bouček 1964. Detail of partly preserved shell; SMF 37801, lower Khebchia Formation, section Torkoz IIa 22b$_{Top}$. Scale 1 mm.

Tentaculitids

In the Torkoz IIa section, dacryoconarid tentaculitids occur in the middle part of the formation (Unit 21). Besides stratigraphically insignificant taxa (e.g. of the styliolinids), *Viriatellina* cf. *pseudogeinitziana* Bouček 1964 is present (Fig. 7). Similar forms have also been found in the Mdâouer-el-Kbîr section near Foum Zguid by Bultynck & Hollard (1980, p. 22).

Brachiopods and goniatites

Small and smooth-shelled brachiopods occur but are without any biostratigraphic value. In upper parts of the Oui-n'-Mesdoûr Formation, *Mimagoniatites* has been found which has its onset in the upper Zlichovian and ranges up into the Dalejian (Chlupáč & Turek 1983).

Correlation and age

According to conodont, dacryoconarid and goniatite data, the Oui-n'-Mesdoûr Formation approximately represents the Zlichovian, also confirmed by the ages of the under- and overlying strata. In comparison with the Achguig region, the formation at its base, just like the base of the Mdâouer-el-Kbîr Formation, is correlative with the '*Jovellania*' Limestone, in its upper parts with the *Anetoceras* Limestone and blue *Mimagoniatites* Limestone (comp. Bultynck & Walliser 2000).

Khebchia Formation (including 'Rich 4')

Distribution, thickness and facies

The Khebchia Formation again represents a complete 'rich' cycle beginning with limestones and marlstones followed by shales and ending with a thick sandstone unit. It follows above the Oui-n'-Mesdoûr Formation in the SW Dra Valley. The thickness reaches up to c. 300 m. The formation has been subdivided into five members (see R. T. Becker, Jansen *et al.* 2004) which, however, are difficult to recognize in the section at Torkoz due to partly unfavourable exposure (Jansen, G. Becker *et al.* 2004b): *Hollardops* Limestone Member, Brachiopod Marl Member, *Sellanarcestes* Limestone Member, Bou Tserfine Member, 'Rich 4' Sandstone Member.

Conodont fauna C 14

The Khebchia Formation starts in the section Torkoz with a 26 m thick interval of limestone beds within marls. These beds represent a shallow water to mixed neritic–pelagic facies and contain a rather diverse icriodid conodont fauna (Fig. 5, Table 1: conodont fauna C 14) with *Caudicriodus culicellus culicellus* (Bultynck 1976), *Icriodus corniger ancestralis* Weddige 1977, *I. c. leptus* Weddige 1977, *I. c. rectirostratus* Bultynck 1970, *I. fusiformis* Carls & Gandl 1969 and *Icriodus* sp., aff. *I. werneri* Weddige 1977. This fauna shows icriodid species without lateral processes, demonstrating a higher stage of phylogenetic development. Such icriodids are known from the Upper Emsian neritic facies in the Rheinisches Schiefergebirge (e.g. Bultynck 1970; Weddige 1977) as well as from the hemipelagic goniatite facies of the Lezna Mbr. in the Palentinian Domain where they are accompanied by *Latanarcestes* (Jahnke et al. 1983).

However, the stratigraphic entry of such icriodids without lateral processes within the Upper Emsian is—hitherto—insufficiently documented, as pointed out by Weddige & Requadt (1985, pp. 363–366, text-fig. 7) who described forms that demonstrate the phylogenetic loss of lateral processes. These forms are associated with brachiopods of the lower to middle Upper Emsian in the sense of the Rhenish regional stages, e.g. *Arduspirifer arduennensis arduennensis* (Schnur 1853). There is, however, no age assignment directly to the global standard conodont zonation, because of the lack of polygnathids as decisive guide fossils. After indirect correlation, Weddige & Requadt (1985) and consequently Bultynck (2003) as well, dated a probable entry within the *laticostatus–inversus* Zone (see Weddige 1996b, within DCT interval di ~14.15 to 15.85 cm). Because the icriodids from the Khebchia Formation include the evolutionary more advanced form *I. c. rectirostratus*, an age within the interval *serotinus* to *patulus* Zone is also probable for the Khebchia Formation, as well (see Carls & Weddige 1996, DCT interval di ~15.85 to 18.85 cm). The conodont fauna allows a correlation with the upper sandstones of the Mdâouer-el-Kbîr Formation, from where the same conodont association has been reported (Bultynck & Hollard 1980; see above, fauna C 15).

Tentaculitids

In the lower Khebchia Formation (Unit 22) stratigraphically insignificant styliolinids, a badly preserved specimen of ?*Homoctenus* sp. and representatives of viriatellinid dacryoconarids are present. They include *Viriatellina* cf. *pseudogeinitziana* Bouček 1964, *V.* cf. *hercynica* Bouček 1964, and *V.* cf. *gracilistriata* (Hall 1879)—and transitional forms (comp. Jansen, G. Becker *et al.* 2004b). Even taxa unquestionably assigned to these species are difficult to be separated from each other as already pointed out by Lardeux (1969, p. 138). He assigns this group to be typical for Emsian strata preferably close to the Zlichovian/Dalejian boundary (as also indicated for *V.* cf. *hercynica* by Alberti 1998) even when *V.* cf. *hercynica* may range as high as the Emsian/Eifelian boundary level (Lardeux 1969). Due to the occurrence of *V.* cf. *gracilistriata* and forms of *V.* cf. *hercynica* transitional to *V.* cf. *gracilistriata* in the youngest bed with dacryoconarids (Bed 22e), a position in the Dalejian may be indicated. This observation fits with Lardeux (1969, p. 126) who reports *V.* cf. *gracilistriata* from the La Grange Limestone in the French Armorican Massif together with the Dalejian *Nowakia cancellata* (Rh. Richter 1854). In Truyóls-Massoni & García-Alcalde (1994, p. 223), *V.* cf. *hercynica* and *V.* cf. *gracilistriata* are figured to enter the La Ladrona–Cabo La Vela section (Asturia, Spain) in the uppermost parts of the Lower Emsian and reaching well into the Upper Emsian. In Figure 7 specimens of *V.* cf. *pseudogeinitziana* and *V.* cf. *hercynica* are shown. Stratigraphically, the viriatellinid fauna is in accordance with the age assignment by other faunal groups mentioned above and below.

Brachiopod fauna B 9

The Khebchia Formation begins with the Brachiopod Marl Member containing small and smooth-shelled brachiopods and a plectodontid form (Becker, Bockwinkel et al. 2004) that are, according to present knowledge, without any biostratigraphic significance. The 'Rich 4' sandstones, however, yielded *Arduspirifer* cf. *steiningeri* (Solle 1953), *Alatiformia* sp., and *Iridistrophia* cf. *hipponyx* (Schnur 1851) suggesting a late Late Emsian or early Eifelian age (brachiopod fauna B 9: Table 1 and Fig. 5).

Trilobites, goniatites and ostracodes

A trilobite fauna has been described from units 22c-e of the section Torkoz II, situated in the lower Khebchia Formation (Schraut 2000); it includes the following taxa: *Hollardops mesocristata* (Le Maître 1952), *Psychopyge elegans* Termier & Termier 1950, and *Diademaproetus holzapfeli praecursor* Alberti 1969. *Psychopyge elegans* is supposed to be confined to the *cancellata* to *Sellanarcestes wenkenbachi* zones of the Upper Emsian (Morzadec 1988, p. 158; Schraut 2000). *D. h. praecursor* occurs in the uppermost part of the *S. wenkenbachi* Zone (Alberti 1969, pp. 215–216), resp. DCT interval 16.4 to 17.5 cm. Bultynck & Hollard (1980) mentioned *Gyroceratites gracilis* (Bronn 1835) from the lower Khebchia Formation which has its onset near the base of the Dalejian (Chlupáč & Lukeš 1999) and characterizes the *G. gracilis* Zone of the lower Dalejian (R. T. Becker 1996a, DCT di 14.3 to 15.1 cm). In Unit 22e of the Torkoz section, *Latanarcestes noeggerathi* (von Buch 1832) indicates the *L. noeggerathi* Zone of the Lower Dalejian (upper *laticostatus* to basal *serotinus* Zone resp. DCT di 15.1 to 16.4 cm, R. T. Becker 1996a). The *Sellanarcestes* Limestone Member contains a goniatite fauna including the genera *Sellanarcestes* and *Anarcestes* corresponding to the *serotinus* Zone of the middle Upper Emsian (R. T. Becker, Jansen et al. 2004; R. T. Becker, Bockwinkel et al. 2004) resp. the DCT interval di 16.4 to 18.7 cm.

According to G. Becker et al. (2003), ostracodes from Unit 22 of the Torkoz section indicate a Late Emsian age.

Correlation and age

Due to the icriodid conodont and goniatite faunas, the Khebchia Formation seems to start in a lower part of the Upper Emsian, i.e. in the *laticostatus* Zone. The *Sellanarcestes wenkenbachi* Limestone Member indicating the *serotinus* Zone is a marker horizon which can be traced from the SW to the NE Dra Valley, where it is present within the Timranrhart Formation (see below). In the Maïder and Tafilalt areas, the same horizon is also present (Bultynck & Walliser 2000). The 'Rich 4' sandstones can only be dated on the base of the brachiopods, indicating a latest Emsian to probably early Eifelian age. It is not clear whether the Lower/Middle Devonian boundary is surpassed.

Timrhanrhart Formation

Distribution, thickness and facies

The Timrhanrhart Formation follows above the Mdâouer-el-Kbîr Formation in the NE Dra Valley, well exposed in the upper part of the Foum Zguid section (Jansen, G. Becker et al. 2004a). The formation starts with trilobite limestones overlain by a solid *Sellanarcestes* Limestone which, finally, is followed by nodular limestones, marls and greenish shales. Unfortunately, the uppermost parts of the formation are lacking in the studied section, because the succession is cut off by a dolerite. At the type locality, the total thickness reaches 37 m (Hollard 1978).

Conodont faunas C 16 to C 17

In some beds of the Timrhanrhart Formation in the Foum Zguid section (16 k-3 to 17top; Fig. 5, Table 1: conodont fauna C 16 to C 17), a few fragments of stratigraphically significant conodonts were found with *Eucostapolygnathus costatus costatus* (Klapper 1971), ?*Linguipolygnathus cooperi cooperi* (Klapper 1971), ?*L. linguiformis pinguis* (Weddige 1977), ?*Polygnathus trigonicus* Bischoff & Ziegler 1957, and ?*P. benderi* Weddige 1977. For the first conodont occurrence (sample 16 k-3) with a questionable subspecies of *E. costatus*, the Middle Devonian *partitus* Zone may be considered. The following assemblages, however, although badly preserved, plead for the *costatus costatus* Zone. The early part of the *c. costatus* Zone can more or less be assumed (see Carls & Weddige 1996, DCT interval dm 0.7 to about 2.3 cm).

Tentaculitids

The two nowakiids *Nowakia* (*Dmitriella*) *sulcata sulcata* (F. A. Roemer 1843) and *N.* (*Maureriana*) cf. *procera* (Maurer 1880) indicate the start of the Eifelian stage in the Foum Zguid section (e.g. Alberti 1993). A specimen of the former taxon from bed 16 k-1 is shown in Figure 7.

Goniatites, trilobites, brachiopods and ostracodes

The *Sellanarcestes* Limestone in the lower part of the Timrhanrhart Formation with its abundant goniatites corresponds to the *serotinus* Zone of the Upper Emsian. Specimens of the goniatites *Fidelites* and ?*Foordites* in Unit 16 k indicate the start of the Eifelian in the Foum Zguid section (Jansen, G. Becker et al. 2004a). Hollard (1978) recorded joint occurrences of *Pinacites* and *Paraspirifer cultrijugatus* (C. F. Roemer 1844) which, however, require modern revision.

The formation yielded a diverse trilobite fauna in the Foum Zguid section (det. G. Schraut, O. Vogel; see Jansen, G. Becker *et al.* 2004*a*). Up to Unit 16j, ?*Treveropyge* sp. occurs still demonstrating a Late Emsian age, whereas *Thysanopeltis* sp. in Unit 16 k already indicates the Eifelian.

Finally, a highly diverse ostracode fauna of E Thuringian provenance is present in the same section (G. Becker *in* Jansen, G. Becker *et al.* 2004*a*). The faunal association shows close relationships to that of the uppermost Emsian to lowermost Eifelian Moniello and Polentinos formations of the Cantabrian Mountains. The overall aspect is Emsian until bed 16 k-3 proved by species of several genera. Distinct *Polyzygia* species found in 16 k samples agree with the Emsian/Eifelian boundary as indicated by other fossil groups.

Correlation and age

The Timrhanrhart Formation of the Foum Zguid section begins slightly below the *Sellanarcestes* Zone in the lower part of the Dalejian or, respectively, in the Upper Emsian. Higher in the section, conodonts, dacryoconarid tentaculitids, trilobites and ostracodes allow to place the Emsian/Eifelian boundary precisely (for details, see Jansen, G. Becker *et al.* 2004*a*, Fig. 7). In the section, the *costatus* Zone is the most probable assignment for the upper parts. According to R. T. Becker, Jansen *et al.* (2004), the formation ranges to the *australis* Zone in sections where its upper part is preserved.

Summary of results

Lower Devonian successions are widely exposed in the Dra Valley (Southern Anti-Atlas, Moroccan Pre-Sahara). Thanks to lateral and vertical facies changes from pelagic or hemipelagic limestones and shales to neritic sandstones, it is possible to correlate bio- and chronostratigraphies of the two contrasting facies realms.

The Lmhaïfid Formation of the el Ayoun section in the NE Dra Valley has an early to middle Lochkovian age in the global sense due to the conodonts and can be attributed by brachiopods to lower and upper parts of the Lower Gedinnian in the sense of the Rhenish subdivision. On the contrary, in the SW Dra Valley the formation seems to document the whole Lochkovian resp. Gedinnian according to literature data (Hollard 1977).

The Oued-el-Mdâouer Formation in the NE Dra Valley has its onset within the Middle Lochkovian and reaches into the Pragian; this assignment in the sense of the Bohemian subdivision is mainly provided by means of conodonts.

The Pragian to Upper Emsian succession generally shows four main sedimentary cycles which are called 'Rich 1' to 'Rich 4', each generally composed of limestones at the base followed by shales in the middle part and sandstones at the top. The Assa Formation resp. the 'Rich 1' in the SW Dra Valley probably starts in the uppermost Lochkovian, but already in its basal part a Pragian age is indicated according to nowakiids. The formation has yielded at its top *Latericriodus steinachensis* indicating an early to middle Pragian age below the *pireneae* Zone. In the sense of the Rhenish stratigraphy, brachiopods in the sandstones of the upper Assa Formation allow an assignment to the upper Lower Siegenian.

The Merzâ-Akhsaï Formation ('Rich 2') is distributed almost in the whole Dra Valley as a typical 'rich' succession. The basal limestone interval still contains *Latericriodus steinachensis* but also questionably determinable specimens of *Caudicriodus curvicauda* which point to a younger age of the *steinachensis* range; that would plead for a later Pragian age (in the GSSP sense) below the *pireneae* Zone. In comparison with the Achguig region (Tafilalt), the basal limestone interval of the Merzâ-Akhsaï Formation is coeval with the 'Pragian' Limestone *sensu* Bultynck & Walliser (2000). Brachiopod faunas from the 'Rich 2' sandstones indicate a Middle to Late Siegenian age in terms of the Rhenish subdivision. The change from shallow-water sediments of the upper Merzâ-Akhsaï Formation to pelagic limestones of the overlying formations may correspond to the Basal Zlíchov Event (Chlupáč & Kukal 1988).

The Mdâouer-el-Kbîr Formation ('Rich 3') is restricted to the NE Dra Valley. Its basal limestones contain dacryoconarids such as *Nowakia praecursor* and the goniatite *Erbenoceras advolvens* demonstrating a Zlichovian age in terms of the Bohemian subdivision. The conodonts include *Eucostapolygnathus excavatus*, *Caudicriodus sigmoidalis* and *Latericriodus celtibericus* which indicate the *excavatus* Zone. In comparison to the Achguig region, the basal limestone well corresponds with the '*Jovellania*' Limestone (Bultynck & Walliser 2000). The brachiopods in the upper sandstones of the formation represent a fauna from the upper Lower Emsian to the lower Upper Emsian in the classical Rhenish sense; the Lower/Upper Emsian boundary can approximately be reproduced by detailed brachiopod comparisons. The same beds have yielded *Mimagoniatites fecundus* indicating the upper Zlichovian to lower Dalejian.

The Oui-n'-Mesdoûr Formation is restricted to the SW Dra Valley and correlates nearly completely with the Mdâouer-el-Kbîr Formation in the NE. It consists of dark limestones and marls with dacryoconarids and conodonts indicating the *excavatus* Zone and the *Nowakia praecursor* Zone. In comparison with the Achguig region, the lower part of the formation, just like the base of the Mdâouer-el-Kbîr Formation, is correlative with the '*Jovellania*' Limestone, in its upper parts with the *Anetoceras* Limestone and blue *Mimagoniatites* Limestone (Bultynck & Walliser 2000).

The Khebchia Formation resp. the 'Rich 4' in the SW Dra Valley contains in its lower part the *Hollardops* Limestone, a goniatite fauna with *Latanarcestes* and a level with *Sellanarcestes* corresponding to the *serotinus* Zone, i.e. a middle Dalejian age in the sense of the Bohemian subdivision. Brachiopods from the 'Rich 4' sandstone document the uppermost Emsian and probably lower Eifelian.

The Timrhanrhart Formation in the NE Dra Valley partly represents an age equivalent of the Khebchia Formation in the SW. *Sellanarcestes* Beds with goniatites corresponding to the *serotinus* Zone are present slightly above the base of the formation. Trilobites, dacryoconarids, goniatites, conodonts and ostracods allow the recognition of the Emsian/Eifelian boundary in a succession of marlstones with intercalated layers of nodular limestones (Foum Zguid section).

According to present knowledge, the following neritic–pelagic correlations can be stated: The Lower Gedinnian correlates with lower and middle parts of the Lochkovian. The upper part of the Lower Siegenian corresponds to the middle Pragian in the GSSP sense. The Middle and Upper Siegenian correlate approximately with the upper Pragian and lowermost Emsian, both stages in the GSSP sense. The Lower/Upper Emsian boundary in the traditional sense corresponds with the Zlichovian/Dalejian boundary in terms of the Bohemian subdivision.

We thank M. Bensaid (Rabat), A. El Hassani (Rabat) and H. Ouanaimi (Marrakech) for help in organizing field trips. Thanks are due to A. M. Murphy (Davis/California) for updating determinations of Lochkovian conodonts and O. H. Walliser (Göttingen) and R. T. Becker (Münster) for goniatite determinations. G. Schraut (Gießen) and O. Vogel (Senckenberg Museum, Frankfurt) determined the trilobites; G. Becker (Senckenberg) provided important ostracode data. U. Jansen and M. Schemm–Gregory thank P. Carls (Braunschweig) for making it possible to work on brachiopod material from his collections. Technical assistance was provided by J. Anger, E. Scheller–Wagner and M. Ricker (Senckenberg). Thanks are also due to P. Carls (Braunschweig) and R. T. Becker (Münster) for critical review. Financial support for the field trip in the year 2000 was granted by the Deutsche Forschungsgemeinschaft (DFG).

References

AÏT MALEK, Z., RACHEBOEUF, P. R. & LAZREQ, N. 2000. Nouveaux brachiopodes Strophomenata du Dévonien inférieur de l'Anti-Atlas occidental, Maroc. *Geobios*, **33**, 309–318.

ALBERTI, G. K. B. 1969. Trilobiten des jüngeren Siluriums sowie des Unter- und Mitteldevons. I. *Abhandlungen der Senckenbergischen Naturforschenden Gesellschaft*, **520**, 1–692.

ALBERTI, G. K. B. 1982. Nowakiidae (Dacryoconarida) aus dem Hunsrückschiefer von Bundenbach (Rheinisches Schiefergebirge). *Senckenbergiana lethaea*, **63**, 451–463.

ALBERTI, G. K. B. 1993. Dacryoconaride und homoctenide Tentaculiten des Unter- und Mittel-Devons. *Courier Forschungsinstitut Senckenberg*, **158**, 1–229.

ALBERTI, G. K. B. 1996. Tentaculiten, planktonische. *In*: WEDDIGE, K. (ed.) Devon-Korrelationstabelle. *Senckenbergiana lethaea*, **76(1/2)**, 275, column B070di96.

ALBERTI, G. K. B. 1998. Planktonische Tentakuliten des Devon. III. Dacryoconarida Fisher 1962 aus dem Unter- und oberen Mitteldevon. *Palaeontographica, Abteilung A*, **250**, 1–46.

ALBERTI, G. K. B. & HOLLARD, H. 1963. *Warburgella rugulosa* (Alth 1874) (Trilobita, Proetidae) dans le Gédinnien inférieur du Sud Marocain. *Notes et Mémoires du Service Géologique du Maroc*, **23**, 125–130.

AL-RAWI, D. 1977. Biostratigraphische Gliederung der Tentaculiten-Schichten des Frankenwaldes mit Conodonten und Tentaculiten (Unter- und Mittel-Devon; Bayern, Deutschland). *Senckenbergiana lethaea*, **58**, 25–79.

ALVAREZ, F. 1990. Devonian athyrid brachiopods from the Cantabrian Zone (NW Spain). *Biostratigraphie du Paléozoique*, **11**, 1–311.

BARRANDE, J. 1865. *Systême Silurien du centre de la Bohême*. I. Vol. II. Cephalopodes. planche 1–107, Prague, Paris.

BAYLE, E. 1878. *Explication de la carte géologique de la France*. Tome **4**, Atlas, premier partie, planches I-CLVIII.

BECKER, G., LAZREQ, N. & WEDDIGE, K. 2003. Ostracods of Thuringian provenance in the Lower Devonian of Eurasia and North Africa with special reference to the Emsian of Morocco. *Courier Forschungsinstitut Senckenberg*, **242**, 39–49.

BECKER, G., LAZREQ, N. & WEDDIGE, K. 2004. Ostracods of Thuringian provenance from the Devonian of Morocco (Lower Emsian–middle Givetian; southwestern Anti-Atlas). *Palaeontographica, Abteilung A*, **271**, 1–109.

BECKER, R. T. 1996a. Ammonoideen-Zonen, globale. *In*: WEDDIGE, K. (ed.) Devon-Korrelationstabelle. *Senckenbergiana lethaea*, **76**, 275, column B060di96.

BECKER, R. T. 1996b. Ammonoideen-Genozonen. *In*: WEDDIGE, K. (ed.) Devon-Korrelationstabelle. *Senckenbergiana lethaea*, **76**, 275, column B070di96.

BECKER, R. T., BOCKWINKEL, J., EBBIGHAUSEN, V., ABOUSSALAM, Z. S., EL HASSANI, A. & NÜBEL, H. 2004. Lower and Middle Devonian stratigraphy and faunas at Bou Tserfine near Assa (Dra Valley, SW Morocco). *Documents de l'Institut Scientifique*, **19**, 90–100.

BECKER, R. T., JANSEN, U., PLODOWSKI, G., SCHINDLER, E., ABOUSSALAM, Z. S. & WEDDIGE, K. 2004. Devonian litho- and biostratigraphy of the Dra Valley area—an overview. *Documents de l'Institut Scientifique*, **19**, 3–18.

BISCHOFF, G. & ZIEGLER, W. 1957. Die Conodontenchronologie des Mitteldevons und des tiefsten Oberdevons. *Abhandlungen des Hessischen Landesamtes für Bodenforschung*, **22**, 1–136.

BISCHOFF, G. & SANNEMANN, D. 1958. Unterdevonische Conodonten aus dem Frankenwald. *Notizblatt des Hessischen Landesamtes für Bodenforschung*, **86**, 87–110.

BOERSMA, K. T. 1974. Description of certain Lower Devonian platform conodonts of the Spanish central Pyrenees. *Leidse Geologische Mededelingen*, **49**, 285–301.

BOUČEK, B. 1964. *The tentaculites of Bohemia*. Czechoslovak Academy of Sciences, 1–215, Czechoslovak Academy of Sciences, Prague.

BRANSON, E. B. & MEHL, M. G. 1933. Conodonts from the Bainbridge (Silurian) of Missouri. *The Universtity of Missouri Studies*, **8**, 39–52.

BRICE, D., CARLS, P., COCKS, L. R. M., COPPER, P., GARCIA-ALCALDE, J. L., GODEFROID, J. & RACHEBOEUF, P. R. 2000. Brachiopoda. *Courier Forschungsinstitut Senckenberg*, **220**, 65–86.

BRONN, H. G. 1835–1837. *Lethaea Geognostica, oder Abbildungen und Beschreibungen der für die Gebirgs-Formationen bezeichnendsten Versteinerungen*. 1–1346, Schweizerbart, Stuttgart.

VON BUCH, L. 1832. Über Goniatiten. *Abhandlungen der physikalischen Klasse der Königlich-Preußischen Akademie der Wissenschaften*, **1830**, 159–187, Berlin.

BULTYNCK, P. 1970. Révision stratigraphique et paléontologique de la coupe type du Couvinien. *Mémoires de l'Institut Géologique de l'Université de Louvain*, **26**, 1–152.

BULTYNCK, P. 1971. Le Silurien supérieur et le Dévonien inférieur de la Sierra de Guadarrama (Espagne Centrale). Deuxième partie: Assemblages de Conodontes à *Spathognathodus*. *Bulletin de l'Institut Royal des Sciences Naturelles de Belgique*, **47**, 1–43.

BULTYNCK, P. 1976. Comparative study of Middle Devonian conodonts from North Michigan (U.S.A.) and the Ardennes (Belgium—France). *The Geological Association of Canada, Special Paper*, **15**, 119–141.

BULTYNCK, P. 2003. Devonian Icriodontidae: biostratigraphy, classification and remarks on paleoecology and dispersal. *Revista Española de Micropaleontología*, **35**, 295–314.

BULTYNCK, P. & HOLLARD, H. 1980. Distribution comparée de conodontes et goniatites dévoniens des plaines du Dra, du Ma'der et du Tafilalt (Maroc). *Aardkundige Mededelingen*, **1**, 9–73.

BULTYNCK, P. & WALLISER, O. H. 2000. Devonian Boundaries in the Moroccan Anti-Atlas. *Courier Forschungsinstitut Senckenberg*, **225**, 211–226.

BULTYNCK, P., LARDEUX, H. & WALLISER, O. H. 2000. On the correlation of middle-Emsian. *SDS Newsletter*, **17**, 10–11.

BULTYNCK, P., COEN-AUBERT, M. & GODEFROID, J. 2004. Ardennes, Dinant Synclinorium. *In*: WEDDIGE, K. (ed.) Devonian Correlation Table. *Senckenbergiana lethaea*, **84**, 397, column R410di04.

CARLS, P. 1969. Die Conodonten des tieferen Unter-Devons der Guadarrama (Mittel-Spanien) und die Stellung des Grenzbereiches Lochkovium/Pragium nach der rheinischen Gliederung. *Senckenbergiana lethaea*, **50**, 303–355.

CARLS, P. 1974. Die Proschizophoriinae (Brachiopoda: Silurium-Devon) der Östlichen Iberischen Ketten (Spanien). *Senckenbergiana lethaea*, **55**, 153–227.

CARLS, P. 1985. *Howellella (Hysterohowellella) knetschi* (Brachiopoda, Spiriferacea) aus dem tiefen Unter-Gedinnium Keltiberiens. *Senckenbergiana lethaea*, **65**, 297–326.

CARLS, P. 1987. Ein Vorschlag zur biostratigraphischen Redefinition der Grenze Gedinnium/Siegenium und benachbarter Unter-Stufen. 1. Teil: Stratigraphische Argumente und Korrelation. *Courier Forschungsinstitut Senckenberg*, **92**, 77–121.

CARLS, P. 1996a. Conodonten, Conodonten-'Schritte' Keltiberien. *In*: WEDDIGE, K. (ed.) Devon-Korrelationstabelle. *Senckenbergiana lethaea*, **76**, 274, column B031di96.

CARLS, P. 1996b. Trilobiten, Trilobiten-'Schritte' Keltiberien. *In*: WEDDIGE, K. (ed.) Devon-Korrelationstabelle. *Senckenbergiana lethaea*, **76**, 276, column B101di96.

CARLS, P. 1996c. Brachiopoden, Brachiopoden-'Schritte' Keltiberien. *In*: WEDDIGE, K. (ed.) Devon-Korrelationstabelle. *Senckenbergiana lethaea*, **76**, 277, column B121di96.

CARLS, P. 1996d. Keltiberien. *In*: WEDDIGE, K. (ed.) Devon-Korrelationstabelle. *Senckenbergiana lethaea*, **76**, 278, column R142di96.

CARLS, P. & GANDL, J. 1969. Stratigraphie und Conodonten des Unter-Devons der östlichen Iberischen Ketten (NE-Spanien). *Neues Jahrbuch für Geologie und Paläontologie, Abhandlungen*, **132**, 155–218.

CARLS, P. & HEDDEBAUT, C. 1980. Les brachiopodes Spiriferida. *In*: PLUSQUELLEC, Y. (coord.) Les Schistes et Calcaires de l'Armorique (Dévonien inférieur, Massif Armoricain). *Mémoires de la Société Géologique et Minéralogique de Bretagne*, **23**, 215–222.

CARLS, P. & VALENZUELA-RÍOS, J. I. 1998. The ancestry of the Rhenish Middle Siegenian brachiopod fauna in the Iberian Chains and its palaeozoogeography (Early Devonian). *Revista Española de Paleontología*, no extraordinario, Homenaje al Prof. Gonzalo Vidal, 123–142.

CARLS, P. & WEDDIGE, K. 1996. Conodonten-Zonen, globale; aktuelle. *In*: WEDDIGE, K. (ed.) Devon-Korrelationstabelle. *Senckenbergiana lethaea*, **76**, 274, column B030di96.

CARLS, P., MEYN, H. & VESPERMANN, J. 1993. Lebensraum, Entstehung und Nachfahren von *Howellella (Iberohowellella) hollmanni* n. sg., n. sp. (Spiriferacea; Lochkovium, Unter-Devon). *Senckenbergiana lethaea*, **73**, 227–267.

CHLUPÁČ, I. 1982. The Bohemian Lower Devonian stages. *Courier Forschungsinstitut Senckenberg*, **55**, 345–400.

CHLUPÁČ, I. 1998a. Barrandian chronostratigraphy. *In*: WEDDIGE, K. (ed.) Devon-Korrelationstabelle. *Senckenbergiana lethaea*, **77**, 293, column A011di97.

CHLUPÁČ, I. 1998b. Barrandian goniatites. *In*: WEDDIGE, K. (ed.) Devon-Korrelationstabelle. *Senckenbergiana lethaea*, **77**, 293, column B051di97.

CHLUPÁČ, I. & KUKAL, Z. 1988. Possible global events and the stratigraphy of the Barrandian Palaeozoic (Cambrian and Devonian). *Sborník Geologických Věd, Geologie*, **43**, 83–146.

CHLUPÁČ, I. & LUKEŠ, P. 1999. Pragian/Zlíchovian and Zlíchovian/Dalejan boundary sections in the Lower Devonian of the Barrandian area, Czech Republic. *Newsletters on Stratigraphy*, **37**, 75–100.

CHLUPÁČ, I. & TUREK, V. 1983. Devonian goniatites from the Barrandian area, Czechoslovakia. *Rozpravy Ústředního ústavu geologického*, **46**, 1–160.

CHLUPÁČ, I., LUKEŠ, P., PARIS, F. & SCHÖNLAUB, H.-P. 1985. The Lochkovian–Pragian Boundary in the Lower Devonian of the Barrandian Area (Czechoslovakia). *Jahrbuch der Geologischen Bundesanstalt Wien*, **128**, 9–41.

CHOUBERT, G. 1956. Lexique stratigraphique du Maroc. *Notes et Mémoires du Service Géologique du Maroc*, **134**, 1–164.

DAHMER, G. 1932. Fauna der belgischen 'Quartzophyllades de Longlier' in Siegener Rauhflaserschichten auf Blatt Neuwied. *Jahrbuch der Preussischen Geologischen Landesanstalt für das Jahr 1931*, **52**, 86–111.

DAHMER, G. 1940. Die Fauna der Siegener Schichten (Unter-Devon) zwischen Bürresheim und Kirchesch in der Südost-Eifel. *Senckenbergiana*, **22**, 77–102.

DREVERMANN, F. 1904. Die Fauna der Siegener Schichten von Seifen unweit Dierdorf (Westerwald). *Palaeontographica*, **50**, 229–287.

DROT, J. 1964. Rhynchonelloidea et Spiriferoidea silurodévoniens du Maroc pré-saharien. *Notes et Mémoires du Service Géologique du Maroc*, **178**, 1–288.

DROT, J. 1971. Rhynchonellida siluriens et dévoniens du Maroc présaharien. Nouvelles observationes. *Notes et Mémoires du Service Géologique du Maroc*, **31**, 65–108.

DROT, J. 1975. Orthida (brachiopodes) du Maroc présaharien. 1. Orthidina. 2. Dalmanellidina du Dévonien inférieur, à l'exclusion du genre Schizophoria. *Annales de Paléontologie*, **61**, 1–99.

EBBIGHAUSEN, V., BOCKWINKEL, J., BECKER, R. T., ABOUSSALAM, Z. S., BULTYNCK, P., EL HASSANI, A. & NÜBEL, H. 2004. Late Emsian and Eifelian stratigraphy at Oufrane (Tata region, eastern Dra Valley, Morocco). *Documents de l'Institut Scientifique*, **19**, 44–52.

ERBEN, H. K. 1960. Primitive Ammonoidea aus dem Unterdevon Frankreichs und Deutschlands. *Neues Jahrbuch für Geologie und Paläontologie, Abhandlungen*, **110**, 1–128.

FUCHS, A. 1915. Der Hunsrückschiefer und die Unterkoblenzschichten am Mittelrhein (Loreleigegend). I. Teil: Beitrag zur Kenntnis der Hunsrückschiefer- und Unterkoblenzfauna der Loreleigegend. *Abhandlungen der Königlich Preußischen Geologischen Landesanstalt, Neue Folge*, **79**, 1–79.

FUCHS, A. 1919. Beitrag zur Kenntnis der Devonfauna der Verse- und der Hobräcker Schichten des sauerländischen Faciesgebietes. *Jahrbuch der Preussischen Geologischen Landesanstalt für das Jahr 1918*, **39**, 58–95.

GANDL, J. 1972. Die Acastavinae und Asteropyginae (Trilobita) Keltiberiens (NE-Spanien). *Abhandlungen der Senckenbergischen Naturforschenden Gesellschaft*, **530**, 1–184.

GARCÍA-ALCALDE, J. L. 1996. El Devónico del Dominio Astur-Leonés en la Zona Cantábrica (N de España). *Revista Esañola de Paleontología*, no extraordinario, 58–71.

GARCÍA-ALCALDE, J. L. 1997. North Gondwanan Emsian events. *Episodes*, **20**, 241–246.

GARCÍA-ALCALDE, J. L. 2004. Lower Devonian Delthyridoidea (Brachiopoda, Delthyridina) of the Cantabrian Mountains. *Bulletin de l'Institut Royal des Sciences Naturelles de Belgique, Sciences de la Terre*, **74**(supplement), 9–38.

GARCÍA-ALCALDE, J. L., ARBIZU, M. A., GARCÍA-LÓPEZ, S., LEYVA, F., MONTESINOS, R., SOTO, F. & TRUYÓLS-MASSONI, M. 1990. Devonian stage boundaries (Lochkovian/Pragian, Pragian/Emsian, and Eifelian/Givetian) in the Cantabric region (NW Spain). *Neues Jahrbuch für Geologie und Paläontologie, Abhandlungen*, **180**, 177–207.

GODEFROID, J. 1994. Le genre *Euryspirifer* Wedekind, 1926 (Brachiopoda, Spiriferida) dans le Dévonien inférieur de la Belgique. *Bulletin de l'Institut Royal des Sciences Naturelles de Belgique, Sciences de la Terre*, **64**, 57–83.

GOSSELET, M. J. 1880. Esquisse géologique du Nord de la France et des Contrées voisines. Fascicule 1, *Terrains primaires*, 1–167.

GOSSELET, M. J. 1888. *L'Ardenne*. Mémoires pour servir à l'explication de la carte géologique detailée de la France. 1–881, Baudry et Cie, Paris.

GOURVENNEC, R. 1988. Nouvelles definition de *Spirifer rousseaui* Rouault, 1846 et description de *Acrospirifer ? rouaulti* n. sp. (Spiriferacea, Brachiopoda) du Dévonien inférieur du Massif Armoricain (France). *Hercynica*, **2**, 149–166.

GOURVENNEC, R. 1989. Brachiopodes Spiriferida du Dévonien inférieur du Massif Armoricain. Systématique, paléobiologie, évolution, biostratigraphie. *Biostratigraphie du Paléozoïque*, **9**, 1–281.

HALL, J. 1879. Descriptions of the Gasteropoda, Pteropoda and Cephalopoda of the Upper Helderberg, Hamilton, Portage and Chemung Groups. *Paleontology of New York*, **5**, 1–492, Albany, N.Y.

HOLLARD, H. 1960. La découverte de goniatites du genre *Sellanarcestes* dans le Dévonien du Draa (Maroc présaharien) et ses conséquences stratigraphiques. *Notes et Mémoires du Service Géologique du Maroc*, **19**, 55–60.

HOLLARD, H. 1963a. Un tableau stratigraphique du Dévonien du Sud de l'Anti-Atlas. *Notes et Mémoires du Service Géologique du Maroc*, **23**, 105–109.

HOLLARD, H. 1963b. Les *Acastella* et quelques autres Dalmanitacea du Maroc présaharien. Leur distribution verticale et ses conséquences pour l'étude de la limite Silurien-Dévonien. *Notes et Mémoires du Service Géologique du Maroc*, **23**, 3–67.

HOLLARD, H. 1963c. Présence d'*Anetoceras advolvens* Erben (Ammonoïdée primitive) dans le Dévonien inférieur du Maroc présaharien. *Notes et Mémoires du Service Géologique du Maroc*, **23**, 131–138.

HOLLARD, H. 1965. Précisions sur la stratigraphie et la repartition de quelques espèces importantes du Silurien supérieur et de l'Éodévonien du Maroc présaharien. *Notes et Mémoires du Service Géologique du Maroc*, **24**, 23–32.

HOLLARD, H. 1967. Le Dévonien du Maroc et du Sahara nord-occidental. *In*: OSWALD, D. H. (ed.) *International Symposium on the Devonian System, Calgary 1967*. Vol. **1**, 203–244.

HOLLARD, H. 1977. Le domaine de l'Anti-Atlas au Maroc. *In*: MARTINSSON, A. (ed.) The Silurian-Devonian boundary. Final report of the Committee on the Silurian–Devonian Boundary within IUGS Commission on Stratigraphy and a state of the art report for Project Ecostratigraphy. *IUGS Series A*, **5**, 168–194.

HOLLARD, H. 1978. Corrélations entre niveaux à brachiopodes et à goniatites au voisinage de la limite Dévonien inférieur–Dévonien moyen dans les plaines du Dra (Maroc présaharien). *Newsletters on Stratigraphy*, **7**, 8–25.

HOLLARD, H. 1981a. Principaux charactères des formations dévoniennes de l'Anti-Atlas. *Notes et Mémoires du Service Géologique du Maroc*, **42**, 15–22.

HOLLARD, H. 1981b. Tableaux de corrélations du Silurien et du Dévonien de l'Anti-Atlas. *Notes et Mémoires du Service Géologique du Maroc*, **42**, 23.

HOLLARD, H. & JACQUEMONT, P. 1956. Le Gothlandien, le Dévonien et le Carbonifère des régions du Dra et du Zemoul. *Notes et Mémoires du Service Géologique du Maroc*, **15**, 7–33.

JAHNKE, H., HENN, A., MADER, H. & SCHWEINEBERG, J. 1983. Silur und Devon im Arauz-Gebiet (Prov. Palencia, N-Spanien). *Newsletters on Stratigraphy*, **13**, 40–66.

JANSEN, U. 1999. Brachiopod fauna and age of the Assa Formation (Early Devonian, Dra Plains, southern Anti-Atlas, Moroccan Pre-Sahara). *Senckenbergiana lethaea*, **79**, 191–207.

JANSEN, U. 2000. Stratigraphy of the Early Devonian in the Dra Plains (Moroccan Pre-Sahara). *Travaux de l'Institut Scientifique, Série Géologie et Géographie Physique*, **20**, 36–44.

JANSEN, U. 2001. Morphologie, Taxonomie und Phylogenie unter-devonischer Brachiopoden aus der Dra-Ebene (Marokko, Prä-Sahara) und dem Rheinischen Schiefergebirge (Deutschland). *Abhandlungen der Senckenbergischen Naturforschenden Gesellschaft*, **554**, 1–389.

JANSEN, U., BECKER, G., PLODOWSKI, G., SCHINDLER, E., VOGEL, O. & WEDDIGE, K. 2004a. The Emsian to Eifelian near Foum Zguid (NE Dra Valley, Morocco). *Documents de l'Institut Scientifique*, **19**, 19–28.

JANSEN, U., BECKER, G., PLODOWSKI, G., SCHINDLER, E., VOGEL, O. & WEDDIGE, K. 2004b. Pragian and Emsian near Aouinet Torkoz (SW Dra Valley, Morocco). *Documents de l'Institut Scientifique*, **19**, 75–84.

JANSEN, U., PLODOWSKI, G., SCHINDLER, E. & WEDDIGE, K. 2004a. Stratigraphy and facies of the Lower Devonian in the Dra Valley (Moroccan Pre-Sahara). *International Meeting on Stratigraphy, Annual Meeting of the Subcommission on Devonian Stratigraphy (SDS)*, Abstracts, 24.

JANSEN, U., PLODOWSKI, G., SCHINDLER, E. & WEDDIGE, K. 2004b. The Pragian at Assa (SW Dra Valley, Morocco). *Documents de l'Institut Scientifique*, **19**, 64–68.

KLAPPER, G. 1971. Sequence within the conodont genus *Polygnathus* in the New York lower Middle Devonian. *Geologica et Palaeontologica*, **5**, 59–72.

KLAPPER, G. 1980. Conodont Systematics. *In*: JOHNSON, J. G., KLAPPER, G. & TROJAN, W. R. (eds) 1980. Brachiopod and conodont successions in the Devonian of the northern Antelope Range, central Nevada. *Geologica et Palaeontologica*, **14**, 77–116.

KLAPPER, G. & JOHNSON, D. B. 1975. Sequence in conodont genus *Polygnathus* in Lower Devonian at Lone Mountain, Nevada. *Geologica et Palaeontologica*, **9**, 65–83.

KLAPPER, G., ZIEGLER, W. & MASHKOVA, T. V. 1978. Conodonts and correlation of Lower-Middle Devonian boundary beds in the Barrandian area of Czechoslovakia. *Geologica et Palaeontologica*, **12**, 103–115.

KLUG, C. 2001. Early Emsian ammonoids from the eastern Anti-Atlas (Morocco) and their succession. *Paläontologische Zeitschrift*, **74**, 479–515.

LARDEUX, H. 1969. Les Tentaculites d'Europe occidentale et d'Afrique du Nord. *Cahiers de Paléontologie*, 12–17, 1–238.

LAZREQ, N. & OUANAIMI, H. 1998. Le Dévonien inférieur de Tizi-n-Tichka (Haut Atlas) et de Laâyoune (Tata, Anti-Atlas, Maroc): Nouvelles datations et implications paléogéographiques. *Senckenbergiana lethaea*, **77**, 223–231.

LE MAÎTRE, D. 1952. La faune du Dévonien inférieur et moyen de la Saoura et des Abords de l'Erg el Djemel (Sud-Oranais). *Materiaux pour la carte géologique de l'Algérie, 1re Série Paléontologie*, **12**, 1–170.

LUKEŠ, P. & CHLUPÁČ, I. 1998. Barrandian planctonic tentaculites. *In*: WEDDIGE, K. (ed.) Devon-Korrelationstabelle. *Senckenbergiana lethaea*, **77**, 293, column B071di97.

MASHKOVA, T. V. 1968. Konodonty roda *Icriodus* Branson, Mehl, 1938, tiz Borshchovskogo i Chortkovskogo gorizontov Podolii. *Doklady Akademii Nauk*, SSSR, **182**, 941–944.

MAURER, F. 1880. Paläontologische Studien im Gebiet des Rheinischen Devon. 4. Der Kalk bei Greifenstein. *Neues Jahrbuch für Mineralogie und Geologie*, Beilagen-Band **I**, 1–112.

MAURER, F. 1886. Die Fauna des rechtsrheinischen Unterdevon aus meiner Sammlung zum Nachweis der Gliederung. *Neues Jahrbuch für Mineralogie, Geologie und Palaeontologie*, **1882**, 3–55.

MÉLOU, M. 1981. Les Brachiopodes Orthida. *In*: MORZADEC, P., PARIS, F. & RACHEBOEUF, P. R. (coords) La tranchée de la Lézais–Emsien supérieur du Massif Armoricain. Sédimentologie, paléontologie, stratigraphie. *Mémoires de la Société Géologique et Minéralogique de Bretagne*, **24**, 135–141.

MITTMEYER, H.-G. 1972. Delthyrididae und Spinocyrtiidae (Brachiopoda) des tiefen Ober-Ems im Mosel-Gebiet (Ems-Quarzit, Rheinisches Schiefergebirge). *Mainzer Geowissenschaftliche Mitteilungen*, **1**, 82–121.

MITTMEYER, H.-G. 1973. Grenze Siegen/Unterems bei Bornhofen (Unter-Devon, Mittelrhein). *Mainzer Geowissenschaftliche Mitteilungen*, **2**, 71–103.

MITTMEYER, H.-G. 1982. Rhenish Lower Devonian biostratigraphy. *Courier Forschungsinstitut Senckenberg*, **55**, 257–269.

MORZADEC, P. 1988. Le genre *Psychopyge* (Trilobita) dans le Dévonien inférieur du nord de l'Afrique et l'ouest de l'Europe. *Palaeontographica, Abteilung A*, **200**, 153–161.

MORZADEC, P. 2001. Les Trilobites Asteropyginae du Dévonien de l'Anti-Atlas (Maroc). *Palaeontographica, Abteilung A*, **262**, 53–85.

MORZADEC, P., PARIS, F. & RACHEBOEUF, P. R. (coords) 1981. La tranchée de la Lézais–Emsien supérieur du Massif Armoricain. Sédimentologie, paléontologie, stratigraphie. *Mémoires de la Société Géologique et Minéralogique de Bretagne*, **24**, 1–313.

MURPHY, M. A. 1989. Lower Pragian Boundary (Lower Devonian) and its application in Nevada. *Courier Forschungsinstitut Senckenberg*, **117**, 61–70.

MURPHY, M. A. 2000. Conodonts first occurrences in Nevada. *In*: WEDDIGE, K. (ed.) Devonian Correlation Table. *Senckenbergiana lethaea*, **80**, 695, column B032di00.

MURPHY, M. A. & VALENZUELA-RÍOS, J. I. 1999. *Lanea* new genus of Early Devonian conodonts. *In*: SERPAGLI, E. (ed.) Studies on conodonts—Proceedings of the Seventh European Conodont Symposium. *Bollettino della Società Paleontologica Italiana*, **37**(2–3), 321–334.

MURPHY, M. A., VALENZUELA-RÍOS, J. I. & CARLS, P. 2004. On classification of Pridoli (Silurian)–Lochkovian (Devonian) Spathognathodontidae (conodonts). *University of California, Riverside, Campus Museum Contribution*, **6**, 1–25.

OEHLERT, D.-P. 1877. Sur les fossiles dévoniens du département de la Mayenne. *Bulletin de la Société Géologique de France*, 3e série, **5**, 578–603.

PLUSQUELLEC, Y. 1980 (coord.). Les Schistes et Calcaires de l'Armorique (Dévonien inférieur, Massif Armoricain). Sédimentologie, Paléontologie, Stratigraphie. *Mémoires de la Société Géologique et Minéralogique de Bretagne*, **23**, 1–317.

RACHEBOEUF, P. R. 1976. Chonetacea (brachiopodes) du Dévonien inférieur du Bassin de Laval (Massif Armoricain). *Palaeontographica, Abteilung A*, **152**, 14–89.

RACHEBOEUF, P. R. 1981. Chonetacés (brachiopodes) Siluriens et Dévoniens du Sud-Ouest de l'Europe (Systématique–Phylogénie–Biostratigraphie–Paléobiogéographie). *Mémoires de la Société Géologique et Minéralogique de Bretagne*, **27**, 1–294.

RENOUF, J. T. 1972. Brachiopods from the Grés à *Orthis monnieri* formation of northwestern France and their significance in Gedinnian/Siegenian stratigraphy of Europe. *Palaeontographica, Abteilung A*, **139**, 89–133.

RICHTER, REINH. 1854. Thüringische Tentaculiten. *Zeitschrift der Deutschen Geologischen Gesellschaft*, **6**, 275–290.

RICHTER, RUD. & RICHTER, E. 1954. Die Trilobiten des Ebbe-Sattels und zu vergleichende Arten (Ordovizium/Gotlandium/Devon). *Abhandlungen der Senckenbergischen Naturforschenden Gesellschaft*, **488**, 1–76.

ROEMER, C. F. 1844. *Das Rheinische Uebergangsgebirge. Eine palaeontologisch-geognostische Darstellung.* 1–96, Hahn'sche Hofbuchhandlung, Hannover.

ROEMER, F. A. 1843 *Die Versteinerungen des Harzgebirges.* 26–32, Hahn'sche Hofbuchhandlung, Hannover.

ROUAULT, F. 1851. Fossiles du terrain silurien. *Bulletin de la Societé Géologique de France*, **8**, 358–399.

SCHEMM-GREGORY, M. & JANSEN, U. 2004. Phylogenie der Gattung *Arduspirifer* (Brachiopoda, Devon). 74. Jahrestagung der Paläontologischen Gesellschaft, Kurzfassungen, Universitätsdrucke Göttingen, 200–201.

SCHEMM-GREGORY, M. & JANSEN, U. 2005. *Arduspirifer arduennensis treverorum* n. ssp., eine neue Brachiopoden-Unterart aus dem tiefen Ober-Emsium des Mittelrhein-Gebiets (Unter-Devon, Rheinisches Schiefergebirge). *Mainzer Geowissenschaftliche Mitteilungen*, **33**, 79–100.

SCHLOTHEIM, E. F. VON 1820. *Die Petrefactenkunde auf ihrem jetzigen Standpunkte durch die Beschreibung seiner Sammlung versteinerter und fossiler Überreste des Thier- und Pflanzenreiches der Vorwelt erläutert.* I–XLII + 1–387, 15 plates in an extra volume; Becker'sche Buchhandlung, Gotha.

SCHNUR, J. 1851. Die Brachiopoden aus dem Uebergangsgebirge der Eifel. *Programm der vereinigten höhern Bürger- und Provinzial-Gewerbeschule zu Trier, Schuljahr 1850–1851*, 1–16.

SCHNUR, J. 1853. Zusammenstellung und Beschreibung sämmtlicher im Uebergangsgebirge der Eifel vorkommenden Brachiopoden nebst Abbildungen derselben. *Palaeontographica*, **3**, 169–247.

SCHRAUT, G. 1998a. Die Gattung *Scabrella* (Trilobita) im Unterdevon von Nord-Afrika, West- und Süd-Europa. *Senckenbergiana lethaea*, **77**, 47–59.

SCHRAUT, G. 1998b. Trilobiten aus dem Unter-Devon des südlichen Anti-Atlas (Marokko)—ihre Bedeutung für die zeitliche Korrelation zwischen Rheinischer und Herzynischer Fazies. *Senckenbergiana lethaea*, **77**, 61–69.

SCHRAUT, G. 2000. Trilobiten aus dem Unter-Devon des südöstlichen Anti-Atlas, Süd-Marokko. *Senckenbergiana lethaea*, **79**, 361–433.

SLAVÍK, L. 2004. A new conodont zonation of the Pragian Stage (Lower Devonian) in the stratotype area (Barrandian, central Bohemia). *Newsletters on Stratigraphy*, **40**, 39–71.

SLAVÍK, L. & HLADIL, J. 2004. Lochkovian/Pragian GSSP revisited: evidence about conodont taxa and their stratigraphic distribution. *Newsletters on Stratigraphy*, **40**, 137–153.

SOLLE, G. 1953. Die Spiriferen der Gruppe *arduennensis-intermedius* im rheinischen Devon. *Abhandlungen des Hessischen Landesamtes für Bodenforschung*, **5**, 1–156.

SOWERBY, J. G. 1842. Description of Silurian fossils from the Rhenish Provinces. *In*: D'ARCHIAC, E. A. J. D. & DE VERNEUIL, E. On the fossils of the older deposits in the Rhenish provinces. *Transactions of the Geological Society of London*, **6**, 408–410.

TERMIER, G. & TERMIER, H. 1950. Paléontologie marocaine. Tome II: Invertébrés de l'ère primaire, Bryozoaires et brachiopodes. *Actualités Scientifiques et Industrielles*, **1093**, fascicule II, 1–253.

TRUYÓLS-MASSONI, M. & GARCÍA-ALCALDE, J. L. 1994. Faune rhéno-bohémienne (Dacryoconarides, Brachiopodes) à la limite Emsien inférieur/supérieur au Cabo la Vela (Asturies, Espagne). *Geobios*, **27**, 221–241.

VALENZUELA-RÍOS, J. I. & MURPHY, A. M. 1997. A new zonation of middle Lochkovian (Lower Devonian) conodonts and evolution of *Flajsella* n. gen. (Conodonta).

In: KLAPPER, G., MURPHY, A. M. & TALENT, J. A. (eds). Paleozoic Sequence Strati-graphy, Biostratigraphy, and Biogeography: Studies in Honor of J. Granville ('Jess') Johnson. *Geological Society of America Special Paper*, **321**, 131–144.

VERNEUIL, E. DE 1850. Note sur les fossiles dévoniens du distrikt de Sabero (Léon). *Bulletin de la Société géologique de France*, **7**, 155–186.

VERNEUIL, E. DE & D'ARCHIAC, E. A. J. D. 1845. Note sur les fossiles du terrain paléozoïque des Asturies. *Bulletin de la Societé Géologique de France*, **2** (série 2), 458–480.

WALTHER, K. 1903. Das Unterdevon zwischen Marburg a. L. und Herborn (Nassau). *Neues Jahrbuch für Mineralogie, Geologie und Palaeontologie*, Beilagen-Band **17**, 1–75.

WALLISER, O. H. 1964. Conodonten des Silurs. *Abhandlungen des Hessischen Landesamtes für Bodenforschung*, **41**, 1–106.

WEDDIGE, K. 1977. Die Conodonten der Eifel-Stufe im Typusgebiet und in benachbarten Faziesgebieten. *Senckenbergiana lethaea*, **58**, 271–419.

WEDDIGE, K. 1985. Systematik von Ober-Emsium-Icriodontiden und Formenentwicklung. *In*: WEDDIGE, K. & REQUADT, H. 1985. Conodonten des Ober-Emsium aus dem Gebiet der Unteren Lahn (Rheinisches Schiefergebirge). *Senckenbergiana lethaea*, **66**, 347–381.

WEDDIGE, K. 1987. The Lower Pragian boundary (Lower Devonian) based on the conodont species *Eognathodus sulcatus*. *Senckenbergiana lethaea*, **67**, 479–487.

WEDDIGE, K. (ed.) 1996a. Beiträge zu Gemeinschaftsaufgaben der deutschen Subkommission für Devon-Stratigraphie, 1: Devon-Korrelationstabelle. *Senckenbergiana lethaea*, **76**, 267–286.

WEDDIGE, K. 1996b. Conodonten-Zonen, glbale; Stand ~ 1985. *In*: WEDDIGE, K. (ed.) Devon-Korrelationstabelle. *Senckenbergiana lethaea*, **76**, 274, column B040di.

WEDDIGE, K. 1996c. Mosel-Trog, Unterstufen (nach Mittmeyer, H.-G., Solle, G.). *In*: WEDDIGE, K. (ed.) Devon-Korrelationstabelle. *Senckenbergiana lethaea*, **76**, 277, column R150di96.

WEDDIGE, K. (ed.) 1998a. Devon-Korrelationstabelle. *Senckenbergiana lethaea*, **77**, 289–326.

WEDDIGE, K. (ed.) 1998b. Devon-Korrelationstabelle. *Senckenbergiana lethaea*, **78**, 243–265.

WEDDIGE, K. 2000a. The Devonian Correlation Table—guidelines and implications. *Senckenbergiana lethaea*, **80**, 685–690.

WEDDIGE, K. (ed.) 2000b. Devonian Correlation Table. *Senckenbergiana lethaea*, **80**, 691–726.

WEDDIGE, K. (ed.) 2001. Devonian Correlation Table. *Senckenbergiana lethaea*, **81**, 435–462.

WEDDIGE, K. (ed.) 2003. Devonian Correlation Table. *Senckenbergiana lethaea*, **83**, 213–234.

WEDDIGE, K. (ed.) 2004. Devonian Correlation Table. *Senckenbergiana lethaea*, **84**, 385–415.

WEDDIGE, K. (ed.) 2005a. Devonian Correlation Table. Supplements 2005, part 1. *Senckenbergiana lethaea*, **85**, 196–205.

WEDDIGE, K. (ed.) 2005b. Devonian Correlation Table. Supplements 2005, part 2. *Senckenbergiana lethaea*, **85**, 379–414.

WEDDIGE, K. & REQUADT, H. 1985. Conodonten des Ober-Emsium aus dem Gebiet der Unteren Lahn (Rheinisches Schiefergebirge). *Senckenbergiana lethaea*, **66**, 347–381.

WEDDIGE, K., JANSEN, U., SCHINDLER, E., WEYER, D., ANDERLE, H.-J., BUCHHOLZ, P., GRIMM, M. C., KRAMER, W., RIBBERT, K.-H., STOPPEL, D., WELLER, H., ZAGORA, K. & SUBKOMMISSION, Devon 2002. Devon. *In*: DEUTSCHE STRATIGRAPHISCHE KOMMISSION (ed.), *Stratigraphische Tabelle von Deutschland* 2002 GeoForschungs Zentrum (GFZ) Potsdam, Potsdam.

YOLKIN, E. A., KIM, A. I., WEDDIGE, K., TALENT, J. A. & HOUSE, M. R. 1997. Definition of the Pragian/Emsian Stage boundary. *Episodes*, **20**, 235–240.

ZEILER, F. 1857. Versteinerungen der älteren Rheinschen Grauwacke. *Verhandlungen des Naturhistorischen Vereins der Rheinlande und Westfalen*, **14**, 45–64.

ZIEGLER, W. 1956. Unterdevonische Conodonten, insbesondere aus dem Schönauer und dem Zorgensis-Kalk. *Notizblatt des Hessischen Landesamtes für Bodenforschung*, **84**, 93–106.

ZIEGLER, W. 1960. Conodonten aus dem Rheinischen Unterdevon (Gedinnium) des Remscheider Sattels (Rheinisches Schiefergebirge). *Paläontologische Zeitschrift*, **34**, 169–201.

Basinwide stratigraphic synthesis and sequence stratigraphy, upper Pragian, Emsian and Eifelian stages (Lower to Middle Devonian), Appalachian Basin

C. A. VER STRAETEN

New York State Museum, The State Education Department, Albany, NY 12230, USA
(e-mail: cverstra@mail.nysed.gov)

Abstract: A new synthesis of the Lower to Middle Devonian (upper Pragian, Emsian and Eifelian) succession across the Appalachian Basin has been developed by high resolution event and sequence stratigraphic analysis. The correlation of numerous marker beds and a hierarchy of cycles in the interval of the Oriskany Sandstone to lower Hamilton Group provide a refined picture of the depositional patterns, faunal changes, formation to member-level (and finer) relationships, and sea-level trends. The succession begins above the Wallbridge Unconformity, or its correlative conformity, which lies beneath the Oriskany Sandstone, not above it as previously thought. The new sequence-stratigraphic framework of Oriskany to lower Hamilton strata comprises nine 'third order' stratigraphic sequences (cycles), though an interval of some 25 million years. At a coarse scale, the eustatic Pragian to Eifelian sea-level curve for Euramerica of Johnson *et al.* (1985) shows broad variance with the Appalachian curve, reflecting the regional influence of the Acadian orogeny. However at the finer sequence-scale, the Euramerican sea-level trends are recognizable in the Appalachian Basin succession.

The application of high resolution event and cyclic stratigraphic methods permits a fine-scale subdivision of the sedimentary record and the correlation of thin units across broad regions (e.g. basin-scale). The methodology yields a refined record of tectonic, sedimentary and faunal events and processes, especially when combined with high resolution geochronological age dating.

These methods were applied to over 350 outcrops across the Appalachian Basin, in New York, New Jersey, Pennsylvania, Maryland, Virginia, West Virginia, Ohio, Tennessee, and Ontario, Canada (Fig. 1). The resulting basinwide correlations yield a new, high resolution stratigraphic synthesis of upper Pragian, Emsian and Eifelian strata in the Appalachian Basin. Building on the new synthesis, the paper then examines the sequence stratigraphic framework of the interval, and compares it to the global Devonian record of sea-level change of Johnson *et al.* (1985, 1996).

Geological setting and background

Paleogeographic reconstructions of the Early to Middle Devonian place Eastern North America approximately 25–35° south of the equator, rotated clockwise on the order of 90° to the south (Scotese & McKerrow 1990; Witzke 1990). Oblique collision of the continental margin with one or more landmasses beginning in the latest Silurian or earliest Devonian (Rogers 1967; Quinlan & Beaumont 1984; Ettensohn 1985; Ferrill & Thomas 1988; Rast & Skehan 1993) resulted in formation of an elongate mountain belt (Acadian Orogen) that extended from Greenland and Maritime Canada to the southern Appalachians. The loading of the continental margin, and associated magmatism, metamorphism, uplift and deformation of the orogen, led to subsidence and reorganization of the Appalachian foreland basin. Unroofing and erosion led to progradation of synorogenic clastics into the basin, with eventual infilling and progradational spillover onto the craton by the Late Devonian.

The upper Lower to lower Middle Devonian rocks of the Appalachian Basin (Pragian, Emsian, and Eifelian stages) comprise a vertical succession of marine siliciclastics and carbonates. Overall, using New York stratigraphic terminology, the succession begins above the Wallbridge Unconformity (or correlative conformity) of Sloss (1963) with Pragian quartz sandstones and carbonates (Oriskany Formation and equivalents). Succeeding lower Emsian shales to sandstones (Esopus Formation) are overlain by mixed clastics and carbonates of the upper Emsian Schoharie Formation. Widespread limestone deposition in the lower Eifelian (Onondaga Formation) is succeeded by a second major influx of clastics in the upper Eifelian to Givetian (Union Springs and Oatka Creek–Mount Marion Formations = Marcellus subgroup of the Eifelian–Givetian Hamilton Group).

The combined Pragian to Eifelian stages represent approximately 25 million years of time, from 413 to 387.5 Ma (Tucker *et al.* 1998; Williams

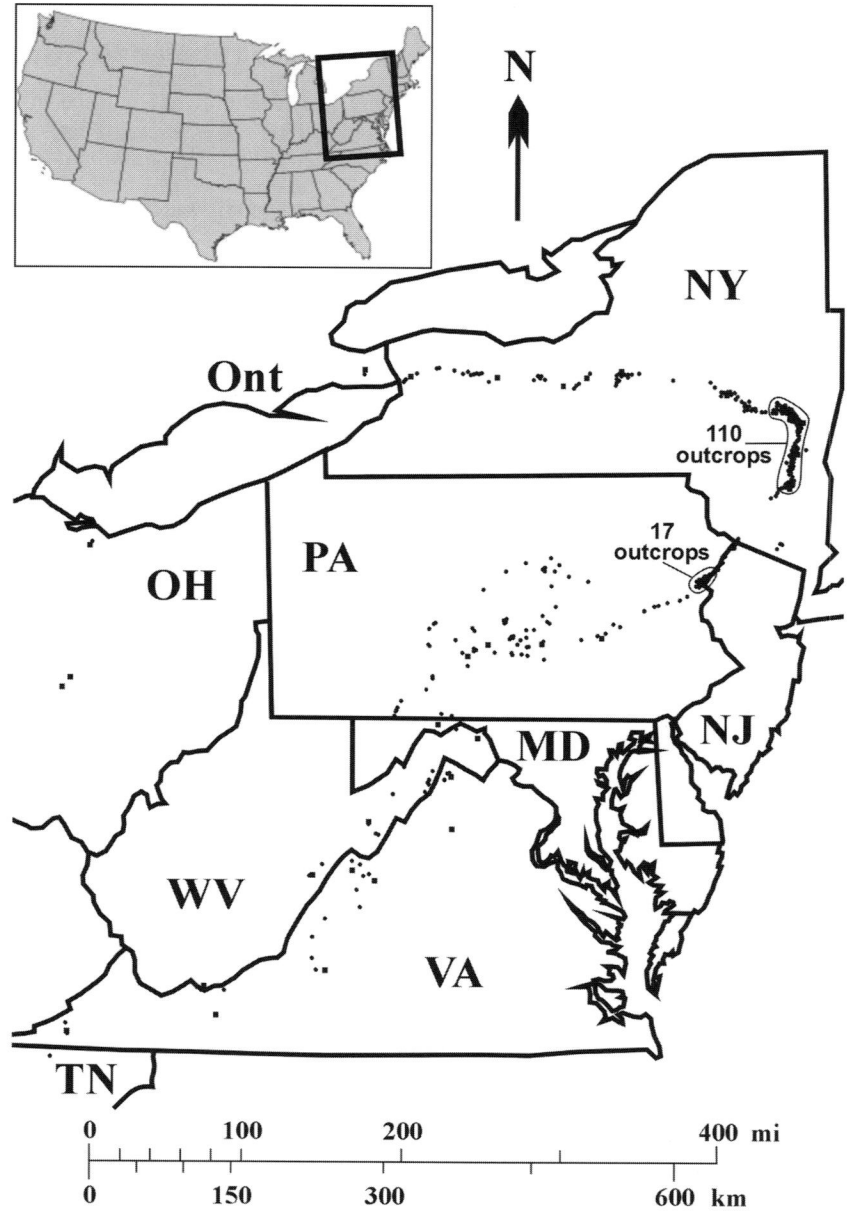

Fig. 1. Map of study localities (>350 sites) of upper Pragian, Emsian and Eifelian (upper Lower and lower Middle Devonian) strata across the Appalachian Basin. MD, Maryland; NJ, New Jersey; NY, New York; OH, Ohio; Ont, Ontario; PA, Pennsylvania; TN, Tennessee; VA, Virginia; WV, West Virginia.

et al. 2000; Kaufmann *et al.* 2005). Previous stratigraphic summaries of the Pragian to Eifelian rocks in the Appalachian Basin have been presented by Oliver *et al.* (1967), Rickard (1975), Berg *et al.* (1983), and Patchen *et al.* (1985) (summarized in Fig. 2). Koch (1978, 1981) discusses Eifelian strata and brachiopod faunas across a broad area of eastern North America. In New York State, the strata have been the subject of numerous theses and papers, including Oliver (1954, 1956, 1966), Johnsen (1957), Johnsen & Southard (1962), Oliver *et al.* (1962), Boucot *et al.* (1970),

Hodgson (1970), Rehmer (1976), Lindemann (1980), Lindemann & Feldman (1987), Griffing & Ver Straeten (1991), Brett & Ver Straeten (1994, 1997), Griffing (1994), Ver Straeten (1994, 1996a), Ver Straeten et al. (1994), and Ver Straeten & Brett (1995). In the central to southern portions of the basin (Pennsylvania, Maryland, Virginia, West Virginia), Emsian and lower Eifelian stratigraphy has been the focus of several studies, most notably Dennison (1960, 1961), Dennison et al. (1972, 1979), Inners (1975), Dennison & Hasson (1976), and Ver Straeten (1996a, b). Epstein (1984) and Ver Straeten (2001a, b) have discussed the correlative strata in the eastern Pennsylvania region, and Ver Straeten et al. (1995) presented data for New Jersey and southeastern New York. Additional work from the Pennsylvania Geological Survey discusses Emsian and Eifelian rocks along various parts of the outcrop belt (e.g. Cate 1963; Epstein et al. 1974; Faill & Wells 1974; Faill et al. 1978). Work from the western part of the basin includes Stauffer (1909), Wells (1947), Chapel (1975), and Sparling (1988).

Pragian, Emsian, and in part, Eifelian biostratigraphy in the Appalachian Basin is generally poorly constrained at present. Available data are chiefly from the New York section, and are summarized in papers in Oliver & Klapper (1981), especially for early to middle Eifelian strata (Onondaga and Union Springs/basal Oatka Creek–Mount Marion Formations). The lower to middle Eifelian has been the focus of biostratigraphic studies of conodonts (Klapper 1971, 1981; Epstein 1984) and goniatites (House 1962, 1978, 1981; Becker & House 1994, 2000). Brachiopod zonation for the entire study interval was established by Boucot & Johnson (1967) and summarized by Dutro (1981); the coral biostratigraphy has been discussed by Oliver & Sorauf (1981).

Precise boundaries for the Pragian, Emsian and Eifelian stages (using conodonts) are also not known at this time. They are interpreted to be close to the base of the Esopus Formation (Pragian–Emsian), within or at the top of the Edgecliff Member of the Onondaga Formation (Emsian–Eifelian) and at some unknown position within the Oatka Creek–Mount Marion Formations (Eifelian–Givetian) (Kirchgasser et al. 1985; Kirchgasser & Oliver 1993; Kirchgasser 2000; D. J. Over, pers. comm.). The present understanding of Pragian to Eifelian biostratigraphy will be presented later in this paper.

Overview of unconformities and sequence stratigraphy

As surfaces of erosion and/or non-deposition, unconformities mark significant breaks in the stratigraphic record. They occur at different temporal and geographical scales, and may be of subaerial or submarine origin (Shanmugan 1988). Their recognition is a crucial part of integrated basinal studies and is important for interpreting changes in relative sea-level, defining depositional sequences, determining the timing of tectonic activity and flexure of the basin, and predicting the occurrence of economic deposits.

The application of sequence stratigraphic methods to sedimentary basin studies provides a powerful tool for the analysis of time–rock relationships (Van Wagoner et al. 1988; Wilgus et al. 1988; Emery & Meyers 1996; Posamentier & Allen 1999; Catuneanu 2002; Coe 2003). It permits chronostratigraphic subdivision of the rock record into cyclic, unconformity-bound, genetically related successions of strata (Van Wagoner et al. 1988). A 'depositional sequence' is the fundamental, meso-scale unit of sequence stratigraphy. A sequence is a coherent package of strata that is bound at bottom and top by unconformities or their correlative conformities (Mitchum et al. 1977). It is formed by a cyclic change in base level (relative sea-level) through the interaction of tectonics, eustatic sea-level change, and sedimentological factors (Allen & Allen 1990). A sequence can be subdivided into 'systems tracts,' composed of smaller scale cycles ('parasequences'), which are deposited during different stages of a transgressive to regressive cycle.

In this paper, four such subdivisions are recognized within each sequence (lowstand, transgressive, highstand, and falling stage systems tracts; following the depositional sequence model 4 of Catuneanu 2002). The base of a sequence is placed at the base of the lowstand systems tract, at the subaerial unconformity or its correlative conformity.

The 'Lowstand Systems Tract', at the base of a cycle, overlies a subaerial unconformity or the correlative marine conformity (Catuneanu 2002). It represents the initiation of base level rise at a time when sedimentation rate is greater than the accommodation space created by the rise. The lowstand systems tract of a sequence is commonly not preserved in shallow, shelf-like margins of epicontinental seas or foreland basins.

The succeeding 'Transgressive Systems Tract' is deposited during the middle stages of a rise in relative sea-level, when sedimentation rate is less than the accommodation space formed due to base level rise. This results in onlap of sea-level and sedimentation over the 'maximum regressive surface'; the latter rests on underlying lowstand deposits, or is amalgamated with the basal sequence-bounding unconformity. Deposition is retrogradational. Condensed sections are common within the transgressive systems tract and the lower part of

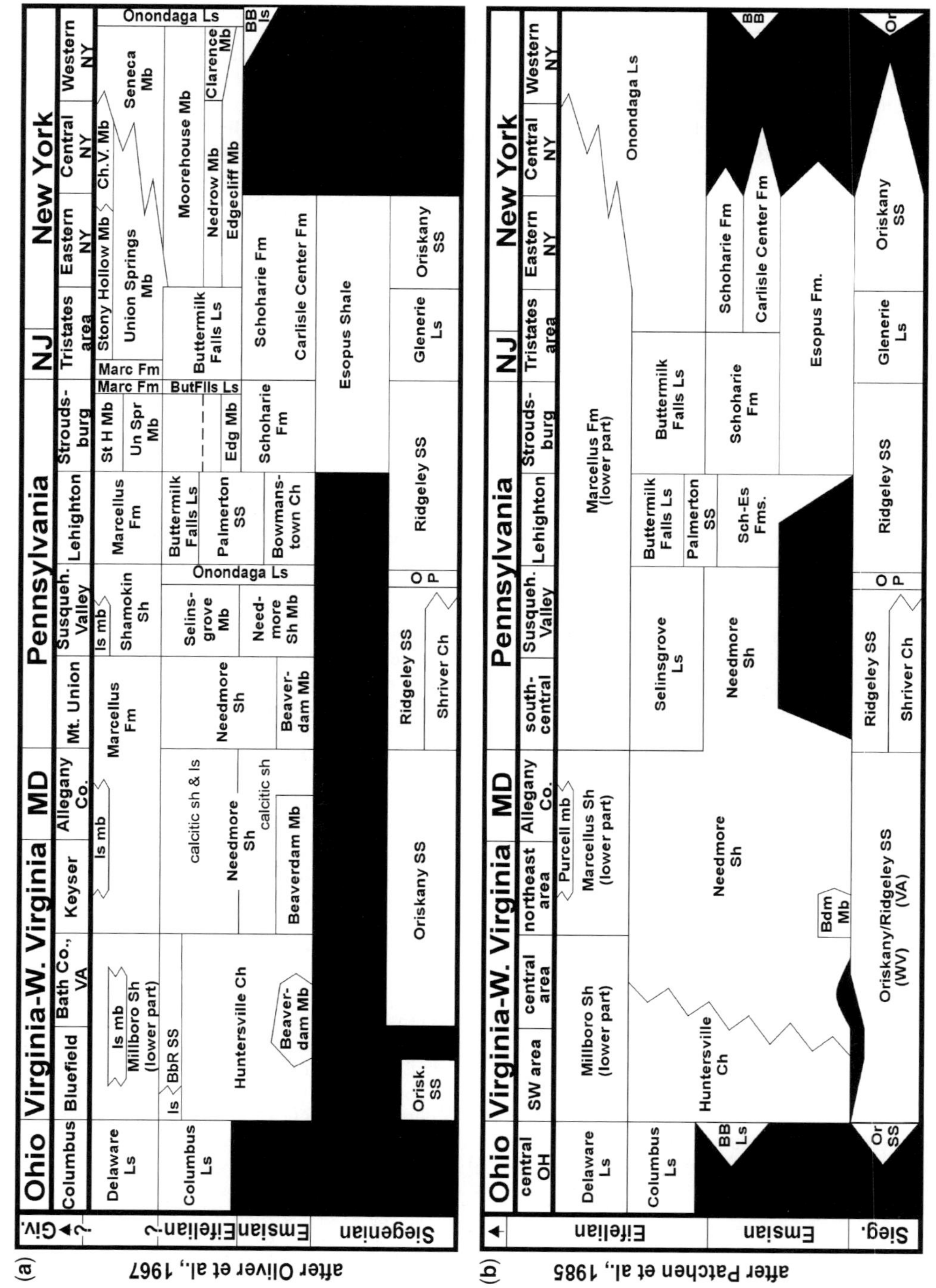
(a)
after Oliver et al., 1967
Ohio
Virginia-W. Virginia
MD
Pennsylvania
NJ
New York
Columbus
Bluefield
Bath Co., VA
Keyser
Allegany Co.
Mt. Union
Susqueh. Valley
Lehighton
Strouds-burg
Tristates area
Eastern NY
Central NY
Western NY
Giv.
Eifelian
Emsian
Siegenian
Delaware Ls
Columbus Ls
Is mb
Millboro Sh (lower part)
Is
BbR SS
Huntersville Ch
Beaver-dam Mb
Orisk. SS
calcitic sh & ls
Needmore Sh
calcitic sh
Beaverdam Mb
Oriskany SS
Marcellus Fm
Needmore Sh
Beaver-dam Mb
Ridgeley SS
Shriver Ch
ls mb
Shamokin Sh
Selins-grove Mb
Need-more Sh Mb
Onondaga Ls
O P
Marcellus Fm
Buttermilk Falls Ls
Palmerton SS
Bowmans-town Ch
Ridgeley SS
St H Mb
Un Spr Mb
Marc Fm
ButFlls Ls
Edg Mb
Schoharie Fm
Marc Fm
Stony Hollow Mb
Union Springs Mb
Buttermilk Falls Ls
Schoharie Fm
Carlisle Center Fm
Esopus Shale
Glenerie Ls
Oriskany SS
Ch.V. Mb
Moorehouse Mb
Nedrow Mb
Edgecliff Mb
Onondaga Ls
Seneca Mb
Clarence Mb
BB ls
(b)
after Patchen et al., 1985
Ohio
Virginia-W. Virginia
MD
Pennsylvania
NJ
New York
central OH
SW area
central area
northeast area
Allegany Co.
south-central
Susqueh. Valley
Lehighton
Strouds-burg
Tristates area
Eastern NY
Central NY
Western NY
Eifelian
Emsian
Sieg.
Delaware Ls
Columbus Ls
BB Ls
Or SS
Millboro Sh (lower part)
Huntersville Ch
Purcell mb
Marcellus Sh (lower part)
Needmore Sh
Bdm Mb
Oriskany/Ridgeley SS (WV) (VA)
Selinsgrove Ls
Needmore Sh
Ridgeley SS
Shriver Ch
O P
Marcellus Fm (lower part)
Buttermilk Falls Ls
Palmerton SS
Sch-Es Fms.
Ridgeley SS
Buttermilk Falls Ls
Schoharie Fm
Glenerie Ls
Esopus Fm.
Schoharie Fm
Carlisle Center Fm
Oriskany SS
Onondaga Ls
BB
Or

the overlying highstand systems tract. Contrary to some interpretations, the transgressive systems tract does not comprise entire black shale successions. Detailed sedimentological, paleobiological and geochemical analyses of numerous Devonian mudrock-dominated sequences indicate that the boundary of the transgressive and overlying highstand systems tracts, marked by a 'maximum flooding surface' (=maximum landward point of the shoreline), occurs at a position down within black shales. In addition, the maximum flooding surface does not generally correspond to the contact of limestones or sandstones with overlying black shales (in the lower part of a sequence), but occurs higher up in the shales.

The 'Highstand Systems Tract' forms during the late stage of a base level rise, as the rate of accommodation space created by the rise becomes less than the sedimentation rate. This results in deposition of aggradational, sediment-starved to progradational strata. The base of highstand deposits occurs above a 'maximum flooding surface', representing the time of the maximum landward position of the shoreline. The maximum flooding surface may be marked by a smaller scale, submarine unconformity during a period of extreme sediment starvation.

The 'Falling Stage Systems Tract' comprises strata deposited during the entirety of a base level fall. Progradation and deposition of offlapping stratal packages characterize the falling stage in the basin, accompanied by the formation of a subaerial unconformity above base level.

The terminology used for key surfaces within sequences in this paper, which also largely follows Catuneanu (2002), is defined as follows: (1) The 'subaerial unconformity' marks a time of erosion and/or non-deposition, which forms above base level during base level fall. It is correlative with a conformity in basinward sections. The two together define the base of a sequence, and underlie the lowstand systems tract. (2) The 'maximum regressive surface', which has also been termed the 'transgressive surface' (Posamentier & Vail 1988). It marks the changeover from coarsening- to fining-upward grain trends and regression to transgression, at the base of the transgressive systems tract. (3) The 'maximum starvation surface', which marks the time of maximum condensation, during the time of the maximum rate of base level rise. This surface is found within the transgressive systems tract. (4) A 'flooding surface', a depositional surface indicative of abrupt deepening (e.g. abrupt shift of limestone to black shale). (5) The 'maximum flooding surface', which can be defined as the maximum landward point of the shoreline. It lies at the contact of the transgressive and highstand systems tracts. Offshore, the maximum bathymetric depth, sometimes also termed 'maximum flooding surface', may occur later in time (Catuneanu 2002). (6) The 'basal surface of forced regression' marks the onset of base level fall in a sequence, and underlies the falling stage systems tract. And (7) the 'regressive surface of marine erosion', where older, previously deposited sediments can be eroded out as by wave processes as base level continues to fall. It may become merged to the previous, underlying surfaces.

Results: correlation of Emsian–Eifelian strata in the Appalachian Basin

High resolution event and cyclic stratigraphic studies across the Appalachian Basin permit basin-wide correlation of upper Pragian, Emsian and Eifelian strata. Cross-sectional figures and photographs (Figs 3–17) outline the correlations at key localities along the outcrop belt.

Emsian correlations

Emsian-age strata in the Appalachian Basin comprise a clastic to mixed clastic and carbonate succession. The clastic component represents the first significant terrigenous materials introduced into the basin during an early stage of the Acadian Orogeny (Ettensohn 1985).

The lateral relationships of Emsian-age strata across the basin have never been well understood. Little faunal information has aided in correlations, especially in the lower part of the succession,

Fig. 2. Upper Pragian, Emsian and Eifelian formation- to member-level stratigraphy after Oliver *et al.* (1967; part **a**) and chiefly formation-level stratigraphy after Patchen *et al.* (1985; part **b**). Geography from left to right is dominantly southwest- to northeast-oriented, except in New York and into Ohio. Turkey Ridge Mbr. in central Pennsylvania (not visible in upper chart) considered by Oliver *et al.* (1967) to be post-Marcellus. Abbreviations: BB, Bois Blanc Fm; Bdm, Beaverdam Member of the Needmore Formation; BbR SS, Bobs Ridge Sandstone Member of Huntersville Fm; ButFlls, Buttermilk Falls Formation; ChV, Cherry Valley Member of the Oatka Creek–Mount Marion Formations; Edg, Edgecliff Member of the Onondaga Formation; Fm, Formation; Ls, limestone; Mb, Member; OP, local area where Ridgeley Fm is included in Old Port Fm; Or, Oriskany Formation; Sch-Es, undifferentiated Schoharie and Esopus Formations; StH, Stony Hollow Member of the Union Springs Formation; Un Spr, Union Springs Formation.

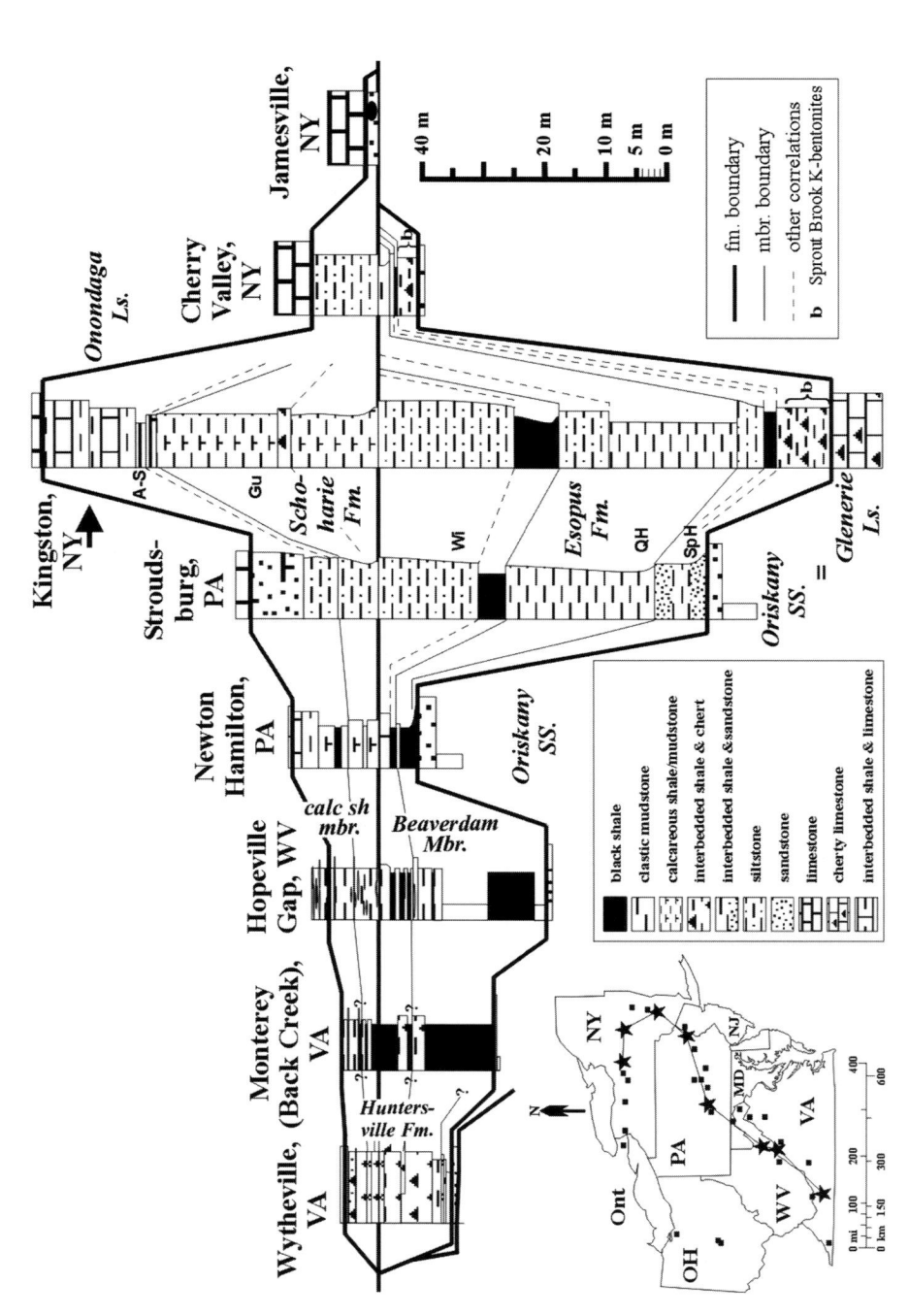

Fig. 3. Stratigraphic sections and correlations of Emsian strata across the Appalachian Basin, New York to Virginia. Dashed and solid lines separate New York formations and correlative horizons across basin. Bold solid lines separate member-level units of the Esopus (Spawn Hollow, SpH; Quarry Hill, QH; Wiltwyck, Wi Members) and Schoharie (Gumaer Island, Gu, and combined Aquetuck–Saugerties, A–S, Members) Formations and equivalents. Datum = contact of Esopus and Schoharie Formations and equivalent horizon basinwide.

where fossils are generally rare to absent. Few correlatable stratigraphic markers, including event marker beds, were known. In addition, the character of the unconformity overlying the late Pragian Oriskany Formation was unclear. The new interpretations presented here are based on the correlation of significant marker units, the correlation of basinwide, sequence-scale cycles (Fig. 3).

Throughout most of the basin, early Emsian mudrocks (Esopus Formation, Beaverdam Member of the Needmore Formation, lower shale and chert of Huntersville Formation) paraconformably overlie the Oriskany Sandstone or equivalent limestone to chert facies (Fig. 3). This contact was interpreted by some workers to be the Wallbridge Unconformity at the base of Sloss's (1963) Kaskaskia Supersequence. The true Wallbridge break, associated with a major sea-level lowstand, however underlies the Oriskany except in the central part of the basin where deposition was continuous through the Pragian (discussed below). Toward the margins of the basins (e.g. central to western New York, central Ohio, southwestern Virginia), the Wallbridge becomes amalgamated with additional younger (+/− older) unconformities (Brett & Ver Straeten 1994; Brett et al. 2000). Throughout the proximal and deeper, central areas of the basin, however the post-Oriskany surface, which is marked locally by phosphate-rich deposits, comprises a significant drowning surface.

Fewer widespread, correlatable marker beds occur through the Emsian succession than in overlying strata. However, the correlation of the marker beds is supported by correlation of five major ('third order') sedimentary cycles that comprise the Emsian succession along the outcrop belt, from eastern New York to southwestern Virginia. These cycles permit widespread subdivision of Emsian strata and a new characterization of their relationships.

In the northeastern part of the basin, three major cycles within lower Emsian strata correspond to three members of the Esopus Formation: in succession, Spawn Hollow (new), Quarry Hill (revised), and Wiltwyck (new) Members (Figs 4a–d). Two additional cycles in upper Emsian rocks comprise the Schoharie Formation: Gumaer Island (new), and Aquetuck and Saugerties Members, respectively (Figs 6b–c, 7a). These five cycles can be correlated from eastern New York into the lower to middle Needmore Formation in central Pennsylvania, where they comprise three cycles in the Beaverdam Member and submember A of the overlying calcareous shale member (Fig. 5a) and two cycles in the remainder of the calcareous shale member (submembers B–D; Fig. 7b), respectively. Strata in central Pennsylvania correlative with the top of the Esopus of eastern New York are characterized by a few to several thin argillaceous to silty limestone beds in the lower part of the calcareous shale member (submember A). From southern Pennsylvania into Maryland–Virginia–West Virginia outcrops, Dennison & Hasson (1976) reported a thin, extensive silt and chert-rich interval (see Fig. 5b, to right of arrow). Field study indicates that the unit represents a coarse cap on top of finer underlying strata, and is correlative with the relatively coarse cycle cap of submember A of the calcareous shale member of the Needmore Formation in central Pennsylvania, and the top of Wiltwyck Member of the Esopus Formation in eastern New York.

Further south in the basin, along the Virginia–West Virginia border, the Esopus and Schoharie-equivalent lower members of the Needmore Formation undergo a transition into the chert-dominated facies of the Huntersville Formation. The facies change occurs between two outcrops spaced 30 km apart (one west of Hightown, Highland Co., VA, Fig. 5b and one northwest of Frost, Pocahontas Co., WV, Figs 5c, 7c). The five cycles, especially the upper four, can be distinguished even in massive cherty facies (e.g. at Wytheville, VA, Figs 5d, 7d), where cycle caps appear as lighter, brown- to grey-coloured intervals amid dark-coloured cherts.

On the very fringe of the basin during late Pragian to Emsian time, in the southwestern tip of Virginia and adjacent Tennessee, strata equivalent to the Oriskany, Esopus and Schoharie Formations of New York occur in sand-dominated facies of the Wildcat Valley Formation (Miller et al. 1964). A coral-rich limestone near the top of the formation features species of late Emsian-age, which also occur in the Schoharie Formation (Oliver 1976).

A cluster of up to 15 K-bentonite beds occurs in the lower Spawn Hollow Member of the Esopus Formation in eastern New York (Ver Straeten 2004a, b). This cluster is termed the Sprout Brook K-bentonites, named for exposures near Cherry Valley, New York (Fig. 3; see Fig. 4a). Through the rest of the basin, a few beds of mixed detrital and volcanic origin are found locally at the same position, but in general discrete beds of volcanic origin are not found (Ver Straeten 2004a, b).

Correlation of the Eifelian Onondaga Formation across the Appalachian Basin

Emsian-age clastic to mixed clastics and carbonates are succeeded across much of the Appalachian Basin and eastern North America by a carbonate-dominated succession. Within the basin, the carbonates are variously termed the Onondaga Formation, Columbus Formation, Selinsgrove Member

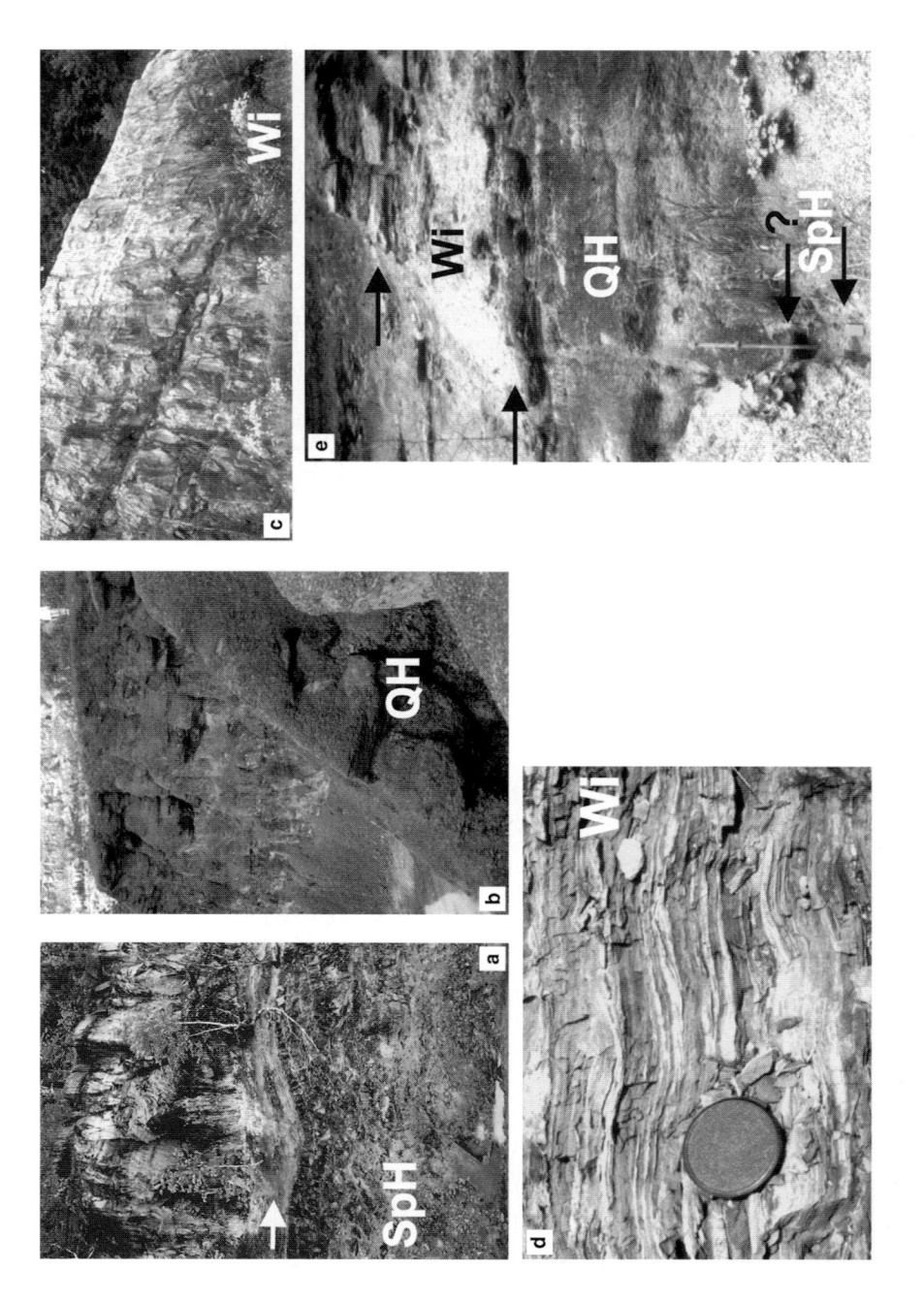

Fig. 4. Photographs of the Esopus Formation and equivalent strata across the northern to central Appalachian Basin (New York, Pennsylvania). New York member-level subdivisions labelled for all localities. Arrows, unless otherwise noted, demarcate formation/member boundaries. SpH, Spawn Hollow Member; QH, Quarry Hill Member; Wi, Wiltwyck Member. (**a**) Spawn Hollow Member, Esopus Fm, Rte. 23a, Catskill, NY. Lower part of member includes the Sprout Brook K-bentonites of Ver Straeten (2004*b*). Arrow points to thickened K-bentonite layer at the top of the Sprout Brook K-bentonites. Visible section = *ca.* 8 m. (**b**) Quarry Hill Member, Esopus Fm, Becraft Mountain, Hudson, NY. Person for scale in upper right. (**c**) Wiltwyck Member, Esopus Fm Rte. 199, Kingston, NY. (**d**) Laminated marker unit, lower part of Wiltwyck Member, Rte. 23, Catskill, NY. (**e**) Beaverdam Member. and lower calcareous shale member, Needmore Formation, Newton Hamilton, PA. 1.5 m staff for scale.

Fig. 5. Photographs of the Esopus Formation and equivalent strata across the Appalachian Basin (New York, Virginia, West Virginia). New York member-level subdivisions labelled for all localities. Arrows, unless otherwise noted, demarcate formation/member boundaries. SpH, Spawn Hollow Member; QH, Quarry Hill Member; Wi, Wiltwyck Member. (**a**) Possible Esopus-equivalent Springvale Member, Bois Blanc Formation, Rte. 88, Phelps, NY. Arrow points to basal contact with Silurian strata. Field book on left for scale. (**b**) Beaverdam Member, Needmore Formation, Back Creek section, west of Monterey, VA. Shovel for scale. (**c**) Lower part of Huntersville Formation, Frost, WV (photo tilted to bedding-horizontal). Section is ca. 16 m thick. (**d**) Lower part of Huntersville Formation, Wytheville, VA. Field book for scale in lower right.

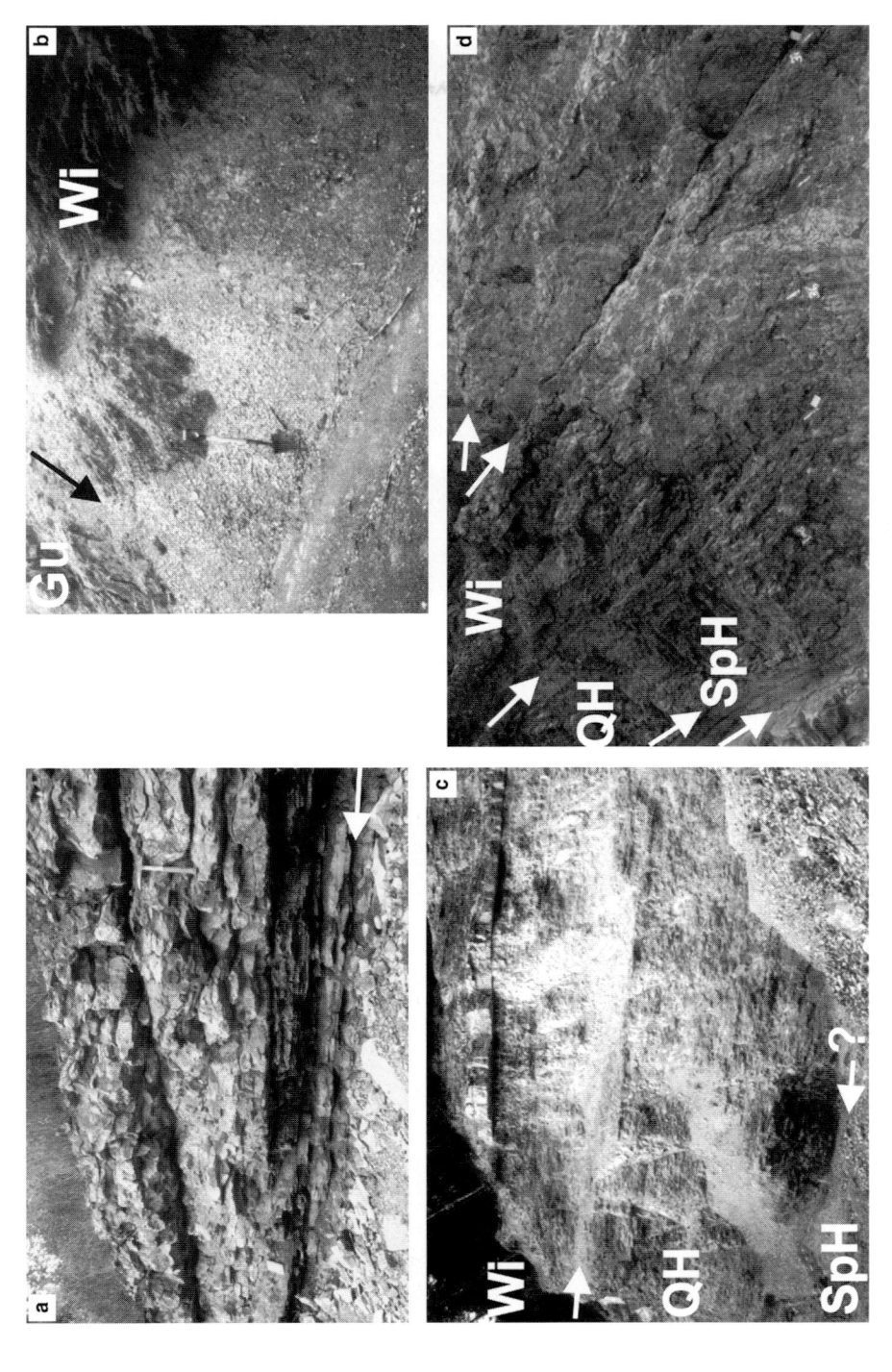

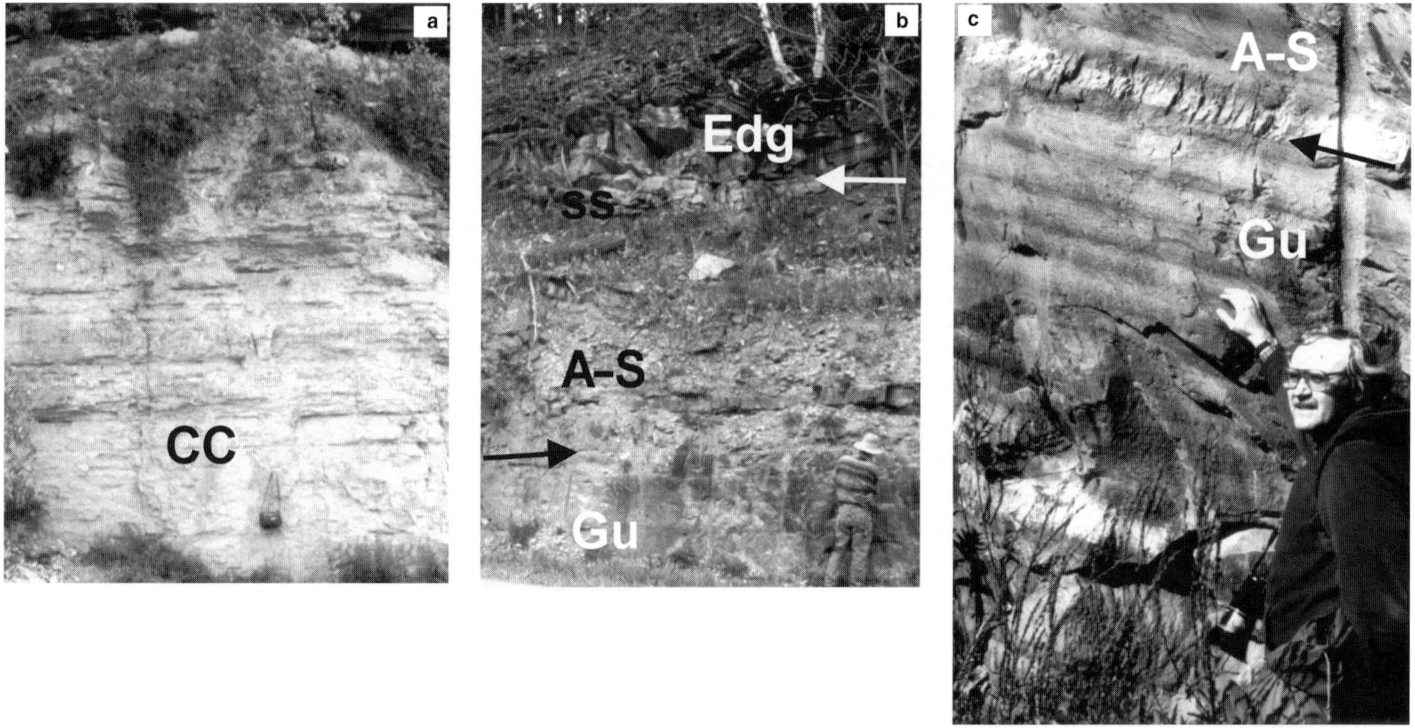

Fig. 6. Photographs of the Schoharie Formation in the northern Appalachian Basin (New York). New York member-level subdivisions labelled for all localities (CC, Carlisle Center Member; Gu, Gumaer Island Member; A–S, combined Aquetuck and Saugerties Members; Edg, Edgecliff Member of the Onondaga Formation; ss, local sandstone facies formerly termed 'Rickard Hill Member' (abandoned)). Arrows demarcate formation/member boundaries. Photos. (**a**) Undifferentiated Schoharie strata (=Carlisle Center Member), Rte. 20, Cherry Valley, NY. Camera bag for scale. (**b**) Schoharie Formation, Rte. 85, Clarksville, NY. (**c**) Close-up of contact interval of Gumaer Island and Aquetuck Members, Rte. 100, Kingston, NY.

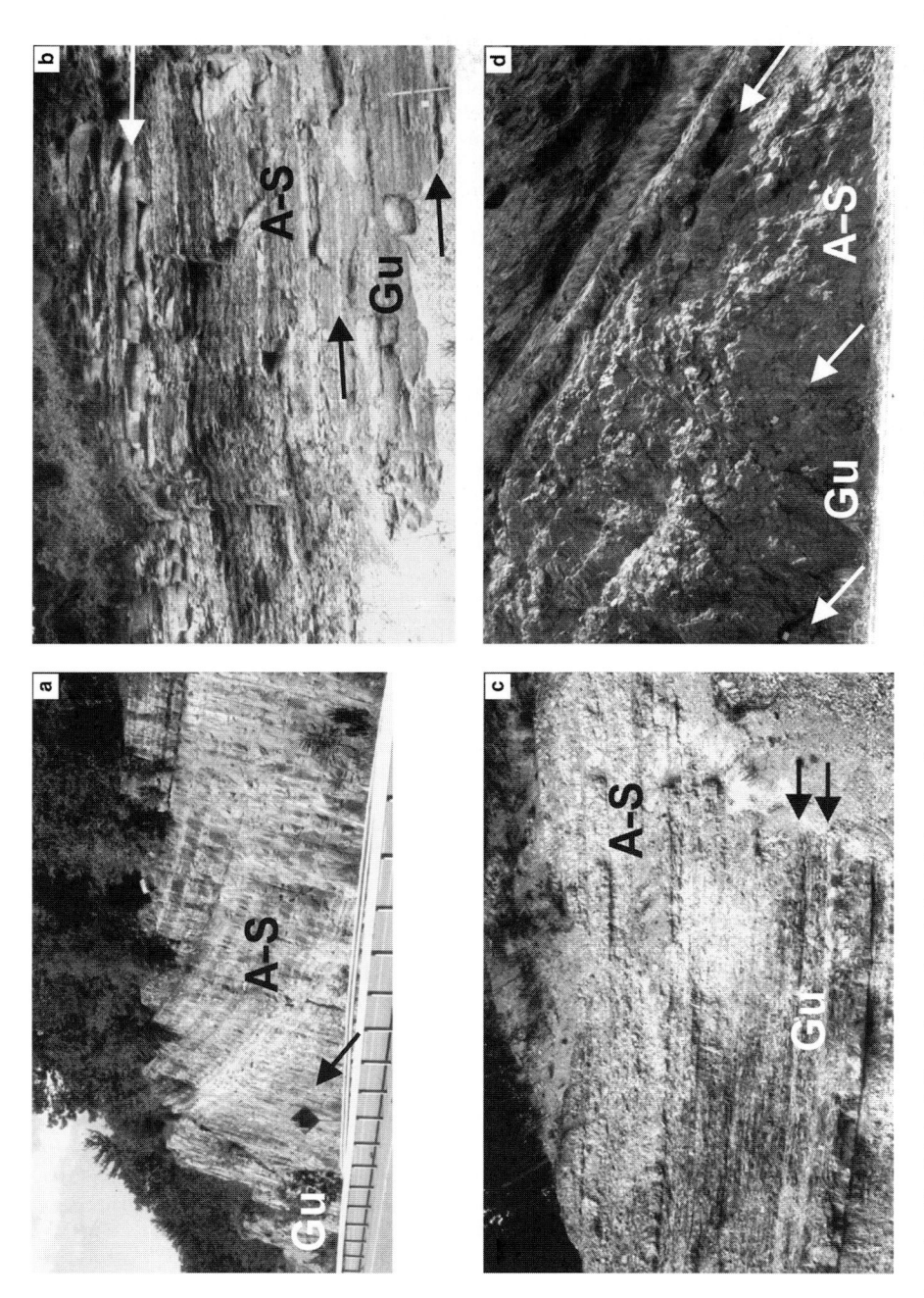

Fig. 7. Photographs of the Schoharie Formation and equivalent strata across the Appalachian Basin (New York, Pennsylvania, West Virginia, Virginia). New York member-level subdivisions labelled for all localities (Gu, Gumaer Island Member; A–S, combined Aquetuck and Saugerties Members). Arrows demarcate formation/member boundaries. (**a**) Schoharie Formation, Rte. 199, Kingston, NY. (**b**) Calcareous shale member and lower part of Selinsgrove Member, Needmore Formation, Newton Hamilton, PA. 1.5 m staff in lower right. (**c**) Upper part of Huntersville Formation, Frost, WV (photo tilted to bedding-horizontal). Section is ca. 7 m thick. (**d**) Upper part of Huntersville Formation and overlying strata, Wytheville, VA. Field book at lower left contact for scale.

(Needmore Formation), and the calcitic shale and limestone member (Needmore Formation). Another named unit, the Buttermilk Falls Formation (of eastern Pennsylvania), has recently been abandoned due to synonymy with the Onondaga Formation (Ver Straeten & Brett, 2006). Extensive fieldwork through the succession revealed a number of widespread, distinctive marker beds. These include K-bentonites (including the widely known Tioga K-bentonites), massive limestones, argillaceous intervals, black shale beds, and pyrite nodule-rich horizons, along with small- to medium-scale cycles (parasequences to parasequence sets). The combination of these multiple event and cyclic markers permits microstratigraphic correlation of the carbonate succession across the entire Appalachian Basin, from New York to Virginia to central Ohio (Fig. 8).

The relatively fine-scale correlations, at a higher resolution than possible in the underlying Emsian rocks, permit recognition of a number of subdivisions of the strata on a basinwide scale. Throughout most of the basin, the basic trends that characterize the four members of the Onondaga Formation in New York (Figs 9a–d, 10a–b) can be seen. Initial relatively shallow facies (Edgecliff Member) are succeeded by finer-grained, more argillaceous strata (Nedrow Member). A return to greater limestone content is marked by a general coarsening up to the upper part of a third unit (Moorehouse Member), followed by a gradual fining-upward (Seneca Member) to the top of the carbonate succession. Again, these basic trends characterize the succession basinwide, except in two areas. These trends break down in the southernmost portion of the basin, in the southern part of the Virginia–West Virginia outcrop belt, where Moorehouse to Seneca equivalent strata occur as black shale dominated facies in southwestern Virginia (at Wytheville; Fig. 12a, c), which grade to the southern terminus of outcrop into carbonates (Duffield, Virginia) and sand-dominated strata (Little War Gap, Tennessee). General basinwide trends for the lower two member-level units are also disrupted to some degree in the central to western New York region, due to flexure of a migrating, bulge-like topographic high and cratonward low (Ver Straeten & Brett 2000).

In central Pennsylvania to northern Virginia and West Virginia, the base of Onondaga equivalent strata can be recognized by thicker limestone beds (Figs 10d, 11a, c) with a relatively coarser lithology and increased faunal diversity (including small rugose corals in some areas). Southward along the Virginia–West Virginia outcrop belt, the base of the succession can be difficult to recognize, until the appearance of a thin sandstone (Bobs Ridge Member) at the top of the Huntersville Formation (Fig. 12a, b). However, the three medial scale cycles which characterize Edgecliff and lowest Nedrow-equivalent strata throughout the New York to Virginia outcrop belt, can be used for correlation into this difficult region, and other areas around the basin.

A number of key Onondaga marker beds can be correlated from the main body of the Appalachian basin into the Columbus Formation of central Ohio (Fig. 8). These include K-bentonite beds in Nedrow, Moorehouse and Seneca strata (solid arrows in Fig. 8), finer-grained facies of the Nedrow Member, and others (Fig. 13a, b). One particularly distinctive bed comprises a fine-grained, argillaceous unit termed the 'false Nedrow' bed by Brett & Ver Straeten (1994), in the upper middle part of the Moorehouse Member (see Fig. 8). This bed is widely correlatable across the basin, and is here termed the 'Stroud Bed' (named for Stroud Township, PA; type section is a classic rock cut on the Norfolk Southern Railroad at East Stroudsburg, where the bed consists of one metre of calcareous, dark grey shale). The framework of correlated marker beds leads to new interpretations of the stratigraphic relationships of the Onondaga and Columbus Formations, most notably in the lower and upper parts of the Columbus (Fig. 8). In this paper, lower to middle parts of the Bellepoint Member of the Columbus are correlated with the Edgecliff Member of the Onondaga. Parts of upper Moorehouse and Seneca equivalent strata, including key K-bentonites of the Tioga A–G zone, are interpreted as missing at unconformities in upper Columbus strata (discussion below).

One of the key revelations in the new stratigraphic correlations is the recognition that the Onondaga Formation succession features two major clusters of volcaniclastic sediments. The extension of the high resolution correlations, including the Tioga A–G K-bentonites in upper Onondaga and equivalent units, indicates that the Tioga Middle Coarse Zone (MCZ) K-bentonites of Dennison & Textoris (1970) underlie the Tioga A–G K-bentonites (Ver Straeten 2004a; Fig. 8). The two were previously correlated; now it is seen that the MCZ zone occurs in the middle of the formation, apparently in strata equivalent to the lower part of the Moorehouse Member of New York (above a key black shale marker bed at the top of the Nedrow Member; Fig. 12a), and is found only in the southern part of the Appalachian Basin (central to southern portion of the VA–WV outcrop belt; Ver Straeten 2004a). The Tioga A–G K-bentonites are found in the upper Moorehouse and Seneca Members and in lowest strata of the overlying Union Springs Formation (Bakoven Member) and correlative units (Figs 9c–d, 10c, 11c, and 12c).

Correlations of the Eifelian Lower Hamilton Group across the Appalachian Basin

Eifelian-age strata in the lower part of the Hamilton Group include rocks that in New York occurred in the lower part of what has long been termed the 'Marcellus Formation'. During the course of this study, it has become apparent that the Marcellus 'Formation' in New York, which ranges in thickness from approximately 15 m on the Lake Erie shore (western New York) to over 580 m in the Catskill Front (eastern New York; Rickard 1989), represents a more complex unit than has been previously recognized. Following the informal recommendations of Ver Straeten *et al.* (1994), Ver Straeten & Brett (2006) have proposed to divide the Marcellus 'Formation' into three formation level units, a lower Union Springs Formation and higher, time-equivalent Oatka Creek and Mount Marion Formations. The latter two formations are distinguished by basinal, black shale-dominated facies (Oatka Creek Formation) and basinal black shale to shoreface, sand-dominated strata associated with progradational infilling to sea-level (Mount Marion Formation). The Union Springs and the laterally equivalent Oatka Creek–Mount Marion Formations comprise two successive, major 'third-order' stratigraphic sequences. The term 'Marcellus' in New York is now assigned to an informal 'subgroup' status within the Hamilton Group (following US Geological Survey guidelines).

The Union Springs and lower part of the Oatka Creek–Mount Marion Formations include several distinctive, widely correlatable marker beds (Fig. 14). Beds in the Union Springs Formation include the widespread Tioga F K-bentonite just above the base of the Bakoven Member, with one to two thin additional Tioga beds closely above; and the mid-Union Springs K-bentonite, found widely in eastern New York, Pennsylvania, Virginia, West Virginia and central Ohio. Another distinctive, if more locally occurring, marker bed in the Union Springs Formation is the *'Cabrieroceras* bed' (formerly *'Werneroceras* bed'; Anderson *et al.* 1988). This goniatite-rich, nodular limestone bed passes laterally from the upper Bakoven Member in east-central into the lower part of the Stony Hollow Member in eastern New York. Marker beds in the succeeding Oatka Creek–Mount Marion Formations include the fossiliferous Chestnut Street submember of the Hurley Member, a limestone or series of limestones recognized throughout much of the basin, except in deep water, basin interior facies; and styliolinid–cephalopod limestones and locally equivalent detrital-rich lithosomes of the Cherry Valley Member.

In deeper parts of the basin, the lower Hamilton Group strata are dominated by black shales, with minor limestones at the base of each formation. In areas more proximal to the basin margin, the black shales of the middle to upper part of the Union Springs Formation undergo lateral facies changes. In eastern New York, the black shale equivalents comprise the calcareous shales, siltstones and fine sandstones of the Stony Hollow Member (Figs 15b, c, 16a). In central Pennsylvania, this interval is represented by the Turkey Ridge Sandstone (Fig. 17a, b). In more proximal sections in eastern New York (e.g. Kingston) and central Pennsylvania (e.g. Harrisburg area), the facies transition occurs a short distance above the mid-Union Springs K-bentonite bed. Similar, correlative facies of the Stony Hollow also appear in areas of central Pennsylvania (e.g. Selinsgrove Junction) and in southwestern Virginia (e.g. Jordan Mines, Fig. 17c).

The lowest strata of the Oatka Creek and Mount Marion Formations in New York are assigned to the Hurley Member (Ver Straeten & Brett, 2006). This differs from an earlier proposal by Ver Straeten *et al.* (1994) to place the Hurley at the top of the Union Springs Formation. In its type area (near Kingston, NY), the Hurley Member is comprised of calcareous shales, thin sandstones, and thin shelly layers. In more basinward areas, the unit consists of fossiliferous limestones (Chestnut Street submember) and overlying black shales (Lincoln Park submember), with local thin siltstones to sandstones (Figs 15a, c, 16b, d and 17c). The Hurley Member and its submembers are recognizable across most of Pennsylvania. In deep basinal facies in the northern part of the Virginia–West Virginia outcrop belt into southernmost Pennsylvania, the member was not recognized. Toward the southern margin of basin (southern VA), the Hurley Member appears as interbedded, impure limestone and shale facies somewhat similar to eastern New York (Fig. 17c).

The Cherry Valley Member, in the lower part of the Oatka Creek–Mount Marion Formations, comprises two main facies. The first is a classic, deep water 'cephalopodenkalk' limestone facies (Griffing & Ver Straeten 1991; Ver Straeten *et al.* 1994; Fig. 15a). The limestone facies commonly contains thin and wispy, to thick black shale interbeds. In more proximal areas, the Cherry Valley Member is a clastic-dominated facies of a calcareous shale (southern PA, southern VA; Fig. 16b, d), nodular limestone in shales (VA, WV), bioturbated sandstone and dark grey mudstone (eastern NY; Fig. 15d) or sandstone (top of Turkey Ridge Sandstone, central PA).

In central Ohio (Columbus–Delaware area), a number of the widespread marker units are found in the Delaware Formation (Fig. 14). Correlatable units include the mid-Union Springs K-bentonite near the middle of the formation, and higher up in the section, a richly fossiliferous bed with

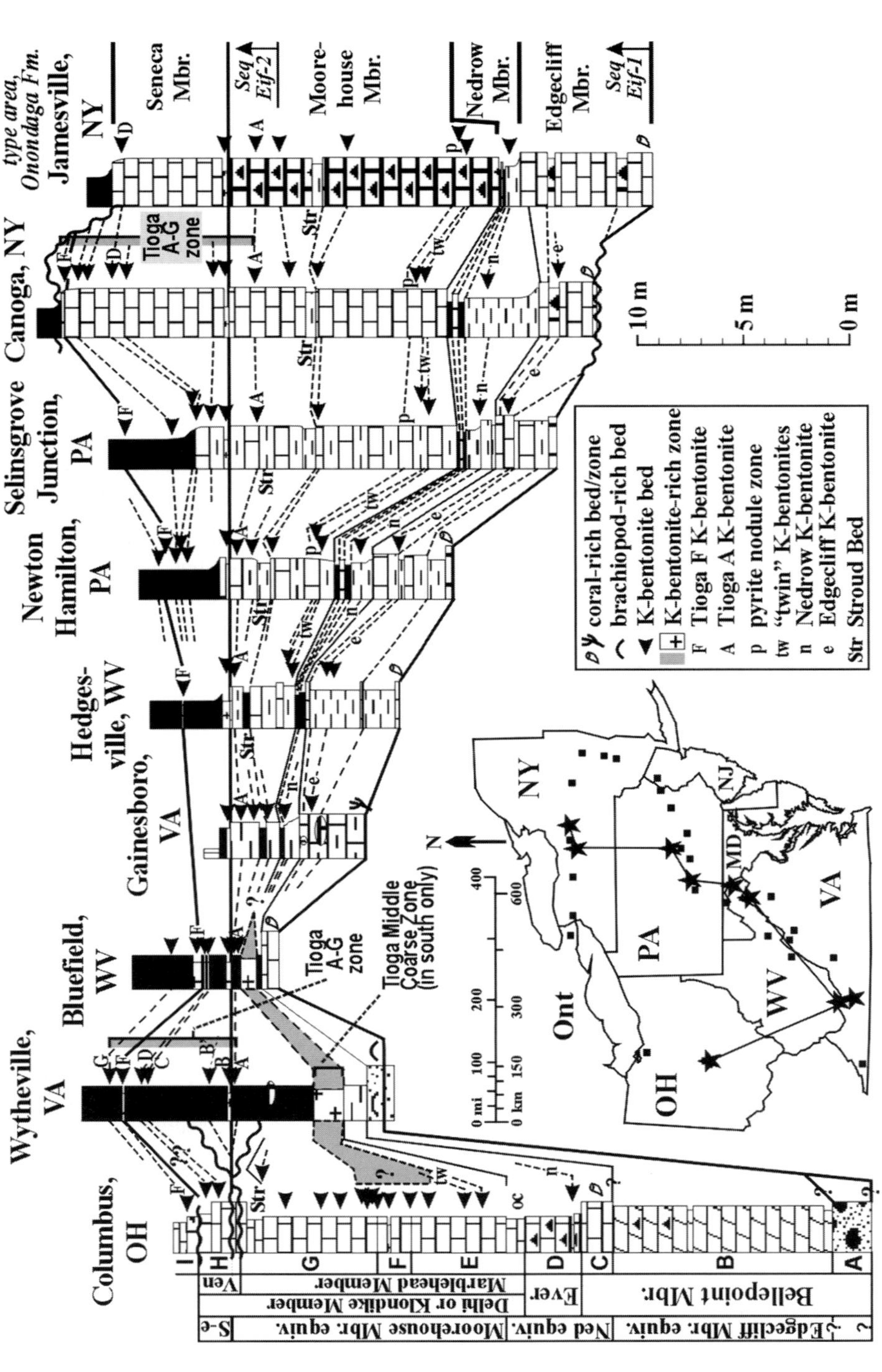
Columbus, OH
Wytheville, VA
Bluefield, WV
Gainesboro, VA
Hedgesville, WV
Newton Hamilton, PA
Selinsgrove Junction, PA
Canoga, NY
Jamesville, NY
type area, Onondaga Fm.
Seneca Mbr.
Seq Eif-2
Moorehouse Mbr.
Nedrow Mbr.
Edgecliff Mbr.
Seq Eif-1
Tioga A-G zone
Tioga Middle Coarse Zone (in south only)
Edgecliff Mbr. equiv.
Bellepoint Mbr.
Ned equiv.
Ever.
Moorehouse Mbr. equiv.
Delhi or Klondike Member
Marblehead Member
S-e
Ven
10 m
5 m
0 m
coral-rich bed/zone
brachiopod-rich bed
K-bentonite bed
K-bentonite-rich zone
F Tioga F K-bentonite
A Tioga A K-bentonite
p pyrite nodule zone
tw "twin" K-bentonites
n Nedrow K-bentonite
e Edgecliff K-bentonite
Str Stroud Bed
Ont
OH
PA
NY
NJ
MD
WV
VA
N
0 mi 100 200 400
0 km 150 300 600

brachiopods, corals and *Dechenella* trilobites, which indicates the presence of the Chestnut Street submember of the Hurley Member within the Delaware Limestone. As discussed further by Desantis *et al.* (this volume), that bed is succeeded by two overlying fauna-rich beds: (1) with numerous button-shaped rugose corals (*Hadrophyllum*); and (2) with auloporid corals and a variety of brachiopods. The two beds appear to correlate in New York with, respectively, the Cherry Valley Member and the Halihan Hill Bed at the base of the Otsego Member, immediately overlying the East Berne Member. These and other correlations indicate that the limestones and minor shales of the Delaware Formation are lateral equivalents of the Union Springs and lower part of the Oatka Creek–Mount Marion Formations of New York, and their equivalents basinwide.

Discussion: stratigraphic synthesis of upper Pragian, Emsian and Eifelian strata across the Appalachian Basin

The data and correlations outlined above present a new, high resolution picture of the stratigraphic relationships through the upper Pragian to Eifelian succession in the Appalachian Basin. Formations, and even many member-level and finer-scale units, are correlatable throughout the basin. The resulting stratigraphic synthesis is summarized in Fig. 18. When compared to earlier correlations (Fig. 2), some significant changes are apparent. These include the following. (1) Basinwide recognition of lower Emsian Esopus-equivalent strata. (2) The Huntersville Formation along the outcrop belt is not equivalent to the Eifelian Onondaga Formation of New York, as has often been claimed, but rather correlates with the underlying Emsian strata of the Esopus and Schoharie Formations. (3) The lower Eifelian K-bentonites of the Tioga Middle Coarse Zone are restricted to the southern (and western?) part of the Appalachian Basin (in lower Moorehouse Member equivalents), and underlie the Tioga A–G K-bentonites (found basinwide in upper Moorehouse, Seneca and overlying Union Springs and equivalent strata). These clusters of K-bentonites were previously thought to be correlative. (4) The middle to upper Onondaga-equivalent strata in the deep, central basin area and in the south are predominantly black shale facies. (5) The Columbus Formation of central Ohio, including lower strata (Zones A and B of Stauffer, 1909; = Bellepoint Member), is the lateral equivalent of the Onondaga Formation. Notably, however, upper parts of the Onondaga succession recognized basinwide (including parts of the upper Moorehouse and Seneca Members, including the Tioga B/'Onondaga Indian Nation' K-bentonite) are missing at unconformities within the upper part of the Columbus (in the lower H zone). And (6) strata of the classic 'Marcellus Shale' of New York represent two major, formation-level packages of strata, with a number of finer-scale units that are widely recognizable through the basin.

Biostratigraphic summary

International biostratigraphic zones for conodonts and goniatites are shown in Figure 18 for Pragian to Eifelian strata. As noted in the introduction, biostratigraphic data for the Pragian to Eifelian is not well constrained, especially for the Pragian and Emsian. This is denoted by dashed lines between biozones in Figure 18. An additional column on the occurrence of key goniatite taxa from the Appalachian Basin is presented based on work by House (1978, 1981), Becker & House (1994, 2000), and discussions with R. T. Becker (pers. comm. 2006). Of note, apparent *Agoniatites vanuxemi nodiferus* goniatites were found during the course of fieldwork in the upper part of the Stony Hollow Member, in a fossil-rich interval ('lower Proetid bed') below the resistant sandstone cap of the Stony Hollow. This would extend the range of *Agoniatites vanuxemi nodiferus* downward from its long recognized position in the Lincoln Park submember of the Hurley Member into the upper part of the Stony Hollow Member.

Proposed stratigraphic nomenclatural changes

Another result of the new, detailed work through the late Pragian to Eifelian succession is a need for

---

Fig. 8. Stratigraphic sections and correlations of the Onondaga Formation and equivalents across the Appalachian Basin (New York to Virginia and central Ohio). Dashed and solid lines show correlations. Bold solid lines separate New York formations and correlative horizons across basin. Non-bold solid lines separate member-level units of the Onondaga Formation and correlative horizons. Datum = base of Tioga B K-bentonite (=base of Seneca Member and equivalent strata) basinwide. Arrows indicate position of K-bentonite beds of Tioga A–G cluster and additional layers; individual Tioga Middle Coarse Zone K-bentonites not marked. Note the correlation of the Tioga Middle Coarse Zone into the Columbus Limestone in central Ohio. Faunal symbols at base indicate key faunal markers for base of Edgecliff Member basinwide. 'oc', 'carbonaceous' interval reported by Conkin & Conkin (1984). Thickness and general facies for the Columbus Formation at Columbus, Ohio, after Conkin & Conkin (1984; their 'old and new Scioto quarries' composite section at Marblecliff).

Fig. 9. Photographs of the Onondaga Formation in northern Appalachian Basin (New York). New York member-level subdivisions labelled for all localities (Edg, Edgecliff Member; Ned, Nedrow Member; Mo, Moorehouse Member; Bak, Bakoven Member of the Union Springs Fm; Hur-ChV, Hurley and Cherry Valley Members of the Oatka Creek–Mount Marion Formations). Abbreviations: bl, 'black beds' at top of the Nedrow Member-equivalent strata; K-b, K-bentonite marker beds; py, widely correlatable pyrite nodule bed in lower Moorehouse Member-equivalent strata. Bars demarcate formation/member boundaries. (**a**) Lower Onondaga strata, Seneca Stone quarry, south of Seneca Falls, NY. (**b**) Middle Onondaga strata, Seneca Stone quarry. Section is *ca.* 13 m thick. Key correlatable marker beds shown: K-b, K-bentonite layer; bl, black beds at top of Nedrow Member; py, widespread pyrite nodule layer. (**c**) Upper Onondaga strata, with overlying Union Springs and Oatka Creek Formations, Seneca Stone quarry. Section is *ca.* 18 m thick. (**d**) Onondaga Formation, quarry at Jamesville, NY. Arrow points to Stroud Bed marker. Section is *ca.* 24 m thick.

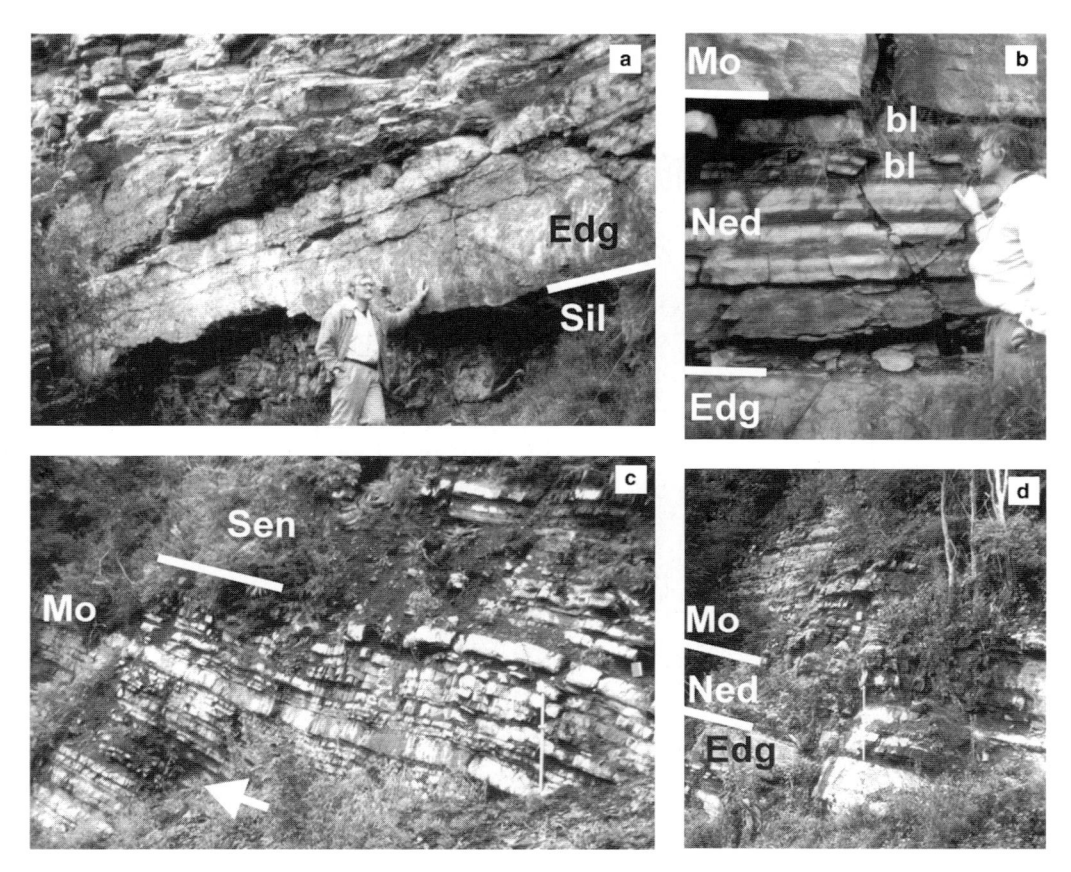

Fig. 10. Photographs of the Onondaga Formation and equivalent strata in the northern to central Appalachian Basin (New York, Pennsylvania). New York member-level subdivisions labelled for all localities (Edg, Edgecliff Member; Ned, Nedrow Member; Mo, Moorehouse Member; and Sen, Seneca Member. Bak, Bakoven Member of the Union Springs Fm; Hur-ChV, Hurley and Cherry Valley Members of the Oatka Creek–Mount Marion Formations. Sil, Silurian). Abbreviation: bl, 'black beds' at top of the Nedrow Member. Bars demarcate formation/member boundaries. (**a**) Irregular unconformity of Onondaga Formation (Edgecliff Member) overSilurian strata, quarry at Oak Corners, NY. (**b**) Nedrow Member and adjacent strata, Interstate 81, Nedrow, NY. (**c**) Upper part of Selinsgrove Member (type locality), Needmore Formation; Selinsgrove Junction, PA. Correlations with Onondaga Formation members of New York shown. Arrow points to the distinctive Stroud Bed. 1.5 m staff in lower right for scale. (**d**) Lower part of Selinsgrove Member (type locality), Needmore Formation, Selinsgrove Junction, PA. Correlations with Onondaga Formation members of New York shown. 1.5 m staff in lower centre for scale.

some changes in stratigraphic nomenclature. The proposed changes, outlined briefly here, are presented in Table 1. They are proposed formally in a separate paper (Ver Straeten & Brett, 2006). For the most part, the changes concern strata in New York State and eastern Pennsylvania, but in a few cases are applicable elsewhere.

In the lower part of the succession, it is proposed that the term 'Oriskany Formation' be applied to all upper Pragian sand-rich strata across the basin, following Ver Straeten's (2001a) recommendation for Pennsylvania; presently the name 'Oriskany' is used in about half the basin (NY, WV, OH). The term 'Ridgeley', used for correlative strata of the same facies, is abandoned (following the US and International stratigraphic codes; North American Commission on Stratigraphic Nomenclature 1983; and Salvador 1994), as it is a junior synonym of the Oriskany Formation.

The Esopus Formation is revised here to recognize two new members (Spawn Hollow and Wiltwyck Members) and to revise an older one (Quarry Hill Member). In addition, two older, rarely used member-level terms (Mountainville

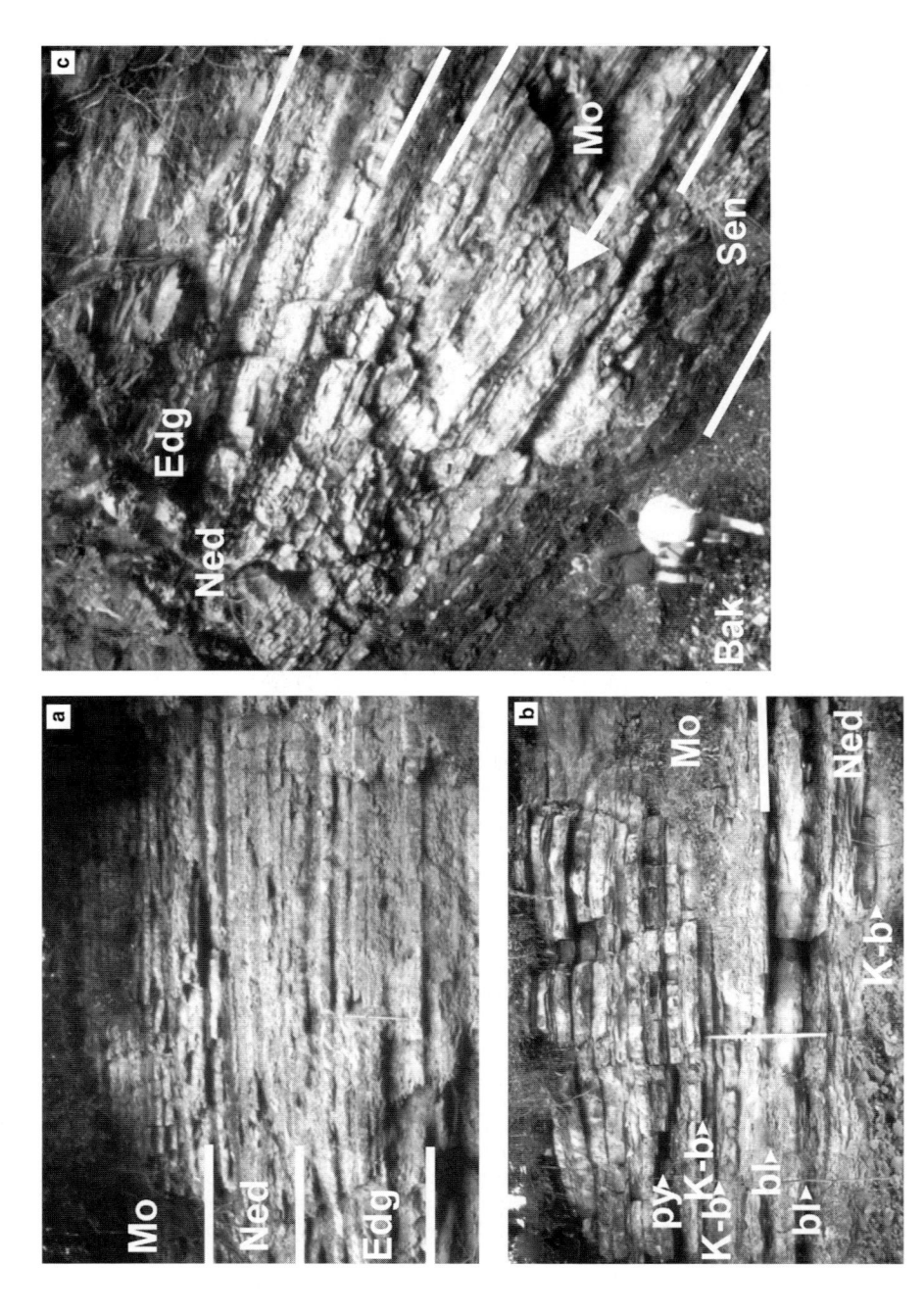

Fig. 11. Photographs of the Onondaga Formation and equivalent strata in the central Appalachian Basin (Pennsylvania, West Virginia). New York member-level subdivisions labelled for all localities (Edg, Edgecliff Member; Ned, Nedrow Member; Mo, Moorehouse Member; Bak, Bakoven Member of the Union Springs Formation). Abbreviations: bl, 'black beds' at top of the Nedrow Member-equivalent strata; K-b, K-bentonite marker beds; py, widely correlatable pyrite nodule bed in lower Moorehouse Member-equivalent strata. Bars demarcate formation/member boundaries. (**a**) Lower to middle Selinsgrove Member, Needmore Formation, Newton Hamilton, PA. 1.5 m staff in lower centre for scale. (**b**) Middle part, Selinsgrove Member, Newton Hamilton, PA. A 1.5 m-thick staff is visible in lower centre for scale. (**c**) Overturned section of Selinsgrove Member, Hedgesville, WV. Arrow points to the Stroud Bed marker.

Fig. 12. Photographs of the Onondaga Formation-equivalent strata in the southern Appalachian Basin (Virginia). New York member-level subdivisions labelled for all localities (Edg, Edgecliff Member; Ned, Nedrow Member; Mo, Moorehouse Member; and Sen, Seneca Member. Bak, Bakoven Member of the Union Springs Formation; Hnt, Huntersville Formation; Sch2, upper Schoharie-equivalent strata of Huntersville Formation; BR, Bobs Ridge Member of Huntersville Formation; 'Marc', Marcellus Formation (as applied outside of New York State). Bars demarcate formation/member boundaries. Photos: **(a)** Onondaga-equivalent strata in uppermost Huntersville and 'Marcellus' Formations, Wytheville, VA. Bobs Ridge Sandstone Member is 1.4 m thick; *MCZ*, Tioga Middle Coarse Zone K-bentonites; **(b)** close-up of upper part of Huntersville Formation, including Bobs Ridge Member, Wytheville, VA. Bobs Ridge Sandstone is 1.4 m thick; **(c)** upper Onondaga equivalent strata, Wytheville, VA. Tioga B (Ti-B) and Tioga F (Ti-F) K-bentonite beds marked by arrows. Person points to Tioga B K-bentonite in left-centre.

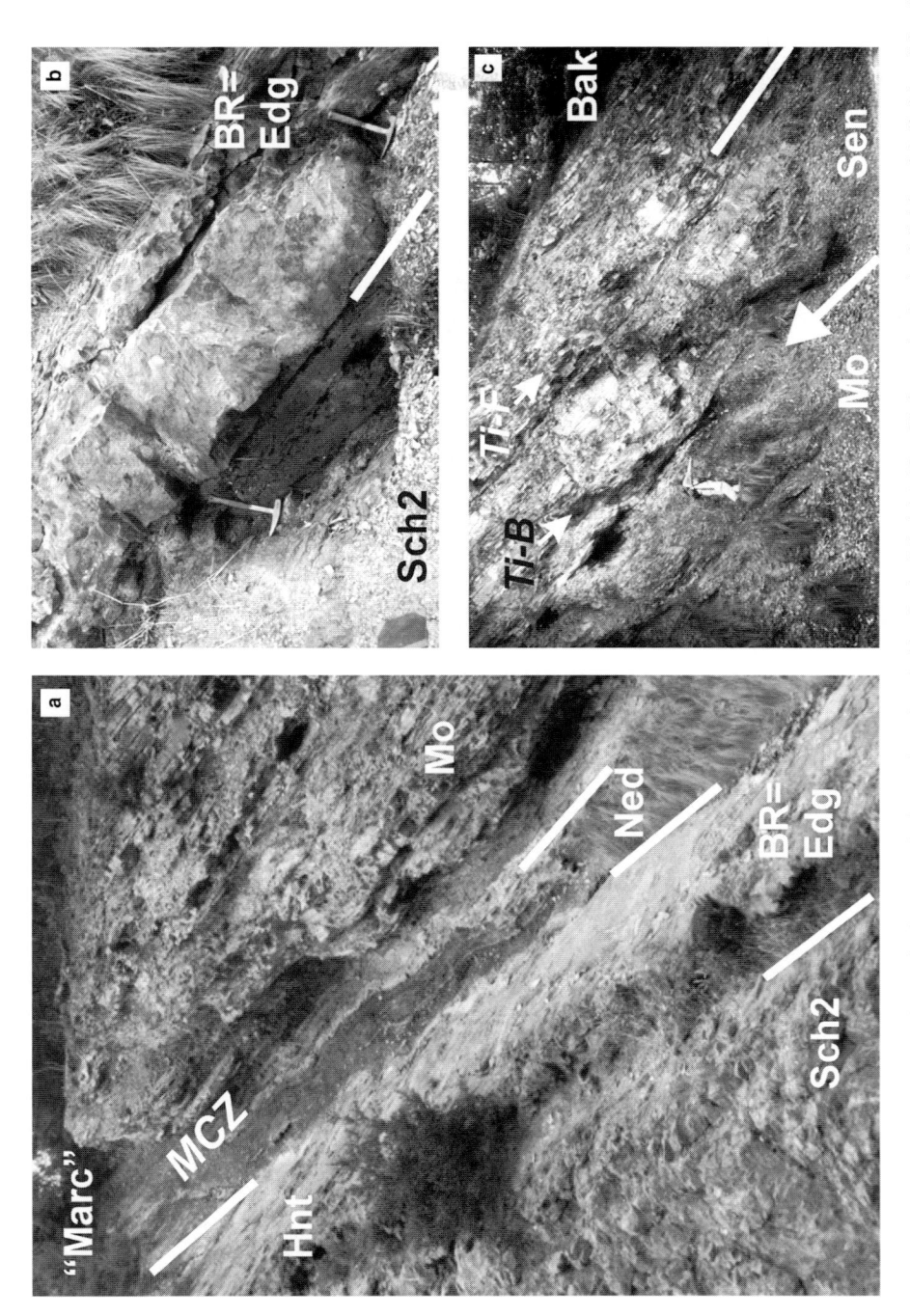

Fig. 13. Photographs of the Onondaga Formation-equivalent strata in the western Appalachian Basin, Ohio. New York member-level subdivisions labelled for all localities (Ned, Nedrow Member; Mo, Moorehouse Member; and Sen, Seneca Member. Bak, Bakoven Member of the Union Springs Formation; BL, Bellepoint Member; Ev, Eversole Member; and Kl, Klondike Member of the Columbus Formation). Bars demarcate formation/member boundaries. (**a**) Upper part of the Columbus Formation, Shaughnessy Dam, NW of Columbus, OH. Arrow points to Stroud Bed marker. (**b**) Middle part of the Columbus Formation, Shaughnessy Dam. Coral bed shown at base. Top ledge of photo = bottom ledge of previous photo.

and Highland Mills), are abandoned. In the Schoharie Formation, the Aquetuck and Saugerties Members are retained with no change. The Gumaer Island Member is proposed as a new member for lower strata of the Schoharie Formation. The Carlisle Center Member is restricted to undifferentiated sandstone to siltstone facies in east-central to central New York. The Rickard Hill Member is abandoned.

In the Eifelian Onondaga Formation, the Clarence Member is abandoned as it only referred to a locally occurring chert-rich limestone facies. The Edgecliff and Nedrow Members are revised to include those strata. In addition, a set of distinctive marker beds ('black beds' of Brett & Ver Straeten 1994) is chosen for the top contact of the Nedrow Member. The overlying Moorehouse and Seneca Members are retained with no change. The name Onondaga Formation is revised to include correlative strata in eastern Pennsylvania, where the term 'Buttermilk Falls' and its four member-level units ('Foxtown', 'McMichael', 'Stroudsburg', and 'Echo Lake') are all abandoned as junior synonyms of the Onondaga Formation and its four members (following Ver Straeten 2001a).

As noted above, the middle to upper Eifelian strata of the classic 'Marcellus Shale' in New York represent two distinctive formation-level packages of strata. A stratigraphic revision for the Marcellus 'Shale' in New York is proposed in which the name 'Marcellus' is raised to subgroup status, as an informal subgroup of the Hamilton Group. It is subdivided into three formation level units: (a) a lower division, the Union Springs Formation (formerly 'Member'); and (b) Oatka Creek formation (formerly 'Member', in central to western NY) and the correlative Mount Marion Formation (in eastern NY). The term 'Marcellus subgroup' is at present only used for strata in New York, although equivalent strata and subdivisions are recognized across the basin. Two members are recognized within the Union Springs Formation in New York (in vertical succession); the Bakoven Member (revised), and Stony Hollow Member (restricted). Above, the lower parts of the time-equivalent Oatka Creek and Mount Marion Formations in New York include, in succession, the Hurley Member (new), the Cherry Valley Member (revised to include correlative clastic strata) and the East Berne Member (new). Two informal submembers are recognized in the Hurley Member: lower, fossiliferous limestones, with or without thin shales (Chestnut Street submember; formerly termed 'proetid bed' by Anderson et al. 1988), and upper black shales, with or without thin sandstone beds (Lincoln Park submember). The name 'Berne Member' is abandoned, and replaced by East Berne Member, as the name Berne Member was previously used for Carboniferous strata in Ohio (Hyde 1915).

Relationship of the Columbus and Onondaga Formations

As previously noted above, a number of key, basinwide Onondaga marker beds recognized in the Columbus Formation of central Ohio (Fig. 8). A combination of recent fieldwork, Chapel's (1975) detailed petrological work on the Columbus

Limestone and Judge's (1998) study of the Bellepoint Member, when integrated into a sequence stratigraphic framework, permit a reinterpretation of the facies and sea-level history of the Columbus. Note that various member-level units have been applied to the Columbus Formation by different workers; they are shown on the left side of the Columbus, OH stratigraphic column of Figure 8.

Overall, the Columbus succession shows an initial, relatively thick transgressive trend. A basal conglomeratic sandstone (Zone A of Stauffer 1909) is succeeded by dolostones of marine to non-marine origin (Judge 1998; = Zone B of Stauffer 1909), a coral-rich limestone (Zone C) and cherty mudstones (Chapel 1975; Zone D). Overlying, a series of wacke- to packstones, with minor mudstones and shales, coarsen upward to coarse grainstones and unconformable surfaces toward the upper part of the formation (Chapel 1975; Zones E–G of Stauffer 1909). The general fining-upward/deepening-upward trend continues into the overlying limestones of the Delaware Formation.

In her study of Bellepoint Member below the coral horizon, Judge (1998) noted that the lower part of the Columbus Formation consists of a set of two shallowing-upward cycles, marked by initial marine transgressive surfaces and thin, overlying shallow marine strata. These are succeeded by tidal and supratidal facies, and terrestrial facies with palaeosols. The base of the coral zone marks a third marine transgressive surface, with is succeeded by increasingly deeper marine facies.

The Bellepoint Member is tentatively interpreted here to represent a coastal facies equivalent of the Edgecliff Member of the Onondaga Formation. Facies patterns through the lower Columbus, close to Findlay arch and the western margin of the Appalachian Basin, indicate overall transgression from below the basal conglomeratic quartz sandstone of Stauffer's (1909) Zone A. The two cycles of the Bellepoint Member below the coral bed are interpreted here to represent the lower two of three parasequence sets in the Edgecliff Member and basal Nedrow Member that are recognized basinwide. The coral horizon at the top of the Bellepoint represents the first time sea-level in the Columbus area rose high enough to lap onto the basin margin, and support a well-developed marine fauna, within a shoal zone setting—in strata correlative with the uppermost Edgecliff Member or lowermost Nedrow Member in the main body of the Appalachian Basin. A bentonite within the overlying lower part of the Eversole Member (Zone D) appears to correlate a mid-Nedrow K-bentonite found at a number of outcrops across the Appalachian Basin. Alternatively, the lower two cycles within the Bellepoint in the Columbus area could represent highly condensed, coastal depositional sequences correlative with the two sequences of the Schoharie Formation (or the partial preservation of minor cycles within them). In that interpretation, the coral bed (average thickness of 1.2 m; Stauffer 1909) would represent a relatively condensed Edgecliff-equivalent, relative to other Onondaga-equivalent strata in the Columbus Formation in the area. Either interpretation appears to be consistent with the little precise biostratigraphic data available. Ramsey (1969) and Judge (1998) found long-ranging spore and conodont taxa that indicated Emsian to Eifelian age, with the exception of one spore taxa (*Emphanisporites schultzi*) she identified, which occurs through lower Emsian strata to the Emsian–Eifelian boundary (Richardson & McGregor 1986). Available biostratigraphic data indicate that the Emsian–Eifelian boundary occurs somewhere within the Edgecliff Member of the Onondaga Formation. Overall, the evidence seems to support a correlation of the Bellepoint Member of the Columbus Formation with the Edgecliff Member of the Onondaga Formation and its correlative strata basinwide.

Correlations from the Onondaga Formation and equivalent strata from the main body of the Appalachian Basin, into the Columbus Formation are much clearer from Zone D upward. The cherty carbonate mudstones of the overlying Eversole Member (Zone D of Stauffer 1909) are interpreted here to represent a relatively deeper offshore facies, associated with transgression above underlying coral-rich facies. Throughout the Devonian carbonates of the Appalachian Basin, cherty limestone facies represent a 'middle shelf' setting (Ver Straeten & Brett 2000). Chapel's (1975) interpretation of them as shallow, 'above wave base' facies, grading upward from coral facies below and grading above into post-Eversole wacke- and packstones is not consistent with a regressive facies shift. The cherty mudstones of the Eversole Member represent a continued transgression from the coral unit to 'middle shelf' depths. The deepening culminated in an organic-rich bed just above the Eversole (noted by Conkin & Conkin 1984, fig. 11), the probable equivalent of the basinwide 'black beds' of the Nedrow, which are the bathymetrically deepest facies of the Onondaga. Overlying wacke- and packstones in the lower part of the Klondike Member (Zones E, F, and G of Stauffer 1909) indicate a minor, initial coarsening/shallowing, correlative with highstand to falling stage systems tract deposits in the lower Moorehouse Member basinwide. A cluster of thin K-bentonites reported by Conkin & Conkin (1984) found between the upper part of Zone E through the lower part of Zone G correlate with the Tioga Middle Coarse Zone in the southern part of the Appalachian Basin. A brief return to finer-grained facies just below a shift to coarse grainstone facies (near the top of Stauffer's

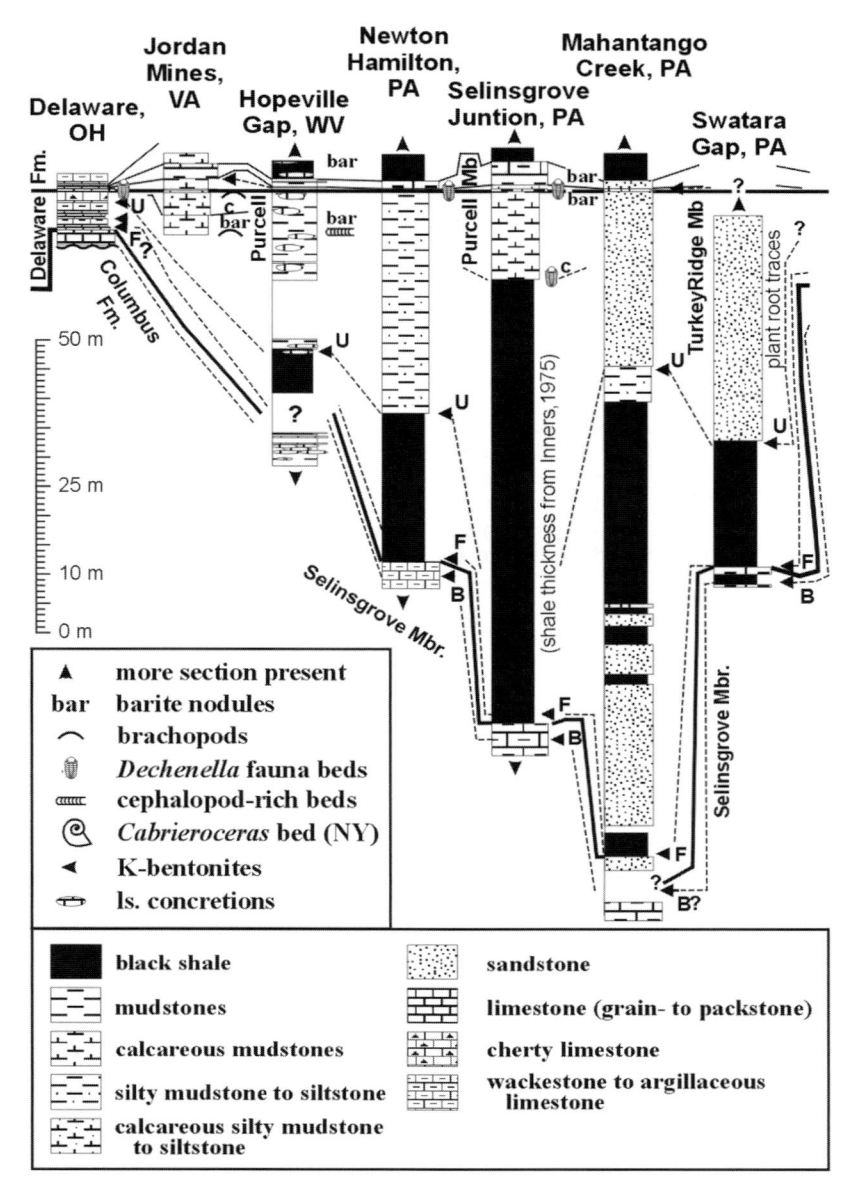

Fig. 14. Stratigraphic sections and correlations of the lower part of the Hamilton Group (Marcellus subgroup = Union Springs and Oatka Creek–Mount Marion Formations of New York) and equivalents across the Appalachian Basin, New York to Virginia and Ohio. Dashed and solid lines show correlations. Bold solid lines separate New York formations and correlative horizons across basin. Non-bold solid lines separate member-level units of the Union Springs and Oatka Creek–Mount Marion Formations and correlative horizons. Datum = base of Hurley Member (H) of the Oatka Creek–Mount Marion Formations and equivalent position basinwide. Abbreviations: B, Tioga B K-bentonite, at base of Seneca Member (Onondaga Fm) in New York; Bak, Bakoven Member of the Union Springs Formation; c, base of fine sandstone cap of Stony Hollow Member of the Union Springs Formation and correlative position; ch, Chestnut Street submember of the Hurley Member; CV, Cherry Valley Member of the Oatka Creek–Mount Marion Formations; EBr, East Berne Member of the Oatka Creek–Mount Marion Formations; F, Tioga F K-bentonite, just above base of Union Springs Formation in New York; H, Hurley Member of the Oatka Creek–Mount Marion Formations; lp, Lincoln Park submember of the Hurley Member; StH, Stony Hollow Member; TR, Turkey Ridge Member of the Mahantango Formation; U, mid-Union Springs K-bentonite. Strata at Delaware, OH, are assigned to the Delaware Formation. Non-black shales in VA and WV and Selinsgrove Junction, PA, are assigned to the Purcell Member of the 'Marcellus Formation' or Millboro Formation.

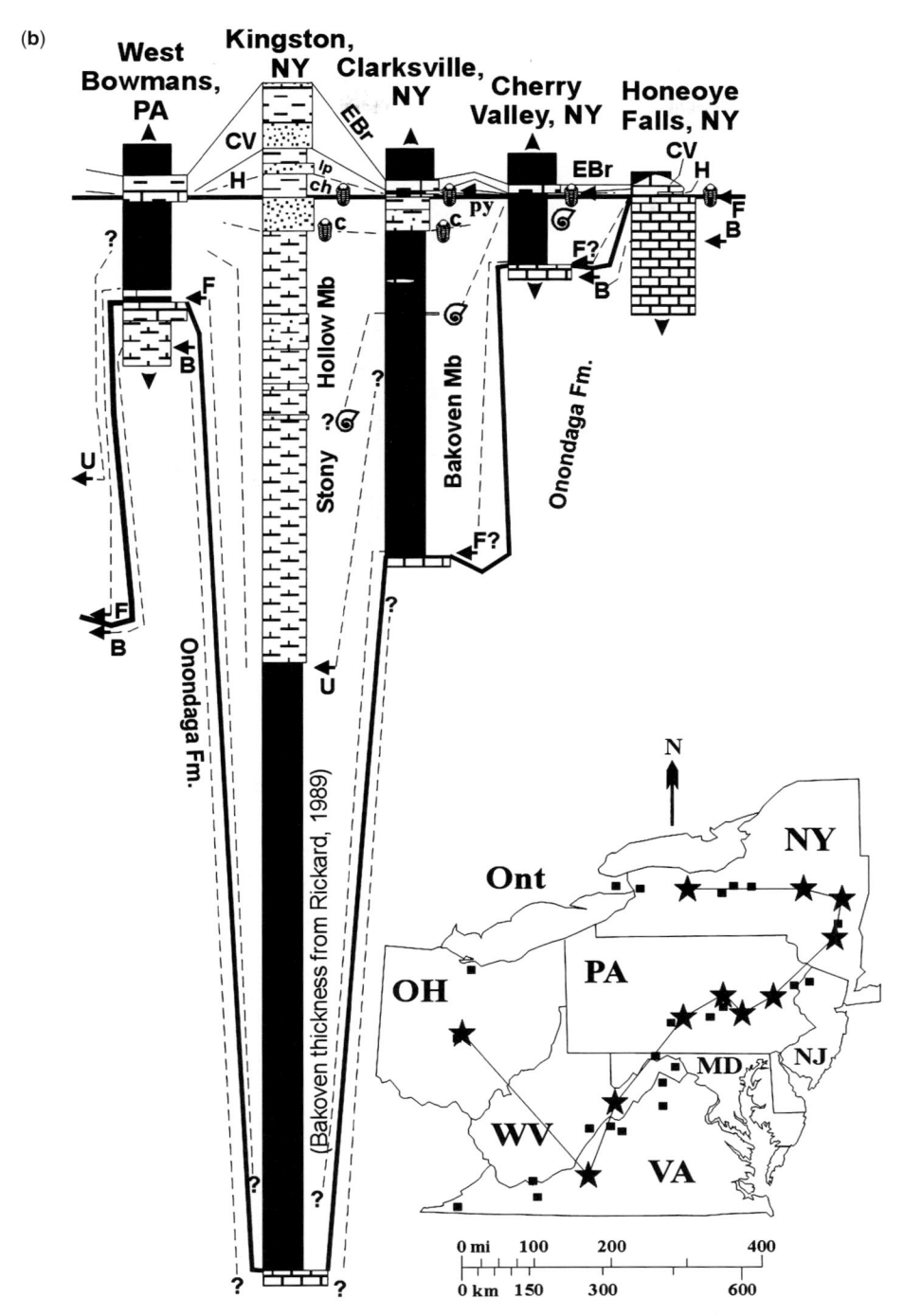

Fig. 14. *Continued.*

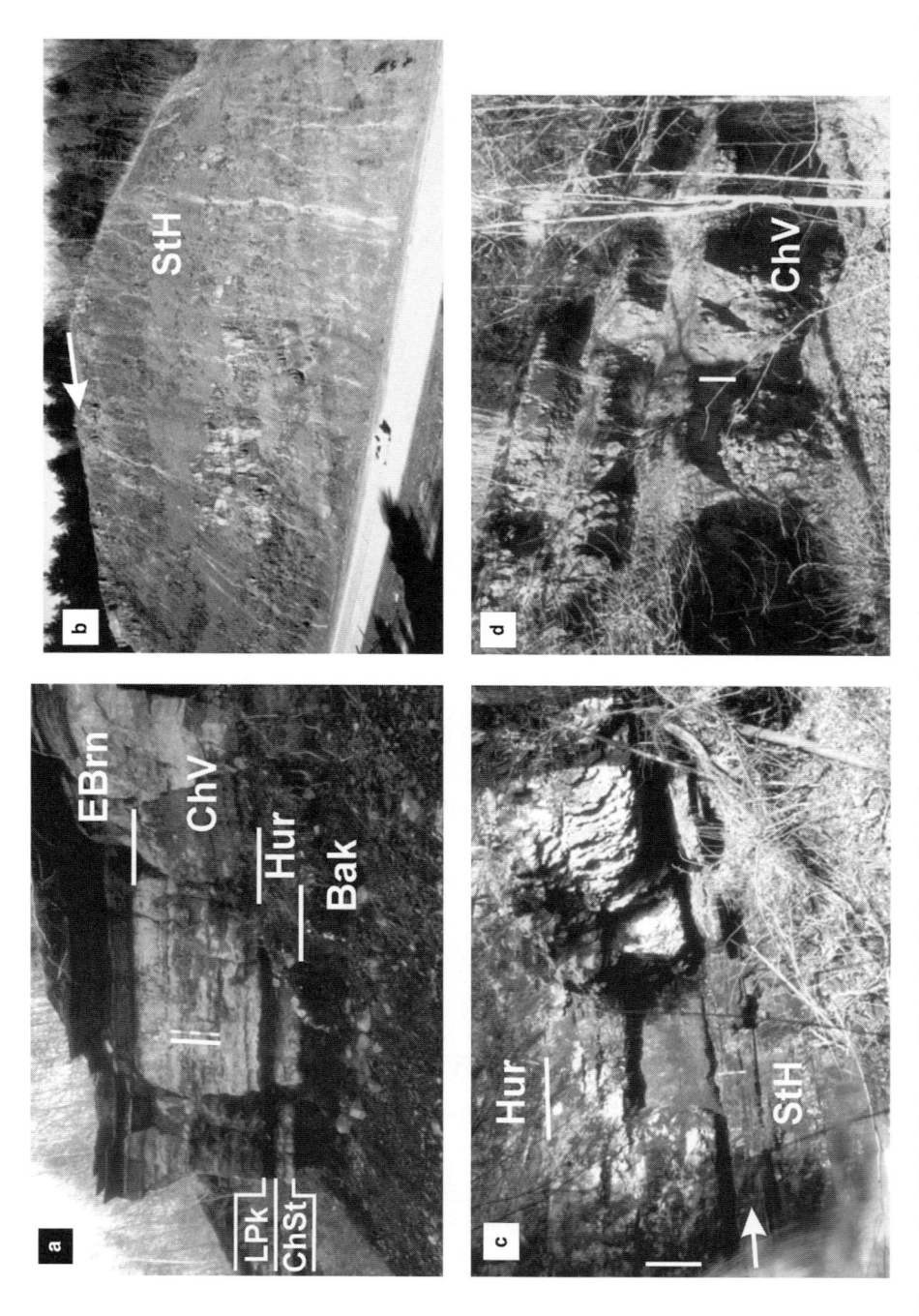

Fig. 15. Photographs of Union Springs and Oatka Creek–Mount Marion Formations and equivalent strata in the northern Appalachian Basin (New York). New York member-level subdivisions labelled for all localities (Bak, Bakoven and StH, Stony Hollow Members of the Union Springs Fm; Hur, Hurley; ChV, Cherry Valley; and EBrn, East Berne Members of the Oatka Creek–Mount Marion Formations of New York). Bars demarcate formation/member boundaries. (**a**) Part of Union Springs and Oatka Creek Formations at Chestnut Street, Cherry Valley, NY. Includes limestone facies of Cherry Valley Member. Scale = 0.5 m. (**b**) Middle to upper Stony Hollow Member, Rte. 28, Kingston, NY. Arrow points to resistant sandstone unit at the top of the Stony Hollow Member. (**c**) Sandstone unit at top of Stony Hollow Member, Buttermilk Falls, Leeds, NY. White bar = 1 m. Arrow points to position of 'lower Proetid bed' of Griffing & Ver Straeten (1991), below the sandstone unit. (**d**) Partial section of Cherry Valley Member in sandstone-mudstone facies, Rte. 28, Kingston, NY. White bar = 0.5 m.

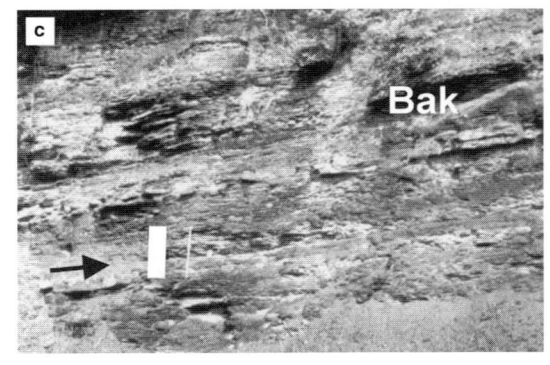

Fig. 16. Photographs of Union Springs and Oatka Creek–Mount Marion Formations and equivalent strata in the northern to central Appalachian Basin (New York, Pennsylvania). New York member-level subdivisions labelled for all localities (Bak, Bakoven and StH, Stony Hollow Members of the Union Springs Formation; Hur, Hurley; ChV, Cherry Valley; and Brn, East Berne Members of the Oatka Creek–Mount Marion Formations of New York). Bars demarcate formation/member boundaries. (**a**) Upper Bakoven and lower Stony Hollow Members, Rte. 28, Kingston, NY. Mid-Union Springs K-bentonite marker bed in upper Bakoven Member. (**b**) Part of Union Springs and Oatka Creek Formations, Palmerton, PA. Note recessive-weathered limestone of Hurley Member. (**c**) Middle part of Union Springs-equivalent strata, Newton Hamilton, PA. Arrow marks mid-Union Springs K-bentonite. Bar = 1.5 m. (**d**) Part of Union Springs and Oatka Creek Formations, Newton Hamilton, PA. Circled stick = 1.5 m.

Fig. 17. Photographs of strata equivalent to the Union Springs and Oatka Creek–Mount Marion Formations central to southern Appalachian Basin. New York member-level subdivisions labelled for all localities (Bak, Bakoven and StH, Stony Hollow Members of the Oatka Creek–Mount Marion Formations of New York). Bars demarcate formation/member boundaries. (**a**) Part of Stony Hollow-equivalent strata of Turkey Ridge Member of the Mahantango Formation, Rte. 333, north of Thompsontown, PA. Section is roughly 3 m thick. (**b**) Overturned section of interbedded sandstone and black shale of Turkey Ridge Member, Rte. 22, Thompsontown, PA. Mid-Union Springs K-bentonite found close below base of Turkey Ridge Member, to right of photo. Arrow points in direction of stratigraphic up in the section. (**c**) Stony Hollow (StH) and Hurley (H) equivalent strata in Purcell Member ('Marcellus Fm.'), Jordan Mines, VA. Arrowed line delineates the equivalent of the top unit of Stony Hollow Member in New York, as seen in Figure 15b & c. Circle marks field book for scale. (**d**) Lower part of Bakoven Member-equivalent (Bak) strata in 'Marcellus Fm.' above Onondaga-equivalents (Mo, Moorehouse and Sen, Seneca Members of the Onondaga Formation of New York), Wytheville, VA. Person for scale in left of centre.

Zone G) is correlative with the shaly Stroud Bed, again found basinwide about two-thirds upsection from the base of the Moorehouse Member throughout the basin. As across the Appalachian Basin, the Stroud Bed in the Columbus–Delaware area of central Ohio is underlain by a thin K-bentonite. This K-bentonite was misinterpreted by Conkin & Conkin (1984) as the Tioga B K-bentonite.

The overlying grainstones of the Klondike Member (topmost Zone G and Zone H) indicate a strong shallowing event, associated with falling stage systems tract deposition. Basinwide this shift, representative of a lowstand systems tract, occurs in upper Moorehouse equivalents. However, in a significant difference from the basinwide trend, upper Moorehouse- and lower Seneca-equivalent strata in the Columbus, representative of lowstand and initial transgressive systems tracts, appear largely absent on the shallow margin of the basin in the Columbus area. These strata, along with the Tioga B and some additional K-bentonites of the Tioga A–G cluster, appear to be cut out at multiple, significant unconformable surfaces above the Stroud Bed interval (the 'smooth surface' of Stauffer 1909, the 'first bone bed' of Wells 1944, and the 'rolling surface'; see fig. 11 of Conkin & Conkin 1984). Upper Seneca-equivalent strata in central Ohio comprise uppermost Columbus and/or lower Delaware Formations (=upper part of Zone H and basal part of Zone I).

To summarize, lower and middle strata of the Bellepoint Member of the Columbus Formation (=Zones A to B of Stauffer 1909) in the Columbus–Delaware area of central Ohio are interpreted here to correlate with the Edgecliff Member of the Onondaga Limestone of New York. Correlations from Zone D upward are more clearly delineated (see Fig. 8). Therefore, lower to middle strata of the Columbus Limestone (Zones A to upper G of Stauffer 1909) are interpreted here to represent the transgressive, lowstand and falling stage systems tracts of a single depositional sequence, and are correlative with the Edgecliff, Nedrow and lower to middle parts of the Moorehouse Members of the Onondaga Limestone of New York, and their equivalents basinwide. Uppermost strata of the G zone, extending up into the H zone, including the unconformities within them, represent the lowstand to transgressive systems tracts of a second sequence, and the development of upper Moorehouse and Seneca strata in central Ohio. The relatively shallower (tidal to terrestrial) facies of the lower to middle Bellepoint Member and the unconformities in the upper Columbus, along with reef development (Zone D) in strata equivalent to the Nedrow Member reflect deposition on the shallow western margin of the basin, adjacent to the Findlay arch.

Sequence stratigraphy of Pragian, Emsian and Eifelian strata in the Appalachian Basin

Emsian and Eifelian strata in the Appalachian Basin represent eight major ('third order') depositional sequences (DSs) (Fig. 19). Five of these are bounded at their base by relatively widespread unconformities; one is bounded across eastern North America by a geographically widespread, conformable sequence boundary, and two others also appear conformable across most of the basin.

The eight sequences comprise the strata of the upper Oriskany and Esopus (three sequences), Schoharie (two sequences), lower to middle Onondaga, upper Onondaga and Union Springs, and Oatka Creek–Mount Marion Formations (= depositional sequences Ems-1, Ems-2, Ems-3, Ems-4, Ems-5, Eif-1, Eif-2, and Eif-3, respectively; Fig. 19). A ninth sequence (depositional sequence Pr-1) is interpreted to occur within underlying strata of the Pragian-age Oriskany Formation and equivalent carbonate facies in the deep central portion of the basin. That sequence is locally bounded at its base by the true Wallbridge Unconformity. Within each sequence, a hierarchy of subsequence-scale packages (e.g. parasequences and parasequence sets) is also recognized.

The Wallbridge Unconformity

The Wallbridge Unconformity represents one of six major unconformities that bound the six Phanerozoic supersequences of North America (Sloss 1963), specifically between the Tippecanoe and Kaskaskia Supersequences. Within the Appalachian Basin, the unconformity may represent only a minor gap, during Sequence Pr-1. Toward the margins of the basin, the Wallbridge becomes progressively amalgamated with other older and younger unconformities, to the point where post-Wallbridge strata may overlie Silurian rocks (e.g. western New York, Brett et al. 2000; and SW Virginia). In deeper portions of the Appalachian Basin (e.g. SE New York, parts of Pennsylvania, Maryland, northern Virginia and northern West Virginia), the rock record is continuous and no unconformity is recorded.

The position of the Wallbridge Unconformity in the Appalachian Basin has been a focus of confusion over the years. As originally defined by Sloss (1963), the base of the Kaskaskia Supersequence underlies quartz arenites of the Oriskany Formation and equivalent strata in the Appalachian Basin. However, later workers (e.g. Dennison & Head 1975) positioned the Wallbridge at a prominent unconformity that occurs, in some parts of the basin, above the Oriskany Sandstone.

66 C. A. VER STRAETEN

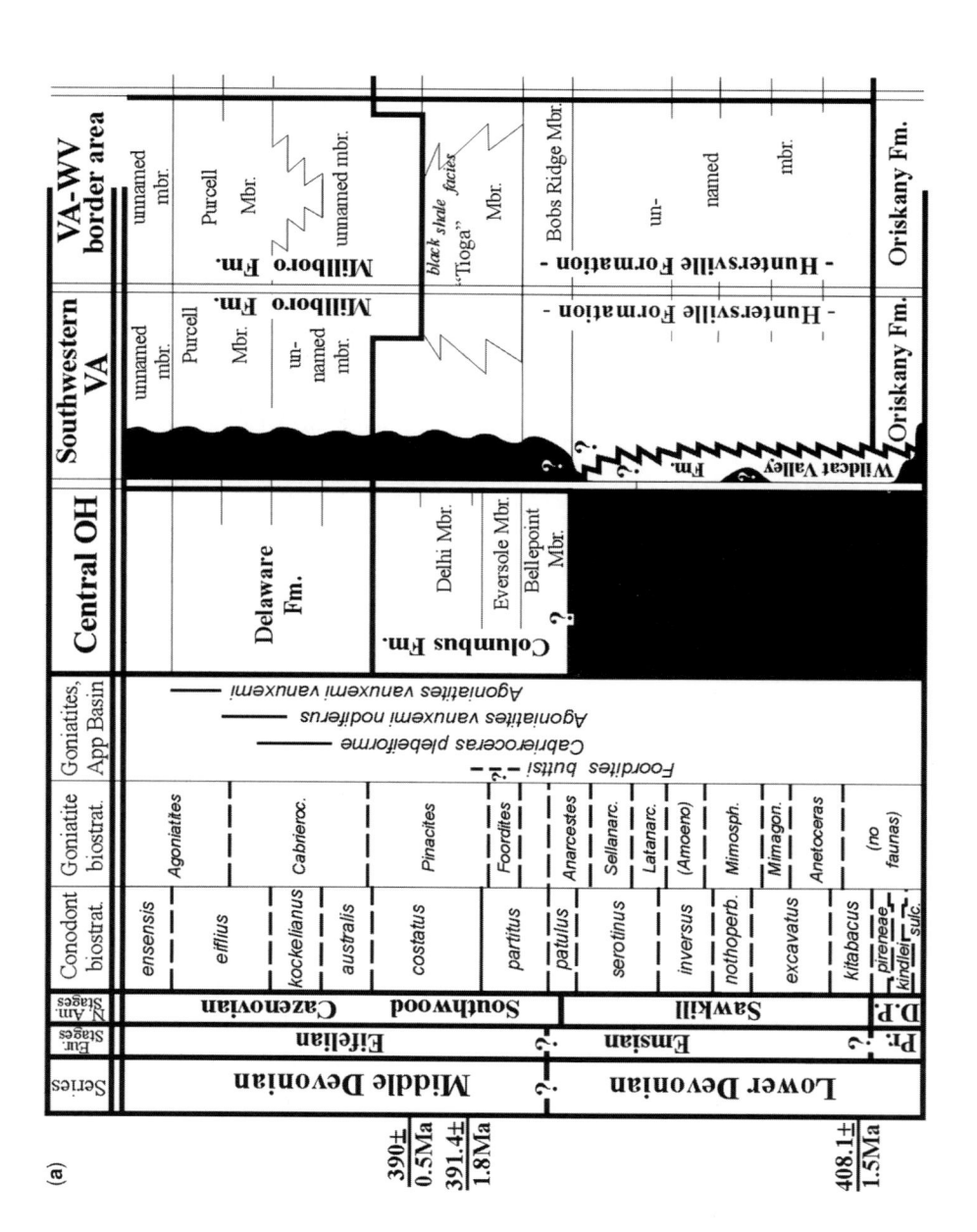

(a)

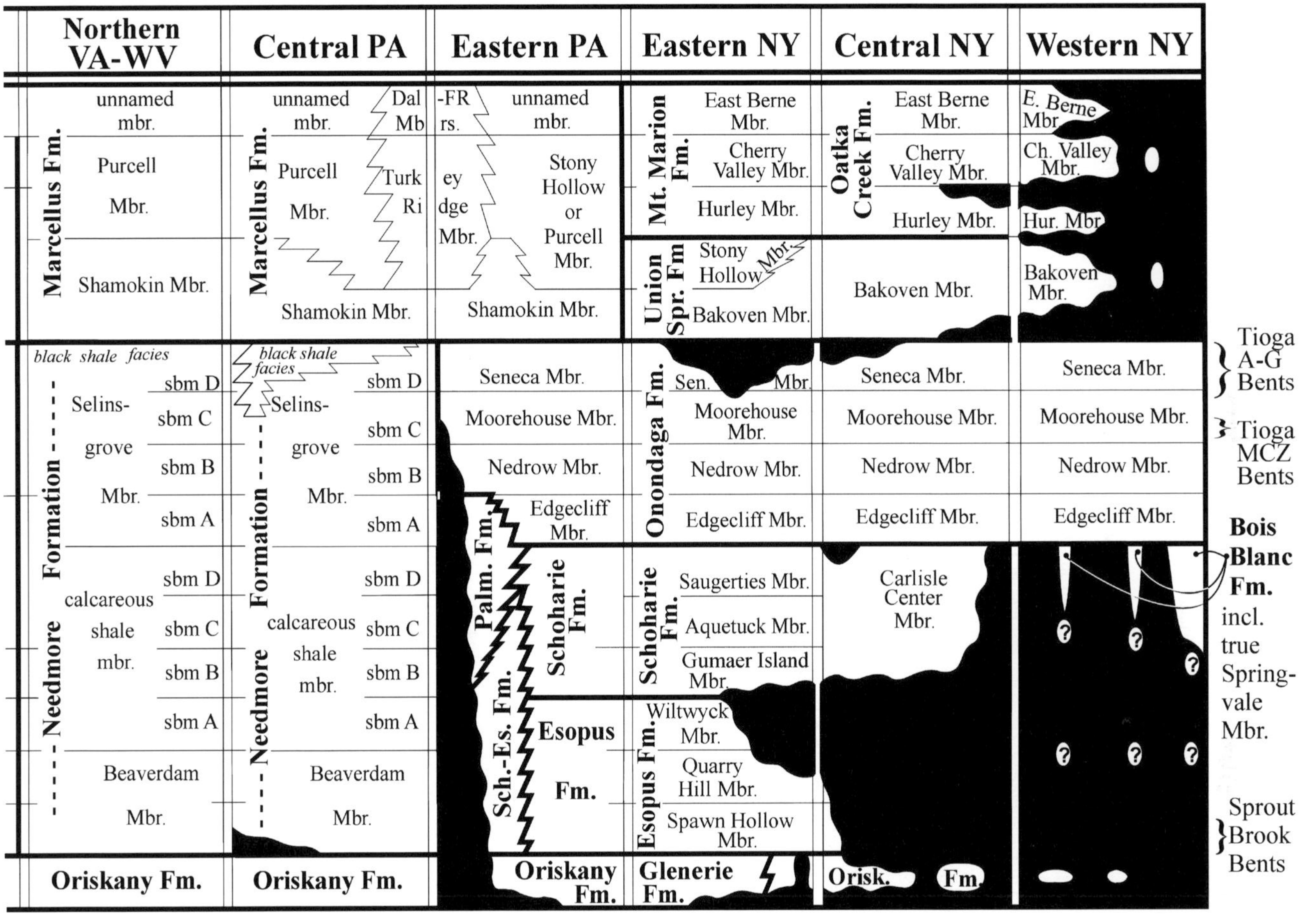

Fig. 18. New stratigraphic synthesis, Late Pragian, Emsian and Eifelian strata, Appalachian Basin. International conodont and goniatite zones shown, along with known range zones of key goniatites from the Appalachian Basin. Dashed lines bounding zones indicate position of biozone bases are unknown. Biostratigraphic data from House (1978, 1981); Becker & House (1994, 2000) and Klapper (1971, 1981). Black areas, unconformities. Geochronologic age dates from Roden et al. (1990) and Tucker et al. (1997). Abbreviations: Ameno., Amenophyllites; bents., K-bentonites; Cabrieroc., Cabrieroceras; D.P., Deerpark Stage; Fm, Formation; Latanarc., Latanarcestes; Mbr., Member; MCZ, Middle Coarse Zone; Mimagon., Mimagoniatites; Mimosph., Mimosphinctes; Mt Mar Fm, Mount Marion Formation; nothoperb., nothoperbonus; Palm. Fm., Palmerton Formation; Pr., Pragian Stage; On., Onondaga Formation; sbm, submember; Sch.-Es Fm., combined Esopus and part (?) of Schoharie Formations; Sellanarc., Sellanarcestes; sulc., sulcatus.

Table 1. *Stratigraphic nomenclature for Pragian, Emsian and Eifelian rocks as outlined in this paper. Changes are formally presented in a separate paper (Ver Straeten and Brett, 2006). Units in bold are new, revised, restricted or retained; units in italics are abandoned; units in normal text remain unchanged*

Upper Pragian strata		
Oriskany Formation	**Vanuxem 1839**	revised
Ridgeley Formation	*Swartz et al. 1913*	*abandoned (junior synonym of Oriskany Fm.)*
Glenerie Formation	Chadwick 1908	no change
Connelly Formation	Chadwick 1908	no change
Emsian strata		
Esopus Formation	**Darton 1894**	
Spawn Hollow Member	**new**	**new**
Quarry Hill Member	**Boucot et al. 1970**	**revised**
Wiltwyck Member	**new**	**new**
Mountainville Member	*Boucot et al. 1970*	*abandoned*
Highland Mills Member	*Boucot 1959*	*abandoned*
Schoharie Formation	**Vanuxem 1839**	
Gumaer Island Member	**new**	**new**
Aquetuck Member	Johnsen & Southard 1962	no change
Saugerties Member	Chadwick 1940	no change
Carlisle Center 'Member'	**Goldring & Flower 1944**	**revised, restricted**
Rickard Hill Member	*Johnsen & Southard 1962*	*abandoned*
Palmerton Member	**Swartz 1939**	**revised**
= undifferentiated Schoharie Formation and Edgecliff Member of the Onondaga Formation		
Eifelian strata		
Onondaga Formation	**Hall 1839**	
Edgecliff Member	**Oliver 1954**	**revised**
Nedrow Member	**Oliver 1954**	**revised**
Moorehouse Member	Oliver 1954	no change
Seneca Member	Vanuxem 1839	no change
Clarence Member	*Ozol 1964; Oliver 1966*	*abandoned*
Buttermilk Falls Formation	*Willard 1938*	*abandoned (junior synonym of Onondaga Fm.)*
Foxtown Member	*Epstein 1984*	*abandoned (junior synonym of Edgecliff Mbr.)*
McMichael Member	*Epstein 1984*	*abandoned (junior synonym of Nedrow Mbr.)*
Stroudsburg Member	*Epstein 1984*	*abandoned (junior synonym of Moorehouse Mbr.)*
Echo Lake Member	*Inners 1975*	*abandoned (junior synonym of Seneca Mbr.)*
Marcellus subgroup (of Hamilton Group)	**Hall 1839**	**revised in NY (from formation to subgroup status)**
Union Springs Formation	**Cooper 1930**	**revised (from member to formation status)**
Bakoven Member	**Chadwick 1933**	**revised**
Stony Hollow Member	**Cooper 1941**	**restricted**
Shamokin Member	*Willard 1935*	*abandoned*
Oatka Creek Formation	**Cooper 1930**	**revised (from member to formation status)**
Mount Marion Formation	**Grabau 1917**	**revised**
Hurley Member	**(Ver Straeten et al. 1994)**	**new**
Cherry Valley Member	**Clarke 1903**	**revised**
Berne Member	*Cooper 1933*	*abandoned*
East Berne Member	**new**	**new**
Turkey Ridge Member	**Willard 1935**	**redefined**
= undifferentiated sandstone facies of Stony Hollow, Hurley, Cherry Valley Members		
Purcell Member	Cate 1963	no change
= undifferentiated Hurley, Cherry Valley +/− Stony Hollow Mbrs.		

(Further member-level nomenclature of Oatka Creek–Mount Marion Formations not addressed in this paper).

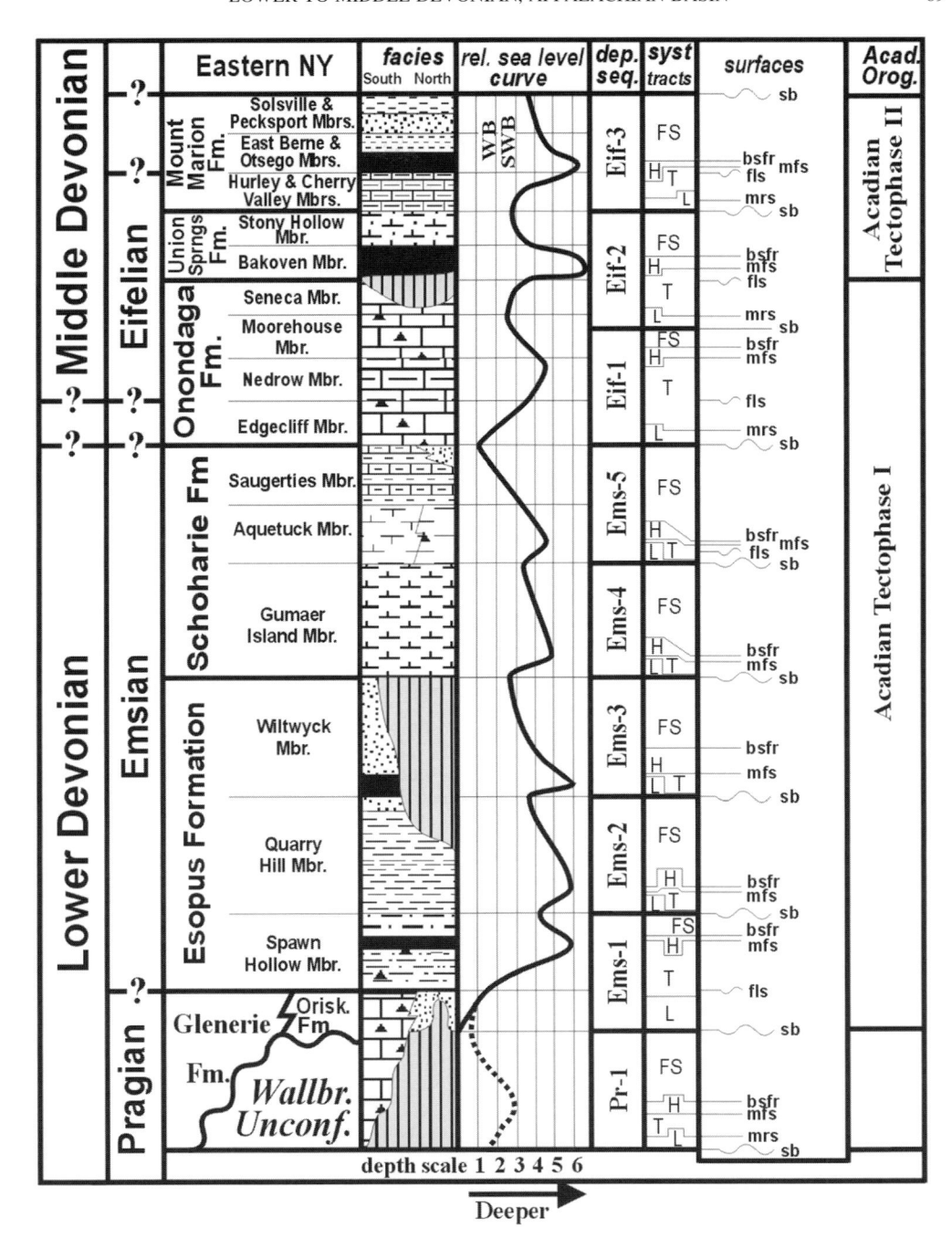

Fig. 19. Sequence stratigraphic framework for Late Pragian, Emsian and Eifelian strata, Appalachian Basin. Stratigraphy based on eastern New York composite section. Acadian Tectophases after Ettensohn (1985). Abbreviations: Acad. Orog., Acadian Orogeny; bsfr, basal surface of forced regression; dep. seq., depositional sequences; fls, flooding surface; H, highstand systems tract; FS, falling stage systems tract; L, lowstand systems tract; mfs, maximum flooding surface; mrs, maximum regressive surface; Orisk., Oriskany Formation; sb, sequence boundary; seq. strat., sequence stratigraphy (systems tracts); SWB, storm wave base; T, transgressive systems tract; Wallbr. Unconf., Wallbridge Unconformity; WB, normal wave base. Depth scale of 1–6, reflecting water depth, based on lithology and benthic assemblages model by Boucot (1982).

Study of Late Pragian to Early Emsian strata across the basin shows that the unconformity above the Oriskany is a major flooding surface, marked locally by deposition of oolitic phosphatic- and/or hematite-rich deposits. This unconformity formed due to subsidence of the foredeep and sediment-starved conditions during the onset of Acadian Tectophase I, combined with an apparent eustatic rise during the maximum rate of sea-level rise in sequence Ems-1. In contrast, the sub-Oriskany unconformity, where developed in the basin, is an unconformity formed during a sea-level fall, and represents a peak time of lowstand and widespread exposure across the North American continent during sequence Pr-1. Therefore, the position of the Wallbridge Unconformity should be placed below the fully developed Oriskany Formation in the deeper areas of the Appalachian Basin, at the base of depositional sequence Pr-1.

Depositional sequence Pr-1

The first sequence at the base of Sloss' (1963) Kaskaskia Supersequence (DS Pr-1) occurs within sandstones in the lower to middle part of the Pragian-age Oriskany Formation, and equivalent carbonate facies in deeper areas of the basin. As previously noted, this sequence is locally bounded at its base by the true Wallbridge Unconformity. In the deeper portions of the basin, deposition was continuous with the underlying strata. The basic framework of the sequence consists of an overall pattern of deepening–shallowing–deepening through lower to middle Oriskany strata, indicated by lithological and faunal changes in western Maryland and adjacent West Virginia (Barrett & Isaacson 1981). The strata of Sequence Pr-1, which were deposited during maximum lowstand of the Kaskaskia Supersequence, are restricted to the deeper portions of the Appalachian Basin.

Depositional sequences Ems-1, Ems-2 & Ems-3

Ver Straeten & Brett (1995) and Ver Straeten (1996a) previously reported two Emsian age sequences, comprising: (1) uppermost Oriskany, and Esopus Formations; and (2) the Schoharie Formation. Subsequent basinwide study, supported by geochronological dating indicating a 17 million year duration for the Emsian (Kaufmann *et al.* 2005), justifies subdividing two Emsian age sequences into five distinct sequences; three lower ones (Ems-1 to Ems-3 associated with the Esopus Formation and equivalent strata across the Appalachian Basin; and two upper ones Ems-4 to Ems-5) in the Schoharie Formation and equivalents.

Fringing the margin of preserved strata of sequence Pr-1, the Wallbridge Unconformity becomes amalgamated with a second, overlying unconformity at the base of sequence Ems-1. It may also combine with additional pre-Wallbridge unconformities. Reworked quartz and/or phosphatic pebbles overlie the basal unconformity in some areas (e.g. eastern NY). The previously noted flooding surface of Ems-1 at the top of the Oriskany Formation is variously marked by a change from: (1) underlying quartz arenites or limestones into shales, siltstones, K-bentonites and cherts (Spawn Hollow Member of the Esopus Formation, eastern NY); or (2) a sharp lithological change from quartz arenites, quartz pebble conglomerates or (locally) limestones to mudstones or shales. (Esopus Formation or Beaverdam Member of the Needmore Formation, PA, MD, VA, WV.)

Across the basin, Esopus-equivalent strata comprise three major sequences, as follows (Figs 18, 19). (1) Ems-1, which consists of upper strata of the Oriskany–Glenerie Formations and the Spawn Hollow Member of the Esopus Formation (eastern NY, eastern PA), upper strata of the Oriskany–Shriver Formations and lowest strata of the Beaverdam Member (Needmore Formation) in the central to southern Appalachian Basin, and upper strata of the Oriskany formation and lowest strata of the Huntersville Chert (southern Appalachian Basin). (2) Ems-2, comprising the Quarry Hill Member of the Esopus Formation (eastern NY, eastern PA), the middle subdivision of the Beaverdam Member (central to southern basin area), and the lower middle strata of the Huntersville Formation (southern basin area). And (3) Ems-3, represented by the Wiltwyck Member of the Esopus Formation (E NY, E PA), upper strata of the Beaverdam Member and submember A of the calcareous shale member (central to southern basin area), and the middle part of the Huntersville Formation (southern basin area).

In some areas of the basin (e.g. deeper, central-basin facies), sequence-capping falling stage systems tract deposits of Ems-1 may be difficult to distinguish due to their fine-grained character. Sequences Ems-2 and Ems-3 commence with dark grey mudstones to black shale and culminate in bioturbated argillaceous siltstone or fine-grained sandstone. In the southern area of the basin, in the cherty Huntersville Formation, falling stage deposits of sequences Ems-1 to Ems-3 are marked by a thickening of beds and increased bioturbation (e.g. Frost, WV) or a shift from black to dark or medium grey colours in the chert (e.g. Wytheville, VA). Smaller-scale cyclicity may be manifest in parts of the Huntersville Formation by alternating 0.3–0.5 m-thick bands of lighter and darker grey-weathering mudstone, or finer-scale chert bands.

On the fringes of the basin (e.g. central to western New York) preservation of the sequences is spotty, apparently represented by local incised valley-fills (e.g. Phelps, NY).

Depositional sequences Ems-4, Ems-5

A depositional break between sequences Ems-3 and Ems-4 is locally marked around the basin by mud firmgrounds, glauconite and scattered quartz pebbles. Evidence of the erosive nature of this unconformity is shown by progressive downward bevelling out of Esopus strata across eastern New York, below the base of the Schoharie Formation. Accompanying finely detailed trace fossils (e.g. *Cruziana*, *Fustiglyphus* and scratch markings) on the basal transgressive surface in east-central New York were produced in semi-consolidated muds previous to deposition of the overlying glauconitic and pebbly sands (Miller & Rehmer 1982). Similar occurrences of delicately sculpted, steep-walled trilobite appendage traces and 'remainie sediments' have been interpreted to indicate that preservation of the traces was a result of excavation into disconformity-related, pre-compacted mud-firmgrounds at major cycle boundaries (Landing & Brett 1987).

The westward, top-downward bevelling of Esopus strata may in part be due some degree of isostatic uplift (e.g. forebulge) in central to western New York during the late, quiescent stage of Acadian Tectophase I. The unconformity, however, is distinctive basinwide, and is found at the correlative position in the top of submember A of the calcareous shale member (Needmore Formation) in the central part of the basin, and at a glauconite- and phosphate-rich interval in the middle of the Huntersville Formation in the south (e.g. Frost, WV).

Depositional sequences Ems-4 and Ems-5 comprise strata of the Schoharie Formation (Gumaer Island; Aquetuck and Saugerties Members, respectively) in New York; the bulk of the calcareous shale member (Needmore Formation) in Pennsylvania, Maryland and the northern portion of the Virginia–West Virginia outcrop belt; and upper strata of the Huntersville Chert (not including the Bobs Ridge Member) in the southern part of the basin. Initial strata of the Gumaer Island Member (Schoharie Formation) and submember B of the calcareous shale member of the Needmore Formation (in the central basin area) comprise the transgressive systems tract of Ems-4. The maximum bathymetric highstand of sea-level in the Ems-4 sequence is represented by a widespread interval of dark grey to black shale to argillaceous strata underlying the highstand systems tract.

A second, less-pronounced unconformity (locally with glauconite and scattered quartz pebbles) occurs at the base of sequence Ems-5, at the contact of the Gumaer Island and Aquetuck Members (eastern NY), and at the correlative position elsewhere in the Appalachian Basin. Lower to middle Aquetuck strata comprise the transgressive to highstand systems tracts of the sequence. A general shallowing-upward trend through the upper part of the Aquetuck and Saugerties Members and equivalents basinwide is indicative of the falling stage systems tract.

The Ems-4 and Ems-5 sequences are recognized in the southern area of the basin (VA, WV), where they comprise the two upper cycles of the cherty Huntersville Formation, below the Bobs Ridge Sandstone Member. At Frost, West Virginia, the sequences are denoted by two well-defined cycles with distinctive, upward thinning to thickening bedding trends. In the more massive chert facies to the south (e.g. Wytheville, VA), dark chert through the transgressive to highstand systems tracts are replaced by a lightening of colour to brown to tan or grey within the falling stage systems tracts. The falling stage of Ems-5 shows increasing sand content upward as it shallows into the Bobs Ridge Sandstone Member (=lowstand of the following sequence).

Depositional sequence Eif-1

The contact of the Schoharie and Onondaga Formations across eastern New York and their equivalents in deeper portions of the Appalachian Basin is relatively conformable and represents a shallowing- to deepening-upward transition. Around the margins of the basin (e.g. central to western New York, Virginia and West Virginia, vicinity of Harrisburg, Pennsylvania), however, there is an unconformity at the contact that represents a major break in deposition which locally becomes amalgamated with the Wallbridge and other post-Wallbridge unconformities (Brett & Ver Straeten 1994; Brett *et al.* 2000). The widespread transition from shallowing- to deepening-upward in equivalent strata across the Appalachian Basin (Brett & Ver Straeten 1994) appears to indicate that the unconformity is associated with a fall and rise in eustatic sea-level. Across central to western New York and the subsurface of northwestern Pennsylvania, development of the unconformity is also associated with uplift and migration of a forebulge or bulge-like feature associated with the Acadian Orogeny (Ver Straeten & Brett 2000).

A subtle break in deposition is also recorded by a flooding surface at the Edgecliff–Nedrow member contact and correlative units basinwide (Fig. 19). This is generally indicated by a lithological break from shallower to deeper water carbonate facies (e.g. medium- to fine-grained limestones to more argillaceous strata; medium-grained, chert-rich to fine-grained, non-cherty limestones). Glauconite and pyritic nodules or pyritic crusts may locally mark the contact.

The first Eifelian sequence (Eif-1) consists of the Edgecliff, Nedrow and lower to middle parts of the Moorehouse Members of the Onondaga Formation of New York State and correlative strata in the Selinsgrove Member of the Needmore Formation and Columbus Formation across the basin. In eastern New York, the base of the sequence is conformable; a laterally equivalent erosive unconformity occurs across central to western New York. Lowstand deposits may be found in the lower part of the Edgecliff Member, associated with initial growth of coral bioherms in shallower water settings, and by relatively shallower water facies in basinal areas. The bulk of the Edgecliff and Nedrow Members and equivalents comprise the transgressive systems tract; the interval of the maximum flooding surface is found at the top of the Nedrow Member, within the widespread Nedrow 'black beds'. Overlying fine-grained, aggradational limestones (basal Moorehouse Member) represent the highstand deposits of Eif-1, which are succeeded by falling stage systems tract deposits of the middle part of the Moorehouse Member and correlative strata across the basin.

Smaller-scale cycles are not as readily recognizable in the Onondaga as in the Schoharie but may be discerned with careful study. Two to three fourth-order subsequences may be recognized in the Edgecliff–Nedrow and lower to middle Moorehouse Members of New York terminology. Parasequence-scale cycles (~1 m-thick) are reported in the Edgecliff Member in western to central New York (Brett & Ver Straeten 1994), and are notable through much of the sequence.

Depositional sequence Eif-2

The base of sequence Eif-2 is widely conformable across the Appalachian Basin, with the exception of the western margin of the basin in central Ohio. The base of sequence Eif-2 is marked by a gradational change from shallowing- to deepening-up lithological and faunal trends. The sequence boundary is found closely above the markedly finer-grained, more argillaceous Stroud Bed approximately two-thirds of the thickness up through the Moorehouse Member and equivalents. Correlation of the Stroud bed and distinctive marker units across the basin show the sequence base to be coeval across much of the basin. Upper Moorehouse strata appear to represent lowstand deposits of Eif-2; fining-upward trends through the overlying Seneca Member are associated with a rise in relative sea-level within the transgressive systems tract. A sequence-bounding unconformity is only developed on the western margin of the basin, in the Columbus Limestone of Ohio.

In contrast, a locally prominent unconformity at the carbonate-black shale contact of the Onondaga and Union Springs Formations and equivalents in some areas of the basin (e.g. NY, OH) represents a major flooding surface within the transgressive systems tract of sequence Eif-2. The upper surface of the Onondaga Limestone in New York is commonly marked by a lag of fish bones and teeth, and/or phosphate pebbles. In New York, the unconformity progressively cuts out upper Onondaga strata (Seneca Member) from central New York and eastern Pennsylvania to the Albany area, in eastern New York (Rickard 1989; also indicated by progressive cutout of marker bed K-bentonites, and top-down thinning of strata). Across the central basin (PA, MD, northern areas of VA and WV) the Onondaga–Union Springs equivalents' contact is relatively conformable, as sediment supply is presumed to have been more continuous in that area.

On the western margin of the basin, in Ohio, the same unconformity at the contact between the Columbus and Delaware Limestones is also marked by a prominent bone bed (Westgate & Fisher 1933; Wells 1944; Conkin & Conkin 1975). Overlying black shales found throughout the rest of the basin are developed only locally in central Ohio. The succeeding strata of sequence Eif-2, in the Delaware Formation, predominantly consist of fine-grained, cherty limestones.

The diachronous character of the unconformity in eastern New York implies some degree of epeirogenic control over the relative sea-level rise in that area of the basin, which can be tied to subsidence of the basin foredeep during the onset of Tectophase II of the Acadian Orogeny. Across most of the basin, including central to western New York, Pennsylvania, Maryland, northern Virginia and West Virginia, and Ohio, event stratigraphic markers (including the Tioga F and other Tioga K-bentonite beds) indicate that flooding of the carbonate platform was relatively synchronous in most areas.

Upper transgressive to highstand systems tracts of Eif-2 in the main body of the foreland basin are composed of black shales of the Union Springs Member (Bakoven Member, NY), equivalent Marcellus strata though the main body of the basin, and carbonates and minor black shales of the lower part of the Delaware Formation in central Ohio. A single, widespread bentonite in the middle of the Union Springs succession (Fig. 14), which is found across much of the basin (eastern NY, PA, OH), provides a time line indicative of the synchronous onset of progradational shallowing during the falling stage across the basin. Falling stage systems tract deposits in sequence Eif-2 include silty black to grey shales, calcareous mudstones, siltstones and sandstones, and carbonates (Stony Hollow Member, eastern NY, eastern PA; and equivalent strata in the lower and middle parts of Purcell and Turkey Ridge Members, central PA; and middle part Delaware Formation, central OH). Sequence Eif-2 is equivalent to Sequence 1 of Brett & Baird (1996), except that the Hurley and Cherry Valley Members have been moved upward into Sequence Eif-3.

Depositional sequence Eif-3

The basal contact of the succeeding depositional sequence Eif-3 (base of the Hurley Member of the Oatka Creek–Mount Marion Formations and equivalents) also appears conformable throughout much of the basin. However, in western New York a sequence-bounding unconformity cuts downward through lowstand deposits of the fossiliferous Hurley Member and upper strata of the underlying Bakoven Member of the Union Springs Formation. Locally (western NY), this unconformity cuts down to the transgressive systems tract of Eif-2 at the top of the Onondaga Formation.

An additional unconformity (flooding surface) is commonly developed overthe top of the Cherry Valley Member of the Oatka Creek–Mount Marion Formations and equivalent strata in the basin. In central to western New York, this surface becomes prominent, and cuts down through the Cherry Valley Member and upper Union Springs Formation (part of transgressive systems tract of Eif-3 and highstand to falling stage systems tracts of Eif-2). In westernmost New York, lower black shales of the Oatka Creek Formation lie directly upon limestones of the Onondaga Formation. In central Ohio, a succeeding unconformity below the Olentangy Shale appears to cut out all but the lower transgressive systems tract of sequence Eif-3 at the top of the Delaware Formation.

Sequence Eif-3 comprises the basinal facies of the Oatka Creek Formation and correlative proximal facies of the Mount Marion Formation in New York, and correlative strata across the basin, of which the middle to upper parts have not been a part of this study. Skeletal limestones of the lower part of the Hurley Member (Chestnut Street submember) represent widely preserved lowstand deposits. Succeeding limestones and the lower part of the succeeding black shales (Cherry Valley and East Berne Members of NY, and lateral equivalents) represent the transgressive systems tract. A drowning surface on the top of the Cherry Valley marks a major stage of deepening and the shutting down of carbonate production.

The stratigraphic position of systems tracts varies across the New York portion of the basin, due to variability in depositional/progradational and subsidence trends during Eif-3. Outside of New York, the upper part of sequence Eif-3 is at present poorly delineated due to limited exposure, black shale-dominated facies, and limited study. Sequence Eif-3 is equivalent to Sequence 2 of Brett & Baird (1996).

Comparison with Devonian T–R cycles model

Johnson *et al.* (1985, 1996) presented the first major model of Devonian eustatic sea-level trends for Euramerica. In this model, they recognized at least 14 transgressive–regressive (T–R) cycles, based on sedimentary successions from two areas in Europe (Belgium, Germany) and five areas in North America (New York, Nevada, Alberta). For the Pragian to Eifelian interval, they recognized five major T–R cycles (cycles Ia–Ie).

The new study of Pragian to Eifelian strata outlined in this paper permits a much more highly refined view of relative sea-level trends across the Appalachian Basin. The refined analysis includes: (1) the application of sequence stratigraphic terminology and interpretations to all units in the succession; (2) recognition of the precise stratigraphic positions of turnaround points of relative sea-level changes; (3) recognition of basinwide relative sea-level trends (without interference of local, tectonically induced variations of water depth curves, e.g. Ver Straeten & Brett 2000).

A comparison of Johnson *et al.*'s (1985) sea-level curve for the Pragian to Eifelian interval with the curve generated in this study for the Appalachian Basin succession is shown in Figure 20. Of some significance in Johnson *et al.*'s (1985) paper, the positions of sea-level turnarounds from regression to transgression and transgression to regression were only presented at a relatively coarse scale. For example, the initiation of transgressions in their Pragian to Eifelian cycles were placed at maximum starvation surfaces or flooding surfaces, both generally well above the true turnaround point of regression to transgression. Therefore, the actual base of each of their cycles occurs at a lower stratigraphic position than they interpreted. The new Appalachian Pragian to Eifelian curve shows the new, refined interpretations.

Johnson *et al.* (1985, 1996) recognized five Pragian to Eifelian T–R cycles (cycles Ia–Ie). In the Appalachian, the succession is shown to comprise nine depositional (third order) cycles. Sequence Pr-1 of this report, developed in the Oriskany Sandstone, basically corresponds to T–R Cycle Ia of Johnson *et al.* (1985, 1996), noting that the transgression of the overlying sequence begins in the upper part of the Oriskany Sandstone.

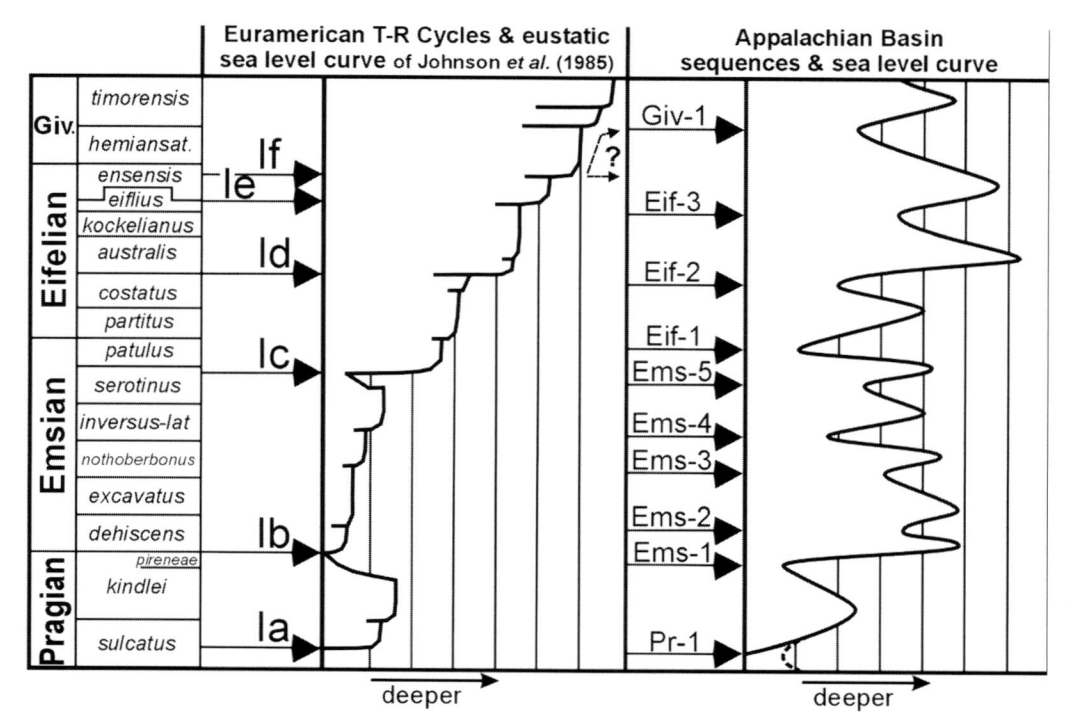

Fig. 20. Comparison of Euramerican eustatic sea-level curve (Johnson et al. 1985) and Appalachian Basin sequences, upper Pragian to Eifelian stages. The sharp Appalachian Basin transgressions at the base of the Emsian and in the middle Eifelian represent flexural subsidence events in the foreland basin during the onset of two tectonically active to quiescent phases of the Acadian Orogeny.

T–R Cycle Ib of Johnson et al. (1985) essentially comprises Sequences Ems-1 to Ems-4 of this paper. T–R cycle Ic represents Sequences Ems-5 and Eif-1. And T–R cycles Id and Ie correspond coarsely to Sequences Eif-2 and Eif-3, respectively, again noting that the positions of the cycle/sequence boundaries from this study occur lower than interpreted by Johnson et al. (1985, 1996).

Cycle by cycle refinements are outlined here. There are few changes to contribute to T–R cycle Ia in this study, as much of the interval was not a focus of this study, and needs more attention. Of significance, however, the upper part of the Oriskany Formation in the Appalachian Basin shows a distinct shallowing, followed by a deepening that is continuous into the overlying Esopus Formation and equivalents across the basin. Johnson et al. (1985) placed the base of T–R Cycle Ib at the position of the major flooding surface at the Oriskany–Esopus contact. Actually, upper Oriskany strata assigned by Johnson et al. (1985) to the top of T–R Cycle Ia represent Lowstand and lower Transgressive systems tracts of overlying Cycle Ib.

Johnson et al.'s (1985) Cycle Ib, which comprises approximately 15 million years of time (based on dates from Tucker et al. 1998) is split into four separate cycles in this paper (sequences Ems-1 to Ems-4). Johnson et al. (1985) did note that 'this cycle was long lasting'. The subdivision of Cycle Ib proposed here is consistent with their recognition of four subcycles (seen in fig. 12 of Johnson et al. 1985).

Johnson et al.'s (1985) T–R Cycle Ic included both the 'Schoharie' and Onondaga Formations. Their sense of the Schoharie Formation in that paper included only the Aquetuck and Saugerties Members, which comprise sequence Ems-5 of this paper. Their inclusion of those strata into Cycle Ic was based on an assumption that reef development in the basal Edgecliff Member of the Onondaga Formation was due to a 'rapid deepening over a Schoharie-Bois Blanc platform'. No such trends are seen in the rocks. In fact, the regressive shallowing into a basal Edgecliff lowstand (or correlative unconformity) basinwide is more significant than the regressive maxima below sequence Ems-5, where Johnson et al. (1985) placed the base of their T–R Cycle Ic. So, Cycle Ic should be split into two distinctive cycles, represented by sequences Ems-5 and Eif-1. Of note, the position of this split occurs at or closely below the Emsian–Eifelian/*patulus-partitus* conodont zones boundaries.

Johnson *et al.* (1985) placed the base of their T–R Cycle Id at the major flooding surface at the contact of the Onondaga and Union Springs Formations and correlative strata. However, the lowstand of the sequence occurs much lower, down in upper strata of the Moorehouse Member of the Onondaga Formation, which is succeeded by transgression beginning in the upper Moorehouse to lower Seneca Members (Onondaga Formation). Therefore, the base of T–R Cycle Id should be moved downward into the middle of the *costatus* conodont zone, at a position correlative with upper strata of the Moorehouse Member. Johnson *et al.* (1985) did recognize the transgressive nature of strata above the upper Moorehouse Member; they reconciled this trend by creating an extra subcycle (comprising essentially the Seneca Member), but placing it down within T–R Cycle Ic. However, those strata clearly are the basal part of the overlying Cycle Id. And the supposed 'Seneca' subcycle does not appear to be of much consequence (if it even exists), at least in the Appalachian Basin, as no significant indicators of shallowing occur below the shift to black shales of the overlying Union Springs Formation.

Similarly, the base of T–R Cycle Ie should be moved down to a position lower in the *kockelianus* conodont zone, from the base of the East Berne Member (and equivalents) to a position correlative with the base of the Hurley Member of New York (to the base of Oatka Creek–Mount Marion Formations).

Poor biostratigraphic control through the Appalachian Basin succession has hindered detailed correlation of Pragian to Emsian strata with other areas globally. The Emsian Esopus and Schoharie Formations of the Appalachian Basin appear by position to correlate with the Zlichov and Daleje cycles of the Barrandian area of the Czech Republic, but no reliable biostratigraphic data at present confirms such a correlation.

More available biostratigraphic data for the Eifelian Stage permits better correlation of Appalachian Basin sequences with successions elsewhere. For example, the base of the lower Eifelian *costatus* conodont zone in the Appalachian Basin (black beds of the Nedrow Member, Onondaga Formation), the Czech Republic (e.g. micritic bed with auloporid corals in basal *Acanthopyge* Limestone, Stop N2-11 of Chlupac and Hladil 2001), the eastern Anti-Atlas mountains of Morocco (e.g. *Pinacites* Limestone at Jebel Mech Irdane; Walliser 2001) is at or closely adjacent to the position of the same widespread, maximum flooding event within the middle subcycle of Johnson *et al.*'s T–R cycle Ic (Sequence Eif-1 of this paper). This event is coincident with the global Chotec (or *jugleri*) bio-event, reported from Europe and Africa (Walliser 1985, 1996; Chlupac & Kukal 1986). Further comparison of Eifelian strata and biostratigraphy between the Appalachian Basin and other areas points to an overall correlation of sequences and relative sea-level trends. The correspondence is evidence of the strong global or eustatic signal in sequence development. Similarly, the transgression of overlying cycle Ie appears to correlate internationally, and is marked by global Kacak bioevent.

However, the Pragian–Eifelian sea-level curves for Euramerica (from Johnson *et al.* 1985) and the Appalachian Basin (Fig. 20) show broad variance in overall relative water depth trends at the coarse scale. This is especially notable in the major transgressions in the early Emsian and middle Eifelian in the Appalachian Basin, each followed by gradual intervals of overall shallowing. These trends are related to the loading and subsidence during separate active to quiescent phases of the Acadian Orogeny (Tectophases I and II of Ettensohn 1985). However, again, at the sequence-/T–R cycles scale, there is a strong correlation between the Appalachian Basin relative sea-level curve and the Euramerican eustatic sea-level curve of Johnson *et al.* (1985, 1996). The eustatic trends are recognizable in the Appalachian Basin, although at times they were enhanced at the larger scale by tectonic overprinting.

Summary and conclusions

Extensive fieldwork across the entire Appalachian basin provides the basis for a new stratigraphic synthesis of the upper Pragian, Emsian and Eifelian interval. A number of distinct, widespread event beds (e.g. K-bentonites, thin black/dark shales, fossil beds, unconformities) and medium- to small-scale cycles are recognizable across the basin. The correlation of unique horizons that maintain stratigraphic relationships through facies and thickness changes provides a high resolution subdivision of the succession, and a more detailed view of stratigraphic relationships throughout the basin.

Among the new interpretations are as follows. The recognition of lower Emsian Esopus-equivalent strata basinwide. The Huntersville Formation along the outcrop belt is not equivalent to the Eifelian Onondaga Formation of New York, but is instead correlative with underlying Emsian strata of the Esopus and Schoharie Formations. The Columbus Formation of central Ohio, including lower strata Zones A and B of the Bellepoint Member, is the lateral equivalent of the Onondaga Formation. However, parts of the upper Onondaga succession seen basinwide (including several of the Tioga A–G K-bentonites) are missing at unconformities within the upper part of the Columbus (in the lower

H zone). Finally, the strata of the classic 'Marcellus Shale' of New York comprise two major, formation-level packages of strata, the Union Springs and Oatka Creek–Mount Marion Formations, whose members are recognizable throughout the basin, including into central Ohio.

The new stratigraphic interpretations indicate a need for some revision of the upper Pragian to Eifelian stratigraphic nomenclature. The revisions mostly relate to New York strata, but in a few cases are applicable elsewhere. Proposed changes include: (1) extending the name 'Oriskany Formation' to all correlative and synonymous late Pragian sandstone lithosomes (i.e. Ridgeley Formation) basinwide; (2) introduction of new, and revision, restriction or abandonment of some older member-level units; (3) the abandonment of the term 'Buttermilk Falls Formation' and its members in eastern Pennsylvania, as junior synonyms of the Onondaga Formation and its members; and (4) raising the term 'Marcellus' in New York to subgroup status and raising the terms 'Union Springs' and 'Oatka Creek' to formation-level status.

The late Pragian, Emsian and Eifelian succession in the Appalachian Basin, deposited overapproximately 25 million years, comprises nine 'third order' stratigraphic sequences. The succession overlies the Wallbridge Unconformity, at the base of Sloss' (1963) Kaskaskia Supersequence. The unconformity, which becomes amalgamated to other pre- and post-Wallbridge unconformities toward the margin of the basin, underlies the Oriskany Formation. A post-Oriskany unconformity, interpreted by some to represent the Wallbridge, is a major drowning surface formed within the transgressive systems tract of the second post-Wallbridge depositional sequence.

Comparison of the nine Pragian to Eifelian sequences from the Appalachian Basin with the eustatic sea-level curve of Johnson *et al.* (1985, 1996) for Euramerica shows a strong correlation at the scale of 'third order' sequences and T–R cycles. However, at the coarser scale, there are notable differences in relative water depth trends in the Appalachian Basin, which are associated with crustal loading and relaxation during separate tectonically active to quiescent phases of the Acadian Orogeny. These are the sea-level rises related to the onset of Acadian Tectophases I (beginning in sequence Ems-1) and II (beginning in sequence Eif-2). The overall pattern appears to indicate that eustatic sea-level trends are superposed over tectonic flexure in the Pragian to Eifelian strata of the Appalachian foreland basin.

The author would like to thank numerous people for discussions of Emsian–Eifelian stratigraphy and depositional systems, especially C. E. Brett, along with J. Inners, J. Dennison, D. Griffing, R. Lindemann, A. J. Boucot, W. Koch, G. Baird, E. Landing, F. Ettensohn, W. A. Oliver Jr., H. Feldman, J. Epstein, J. Day, D. Monteverde and G. Kloc. The manuscript benefited from a critical reading by C. E. Brett and reviews by T. Becker and W. Kirchgasser. D. Allen assisted with the fieldwork. The fieldwork was funded in part by the New York State Museum/State Geological Survey, the Pennsylvania Geological Survey, the Geological Society of America, the Paleontological Society and Sigma Xi.

References

ALLEN, P. A. & ALLEN, J. R. 1990. *Basin analysis: principles and applications.* Blackwell Scientific Publications, Boston.

ANDERSON, E. J., BRETT, C. E., FISHER, D. W., GOODWIN, P. W., KLOC, G. J., LANDING, E. & LINDEMANN, R. H. 1988. Upper Silurian to Middle Devonian stratigraphy and depositional controls, east-central New York. *In*: LANDING, E. (ed.) *The Canadian Paleontology and Biostratigraphy Seminar.* New York State Museum Bulletin, **462**, 111–134.

BARRETT, S. F. & ISAACSON, P. E. 1981. Faunal assemblages developed in a coarse clastic sequence. *In*: GRAY, J. (ed.) *Communities of the Past.* Hutchinson Ross Publishing Co., Stroudsburg, PA, 165–183.

BECKER, R. T. & HOUSE, M. R. 1994. International Devonian goniatite zonation, Emsian to Givetian, with new records from Morocco. *Courier Forschungsinstitut Senckenberg*, **169**, 79–135.

BECKER, R. T. & HOUSE, M. R. 2000. Devonian ammonoid zones and their correlation with established series and stage boundaries. *Courier Forschungsinstitut Senckenberg*, **220**, 113–151.

BERG, T. M., MCINERNY, M. K., WAY, J. H. & MACLACHAN, D. B. 1983 (revised). *Stratigraphic correlation chart of Pennsylvania.* Pennsylvania Topographic and Geologic Survey, Fourth Series, General Geology Report **75**.

BOUCOT, A. J. 1959. Brachiopods of the Lower Devonian rocks at Highland Mills, New York. *Journal of Paleontology*, **33**, 727–769.

BOUCOT, A. J. & JOHNSON, J. G. 1967. Paleogeography and correlation of Appalachian Province Lower Devonian sedimentary rocks. *Tulsa Geological Society Digest*, **35**, 35–87.

BOUCOT, A. J., GAURI, K. L. & SOUTHARD, J. 1970. Silurian and Lower Devonian brachiopods, structure, and stratigraphy of the Green Pond Outlier in southeastern New York. *Palaeontographica*, **135**, 1–59.

BRETT, C. E. & BAIRD, G. C. 1996. Middle Devonian sedimentary cycles and sequences in the northern Appalachian Basin. *In*: WITZKE, B. M., LUDVIGSON, G. A. & DAY, J. (eds) *Paleozoic Sequence Stratigraphy: Views from the North American Craton.* Geological Society of America, Special Paper, **306**, 213–241.

BRETT, C. E. & VER STRAETEN, C. A. 1994. Stratigraphy and facies relationships of the Eifelian Onondaga Limestone (Middle Devonian) in western and west central New York State. *In*: BRETT, C. E. & SCATTERDAY, J.

(eds) *New York State Geological Association, 66th Annual Meeting Guidebook,* 221–269.
BRETT, C. E. & VER STRAETEN, C. A. 1997. *Devonian cyclicity and sequence stratigraphy in New York State.* Subcommission on Devonian Stratigraphy (International Union of Geological Sciences Commission on Stratigraphy), Field Trip Guidebook for Meeting, Rochester, NY, July 20–29.
BRETT, C. E., VER STRAETEN, C. A. & BAIRD, G. C. 2000. Anatomy of a composite sequence boundary. The Silurian–Devonian contact in Western New York State. New York State Geological Association, *72nd Annual Meeting, Field Trip Guidebook*.
CATE, A. S. 1963. *Lithostratigraphy of some Middle and Upper Devonian rocks in the subsurface of southwestern Pennsylvania*. Pennsylvania Topographic and Geological Survey, Fourth Series, General Geology Report **G39**, 229–440.
CATUNEANU, O. 2002. Sequence stratigraphy of clastic systems: Concepts, merits, and pitfalls. *Journal of African Earth Sciences*, **35**, 1–43.
CHADWICK, G. H. 1908. Revision of 'the New York series'. *Science*, **28**, 346–348.
CHADWICK, G. H. 1933. Catskill as a geologic name. *American Journal of Science*, **26**, 479–484.
CHADWICK, G. H. 1940. New York State Geological Association, *16th Annual Meeting Field Guide Leaflets*, 2–3.
CHAPEL, J. D. 1975. *Petrology & depositional history of Devonian carbonates in Ohio*. Unpublished Ph.D. thesis, The Ohio State University.
CHLUPAC, I. & HLADIL, J. 2001. Post-conference Field Trip (N, part 2): Barrandian area (May 20–21, 2001). *In*: JANSEN, U., KONIGSHOF, P., PLODOWSKI, G. & SCHINDLER, E. (eds) *Field Trips Guidebook:* 15th International Senckenberg Conference, Joint Meeting of IGCP project 421 and the Subcommission on Devonian Stratigraphy, 115–151.
CHLUPAC, I. & KUKAL, Z. 1986. Reflection of possible global Devonian events in the Barrandian area, C.S.S.R. *In*: WALLISER, O. H. (ed.) *Global Bio-events, Lecture Notes in Earth Sciences*. **8**, Springer-Verlag, Berlin, 169–179.
CLARKE, J. M. 1903. *Classification of New York Series of Geologic Formations*. New York State Museum Handbook, **19**, Table 2.
COE, A. L. 2003. *The Sedimentary Record of Sea-Level Change*. Cambridge University Press, The Open University, New York.
COOPER, G. A. 1930. Stratigraphy of the Hamilton Group of New York. *American Journal of Science*, **19**, 116–134, 214–236.
COOPER, G. A. 1933. Stratigraphy of the Hamilton Group of eastern New York, part 1. *American Journal of Science*, **26**, 537–551.
COOPER, G. A. 1941. Facies relations of the Middle Devonian (Hamilton) group along the Catskill Front. *Geological Society of America Bulletin*, **52**, 1893.
CONKIN, J. E. & CONKIN, B. M. 1975. Middle Devonian bone beds and the Columbus–Delaware (Onondagan–Hamiltonian) contact in central Ohio. *In*: POJETA, J. & POPE, J. K. (eds) *Studies in Paleontology and Stratigraphy*. Bulletins of American Paleontology, **67**, 99–121.

CONKIN, J. E. & CONKIN, B. M. 1984. *Paleozoic metabentonites of North America: Part 1.—Devonian metabentonites in the eastern United States and southern Ontario: their identities, stratigraphic positions, and correlation*. University of Louisville Studies in Paleontology and Stratigraphy, **16**.
DARTON, N. H. 1894. Report on the relations of the Helderberg limestones and associated formations in eastern New York. *Annual Report of the State Geologist*, **13**, 199–228 *also* New York State Museum Annual Report, **47**, 391–422.
DENNISON, J. M. 1960. *Stratigraphy of Devonian Onesquethaw Stage in West Virginia, Virginia, and Maryland*. Unpublished Ph.D. thesis, University of Wisconsin.
DENNISON, J. M. 1961. *Stratigraphy of Onesquethaw Stage of Devonian in West Virginia and bordering states*. West Virginia Geological Survey Bulletin, **22**.
DENNISON, J. M. & HASSON, K. O. 1976. Stratigraphic cross section of Hamilton Group and adjacent strata along south border of Pennsylvania. *AAPG Bulletin*, **60**, 278–287.
DENNISON, J. M. & HEAD, J. W. 1975. Sealevel variations interpreted from the Appalachian basin Silurian and Devonian. *American Journal of Science*, **275**, 1089–1120.
DENNISON, J. M. & TEXTORIS, C. A. 1970. Devonian Tioga Tuff in Northeastern United States. *Bulletin Volcanologique*, **34**, 289–294.
DENNISON, J. M., DE WITT, W., JR., HASSON, K. O., HOSKINS, D. M. & HEAD, J. W., III 1972. Stratigraphy, sedimentology, and structure of Silurian rocks along the Allegheny Front in Bedford County, Pennsylvania, Allegheny County Maryland, and Mineral and Grant Counties, West Virginia. *37th Annual Field Conference of Pennsylvania Geologists*.
DENNISON, J. M., HASSON, K. O., HOSKINS, D. M., JOLLEY, R. M. & SEVON, W. D. 1979. Devonian shales of south-central Pennsylvania and Maryland. *44th Annual Field Conference of Pennsylvania Geologists, Fieldtrip Guidebook*.
DUTRO, J. T. 1981. Devonian brachiopod biostratigraphy of New York State. *In*: OLIVER, W. A. & KLAPPER, G., JR. (eds) *Devonian Biostratigraphy of New York, Part 1*: International Union of Geological Sciences, Subcommission on Devonian Stratigraphy, 67–92.
EMERY, D. & MEYERS, K. J. 1996. *Sequence Stratigraphy*. Blackwell Science Ltd., Cambridge, MA, 297 pp.
EPSTEIN, J. B. 1984. *Onesquethawan stratigraphy (Lower and Middle Devonian) of northeastern Pennsylvania*. US Geological Survey, Professional Paper, **1337**.
EPSTEIN, J. B., SEVON, W. D. & GLAESER, J. D. 1974. *Geology and mineral resources of the Lehighton and Palmerton Quadrangles, Carbon and Northampton Counties, Pennsylvania*. Pennsylvania Topographic and Geologic Survey, Fourth Series, Atlas **195cd**.
ETTENSOHN, F. R. 1985. The Catskill Delta Complex and the Acadian Orogeny: a model. *In*: WOODROW, D. L. & SEVON, W. D. (eds) *The Catskill Delta*. Geological Society of America, Special Paper **201**, 39–50.
FAILL, R. T. & WELLS, R. B. 1974. *Geology and mineral resources of the Millerstown Quadrangle, Perry, Juniata, and Snyder Counties Pennsylvania*.

Pennsylvania Topographic and Geologic Survey, Fourth Series, Atlas **136**.

FAILL, R. T., HOSKINS, D. M. & WELLS, R. B. 1978. *Middle Devonian stratigraphy in central Pennsylvania—a revision.* Pennsylvania Topographic and Geologic Survey, Fourth Series, General Geology Report **70**.

FERRILL, B. A. & THOMAS, W. A. 1988. Acadian dextral transpression and synorogenic sedimentary successions in the Appalachians. *Geology*, **16**, 604–608.

GRABAU, A. W. 1917. Stratigraphic relationships of the Tully Limestone and the Genesee Shale in eastern North America. *Bulletin of the Geological Society of America*, **28**, 945–958.

GRIFFING, D. H. 1994. *Microstratigraphy, facies, paleoenvironments, and the origin of widespread, shale-hosted skeletal limestones in the Hamilton Group (Middle Devonian) of New York State.* Unpublished Ph.D. dissertation, State University of New York at Binghamton.

GRIFFING, D. H. & VER STRAETEN, C. A. 1991. Stratigraphy and depositional environments of the lower part of the Marcellus Formation (Middle Devonian) in eastern New York State. *In*: EBERT, J. R. (ed.) New York State Geological Association, *63rd Annual Meeting Guidebook*, 205–249.

GOLDRING, W. & FLOWER, R. H. 1944. Carlisle Center Formation, a new name for the Sharon Springs Formation of Goldring and Flower. *American Journal of Science*, **242**, 340.

HALL, J. 1839. Third annual report of the Fourth Geological District of the State of New York. *New York Geological Survey Annual Report*, **3**, 287–339.

HODGSON, E. A. 1970. *Petrogenesis of the Lower Devonian Oriskany Sandstone and its correlates in New York, with a note on their acritarchs.* Unpublished Ph.D. thesis, Cornell University.

HOUSE, M. R. 1962. Observations on the ammonoids succession of the North American Devonian. *Journal of Paleontology*, **36**, 247–284.

HOUSE, M. R. 1978. Devonian ammonoids from the Appalachians and their bearing on international zonation and correlation. *Special Papers in Palaeontology*, **21**, 70 p.

HOUSE, M. R. 1981. Lower and Middle Devonian goniatite biostratigraphy. *In*: OLIVER, W. A., JR. & KLAPPER, G. (eds) *Devonian Biostratigraphy of New York, Part 1*: International Union of Geological Sciences, Subcommission on Devonian Stratigraphy, 33–37.

HYDE, J. E. 1915. Stratigraphy of the Waverly Formation of central and southern Ohio. *Journal of Geology*, **23**, 655–682 and 757–779.

INNERS, J. D. 1975. *The stratigraphy and paleontology of the Onesquethaw Stage in Pennsylvania and adjacent states.* Unpublished Ph.D. dissertation, University of Massachusetts Amherst.

JOHNSEN, J. H. 1957. *The Schoharie Formation: a redefinition.* Unpublished Ph.D. thesis, Lehigh University.

JOHNSEN, J. H. & SOUTHARD, J. B. 1962. The Schoharie Formation in southeastern New York. New York State Geological Association, *34th Annual Meeting Guidebook*, A7–A19.

JOHNSON, J. G., KLAPPER, G. & SANDBERG, C. A. 1985. Devonian eustatic fluctuations in Euramerica. *Geological Society of America Bulletin*, **96**, 567–687.

JOHNSON, J. G., KLAPPER, G. & ELRICK, M. 1996. Devonian transgressive–regressive cycles and biostratigraphy, Northern Antelope Range, Nevada: Establishment of reference horizons for Global Cycles. *Palaios*, **11**, 3–14.

JUDGE, S. A. 1998. *Stratigraphy, sedimentology, and paleontology of the Bellepoint Member of the Columbus Limestone (Devonian), Central Ohio.* Unpublished MS thesis, The Ohio State University.

KAUFMANN, B., TRAPP, E., MEZGER, K. & WEDDIGE, K. 2005. Two new Emsian (Early Devonian) U–Pb zircon ages from volcanic rocks of the Rhenish Massif (Germany): implications for the Devonian time scale. *Journal of the Geological Society*, **162**, 363–371.

KIRCHGASSER, W. T. 2000. Correlation of stage boundaries in the Appalachian Devonian, Eastern United States, *Courier Forschungsinstitut Senckenberg*, **225**, 271–284.

KIRCHGASSER, W. T., OLIVER, W. A., JR. & RICKARD, L. V. 1985. Devonian series boundaries in the eastern United States. *Courier Forschungsinstitut Senckenberg*, **75**, 233–260.

KIRCHGASSER, W. T. & OLIVER, W. A., JR. 1993. Correlation of Stage Boundaries in the Appalachian Devonian, Eastern United States. Subcommission on Devonian Stratigraphy, *Newsletter*, **10**, 5–8.

KLAPPER, G. 1971. Sequence within the conodont genus *Polygnathus* in the New York lower Middle Devonian. *Geologica et Palaeontologica*, **5**, 59–79.

KLAPPER, G. 1981. Review of New York Devonian conodont biostratigraphy. *In*: OLIVER, W. A., JR. & KLAPPER, G. (eds) *Devonian Biostratigraphy of New York, Part 1*. International Union of Geological Sciences, Subcommission on Devonian Stratigraphy, 57–76.

KOCH, W. F., II 1978. *Brachiopod paleoecology, paleobiogeography, and biostratigraphy in the upper Middle Devonian of eastern North America: an ecofacies model for the Appalachian, Michigan, and Illinois Basins.* Unpublished Ph.D. dissertation, Oregon State University.

KOCH, W. F. 1981. Brachiopod community paleoecology, paleobiogeography, and depositional topography of the Devonian Onondaga Limestone and correlative strata in eastern North America. *Lethaia*, **14**, 83–103.

LANDING, E. & BRETT, C. E. 1987. Trace fossils and regional significance of a Middle Devonian (Givetian) disconformity in southwestern Ontario. *Journal of Paleontology*, **61**, 205–230.

LINDEMANN, R. H. 1980. *Paleosynecology and paleoenvironments of the Onondaga Limestone in New York State.* Unpublished Ph.D. thesis, Rensselaer Polytechnic Institute.

LINDEMANN, R. H. & FELDMAN, H. R. 1987. Paleogeography and brachiopod paleoecology of the Onondaga Limestone in eastern New York. New York State Geological Association, *59th Annual Meeting, Field Trip Guidebook*, D2–D30.

MILLER, M. F. & REHMER, J. 1982. Using biogenic structures to interpret sharp lithologic boundaries: An example from the Lower Devonian of New York. *Journal of Sedimentary Petrology*, **52**, 887–895.

MILLER, R. L., HARRIS, L. D. & ROEN, J. B. 1964. The Wildcat Valley Sandstone (Devonian) of southwest

Virginia. *Geological Survey Research, 1964*, US Geological Survey, Professional Paper, **501-B**, B49–B52.

MITCHUM, R. M., JR., VAIL, P. E. & THOMPSON, S., III 1977. The depositional sequence as a basic unit for stratigraphic analysis. *In*: PAYTON, C. E. (ed.) *Seismic Stratigraphy—Applications to Hydrocarbon Exploration*. American Association of Petroleum Geologists, Memoirs, **26**, 53–62.

North American Commission on Stratigraphic Nomenclature, 1983. North American Stratigraphic Code. *American Association of Petroleum Geologists Bulletin*, **67**, 841–875.

OLIVER, W. A., JR. 1954. Stratigraphy of the Onondaga Limestone (Devonian) in central New York. *Geological Society of America Bulletin*, **65**, 621–652.

OLIVER, W. A., JR. 1956. Stratigraphy of the Onondaga Limestone in eastern New York. *Bulletin of the Geological Society of America*, **67**, 1441–1474.

OLIVER, W. A., JR. 1966. The Bois Blanc and Onondaga Formations in western New York and adjacent Ontario. New York State Geological Association Guidebook, *38th Annual Meeting*, 32–43.

OLIVER, W. A., JR. 1976. *Noncystimorph colonial rugose corals of the Onesquethaw and lower Cazenovia stages (Lower and Middle Devonian) in New York and adjacent areas*. US Geological Survey Professional Paper **869**.

OLIVER, W. A. & KLAPPER, G. 1981. *Devonian Biostratigraphy of New York, part 1. Text*. International Union of Geological Sciences, Subcommission on Devonian Stratigraphy.

OLIVER, W. A., JR. & SORAUF, J. E. 1981. Rugose coral biostratigraphy of the Devonian of New York and adjacent areas. *In*: OLIVER, W. A., JR. & KLAPPER, G. (eds) *Devonian Biostratigraphy of New York, Part 1*. International Union of Geological Sciences, Subcommission on Devonian Stratigraphy, 97–105.

OLIVER, W. A., JR., JOHNSEN, J. H. & SOUTHARD, J. B. 1962. The Onondaga Limestone and the Schoharie Formation in southeastern New York. New York State Geological Association, *34th Annual Meeting, Field Trip Guidebook*, A-1–A-25.

OLIVER, W. A., JR., DE WITT, W. JR., DENNISON, J. M., HOSKINS, D. M. & HUDDLE, J. W. 1967. Devonian of the Appalachian Basin, United States. *In*: OSWALD, D. H. (ed.) *International Symposium on the Devonian System*, **1**, Alberta Society of Petroleum Geologists, 1001–1040.

OZOL, M. A. 1964. *Alkali reactivity of cherts and stratigraphy and petrology of cherts and associated limestones of the Onondaga Formation of central and western New York*. Unpublished Ph.D. thesis, Rensselaer Polytechnic Institute.

PATCHEN, D. G., AVARY, K. L. & ERWIN, R. B. 1985. *Correlation of Stratigraphic Units of North America (COSUNA) Project: Northern Appalachian Region*. American Association of Petroleum Geologists, Tulsa, 1 chart.

POSAMENTIER, H. W. & ALLEN, G. P. 1999. *Siliciclastic sequence stratigraphy—concepts and applications*. SEPM Concepts in Sedimentology and Paleontology **7**, Society for Sedimentary Geology, Tulsa.

POSAMENTIER, H. W. & VAIL, P. R. 1988. Eustatic controls on clastic deposition II—sequence and systems tracts models. *In*: WILGUS, C. K., HASTINGS, B. S., KENDALL, C. G., POSAMENTIER, H. W., ROSS, C. A. & VAN WAGONER, J. C. (eds) *Sea-level Changes: An Integrated Approach*. Society of Economic Paleontologists and Mineralogists, Special Publications, **42**, 125–254.

QUINLAN, G. & BEAUMONT, C. 1984. Appalachian thrusting, lithospheric flexure, and the Paleozoic stratigraphy of the Eastern Interior of North America. *Canadian Journal of Earth Science*, **21**, 973–996.

RAMSEY, N. J. 1969. *Upper Emsian—Upper Givetian conodonts from the Columbus and Delaware Limestones and Lower Olentangy Shale of central Ohio*: Unpublished MS thesis, The Ohio State University.

RAST, N. & SKEHAN, J. W. 1993. Mid-Paleozoic orogenesis in the North Atlantic. *In*: ROY, D. C. & SKEHAN, J. W. (eds) *The Acadian Orogeny: Recent Studies in New England, Maritime Canada, and the Autochthonous Foreland*, Geological Society of America, Special Paper **275**, 153–164.

REHMER, J. 1976. *Petrology of the Esopus Shale (Lower Devonian), New York and adjacent states*. Unpublished Ph.D. thesis, Harvard University.

RICHARDSON, J. B. & MCGREGOR, D. C. 1986. *Silurian and Devonian Spore Zones of the Old Red Sandstone Continent and Adjacent Regions*. Geological Survey of Canada, Bulletin, **364**, 79pp.

RICKARD, L. V. 1975. *Correlation of the Silurian and Devonian Rocks in New York State*. New York State Map and Chart, **24**.

RICKARD, L. V. 1989. *Stratigraphy of the subsurface Lower and Middle Devonian of New York, Pennsylvania, Ohio, and Ontario*. New York State Museum Map and Chart, **39**, 59pp, 40 plates.

RODEN, M. K., PARRISH, R. R. & MILLER, D. S. 1990. The absolute age of the Eifelian Tioga Ash Bed, Pennsylvania. *Journal of Geology*, **98**, 282–285.

ROGERS, J. 1967. Chronology of tectonic movements in the Appalachian region of eastern North America. *American Journal of Science*, **265**, 408–427.

SALVADOR, A. 1994. *International Stratigraphic Guide*. International Union of Geological Sciences and the Geological Society of America.

SCOTESE, C. R. & MCKERROW, W. S. 1990. Revised world maps and introduction. *In*: MCKERROW, W. S. & SCOTESE, C. R. (eds) *Paleozoic Paleogeography and Biogeography*. Geological Society of London, Geological Society Memoirs, **12**, 1–21.

SHANMUGAN, G. 1988. Origin, recognition, and importance of erosional unconformities in sedimentary basins. *In*: KLEINSPEHN, K. L. & PAOLA, C. (eds) *New Perspectives in Basin Analysis*. Springer-Verlag, New York, 83–108.

SLOSS, L. L. 1963. Sequences in the cratonic interior of North America. *American Association of Petroleum Geologists Bulletin*, **74**, 93–114.

SPARLING, D. 1988. Middle Devonian stratigraphy and conodont biostratigraphy, north-central Ohio. *Ohio Journal of Science*, **88**, 2–18.

STAUFFER, C. R. 1909. *The Middle Devonian of Ohio*. *Geological Survey of Ohio*, Bulletin **14**, 204pp.

SWARTZ, F. M. 1939. Keyser Limestone and Helderberg Group. *In*: WILLARD, B., SWARTZ, F. M. & CLEAVES, A. B. (eds). *The Devonian of Pennsylvania*. Pennsylvania. Topographic and Geologic Survey, Fourth Series, Bulletin **G19**, 29–91.

SWARTZ, C. K., MAYNARD, T. P., SCHUCHERT, C. & ROWE, R. B. (eds) 1913. The Lower Devonian deposits of Maryland: Stratigraphic and paleontologic characteristics. Maryland Geological Survey, *Lower Devonian*, 84–95.

TUCKER, R. D., BRADLEY, D. C., VER STRAETEN, C. A., HARRIS, A. G., EBERT, J. R. & MCCUTCHEON, S. R. 1998. New U–Pb ages and the duration and division of Devonian time. *Earth and Planetary Science Letters*, **158**, 175–186.

VANUXEM, L. 1839. Third annual report of the geological survey of the third district. *New York Geological Survey Annual Report*, **3**, 241–285.

VAN WAGONER, J. C., POSAMENTIER, H. W., MITCHUM, R. M., VAIL, P. R., SARG, K. G., LOUTIT, T. S. & HARDENBOL, J. 1988. An overview of the fundamentals of sequence stratigraphy and key definitions. *In*: WILGUS, C. K., HASTINGS, B. S., ST., C., KENDALL, C. G., POSAMENTIER, H. W., ROSS, C. A. & VAN WAGONER, H. C. (eds) *Sea-Level Changes: An Integrated Approach*. Society of Economic Paleontologists and Mineralogists, Special Publications, **42**, 39–45.

VER STRAETEN, C. A. 1994. Microstratigraphy and depositional environments of a Middle Devonian foreland basin: Berne and Otsego Members, Mount Marion Formation, eastern New York State. *In*: LANDING, E. (ed.) *Studies in Stratigraphy and Paleontology in Honor of Donald W. Fisher*, New York State Museum Bulletin **481**, 367–380.

VER STRAETEN, C. A. 1996*a*. Stratigraphic synthesis and tectonic and sequence stratigraphic framework, upper Lower and Middle Devonian, Northern and Central Appalachian Basin. Unpublished Ph.D. thesis, University of Rochester.

VER STRAETEN, C. A. 1996*b*. Upper Lower and lower Middle Devonian stratigraphic synthesis, central Appalachian Basin of Pennsylvania. *Open File Report* **96–47**, Pennsylvania Topographic and Geological Survey.

VER STRAETEN, C. A. 2001*a*. Event and sequence stratigraphy and a new synthesis of the Lower to Middle Devonian, eastern Pennsylvania and adjacent areas. *In*: INNERS, J. D. & FLEEGER, G. M. (eds) *66th Annual Field Conference of Pennsylvania Geologists*, Delaware Water Gap, 35–53.

VER STRAETEN, C. A. 2001*b*. The Schoharie Formation in eastern Pennsylvania. *In*: INNERS, J. D. & FLEEGER, G. M. (eds) *66th Annual Field Conference of Pennsylvania Geologists*, Delaware Water Gap, 54–60.

VER STRAETEN, C. A. 2004*a*. K-bentonites, Volcanic Ash Preservation, and Implications for Lower to Middle Devonian Volcanism in the Acadian Orogen, Eastern North America. *Geological Society of America Bulletin*, **116**, 474–489.

VER STRAETEN, C. A. 2004*b*. Sprout Brook K-bentonites: New Interval of Devonian (Early Emsian?) K-bentonites in eastern North America. *Northeastern Geology and Environmental Science*, **26**, 298–305.

VER STRAETEN, C. A. & BRETT, C. A. 1995. Lower and Middle Devonian foreland basin fill in the Catskill Front: Stratigraphic synthesis, sequence stratigraphy, and the Acadian Orogeny. *In*: GARVER, J. I. & SMITH, J. A. (eds) New York State Geological Association, *67th Annual Meeting Guidebook*, 313–356.

VER STRAETEN, C. A. & BRETT, C. E. 2000. Bulge Migration and Pinnacle Reef Development, Devonian Appalachian Foreland Basin. *Journal of Geology*, **108**, 339–352.

VER STRAETEN, C. A. & BRETT, C. E. 2006. Pragian to Eifelian strata (mid Lower to lower Middle Devonian), northern Appalachian Basin—A stratigraphic revision. *Northeastern Geology*, **28**, 80–95.

VER STRAETEN, C. A., GRIFFING, D. H. & BRETT, C. E. 1994. The lower part of the Middle Devonian Marcellus 'Shale,' central to western New York State: Stratigraphy and depositional History. *In*: BRETT, C. E. & SCATTERDAY, J. (eds) New York State Geological Association, *66th Annual Meeting Guidebook*, 270–321 (includes field trip log).

VER STRAETEN, C. A., BRETT, C. A. & ALBRIGHT, S. S. 1995. Stratigraphic and paleontologic overview of the upper Lower and Middle Devonian, New Jersey and adjacent areas. *In*: BAKER, J. E. B. (ed.) *Contributions to the Paleontology of New Jersey*. Geological Association of New Jersey, **12**, 229–239.

WALLISER, O. H. 1985. Natural boundaries and Commission boundaries in the Devonian. *Courier Forschung-Institut Senckenberg*, **75**, 401–408.

WALLISER, O. H. 1996. Global events in the Devonian and Carboniferous. *In*: WALLISER, O. H. (ed.) *Global Events and Event Stratigraphy in the Phanerozoic*, Springer, New York, 225–250.

WALLISER, O. H. 2001. The Jebel Mech Irdane section. *Excursion Guidebook*, Moroccan Meeting of the Subcommission on Devonian Stratigraphy (SDS)—IGCP 421, 57–61.

WELLS, J. W. 1944. Middle Devonian bone beds of Ohio. *Geological Society of America Bulletin*, **55**, 202–217.

WELLS, J. W. 1947. Provisional paleoecological analysis of the Devonian rocks of the Columbus region. *Ohio Journal of Science*, **47**, 119–126.

WESTGATE, L. G. & FISCHER, R. P. 1933. Bone beds and crinoidal sands of the Delaware Limestone of central Ohio. *Geological Society of America Bulletin*, **44**, 1161–1172.

WILGUS, C. K., HASTINGS, B. S., KENDALL, C. G., POSAMENTIER, H. W., ROSS, C. A. & VAN WAGONER, J. C. (eds) *Sea-level Changes: An Integrated Approach*. Society of Economic Paleontologists and Mineralogists, Special Publications, **42**.

WILLARD, B. 1935. Hamilton Group along the Allegheny Front, Pennsylvania. *Geological Society of America Bulletin*, **46**, 1275–1290.

WILLARD, B. 1938. *A Paleozoic section at Delaware Water Gap*. Pennsylvania Geological Survey, Fourth Series, Bulletin **G-11**.

WILLIAMS, E. A., FRIEND, P. F. & WILLIAMS, B. P. J. 2000. A review of Devonian time scales: Databases, construction and new data. *In*: FRIEND, P. F. & WILLIAMS, B. P. J. (eds) *New Perspectives on the Old Red Sandstone*. Geological Society of London, Special Publications, **180**, 1–21.

WITZKE, B. J. 1990. Paleoclimatic constraints for Paleozoic paleolatitutes of Laurentia and Euramerica. *In*: MCKERROW, W. S. & SCOTESE, C. R. (eds) *Paleozoic Paleogeography and Biogeography*. Geological Society of London, Geological Society Memoirs, **12**, 57–74.

Persistent depositional sequences and bioevents in the Eifelian (early Middle Devonian) of eastern Laurentia: North American evidence of the Kačák Events?

M. K. DESANTIS[1], C. E. BRETT[2] & C. A. VER STRAETEN[3]

[1]*Department of Geology, University of Cincinnati, Cincinnati, OH 45221-0013, USA (e-mail: desantmk@email.uc.edu)*

[2]*Department of Geology, University of Cincinnati, Cincinnati, OH 45221-0013, USA*

[3]*New York State Museum, The State Education Department, Albany, NY 12230, USA*

Abstract: The late Eifelian–earliest Givetian interval (Middle Devonian) represents a time of significant faunal turnover in the eastern Laurentia and globally. A synthesis of biostratigraphic, K-bentonite and sequence stratigraphic data indicates that physical and biotic events in the Appalachian foreland basin sections in New York are coeval with the predominantly carbonate platform sections of southern Ontario and Ohio. The upper Eifelian (*australis* to *ensensis* conodont zones) Marcellus subgroup in New York comprises two large-scale (3rd-order) composite depositional sequences dominated by black shale, which are here assigned to the Union Springs and Oatka Creek Formations. The succession includes portions of three distinctive benthic faunas or ecological–evolutionary sub-units (EESUs): 'Onondaga', 'Stony Hollow' and 'Hamilton'. In the northern Appalachian Basin in New York, the boundaries of these bioevents show evidence of abrupt, widespread extinctions, immigration and ecological restructuring. In the Niagara Peninsula of Ontario and from central to northern Ohio, the same sequence stratigraphic pattern and bioevents are recognized in coeval, carbonate-dominated facies.

The correlations underscore a relatively simple pattern of two major sequences and four sub-sequences that can be recognized throughout much of eastern Laurentia. Moreover, the biotic changes appear to be synchronous across the foreland basin and adjacent cratonic platform. However, the degree of change differs substantially, being less pronounced in carbonate-dominated mid-continent sections. Finally, we make the case that the two major faunal changes align with regional sequence stratigraphic patterns as well as with the global Kačák bioevents.

The latest Eifelian to early Givetian of the Middle Devonian was a period of significant environmental change and faunal turnover in the Appalachian Basin. In New York, the interval is marked by the onset of black shales of the Marcellus subgroup (basal Hamilton Group), following the carbonate deposits of the Onondaga Formation. Parts of three major stable, predominantly benthic faunal blocks, or ecological evolutionary subunits (EESUs), occur within the Marcellus subgroup, the lower two formations of the Hamilton Group in the Appalachian Basin of New York, Pennsylvania and elsewhere (Brett & Baird 1995). Concurrent with the influx of black siliciclastics of the basal Marcellus is the replacement of the stable Onondaga Fauna by a short-lived incursion, the Stony Hollow Fauna. A second change associated with the appearance of the stable Hamilton Fauna occurs in the Halihan Hill Bed near the base of the second or Oatka Creek sequence.

On a wider scale, these Appalachian bioevents may also be related, in part, to the latest Eifelian Kačák Event (House 1985), also known as the *otomari* Event of Walliser (1985). This event has been described over a wide region including the Czech Republic (Budil 1995), Germany (Schöne 1997), Spain (Truyóls-Massoni *et al.* 1990), and Northern Africa (e.g. Walliser *et al.* 1995; House 1996. The event as defined by House (1985, 1996*a*) was named for ammonoid extinctions and originations associated with black shales of the Kačák Member of the Sbskro Formation (Czech Republic) and correlative strata elsewhere. Others (e.g. Walliser 1996; Schöne 1997) consider separate lower and upper events within this interval. The lower L'Ei 1 Event (= *otomari* Event) occurs with the onset of black shales, conodont extinctions and entry of the dacryoconarid *Nowakia otomari*. The upper L'Ei 2 Event marks the termination of black shale deposition and extinction of several goniatite genera. Clearly, the broadly defined event shows multiple phases of origination and extinction, primarily affecting pelagic taxa (House 1996*a*, *b*). The Kačák Event occurs just below the Eifelian–Givetian Stage boundary at the Global Stratotype Section and Point (GSSP), which is defined by the entry of the conodont *Polygnathus hemiansatus* (Walliser *et al.* 1995). Thus the event falls wholly within the former Lower *ensensis* conodont zone (Bultynck 1987; Belka *et al.* 1997; Bultynck & Walliser 2000). This level also coincides with the upper part of the *Agoniatites costulatus* ammonoid zone (MD I-F_2 of Becker & House 1994).

To date, the evidence of the faunal turnovers from New York has been from strata chiefly representing dysoxic environments reflecting in part a phase of foreland basin development during the Acadian Orogeny (Ettensohn 1985). This raises the question of whether the local changes were due merely to an extreme shift in environment (i.e. facies tracking on a grand scale). One way of getting at this question is to examine faunas on the nearby carbonate platform to the west, where the magnitude of environmental change was not as extreme. Then we might ask whether biotic changes on the carbonate platform occur synchronously with those in the main body of the Appalachian Basin. Are changes of the same magnitude and do they involve the same taxa in both areas?

Study approach

Determining the precise comparisons of these late Eifelian to early Givetian faunal changes in upramp sections of the Laurentian carbonate platform has previously been hindered by lack of detailed correlations, due in part to the extreme difference in facies. In this context, this paper presents a detailed, regional correlation of upper Eifelian–lowest Givetian strata in New York, Ontario and parts of Ohio. The delineation of coeval stratigraphic units/bioevents requires a synthesis of existing bio- and event stratigraphic data, as well as new insights provided by sequence stratigraphy. Data come from a review of existing literature, detailed examination of stratigraphic sections and drill cores, biostratigraphic sampling, and geochemical analysis of regionally significant K-bentonites.

Study area

The area encompassed by this study (Fig. 1) includes sections in the Appalachian Foreland Basin, in a transitional carbonate ramp in central Ohio and southern Ontario, and in the Chatham Sag, an area of subsidence that separates the Algonquin and Findlay Arches. Appalachian Basin sections are primarily along the east–west outcrop belt in New York State from the Hudson Valley, westward to Buffalo. The New York sections are supplemented by a series of outcrops in the central Appalachians of Pennsylvania. Two areas of southern Ontario are also considered, including four drill cores from the southwestern area from London to Sarnia, and a series of small outcrop exposures reported as 'Delaware Formation' in the vicinity of Selkirk about 85 km west of Buffalo. A third major study area encompasses the north–south outcrop belt along the eastern flank of the Findlay Arch, in central to northeastern Ohio, roughly from Columbus northward to Sandusky.

Units examined

Exposed units (Fig. 2) examined in New York include the upper Onondaga Formation, the Union Springs Formation (Bakoven and Stony Hollow Members), and the coeval Oatka Creek/Mount Marion Formations. Units examined in the subsurface of Ontario include the Dundee Formation and basal Hamilton-equivalent strata ('Marcellus Shale'). In Ohio, exposed units examined include the upper Columbus and Delaware Formations in the central outcrop belt, and the Dundee Formation in the northwest outcrop belt. These units were also examined in the subsurface.

Biostratigraphy

Biostratigraphic data, particularly of conodonts, provide the first-order age constraints on strata across the region (Fig. 2). Ammonoid and dacryoconarid data are of more limited use, but offer important ties to the Kačák bioevents reported in Europe and Northern Africa. Despite extensive conodont sampling, detailed correlations remain poorly resolved, in part because key index taxa have not been found in all areas and because multiple bioevents occur within the span of a single biostratigraphic zone.

Conodont biostratigraphy

The conodont zonation for the study interval and region is based primarily on data from Klapper (1971, 1981) for New York, Uyeno *et al.* (1982) for Ontario, and the data of Sparling (1984, 1988, 1999) for Ohio. The zones for the upper Eifelian are the *Polygnathus costatus costatus*, *Tortodus kockelianus australis*, *Tortodus kockelianus kockelianus* and *Po. xylus ensensis* Zones, and for the earliest Givetian the *Po. hemiansatus* (formerly 'Upper' *ensensis*) Zone. At present, the precise boundaries of the *ensensis* and *hemiansatus* Zones remain poorly defined in North America (Sparling 1999).

costatus Zone. The base of the *costatus* Zone is marked by the appearance of *Polygnathus costatus costatus*. In New York, this conodont first occurs in the middle part of the pre-Moorehouse Nedrow

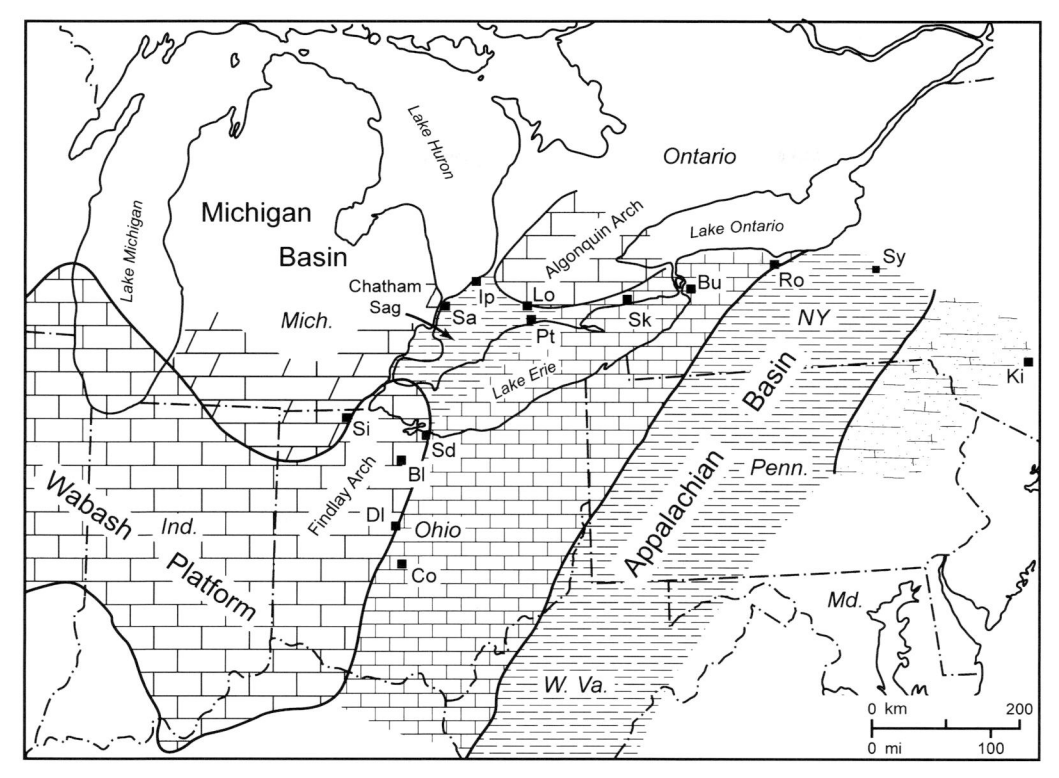

Fig. 1. Map showing late Eifelian palaeogeography of the Appalachian Basin to Wabash Platform region, eastern Laurentia. Locations referred to in text—New York: Bu, Buffalo; Ro, Rochester; Sy, Syracuse; Ki, Kingston. Ontario: Ip, Ipperwash Point; Lo, London; Pt, Port Stanley; Sa, Sarnia; Sk, Selkirk. Ohio: Co, Columbus; Dl, Delaware; Bl, Bloomville; Sd, Sandusky; Si, Silica.

Member of the Onondaga Formation and ranges upward into interbedded limestone and shale in the base of the Bakoven Member of the Union Springs Formation (Fig. 2). In southwestern Ontario, *Po. costatus costatus* ranges through the lower half of the Dundee Formation. In north-central Ohio, it ranges from the lower part of the Marblehead Member of the Columbus Formation to ~1.5 m below the top of the Venice Member of the Columbus (Sparling 1984). In central Ohio, *Po. costatus costatus* ranges from Stauffer's (1909) zone E in the middle of the Columbus Formation upward at least to the contact with the Delaware Formation (Ramsey 1969), and may also occur in the lower Delaware (Uyeno *et al.* 1982). The implications of these ranges are discussed below.

australis Zone. The index species of the *australis* Zone, *Tortodus kockelianus australis*, has not been found in the New York succession. Klapper (1981) referred the basal Union Springs shale and limestone beds at Union Springs, New York, to the *australis* Zone based on the co-occurrence of *Po. linguiformis linguiformis* with *Po. costatus costatus*, but observed that this association could also indicate the *costatus* Zone. Diagnostic *australis* elements have not been found in central Ohio either, but an association similar to that of the basal Union Springs is reported by Uyeno *et al.* (1982) from the lowest 1.5 m of the Delaware (Dublin Shale) at the Marble Cliff quarries in Columbus. This range may imply that most of the Dublin is no younger than the *australis* Zone. *Tort. kockelianus australis* is documented in the upper part of the Venice Member in north-central Ohio, where it occurs with *Po. linguiformis linguiformis*, *Po. pseudofoliatus* and *Po. costatus costatus*, about 2.0 to 2.5 m below the top of the member (Sparling 1984). Sparling notes that the lowest bed of the Delaware at Venice Quarry carries a conodont assemblage identical to that of the upper Venice and implies that both belong to the *australis* Zone. This indicates that the Venice belongs to the *australis* and possibly the lower *australis* Zone. As such, it may be equivalent to the

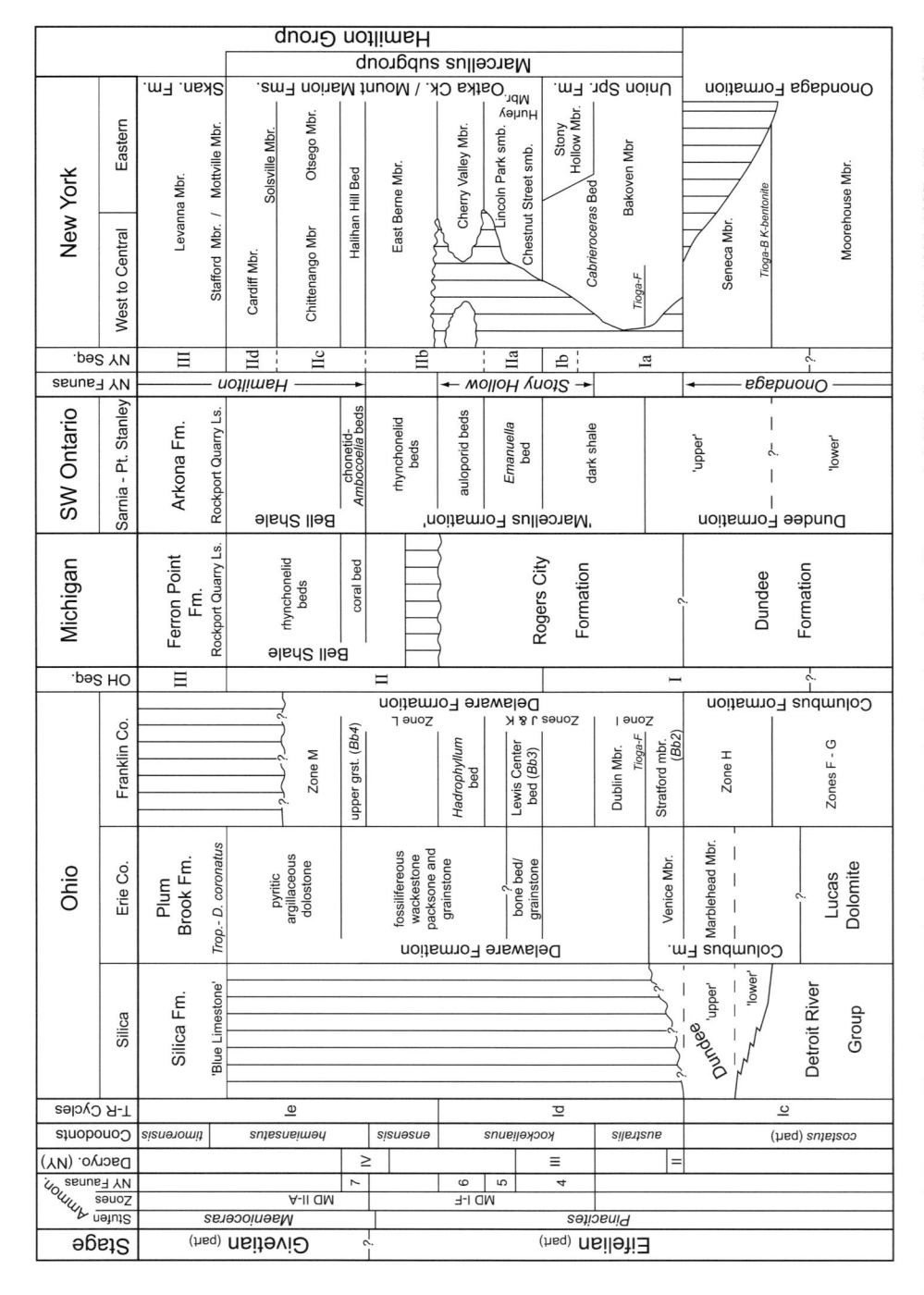

Fig. 2. Time–rock diagram calibrated to existing bio- and event stratigraphy. Conodont data from Uyeno *et al.* (1982); Sparling (1984, 1995, 1999); Klapper (1971, 1981) and Ramsey (1969). Ammonoid zones from House (1978, 1981) and Becker & House (1994, 2000). Dacryoconarid events from Lindemann (2002). TR cycles are those of Johnson *et al.* (1985).

basal Union Springs beds in central New York and to the middle Dundee in Ontario.

kockelianus/eiflius Zones. The base of the *kockelianus* Zone is marked by the first appearance of *Tortodus kockelianus kockelianus*. In New York, this and associated elements comprise the *Po. pseudofoliatus—Po.* aff. *Po. eiflius* fauna of the Chestnut Street submember (Hurley Member) and Cherry Valley Member in central New York (Klapper 1971), indicating placement of both of these units and probably the entire Hurley Member, in the *kockelianus* Zone. Although *Tort. kockelianus kockelianus* has not been found in southwestern Ontario or in Ohio, parts of the Delaware and Dundee Formations are indirectly assignable based on associated taxa.

The Delaware Formation has generally been referred to the *kockelianus* Zone based on *Polygnathus eiflius* reported by Ramsey (1969) from the lower part of the formation in central Ohio. Unfortunately, the exact stratigraphic levels of the samples were not reported. However, all of the *Po. eiflius* specimens apparently came from the lower 26 feet of the Delaware and, thus, somewhere within Stauffer's (1909) Zones I, J or K. As noted above, there is reason to believe that Zone I, including the Dublin Shale, is of *australis* Zone age. Moreover, Rickard (1984, 1985) expressed concern with Ramsey's interpretation, citing that the range of *Po. eiflius* and its differentiation from a similar subspecies, *Po.* aff. *Po. eiflius*, are uncertain. *Po.* aff. *Po. eiflius* is known from the *australis* Zone (Klapper & Ziegler 1979, Figs 3 and 4). However, in Morocco true *Po. eiflius* enters in the upper part of the *kockelianus* Zone and defines a separate *eiflius* Zone (Bultynck 1987; Belka *et al.* 1997). Sparling (1988) assigns all of the Delaware in north central Ohio to the *kockelianus* Zone, apparently based on correlations with the central Ohio sections. However, we suggest that some of the north–central Ohio Delaware is older than this based on our correlation with central Ohio stratigraphic sequences (discussed below). Because of these uncertainties, conodont material from central Ohio needs re-examination.

ensensis/hemiansatus Zones. The base of the *Polygnathus hemiansatus* Zone was designated as the base of the Givetian Stage (Walliser *et al.* 1995). At present, conodonts diagnostic of the preceding 'Lower' *ensensis* Zone (containing the Kačák Event interval) have not been found in eastern North America. *Polygnathus ensensis* elements have recently been preliminarily identified from the *Hadrophyllum* bed in the upper Delaware Formation of Ohio (Over, pers. comm. 2005) but this needs further investigation. The index species *hemiansatus* is also unknown. However, the *hemiansatus* Zone is approximately synonymous with the Upper *ensensis* Zone (*sensu* Bultynck 1987), which has been recognized in several units across the region. A single element recently obtained from the Oatka Creek Formation 90 cm above the Halihan Hill Bed in central New York, identified as long-ranging taxon *Po. pseudofoliatus* (Over, pers. comm. 2004), is permissive of the interval but not particularly constraining. Klapper (1981) reported conodont assemblages indicative of the Upper *ensensis* Zone in the Mottville Member of the Skaneateles Formation in central New York. Uyeno (reported in Sparling 1988) identified Upper *ensensis* conodonts in a portion of the Bell Shale and overlying Rockport Quarry Limestone–Arkona Formation of Ontario. Sparling (1995) reports 'phyletically late forms' of *Polygnathus xylus ensensis* as well as *Icriodus brevis* and early forms of *Po. xylus xylus* from the lowest 1 m of the Plum Brook Formation, and on that basis assigned that unit to the Upper *ensensis* Zone (i.e. *hemiansatus* Zone). However, *Icriodus brevis* and *Po. xylus xylus* both enter within the *timorensis* Zone (lowest part of the Lower *varcus* Zone) in Morocco (Bultynck 1987; Belka *et al.* 1997), suggesting that the Plum Brook may be younger. This interpretation is supported by the Plum Brook goniatites, which are the same as in the upper Arkona Formation, which also has *Icriodus brevis* (Landing & Brett 1987).

Ammonoid biostratigraphy

House (1978, 1981) outlined an ammonoid biostratigraphy for the Middle Devonian of the Appalachian Basin. The *Cabrieroceras* Bed (formerly *Werneroceras* Bed) in the uppermost Bakoven Member (Stony Hollow Member equivalent) in central New York contains fauna 4 of House (1978), including *Cabrieroceras crispiforme plebeiforme*, *Parodiceras discoideum* and *Holzapfeloceras* (reported as *Subanarcestes* cf. *micromphalus* in House 1978; revised in Becker and House 1994). *Cabr. crispiforme plebeiforme* is the index species of the late Eifelian *Cabrieroceras* Genozone and would ordinarily indicate the MD I-E *Cabrieroceras crispiforme plebeiforme* Zone of Becker and House (1994, 2000). However, the occurrence of *Parod. discoideum* in this fauna is more constraining, as *Parodiceras* does not enter before the MD I-F Zone (within the *kockelianus* conodont Zone) elsewhere.

The genus *Agoniatites* enters the New York section with *Agon. vanuxemi nodiferus* (fauna 5 of House 1978) in the Lincoln Park submember of the Hurley Member. This is overlain by the Cherry Valley Member (long known as the '*Agoniatites* Limestone') with its fauna of *Agon. vanuxemi vanuxemi* and other subspecies, as well

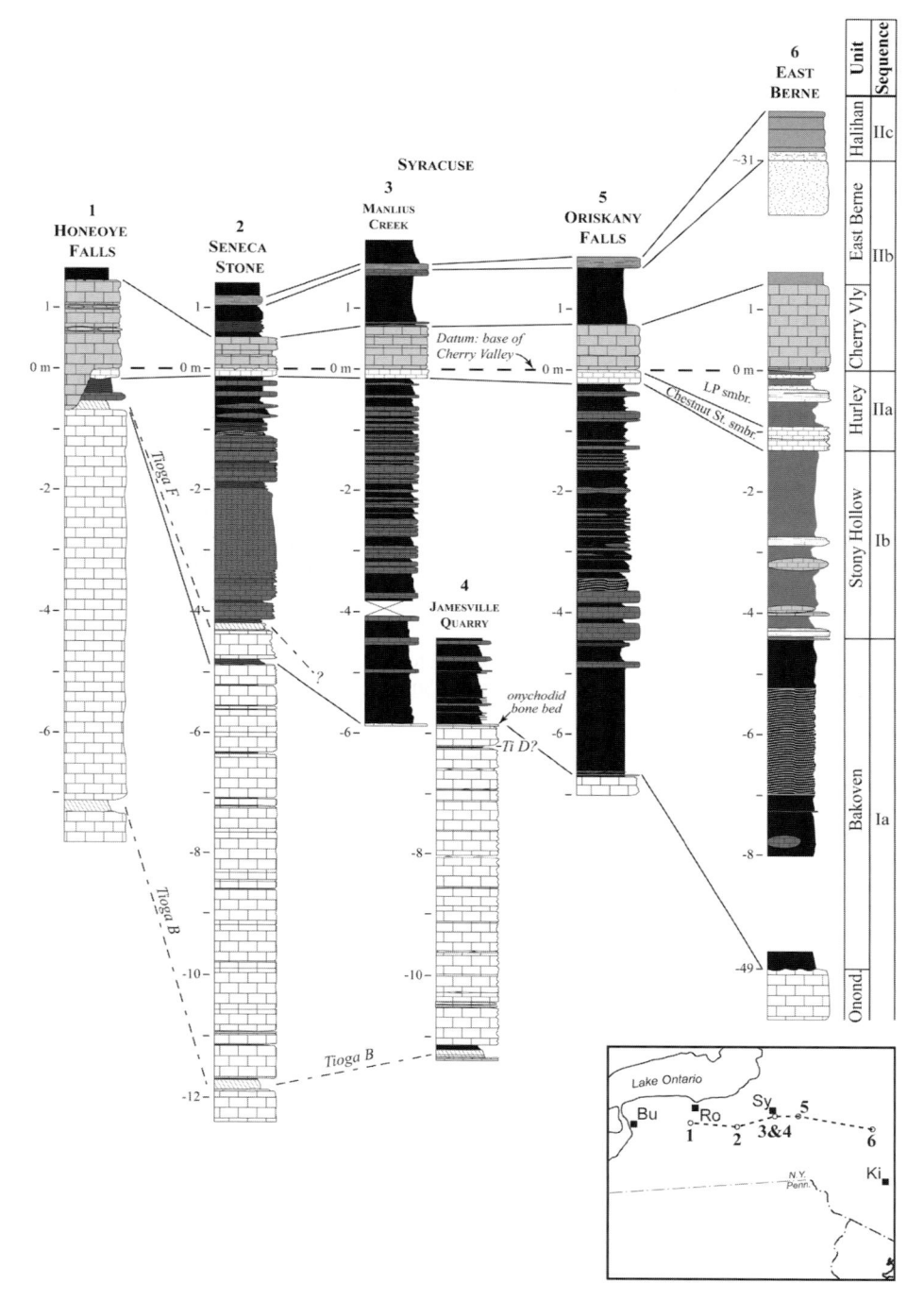

Fig. 3. New York State stratigraphic sections and sequence stratigraphy. Sequences are the same as outlined in Figure 2. Note thinning of units from east to west across the basin.

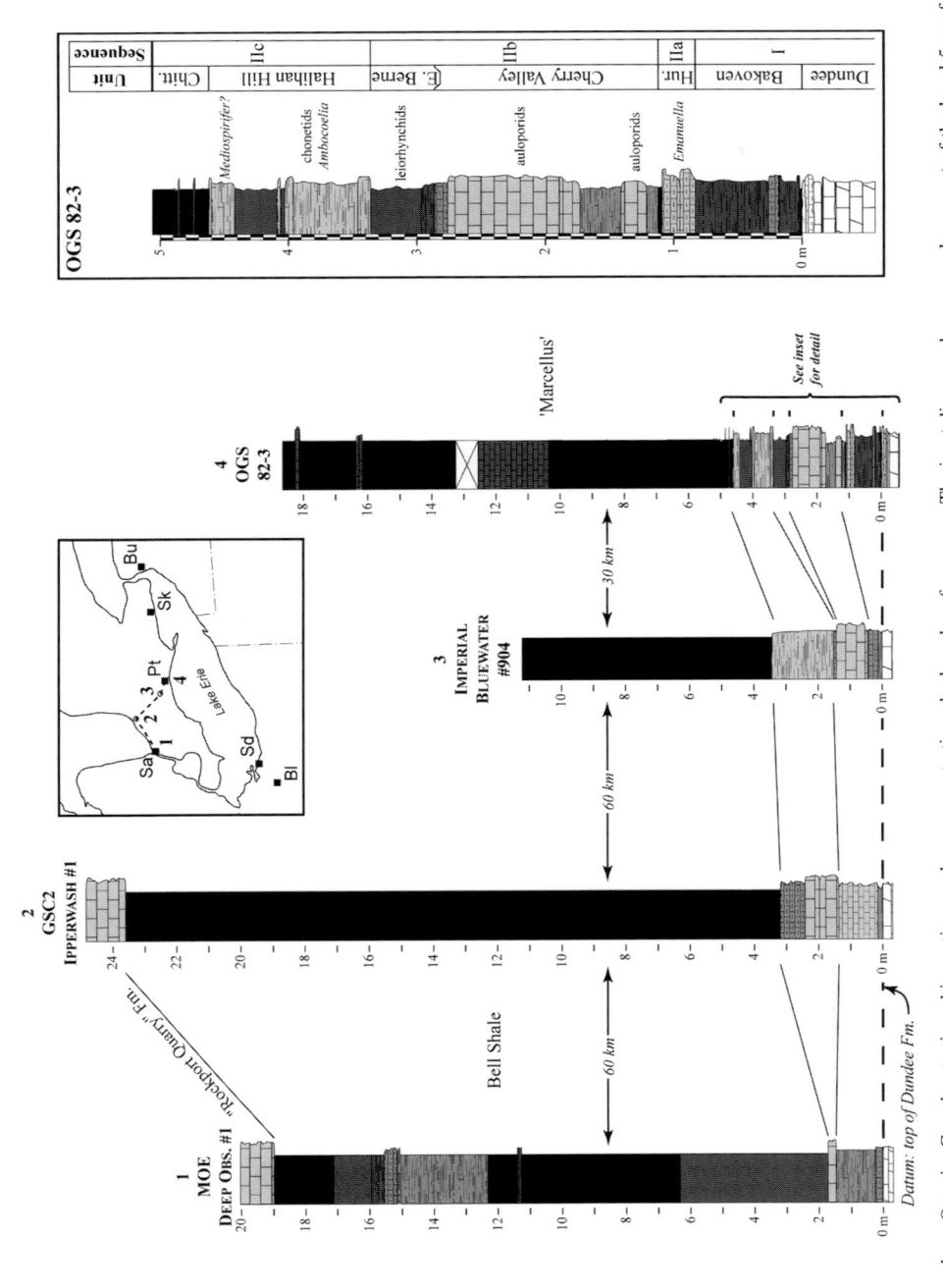

Fig. 4. Southern Ontario, Canada, stratigraphic sections and sequence stratigraphy based on four cores. The inset diagram shows an enlargement of the basal 5 m of OGS 82-3 at Port Stanley with a proposed correlation to New York units and depositional sequences. Cores are stored at the Ontario Oil, Gas and Salt Resources Library (OGSL) in London, Ontario. Core locations: MOE Deep Obs #1, Lambton Co., Sarnia Twp., OGSL core #954; GSC2 Ipperwash #1, Lambton Co., Bosanquet Twp., OGSL core #580; Imperial Bluewater #904, Elgin Co., Dunwich Twp., OGSL core #166; OGS 82-3, Elgin Co., Yarmouth Twp., OGSL core #861.

as *Parod. discoideum* (fauna 6 of House 1978). These forms of *Agon. vanuxemi* in the Oatka Creek/Mount Marion mark the late Eifelian *Agoniatites* Genozone, MD I-F *Agoniatites. costulatus* Zone. The upper part of this interval (MD I-F2) corresponds to the goniatite record of the Kačák Event, as currently defined (Becker & House 2000).

Agoniatites vanuxemi occurs elsewhere in the Appalachian Basin, including the Purcell Limestone in Pennsylvania (Ver Straeten *et al.* 1994), and has been reported from the upper Dundee ('Delaware') Formation in Ontario (Best 1953). *Cabr. crispiforme plebeiforme, P. discoideum, Holzapfeloceras* and *Agoniatites vanuxemi* are all reported from the lower Millboro Shale in Virginia (House 1978, 1981; Griffing & Ver Straeten 1991).

The precise position of the succeeding Givetian *Maenioceras* Genozone, MD II-A, is not known in eastern North America, but is believed to be represented by the early occurrence of *Tornoceras* in the form of *Torn.* aff. *mesopleuron* which with *Parodiceras* sp. nov. comprise fauna 7 reported by House (1965, 1978). According to House (1978, p. 56), this fauna was obtained from the Chittenango Member (Oatka Creek Formation) an estimated 18.3 m above the Cherry Valley Member in a quarry about 4.3 km northwest of Cherry Valley, New York. Given House's description, this specimen probably came from a small quarry exposing the East Berne Member and the Halihan Hill Bed. We suggest that the sample was probably collected from the latter unit (poorly preserved possible tornoceratids occur within the Halihan Hill Bed at other localities), which would place this bed in the earliest Givetian.

Tornoceras arkonense (fauna 8 of House 1978) occurs in the Skaneateles Formation in New York, the Arkona Formation of Ontario, and the Plum Brook and Silica Shales of Ohio (House 1981; Becker & House 2000).

Dacryoconarid biostratigraphy

Dacryoconarid biostratigraphy in the Appalachian Basin is still in its infancy and has not been correlated to the international nowakiid zonation. Lindemann (2002) documents four dacryoconarid events in the Onondaga through lower Oatka Creek Formations at Cherry Valley, New York. Events II–IV occur within the units of interest in this paper. Dacryconarid Event II, marked by the first appearance of *Striatostyliolina*, occurs at the base of the Bakoven Member of the Union Springs Formation. Event III occurs in the upper Union Springs and is concurrent with ammonoid fauna 4 of House (1978). Genera include *Styliolina, Viriatellina, Striatostyliolina, Costulatostyliolina* and *Nowakia*. Lindemann (2002) observes that this is the first reported occurrence of the latter two genera in the Devonian of the Appalachian Basin. He also suggests that Event III may correlate to the Lower Kačák Event *sensu* Walliser (1996). The Kačák interval occurs within younger (*ensensis* Zone) strata, so this cannot be the case. Event III does however coincide with the incursion of the 'Stony Hollow Fauna'. Finally, Lindemann's Event IV in the upper part of the East Berne Member and the Halihan Hill Bed is marked by the occurrence of abundant *Styliolina* and *Nowakia*. Unfortunately, *Nowakia otomari*, the dacryoconarid index fossil of the Kačák-*otomari* Event, remains unknown in New York (Lindemann 2002).

Dacryoconarids have not been reported from the Columbus or Delaware Formations in Ohio. However, a tentaculitid, *Tentaculites scalariformis*, is extremely abundant in certain horizons in the Delaware and in the Venice Member of the Columbus Formation.

K-bentonite event stratigraphy

The Middle Devonian Tioga K-bentonites have been recognized as valuable marker and regional correlation tools for the past half-century. When originally documented by Fettke (1952) from subsurface occurrences in the Pennsylvania Tioga gas field, 'the Tioga' was thought to be a single bed. Later work by Dennison (Dennison 1961; Dennison & Textoris 1970) showed that the Tioga is in fact a cluster of K-bentonite beds. Way *et al.* (1986) identified at least seven major horizons (Tioga A–G) in Pennsylvania. In New York, multiple Tioga beds occur in the upper part of the Onondaga Formation (Conkin & Conkin 1984; Brett & Ver Straeten 1994; Ver Straeten 2004), however only the beds correlated with the Pennsylvania Tioga B and F beds are consistently well preserved. These beds occur at the contact of the Seneca and Moorehouse Members and in the lower part of the Bakoven Member of the Union Springs Formation (Fig. 2).

In Ohio, a K-bentonite immediately overlying the contact between the Columbus (Venice Member) and Delaware Formations in the quarry at Venice has been equated with the Tioga B of New York (e.g. Oliver 1967; Oliver *et al.* 1968). This correlation was disputed by Conkin & Conkin (1984) and Sparling (1985, 1988) because it does not fit with the conodont biostratigraphy. Moreover, recent sampling indicates that the bed may not actually be a K-bentonite, although one highly weathered exposure does resemble Tioga beds seen elsewhere. However, other exposures of the horizon within the same quarry are comprised of dark, laminated calcareous shale. The clay fraction of the weathered material contains illite with no apparent mixed-layered component. Separated mineral residues contain detrital grains (e.g. quartz, garnet, tourmaline and zircon) but no apatite, biotite or other phenocrysts suggestive of a K-bentonite. Even if this bed is, in fact, one of the Tioga beds, we suggest that it is probably not the B horizon, and is more likely a higher bed (Tioga F?) as has been suggested by others (Conkin & Conkin 1984; Sparling 1985).

A recently discovered K-bentonite with moderately abundant zircon and apatite phenocrysts that immediately underlies the Dublin Shale at Slate Run in Columbus, Ohio, is currently being analysed. Preliminary data suggest that the bed at Slate Run does not correlate to the Tioga B. The phenocrysts appear to be chemically similar to those of the Tioga F in Pennsylvania and New York (DeSantis 2005). We are in the process of geochemically fingerprinting apatites from a number of Tioga K-bentonite horizons in the Appalachian Basin sections.

Regional stratigraphy

New York

The Marcellus subgroup (Hamilton Group) of New York as revised by Ver Straeten et al. (1994) is a strongly westward-thinning wedge of dominantly black and dark grey, organic-rich shale with minor siltstone, sandstone and thin, widespread, persistent carbonates. The Marcellus thins from over 580 m near the Catskill front to about 6 m in western New York (Rickard 1989). Despite this nearly hundred-fold thinning, several key marker beds, consisting of concretionary and styliolinid-rich limestones, can be traced across the entire outcrop belt (Ver Straeten et al. 1994). The four most distinctive of these are the closely spaced Chestnut Street submember, the Cherry Valley Member, the Halihan Hill Bed (Ver Straeten 1994), and the capping Stafford-Mottville Limestone of the Skaneateles Formation which overlies the Marcellus subgroup (Figs 2 and 3).

The Marcellus interval was subdivided into two major units, the Union Springs and Oatka Creek/Mount Marion Formations, in the sense of Cooper's (1930, 1933) Hamilton Group stratigraphy. Near its type locality, the Union Springs commences with a transitional series of dark shales and limestones that overlie the Seneca Member of the Onondaga Formation, and passes upward into black, laminated calcareous shales of the Bakoven Member. At most locations, the base of the Bakoven is sharp and marked by a bone bed of onychodid fish teeth and other phosphatic debris indicative of a condensed horizon. In western to central New York, the Tioga F K-bentonite (sometimes termed 'true Tioga') occurs in the lower part of the Bakoven Member, either within the basal transitional shales and limestones or directly overlying the basal bone bed. The Tioga F has not been located in outcrops east of Oakwood (~12 km northeast of Seneca Stone Quarry). The lower Bakoven beds yield a low diversity fauna characterized by styliolines and a few thin-shelled brachiopods, notably the rhynchonellid *Cherryvalleyrostrum limitare*, '*Lingula*' and *Orbiculoidea*. In the dark shales near the top of the Bakoven Member is a concretion horizon with the goniatites *Parodiceras discoideum*, *Cabrieroceras crispiforme plebeiforme* and *Holzapfeloceras* (fauna 4 of House 1978).

In the Hudson Valley region, the Bakoven thickens and passes upward to as much as 68 m of siltstone and silty calcareous mudstones assigned to the Stony Hollow Member (Cooper, in Goldring 1943; Griffing & Ver Straeten 1991; Ver Straeten et al. 1994). The Stony Hollow possesses a distinctive fauna (Stony Hollow Fauna of Ver Straeten et al. 1994; Brett & Baird 1996), including the brachiopods *Variatrypa arctica*, *Emanuella* sp., *Leptaena* '*rhomboidalis*', *Pentamerella* cf. *P. wintereri*, *Warrenella maia* and *Tentaculites* sp. This fauna differs markedly from the fauna of the underlying Onondaga Formation and overlying Hamilton Group.

The Stony Hollow is overlain by 1 to 7 m of fossiliferous, calcareous beds (Proetid Package of Griffing & Ver Straeten 1991), assigned by Ver Straeten et al. (1994) to the Hurley Member and presently considered by Ver Straeten and Brett (2006) to represent the basal member of the redefined Oatka Creek Formation. Careful tracing indicates that the Hurley Member condenses northwestward from East Berne to Oriskany Falls and farther westward into a single, thin interval of fossiliferous limestones, termed the Chestnut Street submember (Griffing & Ver Straeten 1991). The Chestnut Street typically weathers as a single thin ledge (10–43 cm) of fossiliferous wackestone that contains a somewhat less diverse Stony Hollow fauna dominated by sponge spicules, *Emanuella*, small rhynchonellids, *Dechenella haldemanni*, *Guerichophyllum*, and the microcrinoid *Haplocrinites*.

Throughout western and central New York, the Chestnut Street submember is either subjacent to or welded to the overlying Cherry Valley Limestone. However, in the Hudson Valley and central Pennsylvania, the beds are separated by up to 3 m of dark grey to black shale, locally rich in barite nodules (Ver Straeten et al. 1994). These shales, termed the Lincoln Park submember of the Hurley Member by Ver Straeten et al. (1994), have the goniatite subspecies *Agoniatites vanuxemi nodiferus*.

The Cherry Valley Member of the redefined Oatka Creek Formation is a 0.4 to 5 m thick interval of styliolinid pack- to grainstone noted for its cephalopod fauna, including various subspecies of *Agoniatites vanuxemi* and the orthocerid nautiloid *Striacoceras*. In eastern New York (Hudson Valley), the Cherry Valley is predominantly clastic and ranges up to 10 m in thickness (Ver Straeten *et al.* 1994). Throughout the New York outcrop belt, the Cherry Valley shows a consistent stratigraphy with lower and upper compact limestones, and a middle shaly nodular zone. Benthic fossils are generally sparse and of low diversity, but include rhynchonellids and thickets of auloporid corals, especially in the middle nodular beds. Cephalopods are particularly rich in the upper bed. Large individuals of the auloporid *Cladochonus* are particularly typical of the shallow-water grainstone facies of the westernmost Cherry Valley in the Genesee Valley south of Rochester. These grainstones show a slightly richer fauna, with abundant crinoid and bryozoan fragments, small euomphalid snails and brachiopods including *Warrenella maia* in the uppermost beds.

Throughout western and central New York, the Cherry Valley is overlain by a thin (generally <1 m) black, sparsely calcareous shale with styliolinid limestones. This interval, the East Berne Member of the Oatka Creek/Mount Marion Formation (formerly the Berne Member of Goldring 1935), thickens abruptly eastward, to about 80 m in the Hudson Valley (Ver Straeten 1994). The East Berne is everywhere overlain by a thin but significant interval of dacryoconarid-bearing limestones and richly fossiliferous calcareous mudstone, the Halihan Hill Bed (Ver Straeten 1994), defined as the basal unit of the Chittenango Member in central and western New York and of the Otsego Member in eastern New York. The Halihan Hill Bed yields a diverse fauna, comprised largely of species that persist into the overlying Hamilton Group (Brett & Baird 1995), especially the brachiopods *Ambocoelia umbonata*, *Mediospirifer* and *Athyris* sp. In addition, a few holdover Onondaga taxa are also found, including abundant *Hallinetes lineatus*, *Coelospira camilla* and *Strophonella* cf. *ampla*. In the Hudson Valley, near Kingston, the bed carries biostromes of rugose and tabulate corals, especially *Heliophyllum*, *Eridophyllum*, *Cystiphylloides* and branching favositids. The Halihan Hill Bed is particularly significant in carrying the initial occurrences of many taxa characteristic of the later Hamilton Group. It may also contain the lowest occurrence of *Tornoceras* aff. *mesopleuron*.

In western–central New York, the Oatka Creek Formation, above the Halihan Hill Bed, includes the Chittenango and Cardiff Members, comprising some 7 to 50 m of fissile, black to dark grey organic-rich shale. Geochemically, the Oatka Creek black shales in the Genesee Valley area show a very high TOC (5–17%), strong enrichment of Mo and V, and DOP (degree of pyritization) values >0.8, indicating that anoxic, euxinic conditions prevailed in western New York during deposition of this unit (Werne *et al.* 2002; Sageman *et al.* 2003). Most of these shales are barren of benthic fossils, although the uppermost dark- to medium-grey mudstones (Cardiff and its lateral equivalents) become increasingly fossiliferous toward the contact with the overlying Stafford-Mottville carbonates. Their eastward equivalents in the Otsego Member and higher units (up to 425 m thick) carry a diverse fauna with only a few species (e.g. *Paraspirifer* cf. *P. acuminatus*, *Schizophoria*) that do not carry over into higher Hamilton beds.

The Stafford Member and its eastern equivalent, the Mottville Member, form the base of the overlying Skaneateles Formation. These units are characteristically composed of two fossil-rich limestones or calcareous siltstones, separated by a middle mudstone interval of varying thickness (0.5 to 10 m). The Mottville beds carry a diverse 'Hamilton' fauna.

Ontario, Canada

Despite the detailed work of Stauffer (1915) and Best (1953), which remain useful sources of data on faunas, detailed understanding of late Eifelian–early Givetian stratigraphic relationships in southern Ontario and their correlations to the east and west, have long been obscure. Uyeno *et al.* (1982) helped clarify the relationships of the 'Delaware Formation' in Ontario. The problem is two-fold. First, this shale-prone succession is poorly exposed, being covered in most areas by a thick mantle of glacial drift. Secondly, units have not been consistently defined and identified, at least not in an allostratigraphic sense.

Strata believed to be age-equivalent to the Marcellus interval in New York have been variably assigned to four lithological units. In the past, the term 'Delaware' was applied to the lower variably argillaceous and cherty limestones overlying either the 'Onondaga' Formation or the Lucas Formation of the Detroit River Group in southwestern Ontario (Stauffer 1915; Best 1953; Uyeno *et al.* 1982). Carbonate strata between the Detroit River Group and overlying shales are now generally included in a single unit termed the Dundee Formation, a designation extended from outcrops of dolomitic carbonate and fossiliferous grainstones in southern Michigan. It is clear from the conodont biostratigraphy of Uyeno *et al.* (1982) and Sparling (1988) that variable amounts of correlative section are included in the Dundee in Ontario and that these do not strictly conform to the type Dundee of Michigan. Moreover, Uyeno's work makes clear that strata included in the upper portion of the Dundee in parts of western Ontario are age-equivalent to the Rogers City Formation, which is said to overlie the Dundee in the Alpena area of Michigan (Fig. 2). In addition, the term 'Marcellus Formation' has been extended from New York into southern Ontario for black and dark grey shale-dominated facies that are probably equivalent to the upper Delaware. In western drill core sections, Uyeno *et al.* (1982) used the term 'Bell Shale,' a Michigan Basin term, for laterally equivalent dark to medium grey and more fossiliferous shales.

Niagara Peninsula–Selkirk Area. Outcrops previously assigned variably to the Marcellus and Delaware occur along the north shore of Lake Erie and on Dry Creek at Cheapside. Stauffer (1915) measured these sections in detail and concluded that they closely resemble the Delaware Formation (primarily the succession from Dublin Shale upward) of central Ohio, particularly in showing a mixture of dark shale and chert with a low diversity fauna (*Tentaculites gracilistriatus* and *Styliolina*) at the base. Higher cherty limestone beds contain a more diverse fauna including *Arachnocrinus*. *Arachnocrinus* is not known to occur above the Moorehouse Member of the Onondaga Formation in New York (G. McIntosh pers. comm.). *Tentaculites* and stylolinids are also reported from the Nedrow Member of the Onondaga (Lindemann 2002). Thus, the Selkirk outcrops probably correlate to the middle part of the Onondaga Formation and are older than the Delaware of Ohio.

London–Sarnia Area. The facies, strata and faunas in the subsurface of the region south and west of London, Ontario (Fig. 4) appear very similar to those of New York State. This area lies on the eastern margin of and within the so-called Chatham Sag, a low saddle between the Findlay and Algonquin Arches (Fig. 1). Here, the upper Eifelian beds have been assigned mainly to the Dundee Formation, although much of the strata time-equivalent to the Marcellus subgroup in New York are assigned to the overlying 'Marcellus' and/or Bell Shale. Stauffer (1915) recognized on faunal grounds that these beds were age-equivalent to the Delaware of the Niagara Peninsula and central Ohio, and this correlation is largely confirmed by conodont biostratigraphy, although, as previously discussed, only the lower Delaware in Ohio has been assigned to the *australis* to *kockelianus* Zones.

A drill core (Fig. 4, core OGS 82-3) from the area of Port Stanley, southwest of London, Ontario, provides a good reference section for equivalents of the Marcellus subgroup strata in this region. Comparable sections, though increasingly condensed, were also recognized westward in the area of Arkona–Ipperwash and Sarnia, Ontario, but the units are best developed in the Port Stanley core. The lowest unit considered is a brownish grey, vuggy dolostone with crinoidal debris and scattered cystiphyllid rugose corals assignable to the Dundee Formation. As noted by Uyeno *et al.* (1982), the upper beds of Dundee in this region belong to the *australis* to *kockelianus* Zones. Based on conodont biostratigraphy, they are probably equivalent to the Venice Member of the Columbus Formation and lower Delaware Formation in Ohio. The Dundee is overlain abruptly by about 1 m of dark brownish grey to black, laminated shales and styliolinid limestones with rare, small rhynchonellid brachiopods. This unit is lithologically similar to the Bakoven Shale of New York and a correlation with that unit is supported by Uyeno *et al.*'s (1982) report of *Po. pseudofoliatus–Po.* aff. *eiflius* (*sensu* Klapper 1971) conodont assemblages from the uppermost Dundee and equivalent Rogers City Formation.

The dark shales are abruptly overlain by a 30 cm micritic limestone, rich in the amboceeliid brachiopod *Emanuella* cf. *E. subumbona. Emanuella* is reported as common in drift boulders of the 'Marcellus' formation in the Port Stanley area (Stauffer 1915), a faunal association also known from the Chestnut Street submember in New York State. Moreover, the '*Emanuella* bed' in Ontario may be equivalent to the *Emanuella* epibole of the lower–middle Rogers City Formation of Michigan reported by Ehlers & Kesling (1970). The Port Stanley '*Emanuella* bed' is separated by about 2 cm of black shale from an overlying ~1.7 m thick interval of medium dark grey, styliolinid wackestones with abundant, large, auloporid corals, especially near the top and middle, as well as chonetids and other brachiopods. We equate this interval with the Cherry Valley Member of New York based on position with respect to other biostratigraphically dated horizons, sequence pattern, and general lithological and faunal similarities. Like the Cherry Valley, this interval is divisible into lower and upper compact limestones, with a middle shaly, nodular zone rich in large auloporids.

In turn, the 'auloporid beds' are overlain successively upward by ~50 cm of dark grey and black shale with sparse rhynchonellid brachiopods (*Cherryvalleyrostrum limitare*?) and ~1 m of shell-rich limestone and calcareous brownish-grey shale. This interval is packed with chonetid brachiopods (*Longispina, Hallinetes*?), *Ambocoelia* and the rhynchonellid *C. limitare*. Together, these beds form a couplet that very closely resembles the East Berne Member (dark shale) and Halihan Hill Bed (chonetid–*Ambocoelia*-rich calcareous mudstone) found throughout central and western New York to the westernmost outcrop at Stony

Point on the Lake Erie shore (Ver Straeten et al. 1994). Finally, the upper portion of the drill core (assigned by drillers to the Marcellus) is some 10.3 m of dark brownish-grey to black shale with a few beds packed with rhynchonellid brachiopods. This interval is of comparable thickness, position and appearance to dark grey to medium grey rhynchonellid-bearing shale assigned to the Bell Shale in cores to the northwest near Sarnia, Ontario. This is in turn overlain by ~2.5 m of more fossiliferous grey mudstone, perhaps transitional into the Rockport Quarry Formation, although the core ends at this level. As noted previously, the upper Bell and Rockport Quarry interval in southwestern Ontario has yielded Upper *ensensis* Zone (=*hemiansatus* Zone) conodont assemblages (Uyeno, in Sparling 1988). It is thus reasonable on all grounds—biostratigraphy, general fauna, relative thickness and sequence pattern (see below)—to equate the Bell Shale–Rockport Quarry Limestone interval with the upper Marcellus Oatka Creek/Mount Marion Formations and Stafford/Mottville Members of the Skaneateles Formation in New York.

In summary, although more detailed biostratigraphic data are needed to confirm all aspects of this correlation, the package of strata in southwest Ontario is constrained above and below by conodont-dated successions, bracketing it as being of Marcellus age-equivalence. Indeed, the detailed pattern of this succession is so similar to the Marcellus subgroup succession in New York as to leave little doubt of the precise unit-by-unit correlation with that area.

Ohio

Ohio strata coeval to the New York Marcellus include part of the uppermost Columbus Formation (the Venice Member in northern Ohio) and the Delaware Formation. The first extensive survey of these formations was by Stauffer (1909) and subsequent work has built on his framework. In the Columbus area, Stauffer divided the Columbus Formation into eight zones (Zones A–H) based on lithological and palaeontological characteristics. The Delaware Formation was similarly divided into five zones (Zones I–M).

The Delaware Formation is a heterogeneous succession of thin to medium-bedded dark grey to brownish-grey limestone with locally abundant chert and calcareous shale. Lithologically, it contrasts with the underlying Columbus Formation by its generally thinner-bedded, finer-grained appearance. The Delaware has been broadly correlated to the lower part of the Marcellus subgroup, based primarily on a lingulid/rhynchonellid brachiopod fauna found in the Dublin Shale Member (upper Zone I) as well as other typical Hamilton taxa occurring throughout the formation. In northern Ohio, early surveys (Newberry 1873) assigned the lithologically similar strata of the Venice and Delaware to a single formation, the Sandusky Limestone. It was subsequently recognized by Swartz (1907) that the lower part (Venice) of the Sandusky contains brachiopods (e.g. *Acrospirifer duodenarius*, *Paraspirifer acuminatus*) comparable to the uppermost Columbus (Zone H) further south, prompting its correlation to that formation. However, conodont data (Sparling 1984) place at least the upper part of the Venice in the *australis* Zone, seemingly younger than upper Columbus strata in central Ohio.

Lateral facies variation (Fig. 5) between central (Columbus/Delaware) and north-central Ohio (Sandusky), coupled with few complete sections in intermediate areas, has long obscured the fine-scale correlation between these areas. Stratigraphic relationships at the base of the central Ohio succession are important for resolving correlations with the succession further to the north. In sections through the southern part of Delaware County, Stauffer (1909) noted that a basal bone bed and succeeding zones, obvious at Columbus, become harder to distinguish northward. The basal dark bituminous shales are seemingly missing and the lowest units instead are thin-bedded, brown limestones, though some similar faunal elements (*Cherryvalleyrostrum limitare*) are present within them. Stauffer assumed that the Dublin Member shales graded laterally into these limestones. As discussed below, careful observation indicates that these lower beds in fact lie **below** the bituminous shales, which occur with their typical fauna, higher in the section. With that recognized, all key Delaware units are traceable throughout the entire outcrop belt for the first time. Details of this south to north transect are discussed in the following sections.

Columbus/Delaware. Although the Delaware Formation was named for exposures near Delaware in central Ohio, most workers have effectively made the type sections those observed 20 to 30 km further south, along the Scioto River west of Columbus, Ohio. One of these, Slate Run, is shown in Fig. 5. This is somewhat unfortunate because, as discussed below, the basal unit, a key element in correlating the central and north-central regions, is reduced to a single bed at Columbus. The central Ohio succession can be divided into five units.

The basal unit (herein informally termed the Stratford member) generally consists of fine-grained, thin-bedded, brown, bioturbated dolomitic limestones. At Columbus, it is condensed to a 15 cm thick bed of crinoidal grainstone containing abundant fish bone material and brachiopods. Although

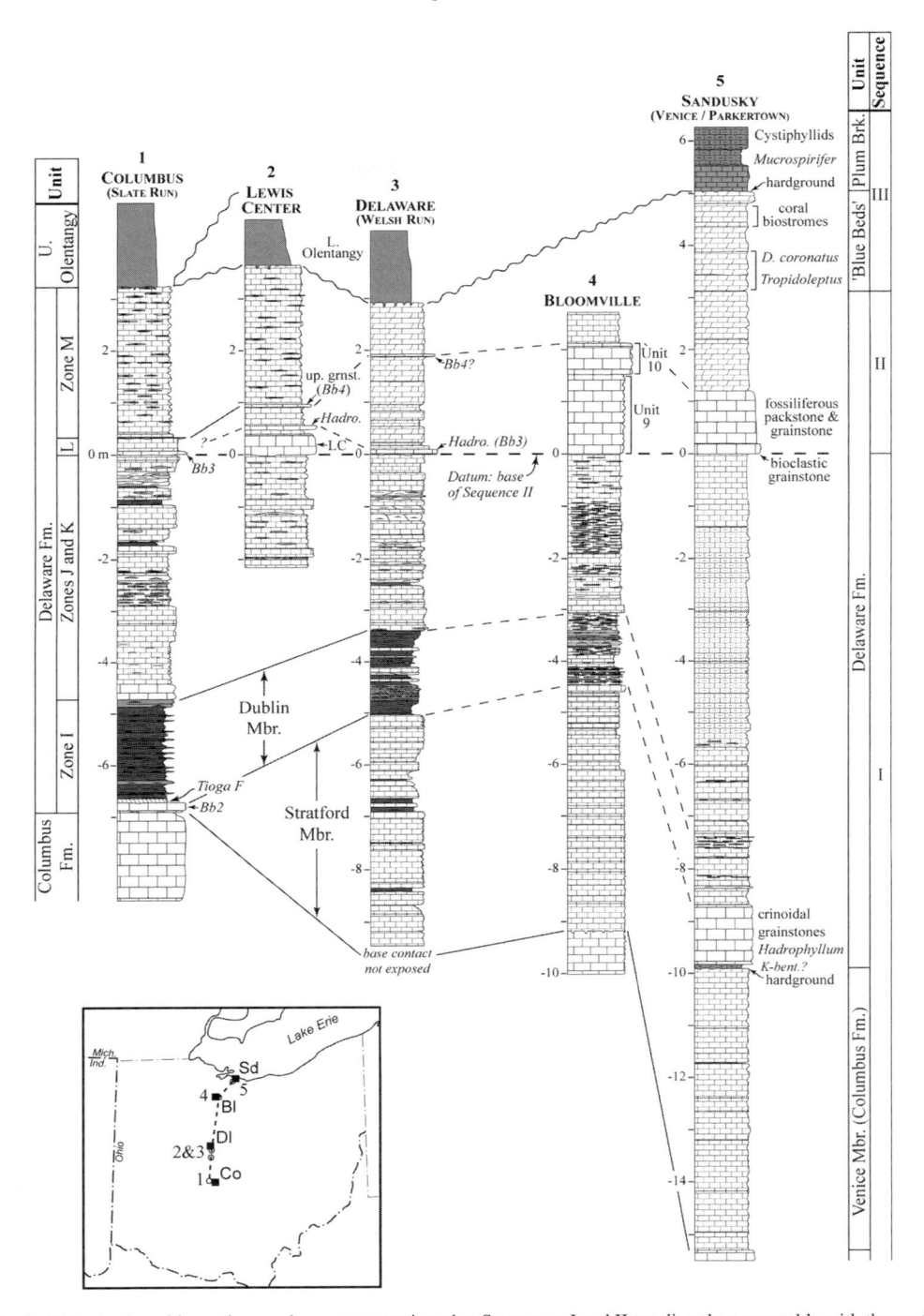

Fig. 5. Ohio stratigraphic sections and sequence stratigraphy. Sequences I and II are directly comparable with those of New York and Ontario. Key horizons are as follows: Bb2, Bb3 and Bb4, bone beds 2, 3 and 4 of Wells (1944); *Hadro.*, *Hadrophyllum* Bed; LC, Lewis Center Bed.

this bone bed (2nd bone bed, Wells 1944), has typically been assigned to the uppermost Columbus (Westgate & Fischer 1933; Wells 1944), Conkin & Conkin (1975) noted that the brachiopod fauna includes a mixture of Columbus (e.g. *Acrospirifer duodenarius*) and Delaware (e.g. *Warrenella maia* and *'Mucrospirifer' consobrinus*) taxa. The Stratford member expands northward, reaching an approximate thickness of 3 to 4 m near Delaware. Thickness differences may be due either to extreme condensation or submarine corrosion in the deeper parts of the basin to the south. The Stratford beds contain a fauna dominated by *'Mucrospirifer' consobrinus* as well as a few representatives of *Cherryvalleyrostrum limitare*, *Leptaena*, *Warrenella*, chonetids, and *Tentaculites*.

At Columbus, the condensed basal unit is overlain by a thin (9 cm), tan, slightly sticky, clay-rich shale, the K-bentonite geochemically correlated to the Appalachian Basin Tioga-F (DeSantis 2005). This bed also contains abundant comminuted fish bones, and also probably represents a highly condensed horizon. The surface between it and the underlying bone bed is an irregular corrosion surface. This K-bentonite has not been definitively located elsewhere in the Ohio outcrop belt. As noted previously, this bed appears to be equivalent to the Tioga-F K-bentonite at the base of the Marcellus subgroup in the Appalachian Basin. The Dublin Shale Member (upper part of Zone I) overlies the K-bentonite at Slate Run. It comprises about 1.5 to 2 m of thin platy, bituminous, calcareous shale or argillaceous limestone interbedded with dark chert. The limited brachiopod-dominated fauna includes *Cherryvalleyrostrum limitare*, *'Lingula'*, *Orbiculoidea*, rare *Ambocoelia*, and *Tentaculites*. The facies suggests dysoxic conditions and stands in sharp contrast to the underlying limestones. At most localities, the faunal transition between the Dublin and overlying units is gradational. As noted above, the Dublin was not previously recognized north of the Columbus/Delaware area, but our field studies show that a similar but slightly more cherty facies persists above the Stratford member to the north (Welsh Run section, Fig. 5).

The middle part of the Delaware Formation (Stauffer's Zones J and K) comprises about 4.5 m of thin- to thick-bedded, greyish-brown, fossiliferous mud- to wackestones, with chert horizons especially prevalent in the upper part. The interval contains a moderately diverse fauna including the brachiopods *Cherryvalleyrostrum limitare* (abundant in lowest beds), *Leptaena*, *Hallinetes*, *Longispina*, *Cupularostrum* and *Warrenella*, the bivalves *Glyptodesma*, *Grammysia*, *Pseudaviculopecten*, and *Tentaculites*. As with other Delaware divisions, fossils are generally concentrated in discrete horizons rather than disseminated throughout the unit.

The middle Delaware units are sharply and erosively overlain by a package of crinoid-rich limestone comprising three grainstones (L Zone of Stauffer 1909) bracketing intermediate intervals of flaggy, slightly cherty wacke- to packstone. The total thickness of this interval expands northward, ranging from 30 cm at Columbus to 120 cm at Delaware. All three grainstone units contain abundant fish bone material and moderately diverse brachiopod assemblages. The lowest unit, herein called the Lewis Center bed (3rd bone bed of Wells 1944) is a pyritiferous, coarsely granular, encrinal limestone up to 39 cm thick. It has a sharp, erosive base, locally incorporating large clasts of the underlying material. The diverse fauna includes fenestrate bryozoans, *sulcoretepora*, *Lichenalia*, brachiopods *Rhipidomella* (particularly abundant), *Mediospirifer*, *Pseudoatrypa*, *Megastrophia*, *Strophodonta demissa*, and proetid trilobites. At Lewis Center (Fig. 5, section 2), where the bed is best developed, the lower part also includes abundant *Favosites*, *Heliophyllum*, *Zaphrentis* and cystiphyllid corals. The middle grainstone (*Hadrophyllum* bed) is 4 to 8 cm thick, and also has an erosive base. The fauna of this bed is generally less diverse than that of the Lewis Center bed, and includes auloporid corals, *Strophodonta demissa*, *Mediospirifer?*, and *Schizophoria?* The small rugose coral *Hadrophyllum* is locally very abundant, primarily occurring in a coarse lag at the base of the bed. At Lewis Center, the two lower grainstones are separated by a thin interval of fine-grained, shaly limestone; however there is some evidence that the *Hadrophyllum* bed may locally erode down to the Lewis Center bed. The latter bed has not been located at Slate Run (Fig. 5, section 3) for example, but the *Hadrophyllum* bed there contains a diverse brachiopod fauna similar to the lower bed at Lewis Center. The 'upper grainstone' (4th bone bed, Wells 1944) is thinner (1 to 5 cm) and often not as well developed as the lower grainstones. It contains a diverse fauna similar that of to the Lewis Center bed, although *Rhipidomella* is not as common. The underlying intervening limestone beds are fine-grained wackestones with locally abundant *Pholidostrophia* and *Cupularostrum* brachiopods.

The grainstones are overlain by up to 3 m of sparsely fossiliferous (rare bivalves and brachiopods), thinly bedded, grey to brown, micritic limestone and chert (Stauffer's M Zone). In places, these upper beds are progressively removed by the unconformity beneath the upper or lower Olentangy shales such that the grainstone beds lie close below the upper contact of Zone M.

Bloomville. A quarry at Bloomville (Fig. 5), Ohio, provides an intermediate exposure between the central and north-central areas. Although the upper part of the section, including the crinoidal grainstones, has been removed by quarrying operations, they were at one time present and described by Stauffer (1909), and it is possible to resolve current exposures with his descriptions. The lower part of the Delaware consists of about 5 m of medium grey, mud- to wackestone which overlies massive, beige, hummocky cross-laminated, pack- to grainstones, the Marblehead Member of the Columbus Formation. The basal contact is marked by large, firm-ground burrows. The lower Delaware beds contain corals and brachiopods comparable to those in the Venice Member to the north and the Stratford member to the south. Above the Stratford member is a 3.7 m interval of laminated calcisiltite with abundant chalky, burrow-replacement chert. The lower part of this interval is a 30 cm thick zone with interbedded seams of black fissile shale, probably equivalent to the Dublin Member. These are the highest beds exposed in the quarry. However, Stauffer (1909) recorded about 11 feet 4 inches (3.5 m) of section above the cherty beds. A large part of this cherty zone is probably comparable to Zones J and K to the south.

Significantly, Stauffer reports a 2 foot (0.6 m) thick coral-rich zone ('Unit 10') about 0.6 m below the top of the section. The reported fauna includes abundant *Cystiphyllum vesiculosum* as well as *Heterophrentis*, the blastoid *Elaeocrinus venustus*, brachiopods including *Rhipidomella*, '*Atrypa*' and *Cupularostrum*, and the ubiquitous Hamilton Group trilobite *Eldredgeops rana*. The immediately underlying 1.5 m interval ('Unit 9') is described as splitting unevenly (bioturbated?) and carrying a rich fauna, including brachiopods *Strophodonta demissa*, *Protoleptostrophia*, *Leptaena*, *Pholidostrophia*, *Brevispirifer*, *Longispina*, *Cupularostrum* and '*Atrypa*', and bivalves *Aviculopecten* and *Glyptodesma*, platycerid snails, and *Pseudaviculopecten*. Although Stauffer did not note grainstone horizons within the interval, the fauna (absent *Hadrophyllum*) suggests that one or both of the lower two grainstones is present within this interval. The coral bed is probably referable to the uppermost grainstone farther south as well as grainstones/bone beds near the top of the Delaware at Sandusky (see below).

Sandusky. In north central Ohio, upper Columbus (Venice Member) and Delaware strata are well exposed in two quarries southwest of Sandusky (Fig. 5). In the first of these, at Venice, recent quarrying has exposed the entire Columbus Formation. However, only the lowermost 1.5 m of strata assigned to 'Delaware' is present, the rest removed by glacial erosion or quarrying. The Venice Member comprises crinoidal mud- to wackestones with *Brevispirifer* and *Tentaculites* in the lower part which pass upward into wacke- to packstones with *Acrospirifer duodenarius*, *Paraspirifer acuminatus*, *Eridophyllum* and cystiphyllid corals.

The Venice-Delaware contact at Venice is a pitted, hardground surface with pyrite-filled burrows and abundant crinoid holdfasts. This horizon is sharply overlain by 9 cm of dark, fissile calcareous shale (possible K-bentonite of Oliver 1967), the lowest bed of a 60 cm interval of wackestones interbedded with calcareous shales. Above this are ~80 cm of thick-bedded non-cherty, crinoidal pack- to grainstone with *Hadrophyllum* corals and *Leptaena*. A section by Stauffer (1909) indicates that an additional ~76 cm of dark, cherty limestone was once present here. Significantly, Stauffer reported the brachiopods *Cherryvalleyrostrum limitare* and *Orbiculoidea* from this limited section. These probably came from the upper cherty interval, which we suggest is correlative with the Dublin Member of central Ohio outcrops.

At Parkertown, the full thickness of the Venice and Delaware are exposed (Fig. 5). Here, the Delaware (not including the Venice/Stratford members) is 14.7 m thick (Sparling 1988). The basal 2.1 m is similar to the Delaware section at the Venice quarry. This interval is overlain, in turn, by 7.7 m of argillaceous limestones with chert in the lower half and abundant carbonaceous layers in the upper portion (the Dublin Member equivalents). At the top of the section is a 4.9 m succession of fossiliferous limestone with lower and upper fossiliferous limestone intervals bracketing finer-grained limestones. A sharply based, 20 cm crinoidal grainstone/bone bed at about 9.8 m above the base of the Delaware is considered to be equivalent to either the Lewis Center bed or the *Hadrophyllum* bed (basal L Zone of Stauffer 1909, 3rd bone bed of Wells 1944) in the Delaware area. The bone bed is immediately overlain by a group of fossiliferous (brachiopods and cystiphyllid corals) wackestone, packstone and grainstone. Overall, this 1.2 m interval of fossiliferous limestones probably correlates with the upper Delaware grainstones to the south and, in turn, to the Hurley and Cherry Valley Members, and the Halihan Hill Bed in New York State. About 2 m of dark, pyritic, argillaceous dolostone overlying the fossiliferous interval may be equivalent to Stauffer's Zone M.

The upper 3 m presently exposed in the quarry consist of fossiliferous packstone and shaly limestone that we correlate with the 'Blue beds' (units 1–6) of the lower Silica Shale of the Toledo, Ohio area. Bartholomew & Brett (2007) correlate this interval with the Stafford/Mottville Members of the Skaneateles Formation in New York State, which is consistent with the conodont biostratigraphy (see below). A 1.8 m interval about 13 m above the base of the Delaware section at Parkertown, considered by Sparling (1996) as the uppermost beds of the Delaware, carries abundant *Tropidoleptus carinatus* and *Devonochonetes* cf. *coronatus* and appears to be equivalent to the *Tropidoleptus–D. coronatus* bed (bed 1) of the basal Silica Shale (Conkin 1984) and the Swanville Member of the North Vernon Limestone in Indiana. The *Tropidoleptus–D. coronatus* interval is overlain by packstones containing a diverse Hamilton type fauna of brachiopods, bryozoans and coral biostromes capped by a dark grey dolomitic bed with a corroded (hardground?) upper surface.

The hardground is overlain by a shaly limestone interval with *Zoophycos* bioturbation, abundant small chonetids and *Mucrospirifer*. This highest interval, identified by Sparling (1995) as the lower Plum Brook Shale and by Conkin (1984) as the 'Silica Formation', has yielded abundant conodonts (Sparling 1995), including forms which may indicate the *timorensis* Zone. Cystiphyllid corals are locally abundant in an uppermost 70–80 cm fossiliferous, silty wackestone in the quarry, which may correspond with Silica unit 6 or the upper limestone of the Mottville Member (Bartholomew & Brett 2007).

In summary, the upper Columbus and Delaware Formations in central Ohio comprise parts of two major depositional sequences recognized in the Marcellus subgroup in New York (Fig. 5). The first sequence (Sequence I) includes the basal beds (lower Zone I and Venice Member at least in part), the Dublin Member and Zones J and K of Stauffer. The succeeding grainstones clearly indicate an abrupt shallowing and start of Sequence II. The sharp erosive base of each grainstone is inferred to be at least a minor (4th- or higher order) sequence boundary. We suggest that these grainstone beds are coeval to the Chestnut Street, Cherry Valley and Halihan Hill Bed of New York. An uppermost coral-rich zone seen at Parkertown may be the equivalent of the Stafford-Mottville Members, lower Silica (Blue Limestone), lower Plum Brook, and Rockport Quarry Limestone; beds which mark the start of Sequence III at the base of the post-Marcellus Hamilton Group.

Sequence stratigraphy

In a sequence stratigraphic interpretation, the late-Eifelian to earliest Givetian (*australis* to *hemiansatus* conodont zones) Marcellus subgroup (Hamilton Group) of New York comprises two large scale (3rd-order) depositional sequences: (I) the upper Onondaga and Union Springs Formations, and (II) the Oatka Creek/Mount Marion Formations (Ver Straeten & Brett 1995; Fig. 2). Each sequence is comprised of a sharply based, widespread, transgressive limestone (TST), overlain by a highstand succession (HST) predominantly of black and dark grey shale. The TST units of the upper Onondaga–Union Springs sequence (Sequence I) consist of a retrogradational succession of crinoidal grainstones of the upper Moorehouse Member (*Paraspirifer* Zone), bioturbated wackestones of the Seneca Member, Onondaga Formation, and dark grey styliolinid limestones and shales of the transition zone of the lower Union Springs Formation (Bakoven Member). The fossiliferous wacke- to grainstone of the Hurley Member (Chestnut Street submember in central-western New York) and the styliolinid limestone of the Cherry Valley Member and lower shales of the East Berne Member form the TST of the Oatka Creek/Mount Marion cycle (Sequence II). The sharp, typically corroded contacts with bone-rich lags between the limestones and overlying black shales (HST) are interpreted to represent maximum starvation surfaces (MSS). These starvation surfaces are the flooding surfaces at the base of Johnson *et al.*'s (1985) T–R cycles; thus the MSS of Sequence I (base of Union Springs Formation) corresponds to the base T–R cycle Id and the MSS of Sequence IIb (base of East Berne Member) corresponds to the base of T–R cycle Ie.

The 3rd-order sequences can be further subdivided into a series of smaller scale (4th- or 5th-order) subsequences: Ia) Upper Onondaga Fm.–Bakoven–lower Stony Hollow Member; Ib) middle–upper Stony Hollow Member; IIa) Hurley Member/IIb) Cherry Valley–East Berne Members; IIc) Halihan Hill Bed–upper Solsville; IId) uppermost Solsville–Cardiff Members; and finally IIIa (in part) Stafford–Mottville limestone Members recognized by Ver Straeten *et al.* (1994) and Brett and Baird (1996). Each subsequence, except for Ia, consists of a thin (0.2–2 m), widespread transgressive limestone (Chestnut Street submember, Cherry Valley Member, Halihan Hill Bed and Stafford/Mottville Member), overlain by a thicker succession of highstand black/dark grey shale (Bakoven, Lincoln Park, East Berne and Chittenango Members).

The major (3rd order) depositional sequences recognized in the middle Eifelian to early Givetian in New York can be identified confidently in the age-equivalent strata of southwestern Ontario subsurface and to the margin of the Michigan Basin. Even the smaller-scale subsequences can be traced in drill core through the southern Ontario subsurface and all have apparent equivalents in the London–Sarnia area. Moreover, this area provides a bridging link to stratigraphically equivalent middle Eifelian to earliest Givetian strata in the Michigan basin. There, two 3rd-order sequences, and even some of the higher-order sequences can be recognized. For example, the coral bed and overlying rhynchonellid-bearing strata in the Bell Shale of northern Michigan appears to equate both with equivalent beds in Ontario and with the Halihan Hill Bed and overlying Chittenango Member of New York.

Biostratigraphically-equivalent late Eifelian strata of Ohio also exhibit two 3rd-order sequences, comparable to the upper Onondaga–Union Springs Formations and the Oatka Creek Formation, and the smaller scale sequences also appear to be present, though are not as clearly recognized owing in part to erosional truncation of parts of the section. The Venice Member of the Columbus Formation in northern Ohio and the lower Delaware Formation (Stratford member) in central Ohio are approximately age-equivalent (*australis* to lower *kockelianus* zone) to the transition beds of the lower Bakoven Member of the Union Springs Formation (New York sequence I). The Venice Member (or lower Delaware) exhibits a back-stepping pattern typical of a transgressive systems tract. The sharp, typically corrosional contact with the overlying Dublin Member of the Delaware is a maximum starvation surface, corresponding to the sharp contact of the transition beds with the shales of the lower Bakoven Member in New York. The generally shallowing-upward pattern which follows is typical of a late highstand (regressive) interval. Thus, Zones J and K of the Delaware are the counterpart of the Stony Hollow regression in New York. The sharp eroded contact of the Lewis Center bed in Ohio is interpreted as a sequence boundary probably equivalent to the sharp base of the Chestnut Street Beds in the Appalachian Basin. The *Hadrophyllum* bed and upper grainstone probably correlate, respectively, to the Cherry Valley Member and Halihan Hill Bed.

Faunal comparisons

Given the sequence stratigraphic framework outlined above, it is possible to consider, in detail, coeval faunas of the Appalachian Basin and the Midwestern carbonate ramp and platform. Figure 6 compares faunal composition in terms of holdover and carryover species (*sensu* Brett & Baird 1995) for the Appalachian Basin and in the central Ohio carbonate platform. Carryover the percentage of species of a given fauna that persist into younger fauna(s). Holdover is the percentage of species of a fauna that is derived from the previous fauna(s).

Several patterns can be recognized from these data. The Union Springs Formation records a major faunal turnover from the diverse, endemic Onondaga Fauna to the widespread incursion of new taxa. The majority (84%) of Onondaga species appear to become extinct, at least locally, at the boundary with the Union Springs. Most of the upper Union Springs Formation (Stony Hollow Member) is typified by a low-diversity fauna marking an incursion of species from the tropical Old World Realm (OWR) of northwestern areas of Laurentia (Koch 1979), including the Stony Hollow brachiopods *Variatrypa arctica*, *Pentamerella* cf. *P. wintereri*, *Emanuella* and *Warrenella*. This fauna shows many similarities with the fauna in the coeval upper *australis* to *kockelianus* Zone Rogers City Formation in the Michigan Basin. Approximately 58% of brachiopod genera are shared between these assemblages. A similar pattern is detected in coeval rocks in the subsurface of southwestern Ontario, where all of the condensed limestones and dark shales seen in the Appalachian Basin can now be recognized, and where corresponding changes occur in the faunas. However, detailed faunal comparison cannot be made at this time because of the limited sampling afforded by core samples.

The base of the Delaware (in sequence I) also coincides with a major loss in species diversity. Some 78.5% of taxa found within the underlying Columbus Formation do not carry over into the lower Delaware. This turnover event, at approximately the *costatus*–*australis* zonal boundary, and the base of the Venice Member (and correlative Stratford member to the south), coincides with the major local loss of diversity in the coeval upper Onondaga–Union Springs transition. However, the percent of species carrying over from the upper Columbus (Marblehead Member and Zone H) into the overlying Venice Member and equivalent Stratford member in Ohio is higher. Taxa such as *Acrospirifer duodenarius*, and *Alaliforma varicosa*, which do not appear above the Onondaga Formation anywhere in New York, evidently persisted into the *australis* Zone in the mid-continent, although apparently not higher. In addition, these taxa are mixed with newly appearing forms including *Warrenella maia*, *Schizophoria* sp. and '*Mucrospirifer*' *consobrinus* that typify the overlying upper Delaware.

The dysoxic black Dublin shale (Zone I) facies of the Delaware Formation (the I highstand unit of Sequence I) of the *australis* to *kockelianus* Zone in central Ohio carries a low-diversity, rhynchonellid brachiopod-dominated fauna remarkably similar to the fauna in the correlative Bakoven Member of the Union Springs sequence in New York. The more oxic argillaceous limestone facies of the overlying Zones J and K of the Delaware, however, contains long-ranging Eifelian–Givetian taxa (e.g. *Strophodonta demissa*, *Rhipidomella vanuxemi*), and unique Delaware taxa (e.g. *Warrenella maia*, *Schizophoria* sp., '*Mucrospirifer*' *consobrinus*). Although this Delaware Fauna shows some similarities to the distinct Stony Hollow Fauna of New York (e.g. shared possession of *Leptaena*, *Warrenella* and *Schizophoria*, which do not generally occur in overlying Hamilton strata),

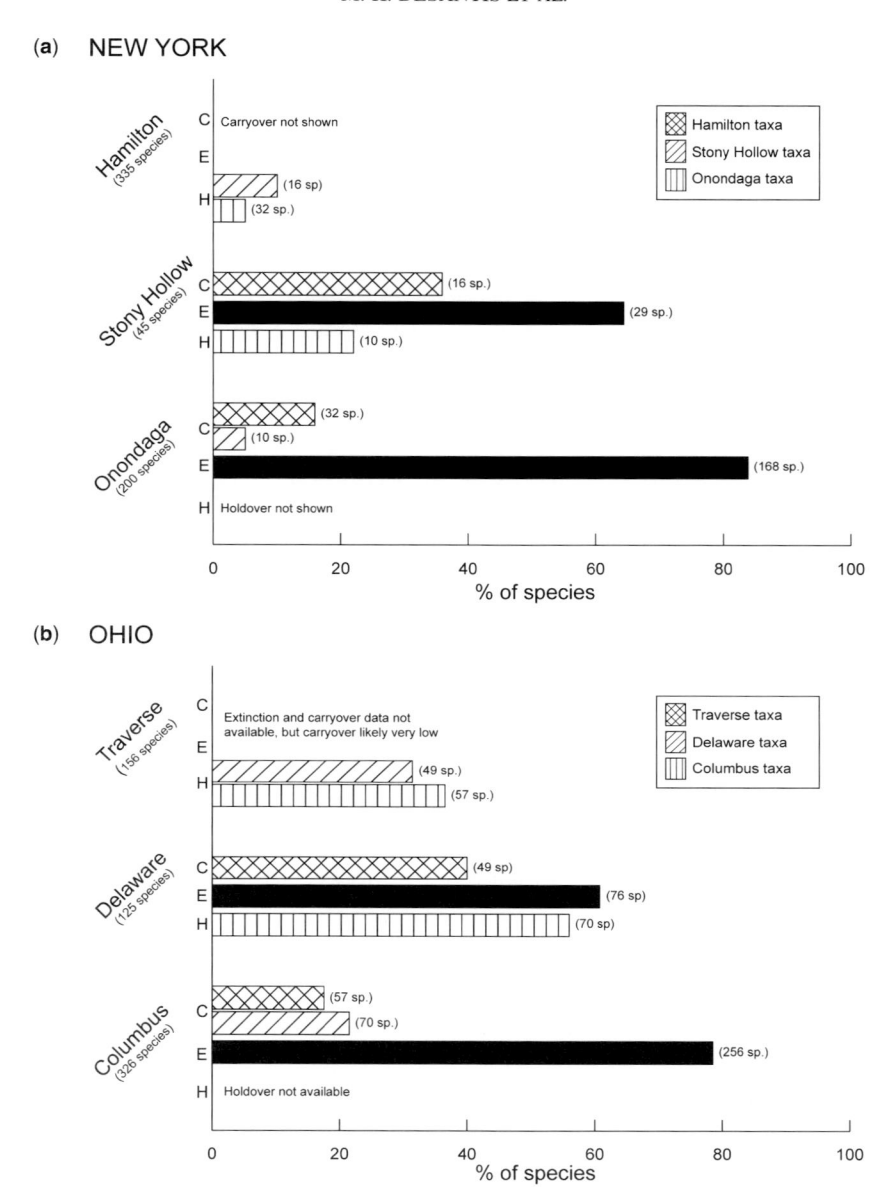

Fig. 6. Comparison of carryover and holdover indices between faunas from (**a**) New York (data from Brett & Baird 1995) and (**b**) Ohio (data from Stauffer 1909). C, carryover species; H, holdover species; E, local extinction of species within or at the top of stratigraphic interval. See explanation in text. Faunas: Hamilton includes taxa starting with the Halihan Hill Bed of the Oatka Creek–Mount Marion Formation and extends to the top of the Hamilton Group. Stony Hollow includes benthic taxa from the Stony Hollow and Hurley Members. Onondaga includes taxa from the Edgecliff through Seneca Members. Ohio data are from Stauffer (1909). Columbus includes taxa of Zones A–H. Delaware includes taxa of Zones I–M. Traverse comprises taxa from platform equivalents of the post-Marcellus Hamilton Group, including the lower Olentangy Formation (central Ohio), and the Plum Brook and Silica Formations of northern Ohio.

surprisingly many typical Stony Hollow taxa (e.g. *Variatrypa, Pentamerella* cf. *P. wintereri, Dechenella haldemani*) are not known from either the Delaware or equivalent Dundee in Ontario. These distributions are problematic given that these Stony Hollow taxa do occur in the more distant Michigan Basin. Moreover, a higher proportion of the older Columbus–Onondaga taxa, including many species (e.g. brachiopods *Rhipidomella vanuxemi, Leptaena 'rhomboidalis', Megastrophia concava, Pholidostrophia, Protoleptostrophia perplana, Protodouvillina inequistriata, Stophodonta demissa*, and corals *Cystiphylloides, Eridophyllum, Heterophrentis*) never observed in the Union Springs of New York, are present in the Delaware of Ohio. These observations suggest that the environmental changes were less severe in the mid-continent than in the adjacent basins and/or that a higher proportion of shallow-water taxa persisted. Many of these Delaware species, particularly the strophomenid brachiopods, carry over into the higher Hamilton Group. The Ohio carbonate ramp environments may have provided a refuge for shallow-water taxa during a time of major extinction in the foreland basin.

A second major faunal turnover in the Appalachian Basin involves a nearly total loss of the Stony Hollow Fauna with the onset of black shales of the East Berne Member. The diverse and relatively stable Hamilton Fauna, which contains a mixture of endemic Appalachian Basin taxa (mainly holdovers from the Onondaga) and newly immigrated species from the OWR Rhenish-Bohemian area (Koch 1979), then makes an abrupt appearance in the Halihan Hill Bed of the Oatka Creek Formation (Brett & Baird 1995, 1996) (Fig. 2).

As discussed previously, the Lewis Center and *Hadrophyllum* beds in the Delaware Formation in Ohio, tentatively placed in the *kockelianus* Zone, are thought to correlate to the Chestnut Street submember and Cherry Valley Member of New York. The overlying upper grainstone (4th bone bed) may represent the Halihan Hill Bed. These upper Delaware beds show a diverse fauna rich in the small rugosan *Hadrophyllum* and other corals, orthid and strophomenid brachiopods, and bryozoans. Nearly all of the species are found in the higher Hamilton Group throughout eastern Laurentia. Moreover, very few lower Delaware or Union Springs taxa are present. This change reflects the incursion of the Hamilton Fauna into the Appalachian Basin. However, in the mid-west, the magnitude of faunal turnover is significantly less than seen in the Appalachian Basin because the lower Delaware fauna is not as discrete as the Stony Hollow Fauna of New York, but rather shows a higher proportion of long-ranging taxa.

Correlation with the Kačák Events

Conodont and ammonoid evidence suggests that the East Berne Member corresponds with the main black shale phase of the Kačák Event observed in Europe and North Africa. The abrupt change from limestone to black shale at the Cherry Valley–East Berne contact (and the loss of the Stony Hollow fauna) would equate to the Lower Kačák Event (L'Ei 1 Event of Walliser 1996; Schöne 1997). The base of the Halihan Hill Bed should correspond to the extinction level (L'Ei 2 Event) near the top of the Kačák Event interval. The Hamilton Fauna would then be a post-event fauna, possibly entering within the topmost event interval.

Interestingly, the major Appalachian Basin benthic faunal overturn (Onondaga–Stony Hollow) entirely precedes the main Kačák Event interval. The local loss of the Onondaga fauna at the base of Union Springs appears primarily to be a local phenomenon related in part to tectonic deepening of the Appalachian Foreland Basin. On the other hand, the Stony Hollow faunal incursion in the *kockelianus* conodont Zone has a widespread signature across eastern Laurentia (Koch & Day 1996), which is associated with the sea-level rise corresponding to TR cycle Ie of Johnson *et al.* (1985). Koch & Boucot (1982) and Boucot (1990) suggested that these incursions were due to changes in water circulation patterns and a lowering of the climate gradient, resulting in migration of warmer-water faunas into the Appalachian Basin. Given that all of these factors are potentially global in scale, it is not unreasonable to expect that they might also have a signature outside of Laurentia. Indeed, this lower 'Stony Hollow Event' roughly correlates to the Kačák–*otomari* level defined by Truyóls-Massoni *et al.* (1990) and the *otomari* 'a' horizon of Walliser (1990), and may correlate to other stages leading up to the Kačák Event interval.

In summary, we return to questions posed in the introduction. With respect to the issue of synchroneity, it does appear that synchronous changes occur at two levels, the bases of two major sequences, throughout the Appalachian Basin and westward onto the carbonate platform. The first is near the *costatus–australis* Zone boundary and the second is in the upper part of the *kockelianus* Zone. However, the change between the lower and upper Delaware faunas is not as discrete as between the coeval Stony Hollow and Hamilton faunas, both because fewer of the distinct OWR taxa appear in the Delaware and because a higher proportion of endemic taxa persisted there than in the Appalachian Basin. As a result, the distinctive ecological–evolutionary sub-units observed in the Appalachian Basin are not as clearly observed in the adjacent cratonward carbonate platform. That said, the widespread and relatively abrupt nature of these faunal changes suggests that they reflect regional if not global events.

In western-central New York, new faunas appear abruptly within widespread condensed beds at the TSTs of some 4th- or higher order sequences, although correlations with thicker/shallower sections in eastern New York indicate that faunal overturns actually occurred during periods of widespread dysoxia associated with preceding highstands. For example, the Stony Hollow Fauna comes in to the Appalachian Basin during the highstand of New York Sequence Ib, but only becomes widespread in the Chestnut Street Beds (TST of Sequence IIa). It is notable that thin intervals of dysoxic sediment also persist into the mid-continent and are associated with similar faunal changes. A combination of dysoxia and climatic change are most probable causes of these widespread changes.

We would like to thank Dr. R. T. Becker for inviting us to submit a paper for this Volume. Reviews by Dr Becker and Dr. W. Kirchgasser greatly improved the final version of this paper. We would also like to thank Dr. D. J. Over (SUNY Geneseo) for processing and identifying conodonts from several central Ohio samples; and the staff (particularly R. Rhea) of the Ohio Geological Survey's Horace R. Collins Laboratory and of the Ontario Oil, Gas and Salt Resources Library for access to drill cores.

References

BARTHOLOMEW, A. J. & BRETT, C. E. 2007. Correlation of Middle Devonian Hamilton Group-equivalent strata in east-central North America: implications for eustasy, tectonics and faunal provinciality. *In*: BECKER, R. T. & KIRCHGASSER, W. T. (eds) *Devonian Events and Correlations*. Geological Society of London, Special Publications, **278**, 105–131.

BECKER, R. T. & HOUSE, M. R. 1994. International goniatite zonation, Emsian to Givetian, with new records from Morocco. *Courier Forshungsinstitut Senckenberg*, **169**, 79–135.

BECKER, R. T. & HOUSE, M. R. 2000. Devonian ammonoid zones and their correlation with established series and stage boundaries. *Courier Forshungsinstitut Senckenberg*, **220**, 113–151.

BELKA, Z., KAUFMANN, B. & BULTYNCK, P. 1997. Conodont-based quantitative biostratigraphy for the Eifelian of the eastern Anti-Atlas, Morocco. *Geological Society of America Bulletin*, **109**, 643–651.

BEST, E. W. 1953. Pre-Hamilton Devonian stratigraphy southwestern Ontario, Canada. PhD dissertation, University of Wisconsin.

BOUCOT, A. J. 1990. Silurian and pre-Upper Devonian bio-events. *In*: KAUFFMAN, E. G. & WALLISER, W. T. (eds) *Extinction Events in Earth History*. Lecture Notes in Earth Sciences, **30**, 125–132.

BRETT, C. E. & BAIRD, G. C. 1995. Coordinated stasis and evolutionary ecology of the Silurian to Middle Devonian faunas in the Appalachian Basin. *In*: ERWIN, D. H. & ANSTEY, R. L. (eds) *New Approaches to Speciation in the Fossil Record*. Columbia University Press, New York, 285–315.

BRETT, C. E. & BAIRD, G. C. 1996. Middle Devonian sedimentary cycles and sequences in the northern Appalachian Basin. *In*: WITZKE, B. J., LUDVIGSON, G. A. & DAY, J. (eds) *Paleozoic Sequence Stratigraphy: Views from the North American Craton*. Geological Society of America Special Paper, **306**, 213–241.

BRETT, C. E. & VER STRAETEN, C. A. 1994. Stratigraphy and facies relationships of the Eifelian Onondaga limestone (Middle Devonian) in western and west central New York State. *In*: BRETT, C. E. & SCATTERDAY, J. (eds) *Fieldtrip Guidebook, New York State Geological Association 68th Annual Meeting*, 221–270.

BUDIL, P. 1995. Demonstration of the Kačák event (Middle Devonian, uppermost Eifelian) at some Barrandian localities. *Věstník Českeho geologického ústavu*, **70**, 1–24.

BULTYNCK, P. 1987. Pelagic and neritic conodont successions from the Givetian of pre-Sahara Morocco and the Ardennes. *Bulletin de l'Institut Royal des Sciences Naturelles de Belgique, Sciences de la Terre*, **57**, 149–181.

BULTYNCK, P. & WALLISER, O. H. 2000. Devonian boundaries in the Moroccan Anti-Atlas. *Courier Forshungsinstitut Senckenberg*, **225**, 211–226.

CONKIN, J. E. & CONKIN, B. M. 1975. Middle Devonian bone beds and the Columbus–Delaware (Onondagan–Hamiltonian) contact in central Ohio. *Bulletins of American Paleontology*, **67**, 99–122.

CONKIN, J. E. & CONKIN, B. M. 1984. *Paleozoic metabentonites of North America: Part I – Devonian metabentonites in the eastern United States and southern Ontario: Their identities, stratigraphic positions, and correlation*. University of Louisville Studies in Paleontology and Stratigraphy, **16**.

COOPER, G. A. 1930. Stratigraphy of the Hamilton Group of New York. *American Journal of Science*, **19**, 116–134, 214–236.

COOPER, G. A. 1933. Stratigraphy of the Hamilton Group, eastern New York, Part I. *American Journal of Science*, **26**, 537–551.

DENNISON, J. M. 1961. *Stratigraphy of the Onesquethaw Stage of Devonian in West Virginia and bordering states*. West Virginia Geological Survey Bulletin, **22**.

DENNISON, J. M. & TEXTORIS, D. A. 1970. Devonian Tioga tuff in northeastern United States. *Bulletin Volcanologique*, **34**, 289–294.

DESANTIS, M. K. 2005. Regional correlation of the Middle Devonian Tioga K-bentonites using apatite trace element fingerprinting. Geological Society of America, Abstracts with Programs, **37**, 73.

EHLERS, G. M. & KESLING, R. V. 1970. *Devonian strata of Alpena and Presque Isle Counties, Michigan. Guidebook for the North-Central Section Meeting*. Michigan, Michigan Basin Geological Society.

ETTENSOHN, F. R. 1985. The Catskill Delta Complex and the Acadian Orogeny: A model. *In*: WOODROW, D. L. & SEVON, W. D. (eds) *The Catskill Delta*. Geological Society of America, Special Paper, **201**, 39–50.

FETTKE, C. R. 1952. Tioga bentonite in Pennsylvania and adjacent states. *American Association of Petroleum Geologists, Bulletin*, **36**, 2038–2040.

GOLDRING, W. 1935. *Geology of the Berne quadrangle*. New York State Museum Bulletin, **301**.

GOLDRING, W. 1943. *Geology of the Coxsackie quadrangle, New York*. New York State Museum Bulletin, **332**.

GRIFFING, D. H. & VER STRAETEN, C. A. 1991. Stratigraphy and depositional environments of the Marcellus Formation (Middle Devonian) in eastern New York. *In*: EBERT, J. R. (ed.) *Fieldtrip Guidebook, New York State Geological Association 63rd Annual Meeting*, 205–249.

HOUSE, M. R. 1965. A study in the Tornoceratidae: the succession of *Tornoceras* and related genera in the North American Devonian. *Philosophical Transactions of the Royal Society, London*, **B250**, 79–130.

HOUSE, M. R. 1978. *Devonian ammonoids from the Appalachians and their bearing on international zonation and correlation*. Special Papers in Palaeontology, **21**.

HOUSE, M. R. 1981. Lower and Middle Devonian goniatite biostratigraphy. *In*: OLIVER, W. A., JR. & KLAPPER, G. (eds) *Devonian Biostratigraphy of New York: Part I*. International Union of Geological Sciences, Subcommission on Devonian Stratigraphy, 33–36.

HOUSE, M. R. 1985. Correlation of mid-Palaeozoic ammonoid evolutionary events with global sedimentary perturbations. *Nature*, **313**, 17–22.

HOUSE, M. R. 1996a. The Middle Devonian Kačák Event. *Proceedings of the Ussher Society*, **9**, 79–84.

HOUSE, M. R. 1996b. Juvenile goniatite survival strategies following Devonian extinction events. *In*: HART, M. B. (ed.) *Biotic Recovery from Mass Extinction Events*. Geological Society Special Publication, **102**, 163–185.

JOHNSON, J. G., KLAPPER, G. & SANDBERG, C. A. 1985. Devonian Eustatic fluctuations in Euramerica. *Geological Society of America Bulletin*, **96**, 567–587.

KLAPPER, G. 1971. Sequence within the conodont genus *Polygnathus* in the New York lower Middle Devonian. *Geologica et Paleontologica*, **5**, 59–79.

KLAPPER, G. 1981. Review of New York Devonian conodont biostratigraphy. *In*: OLIVER, W. A., JR. & KLAPPER, G. (eds) *Devonian Biostratigraphy of New York: Part I*, International Union of Geological Sciences, Subcommission on Devonian Stratigraphy, 57–66.

KLAPPER, G. & ZIEGLER, W. 1979. *Devonian conodont biostratigraphy*. Special Papers in Palaeontology, **23**, 199–224.

KOCH, W. F., II. 1979. *Brachiopod paleoecology, paleobiogeography, and biostratigraphy in the Upper Middle Devonian of eastern North America: An ecofacies model for the Appalachian, Michigan, and Illinois basins*. PhD dissertation, Oregon State University.

KOCH, W. F. & BOUCOT, A. J. 1982. Temperature fluctuations in Devonian Eastern Americas Realm. *Journal of Paleontology*, **56**, 240–243.

KOCH, W. F. & DAY, J. 1996. Late Eifelian–early Givetian (Middle Devonian) brachiopod paleobiogeography of eastern and central North America. *In*: COPPER, P. & JISUO, J. (eds) *Proceedings of the Third International Brachiopod Conference*, 135–143.

LANDING, E. & BRETT, C. E. 1987. Trace fossils and regional significance of a Middle Devonian (Givetian) disconformity in southwestern Ontario. *Journal of Paleontology*, **61**, 205–230.

LINDEMANN, R. H. 2002. Dacryoconarid bioevents of the Onondaga Formation and the Marcellus Subgroup, Cherry Valley, New York. *In*: MCLELLAND, J. & KARABINOS, P. (eds) *Fieldtrip Guidebook, New York State Geological Association 74th Annual Meeting*, p. B7-1–B7-17.

NEWBERRY, J. S. 1873. *The geological structure of Ohio*. Geological Survey of Ohio, **1** (pt. 1, Geology), 89–139.

OLIVER, W. A., JR., 1967. Contact of Delaware and Columbus Limestones in northern Ohio. *Geological Survey Research 1967*. United States Geological Survey Professional Paper, **575-A**, 76.

OLIVER, W. A., JR., DEWITT, W., JR., DENISON, J. M., HOSKINS, D. M. & HUDDLE, J. H. 1968. Devonian of the Appalachian Basin, United States. *In*: OSWALD, D. H. (ed.) *International Symposium on the Devonian System, Calgary*. Alberta Society of Petroleum Geologists, **1**, 1001–1040.

RAMSEY, N. J. 1969. *Upper Emsian–Upper Givetian conodonts from the Columbus and Delaware Limestones and Lower Olentangy Shale of central Ohio*. MS thesis, Ohio State University.

RICKARD, L. V. 1984. Correlation of the subsurface Lower and Middle Devonian of the Lake Erie region. *Geological Society of America Bulletin*, **95**, 814–828.

RICKARD, L. V. 1985. Correlation of the subsurface Lower and Middle Devonian of the Lake Erie region: Alternative interpretation and reply—Reply. *Geological Society of America Bulletin*, **96**, 1218–1220.

RICKARD, L. V. 1989. *Stratigraphy of the subsurface Lower and Middle Devonian of New York, Pennsylvania, Ohio and Ontario*. New York State Museum Map and Chart **39**.

SAGEMAN, B. B., MURPHY, A. E., WERNE, J. P., VER STRAETEN, C. A., HOLLANDER, D. J. & LYONS, T. W. 2003. A tale of shales: the relative roles of production, decomposition, and dilution in the accumulation of organic-rich strata, Middle–Upper Devonian, Appalachian Basin. *Chemical Geology*, **195**, 229–273.

SCHÖNE, B. R. 1997. *Der otomari-Event und seine Auswirkungen auf die Fazies des Rhenherzynischen Schelfs (Devon Rheinisches Schieferegeberge)*. Göttinger Arbeiten zur Geologie und Paläontologie, **70**.

SPARLING, D. R. 1984. Paleoecologic and paleogeographic factors in the distribution of lower Middle Devonian conodonts from north-central Ohio. *In*: CLARK, D. L. (ed.) *Conodont Biofacies and Provincialism*. Geological Society of America, Special Paper, **196**, 113–125.

SPARLING, D. R. 1985. Correlation of the subsurface Lower and Middle Devonian of the Lake Erie region: Alternative interpretation and reply—Alternative interpretation. *Geological Society of America Bulletin*, **96**, 1213–1218.

SPARLING, D. R. 1988. Middle Devonian stratigraphy and conodont biostratigraphy, north-central Ohio. *Ohio Journal of Science*, **88**, 2–18.

SPARLING, D. R. 1995. Conodonts from the Middle Devonian Plum Brook Shale of north-central Ohio. *Journal of Paleontology*, **69**, 1123–1138.

SPARLING, D. R. 1999. Conodonts from the Prout Dolomite of north-central Ohio and Givetian (upper Middle Devonian) correlation problems. *Journal of Paleontology*, **73**, 892–907.

STAUFFER, C. R. 1909. *The Middle Devonian of Ohio*. Geological Survey of Ohio, Fourth Series, Bulletin **10**.

STAUFFER, C. R. 1915. *Devonian of SW Ontario*. Geological Survey of Canada Memoir **34**.

SWARTZ, C. K. 1907. *The relation of the Columbus and Sandusky formations of Ohio*. Johns Hopkins University Circular, n.s.7, **199**, 56–65.

TRUYÓLS-MASSONI, M., MONTESINOS, R., GARCIA-ALCALDE, J. L. & LEYVA, F. 1990. The Kacák-*otomari* event and its characterization in the Paletine domain (Cantabrian Zone, NW Spain). *In*: KAUFFMAN, E. G. & WALLISER, O. H. (eds) *Extinction Events in Earth History*, Lecture Notes in Earth Sciences, **30**, 133–143.

UYENO, T. T., TELFORD, P. G. & SANFORD, B. V. 1982. Devonian conodonts and stratigraphy of southwestern Ontario. *Geological Survey of Canada Bulletin* **332**.

VER STRAETEN, C. A. 1994. Microstratigraphy and depositional environments of a Middle Devonian foreland basin: Berne and Otsego Members, Mount Marion Formation, eastern New York State. *In*: LANDING, E. (ed.) *Studies in Stratigraphy and Paleontology in Honor of Donald W. Fisher*. New York State Museum Bulletin **481**, 367–380.

VER STRAETEN, C. A. 2004. K-bentonites, volcanic ash presentation, and implications for lower and middle Devonian Volcanism in the Acadian Orogen, eastern North America. *Geological Society of America Bulletin*, **116**, 474–489.

VER STRAETEN, C. A. & BRETT, C. E. 1995. Lower and Middle Devonian foreland basin fill in the Catskill Front: Stratigraphic synthesis, sequence stratigraphy, and the Acadian Orogeny. *In*: GARVER, J. I. & SMITH, J. A. (eds) *Fieldtrip Guidebook, New York State Geological Association 67th Annual Meeting*, 313–356.

VER STRAETEN, C. A. & BRETT, C. E. 2006. Pragian to Eifelian strata (middle Lower to lower Middle Devonian), northern Appalachian basin–stratigraphic nomenclatural changes. *Northeastern Geology*, **28**, 80–95.

VER STRAETEN, C. A., GRIFFING, D. H. & BRETT, C. E. 1994. The lower part of the Middle Devonian Marcellus 'shale,' central to western New York State: Stratigraphy and depositional history. *In*: BRETT, C. E. & SCATTERDAY, J. (eds) *Fieldtrip Guidebook, New York State Geological Association 66th Annual Meeting*, 271–306.

WALLISER, O. H. 1985. Natural boundaries and Commission boundaries in the Devonian. *Courier Forshungsinstitut Senckenberg*, **75**, 401–408.

WALLISER, O. H. 1990. How to define 'Global bio-events'. *In*: KAUFFMAN, E. G. & WALLISER, O. H. (eds) *Extinction Events in Earth History*, Lecture Notes in Earth Sciences, **30**, 1–3.

WALLISER, O. H. 1996. Global events in the Devonian and Carboniferous. *In*: WALLISER, O. H. (ed.) *Global Events and Event Stratigraphy*, 225–250.

WALLISER, O. H., BULTYNCK, P., WEDDIGE, K., BECKER, R. T. & HOUSE, M. R. 1995. Definition of the Eifelian–Givetian Stage boundary. *Episodes*, **18**, 107–115.

WAY, J. H., SMITH, R. C., II & RODEN, M. K. 1986. Detailed correlations across 175 miles of the Valley and Ridge of Pennsylvania using 7 ash beds in the Tioga Zone. *In*: SEVON, W. D. (ed.) *Selected Geology of Bedford and Huntington County*, 51st Annual Field Conference of Pennsylvania Geologists, 55–72.

WELLS, J. W. 1944. Middle Devonian bone beds of Ohio. *Geological Society of America Bulletin*, **55**, 273–302.

WERNE, J. P., SAGEMAN, B. B., LYONS, T. W. & HOLLANDER, D. J. 2002. An integrated assessment of a 'type euxinic' deposit: evidence for multiple controls on black shale deposition in the Middle Devonian Oatka Creek Formation. *American Journal of Science*, **202**, 110–143.

WESTGATE, L. G. & FISCHER, R. P. 1933. Bone beds and crinoidal sands of the Delaware Limestone of central Ohio. *Geological Society of America Bulletin*, **44**, 1161–1172.

Correlation of Middle Devonian Hamilton Group-equivalent strata in east-central North America: implications for eustasy, tectonics and faunal provinciality

A. J. BARTHOLOMEW & C. E. BRETT

Department of Geology, University of Cincinnati, Cincinnati, Ohio 45221-00013, USA
(e-mail: alexbartholomew-geo@hotmail.com)

Abstract: An integrated approach, involving all available biostratigraphic data, event and sequence stratigraphy, has been utilized in correlation of the Middle Devonian (latest Eifelian–Givetian) Hamilton Group and equivalent strata in north-central North America. This approach permits high-resolution correlation of strata equivalent to the Oatka Creek (upper Marcellus), Skaneateles, Ludlowville and Moscow Formations from New York into sections bordering the Michigan Basin in Ontario, Canada, as well as southern Michigan, northern Ohio and Indiana, USA. Most member and submember-scale units, herein slightly redefined and interpreted as 3rd and 4th order sequences, respectively, and their bounding condensed beds can be correlated regionally. Moreover, many faunal patterns also persist across this region, which, together with sequence stratigraphy, provides a bridge for correlation into the Michigan Basin. The detailed stratigraphy presented herein permits a more-resolved understanding of far-field tectonics, eustasy and biotic responses during the Middle Devonian. Allocyclic processes, primarily eustasy, played a key role in generating persistent sedimentary cycles. Episodes of rapid mud sedimentation occurred over large areas of the cratonic interior, distal to Acadian source terrains. The major Algonquin–Findlay Arch, which presently separates the Michigan Basin from the Appalachian foreland basin, was not present during deposition of these strata. Conversely, a roughly north–south trending region, running approximately through present-day Cleveland, Ohio, was first a local subsiding area during late Eifelian–early Givetian time and then underwent topographic inversion to form a local arch at which upper Hamilton units were condensed and then bevelled during the later Givetian; we infer that this feature may represent a migrating forebulge. Finally, fossil biotas do not show strong partitioning into Appalachian and Michigan basin faunal subprovinces during the early Givetian, as there appears to have been no physical barrier to migration at least in the study area. However, Hamilton-equivalent strata in the most proximal portion of the Appalachian Basin do show a relatively minor admixture of typical Michigan Basin taxa with normal Hamilton forms.

The Middle Devonian Hamilton Group of western and central New York has been studied extensively for over a century (Vanuxem 1842; Grabau 1899; Cooper 1930, 1933). These rocks form a 'natural laboratory' for examining aspects of sedimentary cycles and events, foreland basin dynamics and evolutionary ecology of the diverse faunas (see papers in Brett 1986; Landing & Brett 1991; Brett & Baird 1995, 1996). A number of cyclic sedimentary and faunal patterns have proven to be widespread within the Appalachian Foreland Basin. However, it has remained unclear whether such patterns can be recognized more broadly. For example, are the sedimentary cycles related to local tectonics of the foreland basin, or do they record more widespread eustatic or climatic processes? Similarly, are episodes of faunal change unique features of the Appalachian Basin region, or are they synchronous with changes seen elsewhere? Of particular interest is whether faunal patterns can be correlated into the adjacent cratonic areas and into the Michigan Basin.

Answers to these questions require detailed correlations of regions to the west of the Appalachian Foreland Basin in the cratonward areas in Ohio and Ontario, Canada. Hamilton-equivalent strata in these regions also provide a critical 'bridge' between the Appalachian and Michigan Basins, as they are the parts of two different palaeobiogeographical subprovinces. The relationship of these areas to the bordering Appalachian and Michigan Basins remains poorly known due to the lack of detailed correlations with those of the classic basinal sections in New York and Pennsylvania.

Correlations out of the Appalachian Basin proposed in the past have been based on a single criterion, either biostratigraphic or gamma-ray evidence (Cooper & Warthin 1942; Rickard 1984; Sparling 1985). This study synthesizes all available evidence including, for the first time, sequence stratigraphy. Sequence stratigraphy not only provides another line of evidence for correlations, but also allows for precise correlations, in some cases at the bed level, across wide areas. Correlations at this scale provide a highly detailed framework in which to test various sedimentological and biogeographical hypotheses.

Geological setting: study area

The area encompassed by this study includes the northern and western portions of the Appalachian Basin that lie east of the Findlay Arch (New York, Ontario, central and north-central Ohio), the southern portion of the Michigan Basin that lies west of the Findlay Arch in northwestern Ohio, particularly outcrops near Sylvania, west of Toledo, and the eastern portion of the Illinois Basin in southeastern Indiana, north of Jeffersonville (Fig. 1). Exposures in west-central New York represent medial portions of the Appalachian Basin, whereas exposures in Ohio east of the Findlay Arch and Ontario represent the very distal portions of the basin. It should be noted that upper Hamilton Group exposures in Ontario (upper Oatka Creek Formation and above) and Ohio (Ludlowville Formation and above) cannot be directly connected with those in New York through the subsurface, as post-Middle Devonian erosion has removed these beds along an arch, roughly from Cleveland, Ohio, northward into Ontario (Fig. 2). The exposures in northwestern Ohio technically represent the southern portions of the Michigan Basin as they are flanked by the Findlay Arch to the south, and exposures in southeastern Indiana have long been considered to lie within the easternmost portion of the Illinois Basin. Although apparently no such structural high separated the eastern Findlay Arch and western Findlay Arch depositional areas during the Middle Devonian (see below), subsequent development of the Findlay Arch has resulted in post

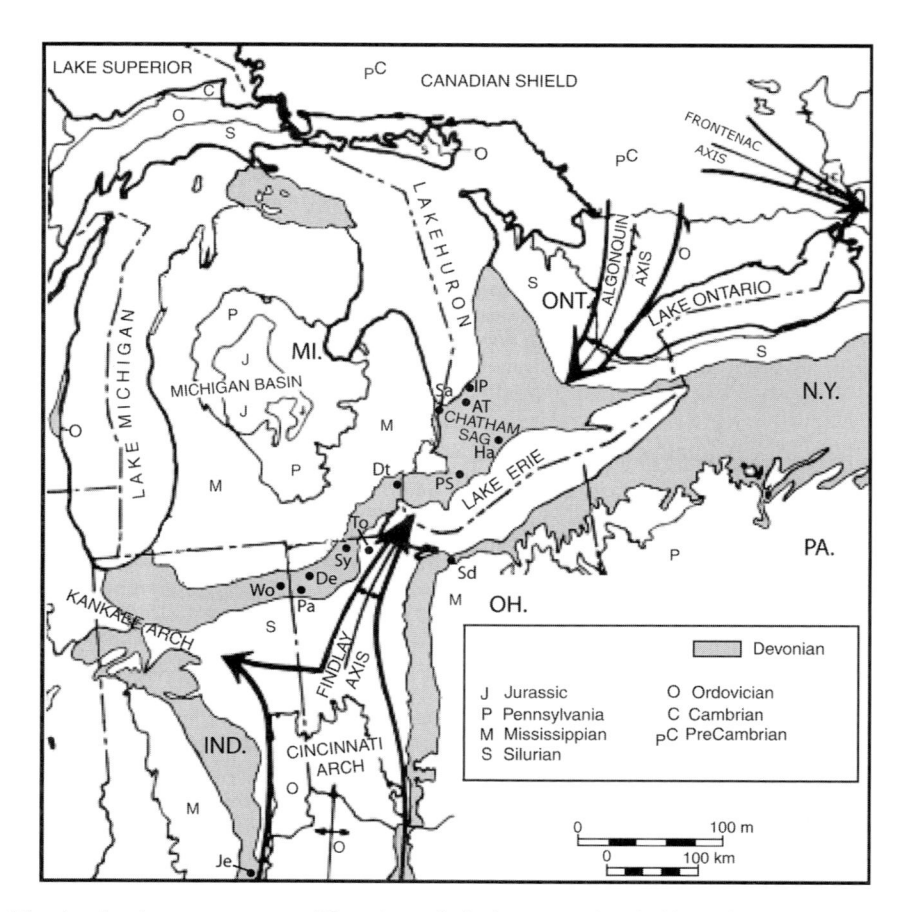

Fig. 1. Map showing the outcrop pattern of Devonian rocks in the western Appalachian and Michigan Basins. (N.Y., New York; PA, Pennsylvania; OH, Ohio; ONT, Ontario; IND, Indiana; MI, Michigan; Sd, Sandusky; Sy, Sylvania; To, Toledo; De, Fort Defiance; Pa, Paulding; Wo, Woodburn; Dt, Detroit; PS, Port Stanley; Ha, Hartwick; AT, Arkona/Thedford; IP, Ipperwash Point Provincial Park; Sa, Sarnia; Je, Jeffersonville.)

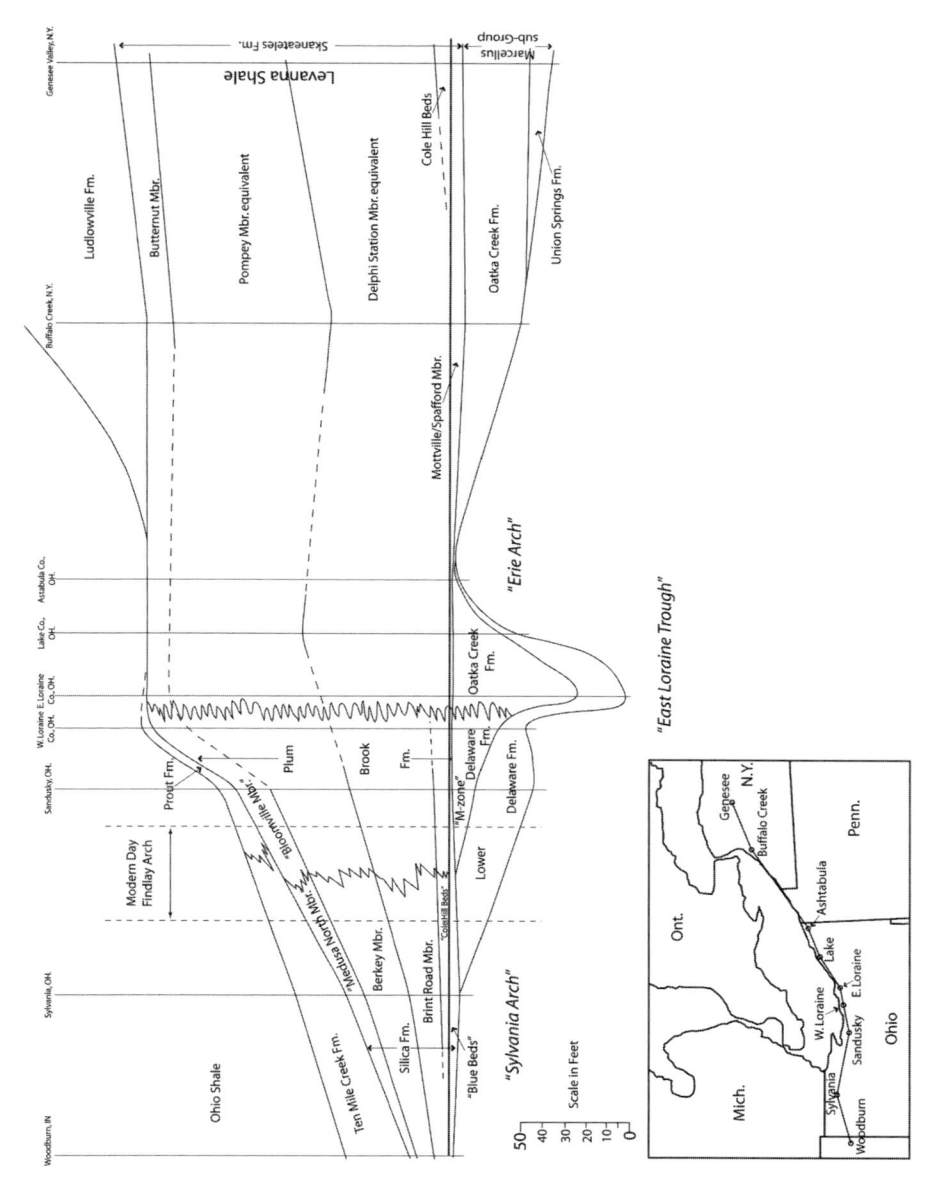

Fig. 2. Cross-section from west-central New York State into northeastern Indiana displaying the relative thickness of units of the lower portion of the Hamilton Group. The section line runs from near the deepest portion of the Appalachian Basin in central New York State up onto the distal western flank of the basin. Note the dramatic changes in thickness in the units of the Marcellus sub-Group about the area labelled the 'Erie Arch'. The units of the Marcellus sub-Group are missing west of Sylvania, Ohio, being removed by erosion at the base of the overlying Skaneateles sequence. Note also the relatively narrow zone of lithofacies change between western and eastern Lorain county, Ohio, during the time of deposition of both the Marcellus sub-Group and the overlying Skaneateles sequence. The overlying units of the Ludlowville sequence are missing west of western New York State and east of Lorain county, Ohio, due to erosion below the overlying upper Devonian shales.

Palaeozoic removal of these strata in the region from Sandusky to Toledo, Ohio (Fig. 2).

Four basic study areas were included, outside New York State (Fig. 1). First outcrops were examined in the vicinity of Arkona and Thedford, Lambton County, Ontario, including the Ausable River and its tributaries and foreshore exposures along the Lake Huron shore near Ipperwash Provincial Park. Very critical new data were obtained from two complete drill cores, one near Ipperwash Point and a second near Sarnia, some 40 km to the southwest. These measured cores were supplemented by data on the lower Hamilton Group (only) in a core taken from Port Stanley, about 25 km south of London, Ontario, and the lower and middle Hamilton Group from a core taken at Hartwick, about 75 km south of Ipperwash Point (Fig. 1).

Poor and incomplete exposures of Middle Devonian strata were examined in the vicinity of Plum Brook near Sandusky, Ohio (Fig. 2); fortunately, two complete drill cores were available through the entire Hamilton from near these areas, permitting for the first time a more complete characterization of the details of stratigraphy, especially for the poorly exposed Plum Brook Shale. Minor outcrops of the correlative lower Olentangy Shale were also examined near Delaware, Ohio.

In the western Findlay Arch area in northwestern Ohio, exposures of the Traverse (=Hamilton) Group were studied in the abandoned Medusa North Quarry in Sylvania (Fig. 1). We also examined condensed sections near the Ohio/Indiana state line at Paulding, Ohio, and Woodburn, Indiana. A single drill core through the Traverse Group near Defiance, Ohio, was also measured (Fig. 1).

The Sellersburg Formation of southeastern Indiana comprises a rather condensed section of lower Givetian age. Exposures were examined in outcrop and both active and abandoned quarries in the area to the north of Jeffersonville (Jennings, Jefferson, Scott, and Clark counties; Fig. 1). The Sellersburg Formation displays a complicated stratigraphic pattern with disconformity-bounded units displaying varying degrees of erosion.

Biostratigraphy

Biostratigraphy is critical to establishing a framework of correlations upon which, more detailed sequence stratigraphic interpretations can be based. In the following sections, we summarize previous work and newer unpublished data on virtually all pertinent lines of biostratigraphical data. Although conodonts, the primary zonal fossils for the Givetian Stage, are rare and non-diagnostic in some portions of the studied interval, nearly all of the fossil data sources are mutually supportive and the consilience of these data provides a strong basis for the more precise correlations that follow.

Basal datum of succession

All units examined in this study overlie a well-correlated datum at the top of the *Polygnathus eiflius* Conodont Zone, near the uppermost Eifelian Stage (Fig. 3). Units below this basal datum include the Stony Hollow–Cherry Valley succession in New York, the Delaware–Dundee

Stage	Conodont zones	Goniatite faunas	Unit
Givetian	semialternans	P. amplexum	up. Tully
Givetian	ansatus		mid. Tully
Givetian	ansatus		low. Tully
Givetian	ansatus	T. uniangulare	Windom
Givetian	ansatus		Portland Pt.
Givetian	ansatus		Jaycox
Givetian	rhenanus/ varcus	M. n. sp./S. unilobatus	Wanakah
Givetian	rhenanus/ varcus	T. amuletum/T.u.aldenense	Ledyard
Givetian	rhenanus/ varcus		Centerfield
Givetian	timorensis		Butternut
Givetian	timorensis	T. arkonense	Pompey
Givetian	timorensis		Delphi Stat.
Givetian	timorensis		Stafford
Givetian	hemiansatus	Parodiceras/ T. mesopleuron	Chittenango/ Cardiff
Eifelian	ensensis		Berne
Eifelian	eiflius	A. vanuxemi	Cherry Valley
Eifelian	kockelianus	C. plebeiforme	Hurley
Eifelian	australis		Bakoven

Fig. 3. Chart showing the current biostratigraphical zonation of conodonts and goniatites for eastern North America. Units listed are those of New York State. Conodont zones adapted from Kirchgasser 2000, goniatite zones from House 1978. Goniatite faunas: *Cabrieroceras plebeiforme* (*Cabrieroceras* Genozone MDI-E; NY Regional Zone 5), *Agoniatites vanuxemi* (*Agoniatites* Genozone MDI-F; NY Regional Zone 6), *Parodiceras* sp. nov. aff. *Tornoceras mesopleuron* (*Agoniatites* Genozone MDI-F; NY Regional Zone 7), *Tornoceras arkonense* (*Maenioceras* Genozone MDII-A; NY Regional Zone 8), *Tornoceras amuletum*/ *T. uniagulare aldenensis* (NY Regional Zone 9–10), *Maeneceras* sp./*Agoniatites unilobatus* (*Sellagoniatites* Genozone MDII-C, NY Regional Zone 11), *Tornoceras uniangulare uniangulare* (*Afromaeneceras* Genozone MDII-D; NY Regional Zone 12), *Pharciceras amplexum* (*Pharciceras* Genozone MDIII-A; NY Regional Zone 13).

succession in Ohio, the upper Dundee Formation in Ontario, Speeds Member in southwestern Indiana, and the Rogers City Formation in Michigan. These units are characterized by a unique macrofauna for eastern North America, termed the Stony Hollow Fauna (C. Ver Straeten, pers. comm. 2005, Union Springs Fauna of Brett & Baird 1995), consisting of the brachiopods *Hallinetes*, *Variatrypa*, *Carinatrypa*, *Subrennsselandia*, *Pentamerella* cf. *P. wintereri*, *Brevispirifer lucasensis*, and *Warrenella maia*, (Koch 1981) and the goniatites *Cabrieroceras plebeiforme*, and *Agoniatites vanuxemi* (House 1981). Koch interpreted these taxa to represent a warm-water fauna that immigrated into the Appalachian and Michigan Basins from northern Canada during the late Eifelian (Boucot 1990, see DeSantis *et al*. 2007).

Conodont biostratigraphy

A detailed conodont biostratigraphy exists for the Middle Devonian of eastern North America (Klapper 1981) and yet conodonts from only a few units of this study have been well documented (Fig. 3). The conodont zones into which the examined units fall include the *Polygnathus hemiansatus*, *Polygnathus timorensis*, *Polygnathus rhenanus* and *Polygnathus ansatus* Zones. Many of the conodonts used as guide-fossils for the different zones show a delayed appearance in the Appalachian Basin, most probably due to facies control (Becker, pers. comm. 2006), thus necessitating reliance upon other biostratigraphical markers, such as goniatites.

The base of the *P. hemiansatus* Zone has been designated as the base of the Givetian Stage (Walliser *et al*. 1995). The conodont *Laticriodus latericrescens latericrescens* is thought to have its first appearance near the base of the *P. hemiansatus* Zone and extends upward to the basal Frasnian Stage. The base of the Skaneateles Formation in New York (Stafford/Mottville Member) corresponds to the first appearance of the conodont *Lat. l. latericrescens* in the central portions of the Appalachian Basin (Klapper 1981). This conodont first appears in Ontario in the Arkona Formation (Uyeno *et al*. 1982), in the eastern Findlay Arch region of Ohio in the basal Plum Brook Formation (Sparling 1988), in the western Findlay Arch region of Ohio in the lowest portion ('Blue Beds' or 'Blue Limestone') of the Silica Formation (Klapper & Ziegler 1967), and in Indiana in the Swanville Member of the Sellersburg Formation (Klug 1983) (Figs 2 & 9). Goniatite evidence presented later shows that the base of the *P. hemiansatus* Zone most probably lies lower down, probably at or near the Halihan Hill Bed of the Oatka Creek Formation.

The base of the *P. timorensis* Zone is marked by the first appearance of the diagnostic conodont, *Polygnathus timorensis*. However, *P. timorensis* displays a delayed entry in the strata of the Appalachian Basin and first appears in the New York succession near the top of the Centerfield Member of the Ludlowville Formation (Klapper 1981), in Ontario in the basal Hungry Hollow Formation (Landing & Brett 1987), and in Ohio east of the Findlay Arch in the Prout Formation (Sparling 1988) (Figs 2 & 9). To date, *P. timorensis* has not been found in the western Findlay Arch region of Ohio, although it is present in the basal Beechwood Member of the Sellersburg Formation of southeastern Indiana (Klug 1983).

Other conodonts indicative of the *P. timorensis* Zone occur lower in the section than the first appearance of *P. timorensis*. In the lower Plum Brook Formation at the Parkertown Quarry near Sandusky, Ohio, Sparling (1988) has identified a 'phyletically late form' of *Polygnathus xylus ensensis* and an early form of *P. xylus xylus*, as well as *Icriodus arkonensis*, *I. expansus*, *I. brevis* and *Lat. l. latericrescens*, placing this unit in the lower portion of the *P. timorensis* Zone (formerly upper *ensensis* Zone of Sparling 1995). Uyeno (pers. comm. as reported in Sparling 1988) had also identified a similar assemblage in a drill core from the Chatham Sag of southern Ontario in the 'Rockport Quarry' and Arkona Formations (Fig. 9).

The *P. rhenanus* Zone is designated as beginning with the first occurrence of *Polygnathus rhenanus*, but as with *P. timorensis*, *P. rhenanus* displays a delayed appearance in the Appalachian Basin, first occurring in the basal Moscow Formation. However, the conodont *Polygnathus varcus*, which first appears near the base of the *P. rhenanus* Zone, is found in the Hungry Hollow Member of the Widder Formation in Ontario (Uyeno *et al*. 1982), indicating that the basal portion of the Ludlowville Sequence (IV) lies within the *P. rhenanus* Zone; this is supported by goniatite evidence (see below). *P. varcus* is also reported from the Ten Mile Creek Formation in the subsurface of northwestern Indiana (Orr 1971).

The base of the *P. ansatus* Zone is defined by the first appearance of the conodont *Polygnathus ansatus*. However, much like *P. timorensis* and *P. rhenanus*, the first appearance of the diagnostic conodont of this zone may show a delayed appearance as suggested by goniatite evidence, again probably in response to facies control. *P. ansatus* first appears in the lower portion of the Moscow Formation in New York State and in the Prout Formation of the eastern Findlay Arch region of Ohio (Sparling 1988). In eastern Ohio, however, *P. ansatus* appears in association with conodonts not found above lower zones, indicating reworking and time-averaging within this unit. Sparling (1999) correlated the Prout with the lower portion of the Tully Limestone of New York based on the presence of *P. ansatus* along with *P. ovatinodosus*. He claimed that the Prout represents deposition during the middle of the *P. ansatus* Zone. It is interesting to note that Sparling's samples that produced conodonts indicative of the upper portion of the *P. ansatus* Zone come from very thin portions of the unit that are found geographically between thicker portions of the Prout. Samples from the thicker portions of the unit produced few, if any, diagnostic conodonts (Sparling, pers. comm. 2004). It is our interpretation that the thin, conodont-rich portions of the Prout sampled by Sparling represent a condensed, reworked lag lying in palaeo valleys eroded between older portions of the unit during the time of the Taghanic Unconformity. *P. ansatus* has not been found to date in Ontario or the western Findlay Arch region of Ohio.

Ammonoid biostratigraphy

A well-developed goniatite biostratigraphy exists for the Middle Devonian of the Appalachian Basin (House 1981; Becker & House 2000; Becker 2005; Fig. 3). Goniatite cephalopods with a bearing upon this study include the genera *Agoniatites*, *Cabrieroceras*, *Tornoceras*, *Parodiceras*, *Sellagoniatites*, *Maenioceras* and *Pharciceras*. The lowest ammonoid-bearing beds in the Hamilton Group occur in eastern New York within a concretionary layer in the upper Bakoven Formation (Sequence I) which contains *Cabrieroceras plebeifome*. Farther to the west in New York, this species is also seen in the overlying Chestnut Street Bed of the Hurley Member of the Oatka Creek Formation (base of Sequence II). In eastern New York, a thin (~30 cm) black shale, known as the Lincoln Park Shale (Ver Straeten *et al*. 1994), overlies the Chestnut Street Bed; this shale is removed by erosion below the overlying Cherry Valley Limestone west of Otsego County, New York. Contained within the Lincoln Park Shale is the primary occurrence of the goniatite, *Agoniatites nodiferum*. The overlying Cherry Valley Limestone contains abundant *Agoniatites vanuxemi* along with *Parodiceras discoideum*. *Agoniatites vanuxemi* is also known to occur in west-central Ontario in the Delaware Limestone (Best 1953).

The earliest-known Givetian ammonoid identified in the Appalachian Basin is *Tornoceras (T.)* aff. *mesopleuron* from near Cherry Valley, New York (House 1965, 1978) collected from a level estimated at ~18.3 m above the Cherry Valley Limestone in the Chittenango Member of the Oatka Creek Formation (Sequence II). Upon further investigation of the collection locality, it seems very likely that the level from which *Tornoceras (T.)* aff. *mesopleuron* was collected is the Halihan Hill Bed. If this interpretation is correct, it provides a much-needed age constraint on the Halihan Hill Bed as lying at or near the base of the Givetian Stage. From this level, a new, as yet unnamed, species of *Parodiceras* was also collected.

The overlying Skaneateles Formation (Sequence III) contains a number of important goniatites that aid in correlation of units across the basin. The lowest record of goniatites from the Skaneateles Formation consists of undescribed specimens of the genus *Tornoceras* from the Mottville Member of central New York State (Grasso 1986; Becker 2005). Of greater biostratigraphical importance are the occurrences of *Tornoceras arkonense* and *Tornoceras mesopleuron* from the Pompey Member of New York, the middle Arkona Formation of Ontario, the lower Olentangy and Plum Brook Formations of eastern Ohio, and the Silica Formation of western Ohio (House 1965, 1978). These species occur as pyritized specimens in widely traceable horizons. *Agoniatites* also occurs in the Pompey Member of the Skaneateles Formation in central New York and apparently also in the upper Arkona Shale at Hungry Hollow, Ontario (Prosh 1990).

The basal Centerfield Member of the overlying Ludlowville Formation (Sequence IV) contains only fragmentary specimens of the genus *Tornoceras*, but the overlying shales of the Ledyard and Wanakah Members contain several important goniatite-bearing intervals. The Alden Pyrite bed in the lower Ledyard Member contains the goniatite *Tornoceras (T.) uniangulare aldenense*, while the laterally equivalent lower shales of the Widder Formation in Ontario contain *Tornoceras (T.) uniangulare widderi* (House 1965). Higher up in the Ledyard Member, the distinctly ribbed goniatite '*Tornoceras*' *amuletum* occurs in a concretionary horizon (Kloc 1983) and provides an important link to the goniatite sequence of Morocco, interpreted to correlate with MD II-B of that area where other ribbed tornoceratids have been found (Becker *et al*. 2004). The overlying Wanakah Member also contains pyritic goniatite-bearing horizons that yield *Sellagoniatites unilobatus* and *Maenioceras* sp. (Kloc 1983) as well as several new species of tornoceratids. The presence of these distinctive goniatites in the upper Ludlowville Formation is important as this demonstrates a correlation with European and African faunas that are characteristic of the *P. rhenanus* Conodont Zone (Becker 2005); the lowest occurrence of this conodont is found at the base of the overlying Moscow Formation.

Tornoceratid goniatites are also present in the Moscow Formation (Sequence V) with several new species present in the Windom Member (Becker, pers. comm. 2006). *Tornoceras (T.) uniangulare uniangulare* is present probably as reworked individuals in the Leicester Pyrite Bed, a lag bed that rests unconformably on the upper Windom Member of the Moscow Formation (Baird & Brett 1986). The upper Tully Formation contains the zonally important goniatite *Pharciceras amplexum* as well as *Tornoceras* cf. *arcuatum*. *Pharciceras amplexum* is also known from the upper Tully of Pennsylvania and from the Portwood Member of the New Albany Shale of Kentucky (Brett *et al.* 2004).

Brachiopod biostratigraphy

Brachiopods are by far the most diverse part of the fauna within the study interval and it is not surprising that certain taxa may be employed as biostratigraphic markers. Among such taxa are the genera *Eumetabolotoechia*, *Paraspirifer*, *Fimbrispirifer*, and *Spinocyrtia*. *Eumetabolotoechia* is a rhynchonellid brachiopod often found in the dark, laminated, relatively deep-water shales of sequences in the Givetian Stage in assemblages thought to represent dysoxic conditions. Where *Eumetabolotoechia* is present in dark shales, it is often the dominant taxa in the associations, almost to the exclusion of all other taxa. *Eumetabolotoechia* is present in the upper Oatka Creek Formation of New York, in the Bell Formation in the subsurface of Ontario, in the upper Delaware Formation of central Ohio, and in the Silver Creek Member of the Sellersburg Formation in southeastern Indiana. An epibole of *Eumetabolotoechia* cf. *E. multicostum* occurs in the Butternut Member of the upper Skaneateles Formation in New York, in the uppermost dark grey shale unit of the Arkona Formation in Ontario, in the Plum Brook and lower Olentangy Formation in central Ohio, and the upper Silica Formation in western Ohio (Figs 2 & 9).

Paraspirifer is another brachiopod that occurs in the Appalachian Basin in restricted horizons. *Paraspirifer* is well known from the lower Silica Formation of western Ohio where well-preserved specimens occur in large numbers. In the more proximal portions of the Appalachian Basin (eastern and central New York), *Paraspirifer* occurs in the Mount Marion Member of the Oatka Creek Formation and the Mottville Member of the Skaneateles Formation. The Mount Marion was once correlated to the Silica Formation based primarily on the co-occurrence of *Paraspirifer* (Cooper & Warthin 1942), but is now known to be older than the base of the Silica Formation, while the Mottville Member is the probable equivalent to the basal Silica Formation ('Blue Beds'). These data indicate that *Paraspirifer* first appeared in the more proximal portions of the Appalachian Basin and then migrated to the distal portions basin where it flourished for a time.

Fimbrispirifer venustus is a distinctive brachiopod that appears in specific assemblages found only in the shallow facies of certain sequences. *F. venustus* first appears in low abundance in the Halihan Hill Bed of the Oatka Creek Formation; but it is most abundant in the Centerfield Member of the Ludlowville Formation; it also appears very rarely higher in the Tichenor–Kashong members of the Moscow Formation in New York. Outside of New York, *F. venustus* occurs in the Prout Formation of eastern Ohio (Stumm 1942), the Ten Mile Creek Formation of western Ohio (Stauffer 1909), the Hungry Hollow Formation of Ontario (Stumm & Wright 1958), and the Beechwood Member of the Sellersburg Formation of southeastern Indiana (Campbell 1942).

The Ludlowville Formation of New York and the Widder and Ipperwash Formations of Ontario have long been correlated with each other, although the precise relationships have remained uncertain. Morphologically distinct species of *Spinocyrtia*, including *S. granulosa* and *S. ravenswoodensis*, from the upper Hamilton Group of New York and Ontario (Ludlowville and lower Moscow Formations in New York State; Widder and Ipperwash Formations in Ontario) display a fine-scale stratal pattern of morphology that is similar between the two areas that help to elucidate the finer-scale relationships between these formations (Kloc, pers. comm. 2004). Morphologically, *Spinocyrtia* species fall into two groups: those with a distinct 'notch' in the medial fold (*S. ravenswoodensis*) and those lacking a 'notch' in the medial fold (*S. granulosa*). The stratal pattern of different morphologies is summarized in Table 1.

Bivalve biostratigraphy

Although less abundant than brachiopods in the Middle Devonian of the Appalachian Basin, bivalves were a major part of the megafauna. Certain taxa display patterns that may prove useful for biostratigraphical interpretation: these include the genera *Ptychopteria*, *Gosseletia*, and *Paracyclas*. Ptychopteria is a long-ranging genus that is present in most siliciclastic-dominated portions of the sequences in the study interval, although it is conspicuously absent from the upper-most sequence, having its last appearance in the Givetian in the upper Ludlowville Formation of New York, the Ipperwash Formation of Ontario, the Prout Formation of eastern Ohio, and the Ten Mile Creek Formation of western Ohio. *Gosseletia* and *Paracyclas* are two genera that are found primarily in the Oatka Creek Formation of New York and are also present in the basal portions of the Silica

Table 1. *Details of the stratigraphic distribution of differing morphologies of the brachiopod Spinocyrtia in New York and Ontario. Morphologies differ in the presence or absence of a notch in the medial fold of the brachiopod*

Unit (N.Y.)	Morphology	Unit (Ont.)	Morphology
Kashong	Notch	Up. Ipperwash	Notch
Up. Wanakah	No Notch	Petrolia	??
Low. Wanakah	Notch	Low. Ipperwash	Shallow Notch
Ledyard	No Notch	Up. Widder	No Notch

Formation in western Ohio and the Silver Creek Member of the Sellersburg Formation in southeastern Indiana (Campbell 1942).

Gastropod biostratigraphy

Although less diverse than bivalves, gastropods also play a role as regional biostratigraphical markers. *Bembexia* is an excellent biostratigraphical marker, being found in abundance in the Skaneateles Formation of New York and generally not found above that unit. It is also found in the Plum Brook and lower Olentangy of central Ohio, and in the Arkona Formation in Ontario. *Bembexia* is, however, reported above the Skaneateles horizon in the Ontario area, being present in the Widder Formation (Stumm & Wright 1958), although we did not confirm this occurrence.

Sequence stratigraphic correlation of Middle Devonian strata: overview of third order sequences in the foreland basin

Cooper (1930) laid out the basic framework of Hamilton stratigraphy, defining four formations, the Marcellus, Skaneateles, Ludlowville and Moscow, of which the upper three were delimited at their bases by a thin but widespread limestone or calcareous siltstone member, the Stafford-Mottville, Centerfield and Tichenor Members, respectively (Fig. 4). The formations of the Hamilton Group range from a few metres to several hundred metres thick. Subsequently, Ver Straeten *et al.* (1994; Ver Straeten & Brett 2006) redefined the Marcellus as a sub-group of the Hamilton and divided it into two formations, the lower Union Springs and an upper Oatka Creek, the latter paralleling the higher Hamilton units in being delimited at its base by the thin, widespread Cherry Valley Limestone.

Brett & Baird (1996) noted that, as defined, each Hamilton formation is comparable, in temporal scale and pattern, to a 3rd-order depositional sequence (Fig. 4). The sharply defined, typically erosional, bases of the basal limestone beds represent sequence boundaries, their sharp, corroded and stained tops are flooding surfaces, recording sediment starvation, and the limestones form portions of the transgressive systems tract (TST). In several cases, the lower TST is actually composed of two or more thin limestones, separated by calcareous, silty mudstones (e.g. Mottville A and B Limestones with intervening, unnamed mudstone; Tichenor Limestone, Deep Run Shale, Menteth Limestone). The TST typically continues some distance into overlying, calcareous shales, thin limestones and concretionary beds that exhibit a backstepping pattern. The maximum flooding surface is typically cryptic but may be marked by a phosphatic pebble bed that underlies a sharp change to dark shales that show an aggradational to upward-shallowing pattern. The remainder of the formation comprises thick shales, mudstones and siltstones with minor shell-rich and concretionary limestones representing the highstand systems tract (HST) and falling-stage (or regressive) systems tract (FSST).

On the basis of absolute dates for the Givetian Stage, ranging from 4.4 to 7 million years (House 1991; Tucker *et al.* 1998; Kaufmann 2006), the estimated durations of Hamilton formations are on the order of 1 to 3 million years, comparable to 3rd order sequences identified by seismic stratigraphers (Vail *et al.* 1991; Emery & Meyers 1996). Thus, the Hamilton Group formations can be classified as 3rd-order cycles following the typical definition of Vail *et al.* (1991; see Catuneanu 2002; Coe 2003).

The most apparent and easily identified portion of the 3rd-order sequences is the TST. The HST of the 3rd-order sequences is less discrete, as it contains the component 4th-order sequences. An abrupt, shallowing-upward or falling stage systems tract (FST) of the 3rd-order sequences is usually identifiable, although some of its upper portions may be removed by erosion at the next sequence boundary.

The main siliciclastic highstand portions of each sequence have been subdivided into members that also represent smaller-scale (4th-order) sequences. Each of these can be further divided into a lower condensed, shell-rich, calcareous, bioturbated siltstone, biostrome or minor concretionary limestone, and an upper siliciclastic-rich interval. Previous workers considered the fossiliferous, calcareous beds to represent the 'caps' of coarsening-upward cycles

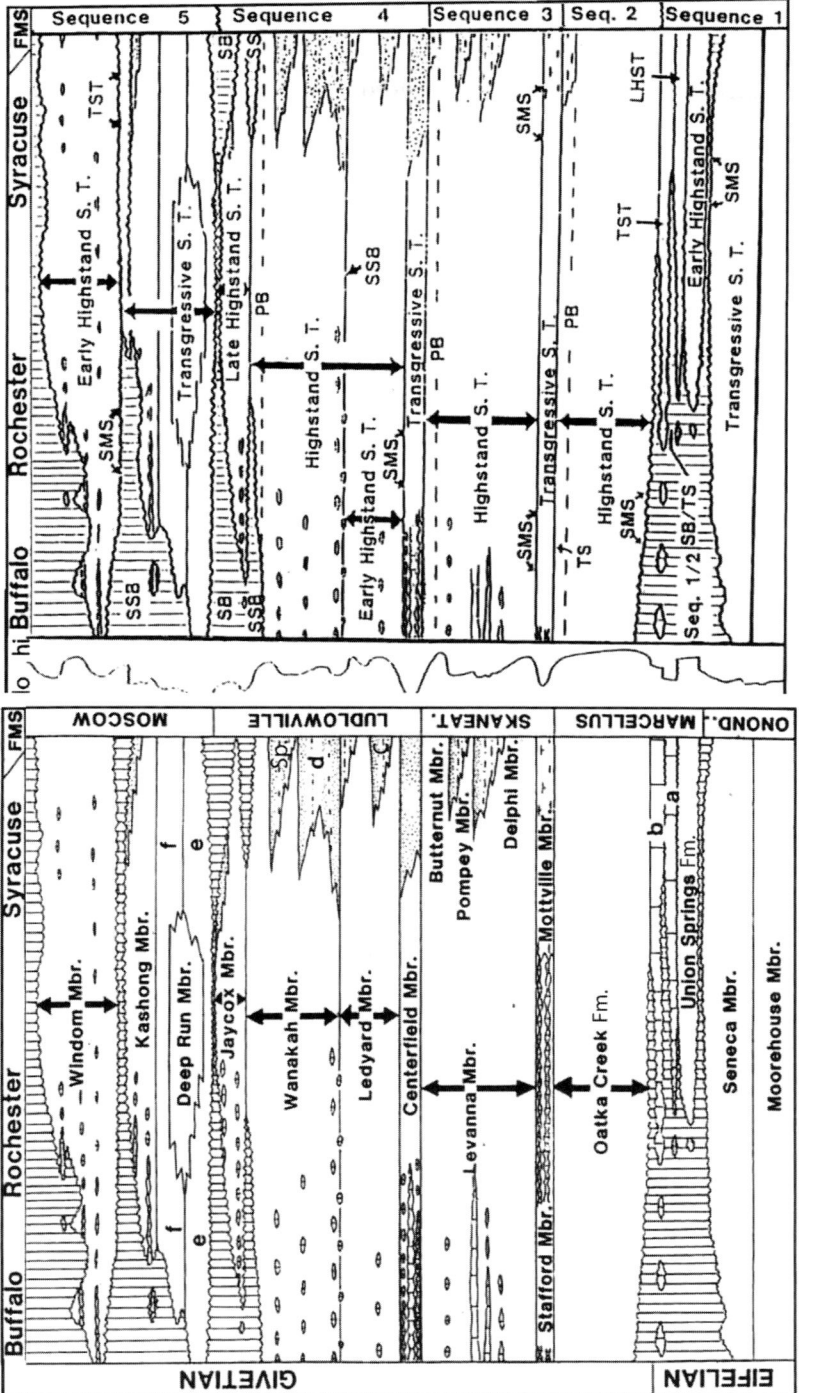

Fig. 4. Diagram showing the exposure of Hamilton units across western and central New York State (left) and the sequence stratigraphic interpretation of the units (right). Between the two diagrams is an interpreted sea-level curve for the Appalachian Basin during the deposition of the Hamilton Group. a, Hurley Member; b, Cherry Valley Member; c, Otisco Member; d, Ivy Point Member; e, Tichenor Member; f, Menteth Member; Mbr., Member; Fm., Formation; Sp., Spafford Member; SMS, surface of maximum starvation; TS, transgressive surface; S.T., systems tract; LHST, late highstand systems tract; TST, transgressive systems tract; PB, precursor bed (analogous to the surface of forced regressions); SB, sequence boundary; SSB, subsequence boundary. From Brett & Baird (1996).

bounded by sharp shifts back to deeper facies, i.e. 'parasequences' (for discussion, see Brett & Baird 1996). However, the limestones are typically sharply set off from underlying siltstones and exhibit a retrogradational (upward-deepening and backstepping) pattern and, thus, these packages are better interpreted as transgressive systems tracts; these decametre-scale cycles do not strictly fit the definition of a 'parasequence' (asymmetrical shallowing-upward cycles bounded by flooding surfaces). Rather we interpret them as small-scale depositional sequences with component systems tracts. In fact, these small-scale sequences are most easily and discretely subdivided into their constituent systems tracts (TST, HST, FST), whereas the delineation of the highstand to regressive portions of the 3rd-order (formation-scale) sequences is somewhat more arbitrary. This reflects their composite nature; the existence of multiple small-scale sequences within each 3rd-order sequence blurs the overall distinction between the different large-scale systems tracts. Based on cyclostratigraphical calibration of Givetian conodont zones (House 1995), these smaller cycles are estimated to represent about 100,000 to 400,000 years and may relate to Milankovitch eccentricity cycles and thus fall in the range of 4th-order sequences.

Correlation of sequences into Ontario, Ohio and Indiana

In the following sections, data will be presented sequence-by-sequence, at the 3rd-order (formational) scale, for all areas beginning with the well-defined New York sequences and extending westward into Ontario, Ohio and Indiana. Evidence linking smaller-scale 4th-order sub-sequences will also be presented within the discussion of each 3rd-order sequence.

This paper deals only with the upper four sequences of the Hamilton Group, in ascending order, the Oatka Creek, Skaneateles, Ludlowville and Moscow sequences (see DeSantis et al. 2007, for discussion of the lowest (Union Springs) sequence of the Hamilton Group). Particularly relevant to this study are the details of the small-scale sequences of the Skaneateles and Ludlowville Formations and these will be discussed briefly in the following sections. More detailed, bed-by-bed correlations are documented in another paper (Bartholomew et al. 2006).

Sequence II: Oatka Creek, Bell and lower Sellersburg Formations

New York–Oatka Creek Formation

The lowest 3rd-order sequence considered here is represented in west-central New York by the Oatka Creek Formation (Ver Straeten et al. 1994). The Oatka Creek sequence in New York begins with a widespread TST, the Hurley–Cherry Valley limestone members; the overlying HST can be divided into three 4th-order subsequences represented by the East Berne, Chittenango and Cardiff Members and their eastern correlatives (Fig. 5). Of particular importance is the Halihan Hill Bed of the Oatka Creek Formation, a condensed fossiliferous, grey, mudstone and limestone interpreted as a 4th-order subsequence TST; this bed marks the first appearance of the Hamilton Fauna in the Appalachian Basin (Brett & Baird 1995).

Southwestern Ontario–Bell Shale Formation

The Oatka Creek Sequence is identifiable in subsurface of Ontario where it is represented by about 20 m of dark grey rhynchonellid-bearing shales assigned to the Bell Shale (Fig. 5). Drill cores in the London to Sarnia area of Ontario reveal a fossiliferous bed rich in chonetid brachiopods, including *Hallinetes* and the athyrid *Coelospira* that, for the Hamilton Group, are uniquely found in the Halihan Hill interval (see DeSantis et al. 2007). This is a key marker bed of the Hamilton Group and its discovery in the Ontario subsurface represents the first report of the Halihan Hill bed outside of the Appalachian Basin. Similar grey, fossiliferous beds, high in the Bell Shale of Ontario, may represent the Cardiff (Solsville-Pecksport) Member (Fig. 5). These beds appear to be traceable into the Bell Shale of Michigan.

Ohio–Delaware Formation

In Ohio, the Oatka Creek sequence is probably represented by the upper portion (M-zone of Stauffer 1909) of the Delaware Formation (eastern Findlay Arch) and the uppermost Dundee Formation (western Findlay Arch) (Fig. 9). The Oatka Creek sequence shows dramatic thinning west of New York State, the nature of which has only recently been elucidated based upon examination of a series of cores in northeastern Ohio in Ashtabula, Lake and Loraine counties (Fig. 2). The Oatka Creek Formation thins to less than a metre in northeastern-most Ohio and then thickens to over 15 m in north-central Ohio while retaining the dark, shaly nature of the unit as observed in more proximal portions of the basin. Just west of the area with the greatest thickness, a rapid lithofacies shift (over a lateral distance of ~25 km) occurs within the unit, the facies changing from dark-grey shale to rusty-brown argillaceous dolostone typical of the upper Delaware Formation in outcrop in the Sandusky area. West of Sandusky, the Oatka Creek Formation is removed beneath

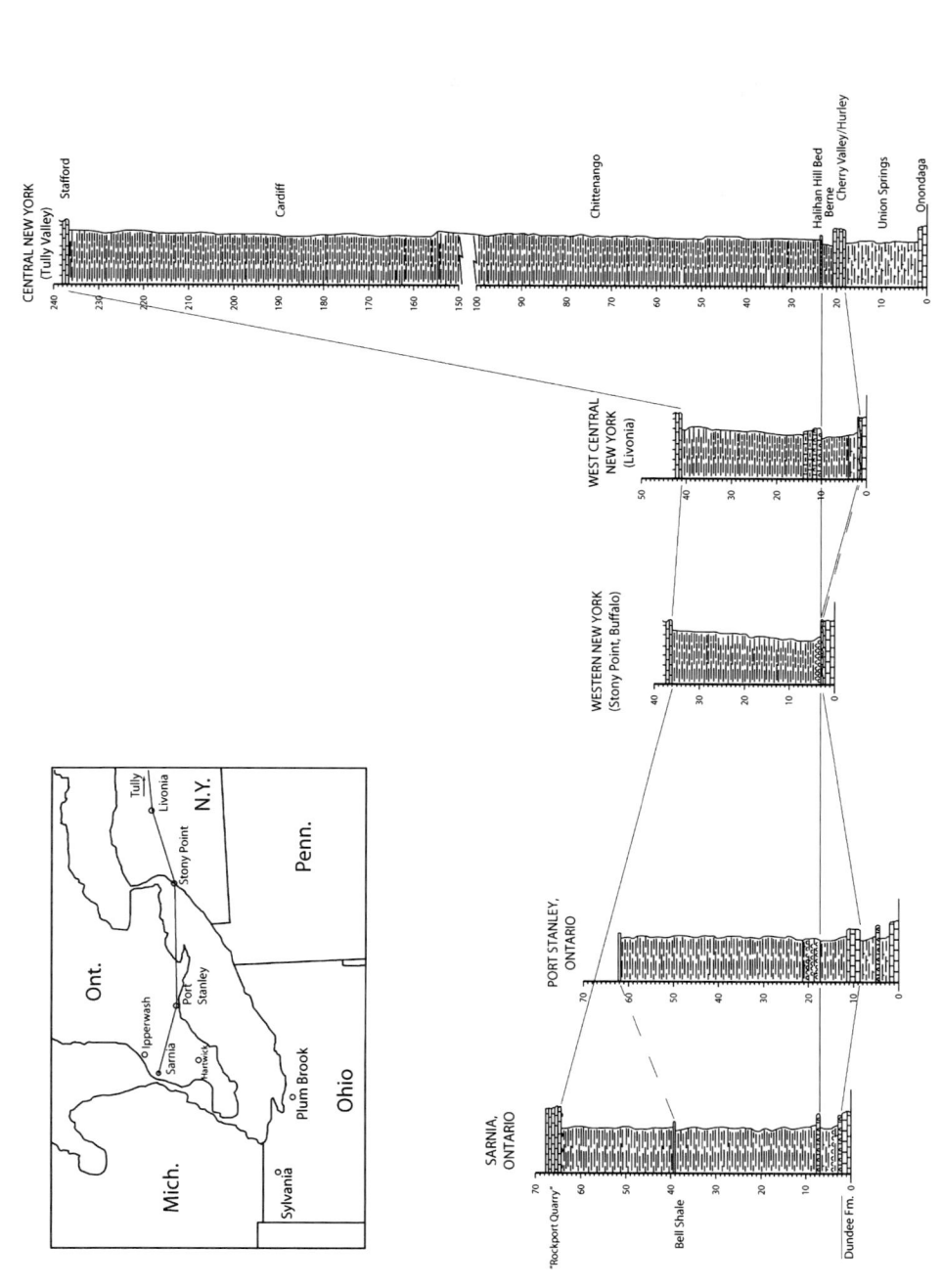

Fig. 5. Cross-section displaying the change in thickness of the different units of the upper Union Springs and Oatka Creek Formations across western New York State into southwestern Ontario through the subsurface. Scale is in feet.

the unconformity at the base of the overlying Skaneateles sequence, here represented by the basal 'Blue Beds' of the Silica Formation.

Southeastern Indiana – Silver Creek Member

The Oatka Creek sequence is represented in southeastern Indiana by the Silver Creek Member of the Sellersburg Formation (Fig. 9). This unit consists of up to 8 m of buff, locally cherty, dolomitic mudstone with abundant chonetid brachiopods. The Silver Creek Member is bounded both above and below by disconformities. The erosive discontinuity at its base locally completely removes the lower units of the Sellersburg Formation to the extent that it rests on the underlying Jeffersonville Formation. The Silver Creek is bounded at the top by the basal discontinuity of the overlying Swanville Member. Faunally, the Silver Creek Member shows greater similarity with the Hamilton Fauna than it does with the Stony Hollow Fauna, although the Halihan Hill interval has yet to be precisely identified in this area. The unit contains many elements indicative of the Oatka Creek sequence, including the bivalve *Paracyclas* and the brachiopod *Eumetabolotoechia*. A shell-rich interval near the top of the unit may be equivalent to the upper Cardiff interval of east-central New York and represents a smaller-order TST within the Oatka Creek sequence. Previous reports (Klug 1983) of younger conodonts from the upper portion of this formation are now known to be derived from lag deposits representing reworked portions of the overlying Swanville Member.

Sequence III: Skaneateles, Arkona, Plum Brook, Silica and middle Sellersburg Formations

New York – Skaneateles Formation

A second 3rd-order sequence to be considered is represented by the Skaneateles Formation of New York State (Vanuxem 1842) (Figs 2 & 6). The Skaneateles sequence begins with a thin (0.5 to 10 m), widespread limestone and calcareous silty mudstone succession, the Mottville/Stafford Members, which contain the first appearance of the conodont *Laticriodus latericrescens latericrescens*, identified by Klapper (1981) as being near the base of the upper *P. ensensis* conodont Zone (presently termed *P. hemiansatus* Zone). The lack of defining polygnathid conodonts does not allow for precise correlation, although goniatites from lower down in the underlying Oatka Creek Formation place the Mottville/Stafford interval well up in the *P. hemiansatus* Zone. Mottville and Stafford Members occupy the position of the 3rd-order TST of the Skaneateles sequence. The Stafford Member is up to ~2.5 m thick and consists of a series of interbedded condensed skeletal limestones and more argillaceous beds in western New York State; this unit thickens eastward into the Mottville Member. The latter consists of two coral-rich, silty, limestone beds, each about 30 to 50 cm thick, separated by an intervening more silty mudstone package that thickens eastward and contains an abundance of the brachiopod *Tropidoleptus carinatus*. The upper Mottville limestone (B) grades upward into dark shales containing a dysoxic fauna; in central New York, these shales form the base of a 7–10 m coarsening-upward cycle, the informal Cole Hill submember.

Overlying the Stafford Member in western New York is a mass (50–100 m) of dark grey and black shales, assigned to the Levanna Member, which is interpretable as a 3rd-order highstand–regressive succession. This interval is traceable eastward into five member-scale units of the Skaneateles Formation in central New York: the Mottville-Cole Hill (informal), Delphi Station, Pompey, Butternut and Chenango Members (Cooper 1930; Baird *et al.* 2000) (Fig. 9). Each of these, as presently defined, commences with thin (0.5 to 5 m) fossiliferous, calcareous siltstones, concretionary limestones or biostromes (Paper Mill/Pole Bridge, Wadsworth, and Marietta beds), which are overlain by a thicker (5–30 m), coarsening-upward shale to siltstone succession. These intervals are interpreted as 4th-order sequences. The thin basal concretionary carbonates of each cycle have been recently traced into the Levanna Shale of western New York (Baird *et al.* 1999, 2000). Although major portions of each member are represented in western New York only by dark shales without obvious coarsening-upward character, the limestone marker beds, interpreted as TSTs, permit correlation of 4th-order sequences to western New York State (Baird *et al.* 1999) (Fig. 6). The Cole Hill cycle is overlain in many areas by concretionary limestones with auloporid coral biostromes, at the base of the relatively thick, sparsely fossiliferous shale of the Delphi Station cycle, with more fossil-rich beds near the top.

A pair of silty carbonate beds, the Paper Mill and Pole Bridge beds, also typically with auloporids and ambocoeliid brachiopods, forms the base of the Pompey Member, which is predominantly dark chonetid-bearing shale and, uniquely for the Skaneateles Formation, carries a zone of well-preserved pyritic fossils that include the goniatites *Tornoceras arkonense* and *Tornoceras mesopleuron*, and the bactritid *Bactrites arkonense* (House 1965). The Pompey also appears to thicken westward in contrast to the Delphi Station, which thins in this direction.

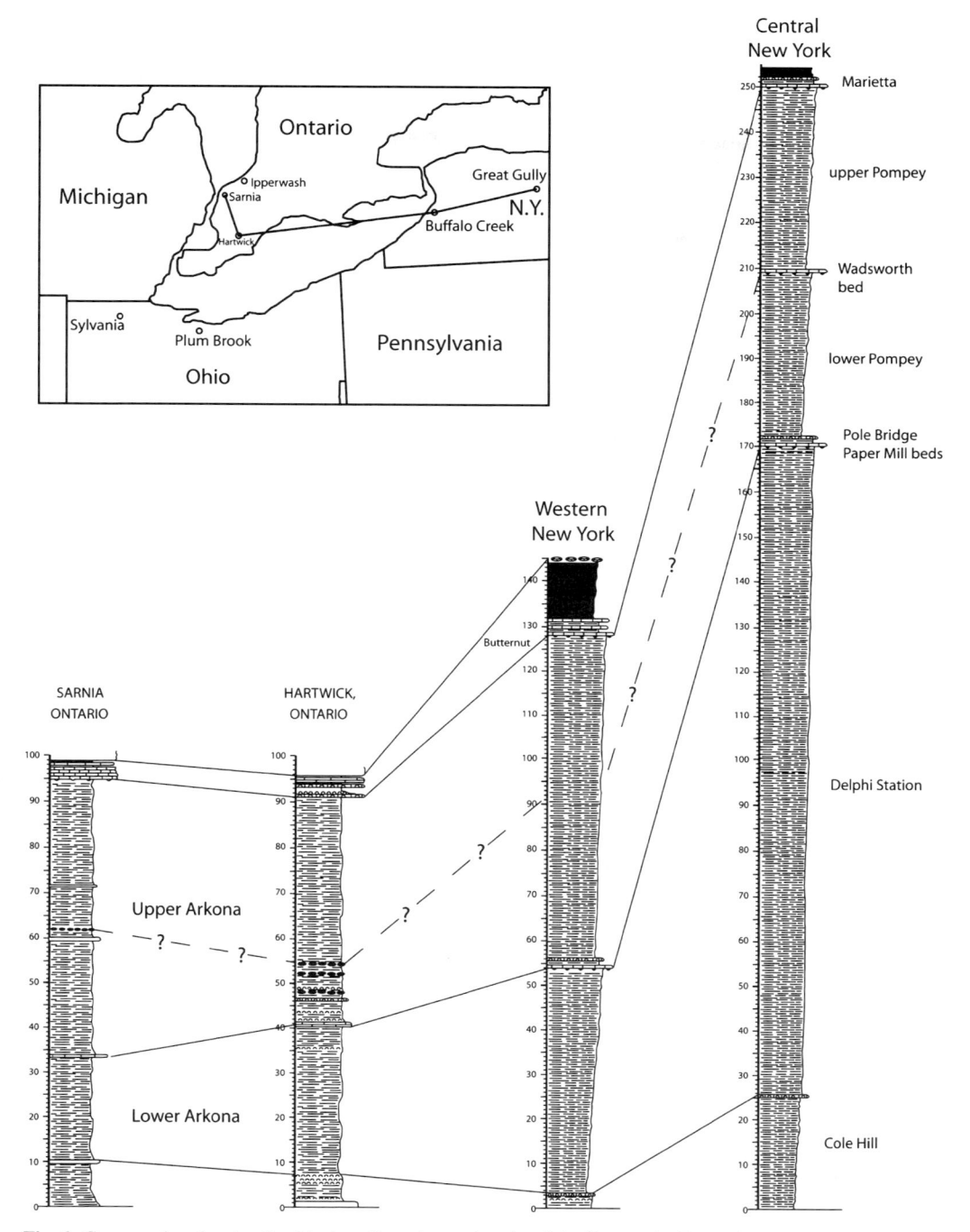

Fig. 6. Cross-section showing the thinning of member scale units of the Skaneateles Formation across New York and the Arkona Shale across southwestern Ontario. Scale in feet. (D.S.TST, Delphi Station transgressive systems tract.)

The Butternut Member commences with a sharply based series of three ambocoeliid–chonetid and auloporid-rich, concretionary limestones (Marietta beds) at its base. Overlying this unit is dark grey to black, slightly silty shale that carries a great abundance (an epibole) of the rhynchonellid brachiopod *Eumetabolotoechia*. The Butternut Shale shows extreme thinning from over 40 m in the Syracuse area to <3 m in western New York.

Collectively, the Cole Hill, Delphi Station, Pompey and Butternut comprise the composite HST of the 3rd-order Skaneateles sequence. However, it is notable that the most dysoxic and perhaps deepest facies (Butternut Member) occurs high in the succession, counter to most sequence stratigraphic models.

A distinct coarsening-upward succession is recorded in the Chenango Siltstone Member in central New York; this sequence is sharply set off from the underlying Butternut Shale by a highly fossiliferous bed, the Peppermill Gulf bed. In western New York, the succession correlative with the Chenango is a thin (1 to 5 m) succession of dark grey to medium grey, calcareous mudstone and limestone succession previously identified as the lower Centerfield, and assigned to the Ludlowville Formation (Gray 1991), but set off from the main Centerfield Limestone by a sequence boundary. The Chenango–lower Centerfield is interpreted as the regressive or falling stage systems tract of the 3rd-order sequence. Its sharp base is a forced regression surface and the siltstones represent progradation induced by sea-level fall.

The Skaneateles sequence is the best developed of the Givetian sequences across the Appalachian Basin, being recognized across northern Ohio where it is represented by the Plum Brook Formation (eastern Findlay Arch) and the Silica Formation (western Findlay Arch), in Ontario, where it is coextensive with the Arkona Formation, and in southeastern Indiana where it is represented by the Swanville Member of the Sellersburg Formation (Figs 2, 6, 7 & 9).

Southwestern Ontario–Arkona Formation

In Ontario, the Skaneateles Formation sequence is represented by the Arkona Formation (Fig. 9). In addition to the recognition of the 3rd-order depositional pattern, the smaller-scale 4th-order subsequences have been resolved for the first time in this area.

Strata equivalent to the 3rd-order TST, Mottville Member, have been identified only in the subsurface where they have been termed 'Rockport Quarry Formation' (Uyeno 1982) (Fig. 5). Although probably coeval with the 'Rockport Quarry Formation' of Michigan, the strata of this interval in Ontario have been found to more closely resemble the Mottville Member of the Skaneateles Formation in central New York State both lithologically and faunally than they do the Rockport Quarry Formation. As in New York, this unit is composed of two grey to brown, fine-grained limestones separated by shales up to 8 m thick. The 'Rockport Quarry' interval and underlying upper Bell Shale of the Ontario subsurface have also been dated on the basis of conodonts as near the base of the *P. timorensis* or 'upper *P. ensensis*' conodont Zone (Uyeno pers. comm. as reported in Sparling 1988) (Fig. 9).

Some 30–40 m of grey mudstone of the Arkona Formation above the 'Rockport Quarry' is equivalent to the 50–75 m Levanna Member in western New York (3rd-order HST), and, as with that unit, it can be subdivided into four intervals by thin, auloporid-rich, concretionary limestones at about 3, 10, and 20 m above the 'Rockport Quarry' in the Sarnia drill core. These appear to be correlative with the Cole Hill 'cap', Paper Mill–Pole Bridge and Marietta beds. The intervening strata are monotonous, very sparsely fossiliferous, medium dark grey mudstones/shales, distinctly less organic-rich and poorly laminated in comparison to coeval shales in western New York State. They are each about half as thick as the western New York equivalents, but appear in the same proportions, except for the Butternut Shale, which is reduced to a very thin remnant. Only the Pompey and Butternut equivalents are well displayed in outcrop. Notably, the upper 9 m of the Arkona Shale carries a pyritic fauna with *T. arkonense* and *Bactrites arkonense*, as seen in the Pompey Shale of New York.

As noted, the putative Butternut interval is highly distinctive and well set off at the top of the Pompey-equivalent upper Arkona. The basal limestone, probably correlative with the upper one or two Marietta limestones, is a highly condensed wacke- to packstone bed that contains phosphatic pebbles and reworked concretions. This bed has a sharply erosive base with deeply incised megaburrows indicative of excavation into overcompacted muds and therefore removal of over a metre of unconsolidated mudstone (Landing & Brett 1987). A very thin (10–30 cm) black shale, rich in rhynchonellid brachiopods, which overlies this limestone, appears to be a feather-edge of the black Butternut Shale.

The regressive or falling stage systems tract of the Skaneateles sequence ('lower Centerfield' transitional equivalents) has evidently been removed at the overlying sequence boundary, as the Hungry Hollow Limestone rests sharply and erosionally on the black shale with no intervening transition. This evidence suggests substantial erosional removal of strata below the Ludlowville sequence boundary.

East Findlay Arch Ohio–Plum Brook Shale

The basal, TST portion, of the Skaneateles sequence has been identified in the eastern Findlay Arch region of Ohio as lying at the base of the Plum Brook Formation formerly exposed at the top of the Parkertown Quarry near Sandusky (Conkin & Conkin 1984). The section described consists of ~1.5 m of limestone containing the brachiopods *Platyrachella*, *Tropidoleptus* and *Devonochonetes* along with various tabulate and rugose corals. This succession very closely resembles the basal portion ('Blue Beds') of the Silica Formation of western Ohio. This portion of the sequence is locally absent in eastern Ohio, having been removed beneath an erosive discontinuity higher up in the sequence (basal Pompey subsequence boundary).

In eastern Ohio, the Delphi Station subsequence is currently identifiable in the subsurface, where it is represented by the basal portion of the Plum Brook Formation (Fig. 2). A basal 30 cm, calcareous siltstone in the area of Sandusky, Ohio, that may represent the TST of the Delphi Station subsequence, yields a conodont assemblage indicative of the Givetian *P. timorensis* Zone (Sparling 1995). Strata representing the upper Delphi Station subsequence consist of ~7 m of fine-grained, dark-grey mudstone with a few thin shell beds, which coarsens upward into more calcareous mudstone with a greater concentration of shell material near the top of the sequence.

The Pompey Member subsequence apparently accounts for the bulk of the thickness (~16 m) of the Plum Brook Formation in the Sandusky area, and probably all of the lower Olentangy Shale near Delaware, Ohio. Both units are dark to medium grey, very sparsely fossiliferous shale with a distinct upper zone of scattered pyritized burrows and fossils, including the cephalopods *Tornoceras arkonense* and *Bactrites arkonense*. This pyritized fauna was correlated between the lower Olentangy and the Plum Brook by Tillman (1970), who pointed to the overall faunal similarity of the two units; this interval is also comparable to the pyritic faunal zone of the upper Arkona Shale in Ontario and that of the Pompey in central New York. We interpret all of these as parts of a widespread biotic and taphonomic event that occurred during the *P. timorensis* Zone. A fossil-rich bed near the top of the Plum Brook Shale appears to represent the Peppermill Gulf bed of New York. It lies slightly below the sharp basal erosion surface of the Prout Limestone.

West Findlay Arch Region–Silica Shale

The Skaneateles 3rd-order sequence is represented in the area of Toledo, Ohio (western Findlay Arch) by the bulk of the Silica Formation (Units 1–29 of Mitchell 1967) (Fig. 2). The basal TST is represented by the 'Blue Beds' of older terminology (Units 1–6 of Mitchell 1967). The 'Blue Beds' consist of ~2.5 m of pack- to grainstone containing numerous rugose corals and large brachiopods including '*Platyrachella*', *Tropidoleptus* and *Devonochonetes* (Fig. 9). A thin, discontinuous layer of calcareous shale up to 15 cm thick is present in the middle of this interval and is thought to represent the medial, argillaceous portion of the Mottville Member of New York State. The conodont *Latricriodus latericrescens latericrescens* has also been identified in these lower beds of the Silica Formation (Klapper & Ziegler 1967), again indicating an early Givetian age and corroborating correlation with the basal Skaneateles Formation. The remainder of the Mottville-equivalent strata is encompassed in Silica unit 7, a 75 cm calcareous shale, thought to be equivalent to the Cole Hill submember in New York.

The higher Silica shows three divisions that are considered to represent the Delphi Station, Pompey and Butternut cycles (Fig. 7). The lowest division, assigned to the Brint Road Member (in part) by Mitchell (1967), consists of Units 8–9. The basal highly fossiliferous beds (Unit 8) pass upward into a 4 m-thick shale containing much of the famous Silica Shale fauna (e.g. pyritized brachiopods, such as *Devonochonetes coronatus* and *Paraspirifer*, large *Eldredgeops milleri*, and the camerate crinoid *Arthroacantha*). This mudstone passes upward into shales and lenticular, bryozoan-rich limestones (Units 10–13) that are locally truncated by a disconformity beneath unit 14.

The Pompey subsequence in western Ohio comprises Units 14–18 of the Silica Formation; the Berkey Member of Mitchell (1967) (Fig. 7). Units 14–17 consist of two thin limestone bands and an intervening shale bed totalling 1.3 m; the lower bed shows a marked discontinuity at its base with hypichnial burrows projecting down ~3 cm into the underlying shale bed. Units 16–17 consist of somewhat shalier limestone than the basal bed and are rich in the brachiopod *Ambocoelia*. Above Unit 17 (limestone) is a mound of auloporid corals which locally extends up to ~5 m into the overlying shales. Units 14–17 are probably the lateral equivalents of the Paper Mill and Pole Bridge 4th-order TST package and display the distinctive back-stepping pattern of these units. Overlying these limestones is the thickest siliciclastic unit of the Silica Formation (Unit 18), consisting of about 6.5 m of medium-grey, very sparsely fossiliferous shale, with the only common fossils consisting of auloporid corals and chonetid brachiopods. This unit, comparable to the bulk of the Plum Brook, lower Olentangy and upper Arkona Shale, has yielded rare pyritized fossils, including

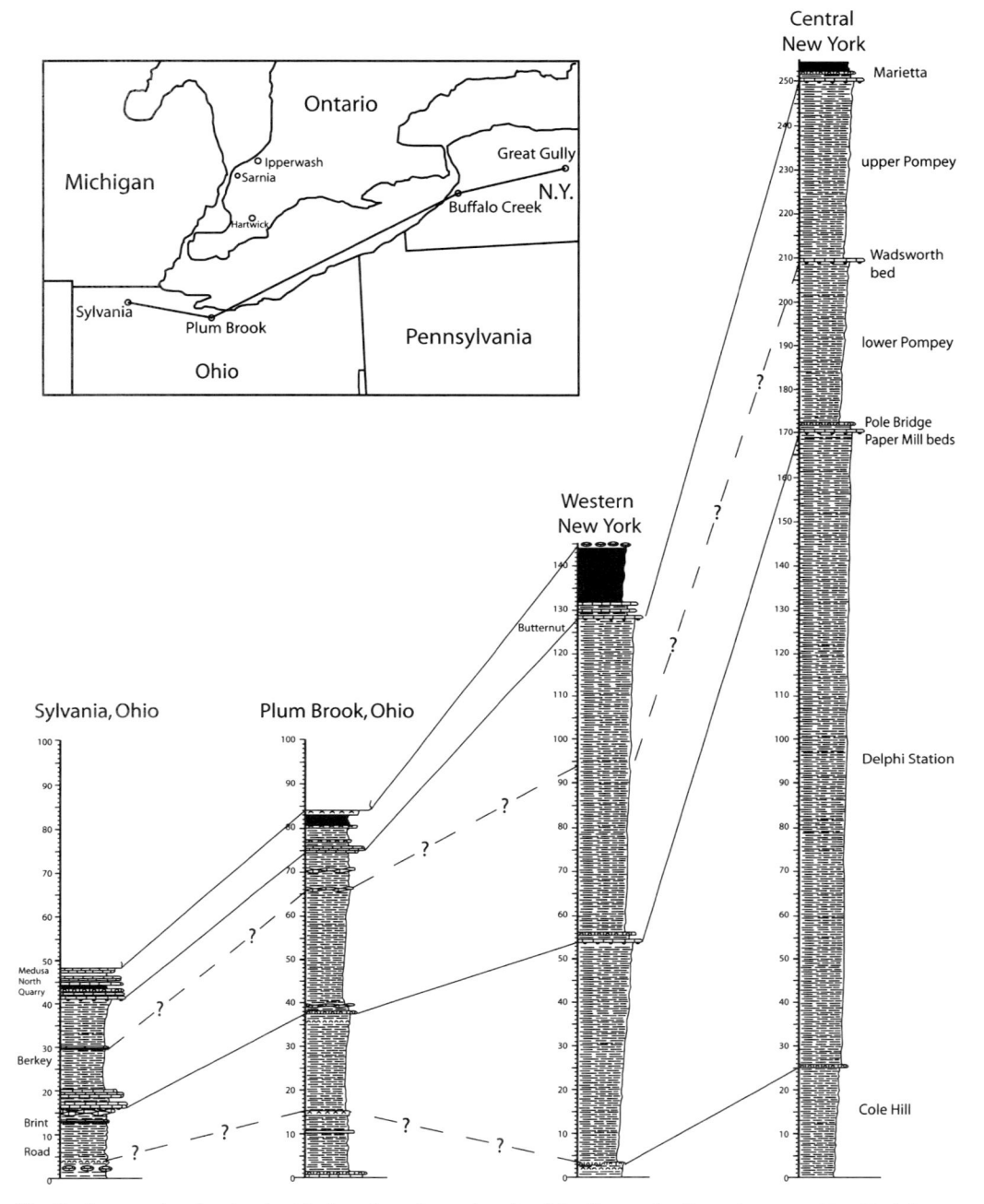

Fig. 7. Cross-section showing the thinning of member scale units of the Skaneateles Formation across New York State and across northern Ohio in the Plum Brook and Silica shales. Scale in feet. Units on the Silica Formation section are from Mitchell, 1967. (D.S.TST, Delphi Station transgressive systems tract.)

Tornoceras arkonense and *T. mesopleuron* (Becker, pers. comm. 2005).

In western Ohio, the Butternut subsequence is represented by beds 19–29 of the upper Berkey Member, Silica Formation (Mitchell 1967) (Fig. 7). The Butternut subsequence here is ~2.25 m thick and consists of a succession of interbedded shell-rich limestones and dark-grey shales, closely resembling the uppermost Arkona. The basal, phosphatic, shell-rich limestone again features incised megaburrow-prods at the base, indicating a discontinuity of the subsequence. The upper beds are shalier and contain an abundance of the rhynchonellid brachiopod *Eumetabolotoechia* in some beds, as in the upper portion of the Plum Brook Formation of eastern Ohio, the Butternut Member shales of New York, and the uppermost Arkona black shales in Ontario. Finally, a thin remnant of the 'lower Centerfield–Chenango' FSST, may be represented in fossiliferous, argillaceous limestone, Units 25–29, which rests sharply upon dark rhynchonellid shales. Again, the upper part of the transition has been removed by erosion prior to deposition of the overlying Ten Mile Creek Dolostone (Fig. 9).

Southeastern Indiana–Swanville Member

The Skaneateles sequence in southeastern Indiana is represented by the Swanville Member of the Sellersburg Formation, preserving only the TST portion of the sequence. The Swanville Member consists of up to ~1.5 m of fossiliferous, crinoidal-grainstone, and is locally absent in the southern portion of the outcrop area in southeastern Indiana (Clark county). Containing an abundance of *Tropidoleptus* and *Devonochonetes* brachiopods near its base along with rare '*Platyrachella*', the fauna of the Swanville Member closely resembles that of the 'Blue Beds' interval at the base of the Silica Formation in northern Ohio. The conodont *Laticriodus latericrescens latericrescens* was reported by Orr & Pollock (1968) as having been extracted from the 'upper Silver Creek' at the Atkins Quarry in Clark county, Indiana. It is now known that the 'upper Silver Creek' of Orr and Pollock represents the Swanville Member of the Sellersburg Formation. The first occurrence of *Lat. l. latericrescens* parallels the first occurrence of this diagnostic element at the base of the Skaneateles sequence as in northern Ohio and New York. The base of the Swanville Member is an erosional disconformity that has removed most of the underlying Silver Creek Member at the Swanville type locality. The upper portions of the Skaneateles sequence have been removed by subsequent erosion beneath the overlying Beechwood Member, which can be seen to rest disconformably upon the Swanville Member at its type locality.

Sequence IV: Ludlowville, Widder, Prout and Ten Mile Formations

New York–Ludlowville Formation

The Ludlowville Formation represents the next higher 3rd-order depositional sequence (Fig. 8) and is a similarly complex interval that, like the Skaneateles, has a thin, persistent carbonate-rich interval, the Centerfield Member at its base, representing the 3rd-order TST of the sequence, the analogue of the Mottville Member (Fig. 4). The Centerfield limestone *sensu stricto* is a thin (0.5 to 2 m), exceptionally fossiliferous pack- to grainstone, with a sharp basal contact, which represents a sequence boundary. This limestone passes upward into calcareous mudstone with thin limestones replete with rugose and tabulate corals in western New York. The top of the Centerfield Member has been drawn at a thin shelly phosphatic pebble bed, the Moonshine Falls Bed, and is interpretable as a maximum flooding surface (Fig. 8).

The remainder of the Ludlowville Formation overlying the Centerfield Member in central New York represents the 3rd-order HST of this sequence and can be subdivided into a series of six 4th-order subsequences in the Ledyard-Otisco (lower and upper submembers), Wanakah-Ivy Point (lower and upper submembers), Spafford, and Jaycox Members (Fig. 8). As with the Skaneateles subsequences, these are represented in central and eastern New York State by thin, shell, coral-rich intervals that are abruptly overlain by coarsening-upward shale-siltstones. The thin, transgressive shell-rich beds and concretionary carbonates are traceable into western New York where they separate intervals of dark to medium-grey shales and mudstones. Pyrite beds contained within the black shales are interpreted as representing the period of maximum water depth in each subsequence.

Southwestern Ontario–Widder Formation

Stauffer (1915) defined the Widder Formation as commencing at the base of the 'Hungry Hollow' Limestone and extending to the base of the Ipperwash Limestone. As such, the Widder parallels Cooper's definition of the Ludlowville Formation. Although subsequent workers have used the term Widder in different senses, we revive Stauffer's definition with only minor modification.

The Hungry Hollow Limestone of southwestern Ontario (redefined as the Hungry Hollow Member of Widder Formation as per Stauffer's, 1915, original definition), was first identified by Cooper & Warthin (1942) as being equivalent to the Centerfield Member on the basis of macrofauna; this was

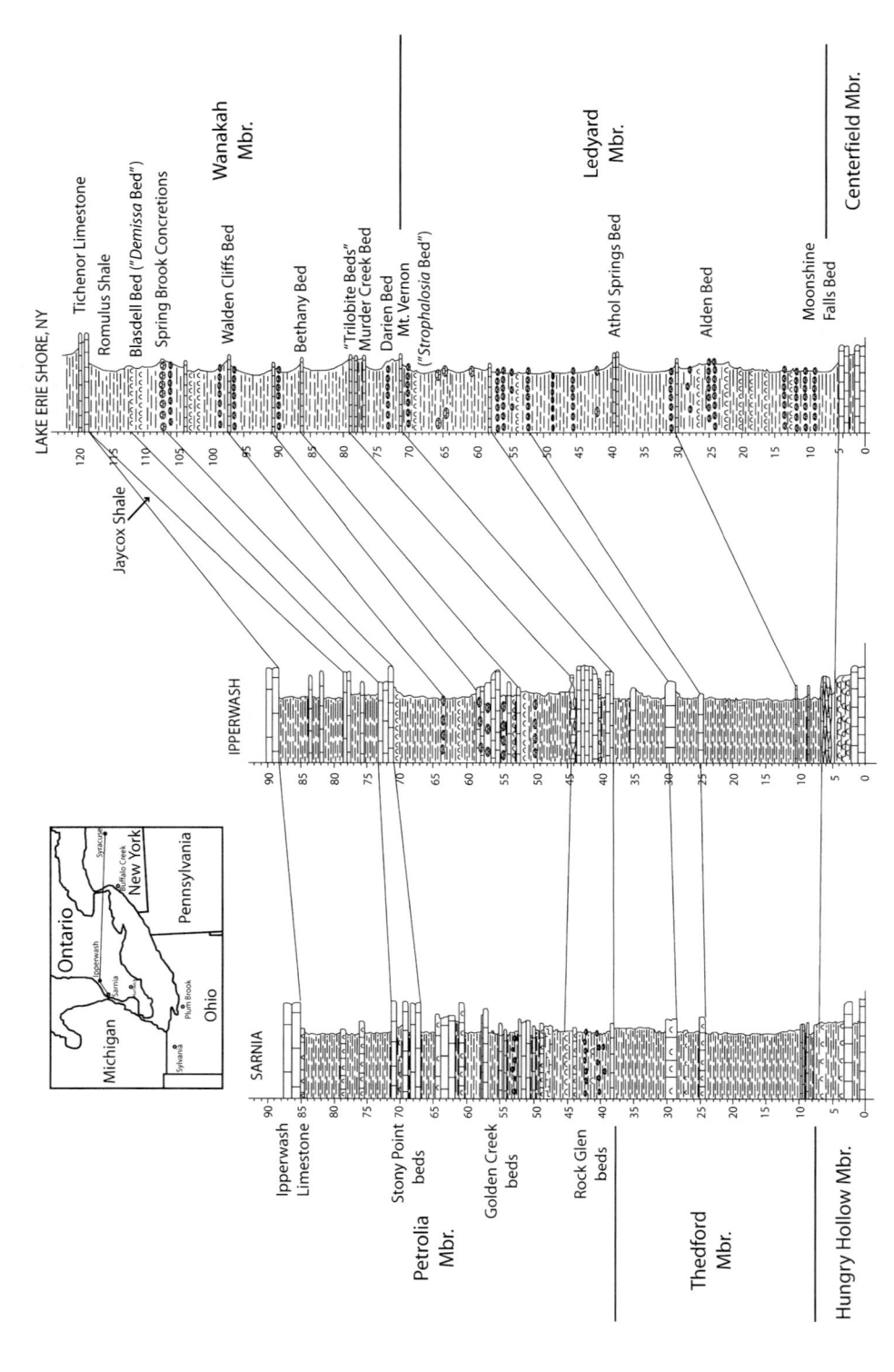

Fig. 8. Cross-section showing change in thickness of units of the Ludlowville and Widder Formations across New York and Ontario. Scale in feet.

corroborated by subsequent conodont studies, which placed both units within the lower *varcus* Zone (Klapper 1971, 1981; Landing & Brett 1987; but see Sparling 1992 for a dissenting opinion) (Fig. 9). Currently, the Centerfield and Hungry Hollow members are placed in the *P. timorensis* Zone but a revision of polygnathids that were previously in the '*P. varcus* group' (Uyeno *et al.* 1982) is desired. As with the Centerfield, the Hungry Hollow is composed of two divisions: a lower crinoidal grainstone bed containing worn fossil material, including corals, and an upper, more argillaceous pack- to wackestone division abounding in rugose and tabulate corals known as the 'coral zone' (Donato 2002). This interval appears to correlate with the upper Centerfield coral biostrome in western New York. A thin phosphatic bed at the top of the 'coral zone' appears to be a local correlative of the Moonshine Falls phosphate bed, interpreted as the maximum flooding surface of the Ludlowville sequence (Fig. 8).

The lower shale unit of the Widder Formation, corresponding to Units 1–14 of Wright & Wright (1961), is ~10 m-thick, approximately half the thickness of the Ledyard in western New York (Fig. 8). This unit, informally termed 'Thedford member' (Bartholomew *et al.* 2006), consists of grey calcareous shale and argillaceous limestone with rhynchonellid brachiopods, pyritized ammonoids, including *T. uniangulare widderi*, and an abundance of the brachiopods '*Mucrospirifer*' *thedfordensis, Arcuaminites scitulus*; *Eumetabolotoechia* cf. *E. multicostum* and the trilobite *Greenops* cf. *G. grabaui*. This morphotype of *Greenops* closely resembles those found in the lower Ledyard Shale of western New York, which has also yielded rare specimens of '*M.*' *thedfordensis* (G. Kloc, pers. comm. 2004). As with the Ledyard–Otisco shales, the Thedford member is characterized by two concretionary limestone intervals separated by a 4 m, soft, grey pyritic shale interval. These limestones presumably represent small-scale TSTs of two 4th-order sequences (Fig. 8).

The remainder of the Widder Formation consists of bluish-grey calcareous shale and medium-grey, argillaceous, nodular limestones, with minor pale-grey to cream-coloured chert. This interval shows a detailed pattern similar to that of the Wanakah, Spafford and Jaycox members of the Ludlowville sequence.

Three major packages of argillaceous limestone and nodular, calcareous shale correspond to three major shallowing-upward cycles in the Wanakah Shale (Fig. 8). Each is bounded by condensed, glauconitic skeletal limestones with sharp erosive bases (4th-order sequence boundaries), marked by incised hypichnial burrow fills. This succession is approximately half as thick as the Wanakah Shale in western New York. The lower cycle, marked near its base by hard, blocky argillaceous limestones, caps falls in tributaries of the Ausable River, including Rock Glen north of Arkona. This interval, identified by Wright & Wright (1961) as Units 14–23, is informally termed the 'Rock Glen member' (Bartholomew *et al.* 2006) (Fig. 8). The Rock Glen member shows moderately diverse faunas, including the large strophomenid brachiopods *Strophodonta* and *Megastrophia*, auloporid corals and large nautiloids. This interval is correlated with the lower Wanakah, Murder Creek and Bidwell beds (Grabau's 1899 *Truncalosia, Pleurodictyum, Nautilus*, and trilobite beds of the classic Eighteen Mile Creek section south of Buffalo, NY).

Higher shales and thin limestones, originally identified as Petrolia Shale, are poorly exposed but may be correlated bed for bed through drill cores (Fig. 8). This succession contains two major skeletal limestone beds that may form bases of two fossiliferous concretionary limestone intervals in the middle Wanakah, equivalent to the Walden Cliffs and Blasdell beds in New York. The predominance of sparsely fossiliferous shale with chonetids and *Mucrospirifer* in this portion of the Petrolia is in line with the generally dysoxic shales of the middle Wanakah.

A 1.5 m-thick interval of glauconitic limestones with abundant strophomenid brachiopods, *Spinocyrtia* and the crinoid holdfast *Ancyrocrinus*, formerly termed the lower Ipperwash Limestone is exposed at Stony Point along the shore of Lake Huron. This interval referred to as the Stony Point beds, appears to record the major upper Wanakah Blasdell shell beds (Grabau's 1899 *Strophodonta demissa* and '*Stictopora*' beds). This correlation is supported by the occurrence of a distinctive morphotype of *Spinocyrtia* (*S. carinatus* Ehlers & Wright 1955), which very closely resembles (and may be synonymous with) a form of *S. granulosa* unique to the Blasdell beds.

The very poorly exposed upper Petrolia beds have been correlated in the subsurface and consist mainly of 10 m of soft bluish-grey, highly fossiliferous shale. The fauna of abundant chonetid brachiopods, especially *Arcuaminites scitulus* and *Longispina? vicinus*, along with *Mucrospirifer*, *Sulcoretepora* and the distinctive *Spinocyrtia ravenswoodensis* Ehlers & Wright, are typical of the highest Ludlowville beds assigned in New York to the Spafford and Jaycox members (Fig. 8).

Ohio–Prout and Ten Mile Creek Formations

Correlation of the Centerfield Member (s.s.) and Hungry Hollow with the basal two units of the Prout Limestone seems nearly certain, not only because of conodont biostratigraphy (see previous discussion for evidence of reworking in upper), but because of the very high degree of macrofaunal similarity (see below), including unique taxa, such as *Fimbrispirifer venustus* and *Callipleura nobilis*, which are nearly restricted to the Centerfield. The basal coral-rich grainstone of the Prout probably links with the main Centerfield Limestone and its sharp basal surface represents the Ludlowville sequence boundary. An overlying, more shaly succession rich in corals probably represents the upper Centerfield and upper Hungry Hollow biostromes.

In western Ohio, the basal units of the Ten Mile Creek Dolostone also show a strong similarity with the lower Prout, Hungry Hollow and Centerfield, and yield abundant rugose and tabulate corals as well as the distinctive crinoid holdfasts of the genus *Ancyrocrinus*. Again, the sharply erosive base of the lowest Ten Mile Creek forms the Ludlowville sequence boundary, though dolomitization has obscured details of most of the higher beds of the Ten Mile Creek. However, a less-dolomitized area of this succession, formerly exposed at the Seacrest Quarry south of Sylvania, Ohio, exhibited an upper shaly coral biostrome with unique blastoids and the trilobite *Basidechenella rowi*, both typical of the upper Hungry Hollow Member (A. Fabian, pers. comm. 2004).

The equivalent of the Ledyard Member is thought to be represented by thin shale and cherty dolowackestone and packstones of the middle to upper Ten Mile Creek Dolostone in Ohio. The nodular, cherty facies somewhat resemble those of the Petrolia Member in Ontario, but the details remain obscure. Equivalents of the higher Ludlowville succession are not securely recognized in the Ten Mile Creek Formation in Ohio, although it is likely that fossiliferous, coral-bearing, cherty, micritic limestones (Units 9–13 of Stauffer 1909) in the upper Ten Mile Creek may be equivalent to portions of the Ludlowville above the Centerfield Member.

Southeastern Indiana–Beechwood Member

The Ludlowville sequence is represented in southeastern Indiana by the Beechwood Member of the Sellersburg Formation. The Beechwood Member has long been considered to be equivalent to the Centerfield Member of New York (Cooper & Warthin 1942). The fauna of the Beechwood Member has been shown to contain all of the distinct elements of the Centerfield Member and also the first appearance of the conodont *Polygnathus timorensis* (Klug 1983). Sparling (1999) suggested a Middle *varcus* Zone age for the Beechwood based on *Polygnathus linguiformis klapperi*. Elsewhere, however, this subspecies is known to occur much lower in the section (Bultynck 1987), making it an unreliable guide fossil for this zone. The presence of *P. timorensis*, along with macrofaunal similarities to the Centerfield and Hungry Hollow members, suggests the Beechwood Member should be placed near the base of the *P. rhenanus* zone.

The Beechwood Member is the most persistent of all the units of the Sellersburg Formation, and is still recognizable in the northernmost exposures of the formation around Vernon, Indiana. The base of the Beechwood Member is distinctly erosional, often containing large pebble- and cobble-sized phosphatic clasts as well as an abundance of glauconite, both indicative of intense sediment starvation and sediment reworking. The upper surface of the Beechwood Member displays evidence for lengthy condensation with pyrite and phosphate encrustation and is overlain by a thin conodont-rich bed at the base of the New Albany Shale that contains elements indicative of the *P. disparilis* Zone (Orr & Klapper 1968); the intervening upper *P. rhenanus* to *P. hermanni* Zones are absent. Locally, the Beechwood Member is absent and the overlying conodont bed at the base of the New Albany Shale rests upon lower strata of the Sellersburg Formation.

Sequence V: Moscow Formation, Ipperwash and unnamed units in Ontario

New York–Moscow Formation

The Moscow Formation of New York has been studied in detail (Cooper 1933; Baird 1979; Brett & Baird 1994). Again, the base of the interval is the sharp lower contact of a thin (30–60 cm) crinoidal grainstone and coral-rich interval, the Tichenor Limestone, analogous to the Centerfield and Stafford-Mottville. The Tichenor is the lowest unit of a complex of calcareous shale and limestone (Deep Run Shale, a hard silty, calcareous mudstone; Menteth Limestone, a silty bioturbated wackestone; and Kashong Shale, a bluish-grey shale rich in *Tropidoleptus*, with middle and upper condensed shell beds). The Tichenor Limestone contains the regional first appearance of *P. rhenanus*, although goniatite evidence from the underlying Ledyard and Wanakah members of the Ludlowville Formation indicates a delayed entry of this species in the region (Becker 2005). The *P. ansatus* Zone is recognized as beginning with the top of Kashong Shale (Ziegler *et al*. 1976). These limestones thin and become stacked into a 0.5 to 3 m interval in both western and central New York, where they and the Tichenor are collectively termed Portland Point Member. The upper Moscow comprises the widespread Windom Shale Member and is divisible into two major shallowing-upward cycles.

Southwestern Ontario – Ipperwash and unnamed uppermost beds

Strata lying above the redefined Widder Formation are herein placed within a newly recognized formation-scale unit apparently correlative with the Moscow Formation (Fig. 9). This unit is best developed in the subsurface in the Sarnia, Ontario, area. The basal portion of this unit consists of the 'upper Ipperwash' of previous workers (Cooper & Warthin 1942; Wright & Wright 1963), herein redefined as the Ipperwash Member (restricted); this was the original unit identified at Ipperwash Point by Stauffer (1915). The Ipperwash Member is the only portion of the Moscow exposed in Ontario in outcrops along the shores of Lake Huron at Silica Point and also present in blocks dredged from boat slips along Lake Huron, four miles to the south and at Smith's Falls on the Seydenham River, near Shetland.

The Ipperwash Member, up to 1 m thick, consists of dark-grey limestone with whole and broken and worn fossils, including the brachiopods *Longispina*? *vicinus*, *Spinocyrtia ravenswoodensis*, *Mucrospirifer*, *Tropidoleptus* and *Protoleptostrophia*; sharply incised hypichnial megaburrow-prods filled with pyritic skeletal debris are present at the base of the unit. The upper portion of this bed is a *Zoophycos*-churned, *Tropidoleptus*-bearing calcistiltite, with black chert nodules. We suggest that the sharp prodded basal surface of the Ipperwash Limestone represents the basal Moscow sequence boundary and that the overlying Ipperwash (s.s.) represents a combination of Tichenor and Menteth Members. The Tichenor often exhibits megaburrows and commonly carries a fauna derived by reworking of underlying muds, as appears to be the case with the base of the Ipperwash Member. Throughout its outcrop, the Menteth is *Zoophycos*-swirled silty carbonate and contains abundant *Tropidoleptus*; it is also the only slightly cherty limestone in the upper Hamilton Group. We would therefore suggest that the upper portion of the Ipperwash Limestone represents the Menteth Member of New York State (Fig. 8).

Beds above the Ipperwash Member in Ontario are only present in the subsurface in the area near Sarnia, Ontario. In its type area, the Ipperwash Member is capped by a pyrite-coated discontinuity at the base of the Upper Devonian Kettle Point Formation. However, in a core from the subsurface near Sarnia, about 6 m of calcareous, fossiliferous, grey shales are observed to overlie the same chert bed at the top of the Ipperwash Member. At the top of this unit is a pyritic discontinuity overlain by the Kettle Point Formation. This unit, herein termed the Sarnia Shale member, has heretofore gone unnoticed in the Middle Devonian of Ontario and helps to identify a top-down removal of units to the east by the overlying unconformity with the Upper Devonian Kettle Point Formation. At this time, the fauna of this unit is poorly understood, but it does contain brachiopods indicative of a Middle Devonian age such as *Rhipidomella penelope*. We tentatively correlate these beds with the upper Kashong to lower Windom members of the Moscow Formation of New York.

The Ipperwash Limestone appears to correlate with upper cherty coral beds at the top of the Prout Formation in the eastern Ohio, and possibly to upper coral beds in the Ten Mile Creek Dolostone of the western Findlay Arch area.

Discussion: implications of detailed correlations

Our ability to correlate all 3rd- and even 4th-order sequences recognized in the New York foreland basin into Ontario, Ohio and Indiana, both east and west of the Findlay Arch, has very important implications for interpreting the processes that generated the cycles, as well as the palaeogeography of the region. The combined biostratigraphical, sequence, and event stratigraphic correlations also provide a well-corroborated, detailed framework within which to examine epeirogenic, far-field tectonic effects, depositional dynamics, patterns of biogeography, faunal migration and biotic turnover, and will facilitate more-refined global correlations.

Implications for relative sea-level

Each major and minor sequence boundary and marker limestone bed (TST) previously identified in the Oatka Creek, Skaneateles and Ludlowville Formations in New York was also recognized in southwestern Ontario (Fig. 9). The same is true for the Skaneateles and lower Ludlowville Formations in northern Ohio, although higher units are either absent at a major post-Middle Devonian unconformity or obscured by condensation and/or dolomitization of the remaining units. Ongoing work by the authors indicates that most of these bracketing carbonates and the sequences they delineate are also recognizable in the Traverse Group of the Michigan Basin proper. This evidence strongly implicates allocyclic, probably eustatic, oscillations in sea-level as the driving mechanism behind these cycles, perhaps surprising in view of the active Acadian tectonism that was ongoing during the Middle Devonian.

The magnitude of the relative sea-level oscillations that produced 3rd- and 4th-order sequences across eastern North America is uncertain. First, there is evidence that biofacies, in particular systems tracts of specific sequences, are not highly different from western New York to Ontario to western Ohio. A good example is the transgressive systems tract of the Ludlowville sequence; the Centerfield Limestone, Hungry Hollow, Prout Limestone and Ten Mile Creek Dolostone share strong similarities in litho- and biofacies (see above). That said, there are subtle and gradual differences in this interval over the study region suggesting that the limestones record very subtle gradients existing during times of low sedimentation rather than markedly diachronous facies migrations. This suggests that topography over the entire region was quite subdued, despite differences in subsidence rates that must have been very nearly balanced by sedimentation.

It is highly important that, despite evidence for minor local shallowing or deepening, the approximate extent of facies change across sequence boundaries and flooding surfaces is similar in all regions considered. The total maximum relative water depth range is approximately represented in the difference between black rhynchonellid-bearing shales and coral-rich biostromes. Previous estimates, based on storm wave base and evidence of the photic zone, including microendolith studies (e.g. Vogel *et al.* 1987, see Brett *et al.* 1993), suggest that the depth range recorded in this biofacies spectrum is on the order of tens of metres, perhaps ranging from about 10 to 50 m. If so, then the larger cycles record about 30 to 50 m of water depth change, while smaller oscillations recorded in 4th-order sequences may reflect 10–20 m. It remains an important and tantalizing question as to how such sea-level variations might be produced during a greenhouse time, such as the Middle Devonian. However, it is increasingly clear that sea-level cycles of comparable amplitude and frequency are recorded in marine strata of many greenhouse intervals including the Ordovician, Jurassic and Cretaceous (Elder *et al.* 1994; Sageman & Hollander 1999; Holland & Patzkowsky 2004). We suggest that these oscillations reflect the synergistic effects of thermal expansion/contraction of seawater, and storage/release of water sequestered in freshwater reservoirs or mountain glaciers (Miller *et al.* 2005).

Sedimentological implications

It is impressive that nearly all condensed, fossiliferous marker horizons that demarcate transgressive systems tracts are traceable from the eastern portions of the Acadian Foreland Basin in New York and Pennsylvania to the margins of the Michigan Basin, despite a 20- to 100-fold thinning of intervening siliciclastic wedges. It is notable that the concretionary limestones interpreted as TSTs show relatively little change in thickness across the entire study area from New York to Michigan, perhaps reflecting their accumulation through autochthonous carbonate sedimentation in the face of

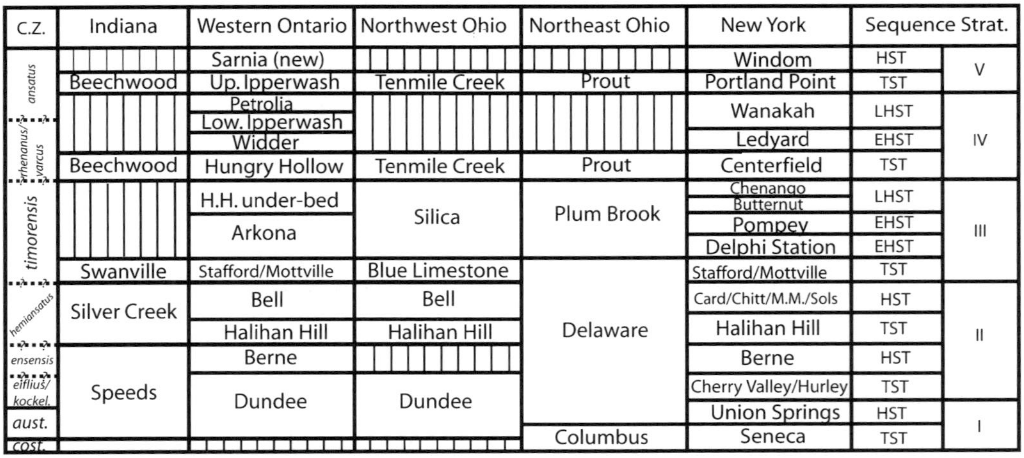

C.Z.	Indiana	Western Ontario	Northwest Ohio	Northeast Ohio	New York	Sequence Strat.	
ansatus		Sarnia (new)			Windom	HST	V
	Beechwood	Up. Ipperwash	Tenmile Creek	Prout	Portland Point	TST	
varcus / *rhenanus*		Petrolia			Wanakah	LHST	IV
		Low. Ipperwash					
		Widder			Ledyard	EHST	
	Beechwood	Hungry Hollow	Tenmile Creek	Prout	Centerfield	TST	
timorensis		H.H. under-bed	Silica	Plum Brook	Chenango / Butternut	LHST	III
		Arkona			Pompey	EHST	
					Delphi Station	EHST	
	Swanville	Stafford/Mottville	Blue Limestone		Stafford/Mottville	TST	
hemiansatus	Silver Creek	Bell	Bell	Delaware	Card/Chitt/M.M./Sols	HST	II
		Halihan Hill	Halihan Hill		Halihan Hill	TST	
ensensis		Berne			Berne	HST	
eiflius / *kockel.*	Speeds	Dundee	Dundee		Cherry Valley/Hurley	TST	
aust.					Union Springs	HST	I
cost.				Columbus	Seneca	TST	

Fig. 9. Time–rock chart showing correlation of units across the study area with vertical lines representing unconformities. Conodont zones are listed on the left-hand side of the diagram. The Tichenor Member is a partial lateral equivalent to the Portland Point Member of the Moscow Formation within New York State. (*cost.*, *costatus* Zone; *aust.*, *australis* Zone; *kockel.*, *kockelianus* Zone; Card, Cardiff Member; Chitt, Chittenango Member; M.M., Mount Marion Member; Sols, Solsville Member; TST, transgressive systems tract; HST, highstand systems tract; EHST, early highstand systems tract; LHST, late highstand systems tract.)

siliciclastic starvation, although facies suggest westward shallowing. However, intervening shale units thin markedly, probably by a combination of lower sedimentation rates and erosional truncation at minor sequence boundaries. We suggest that the source of muds for the Plum Brook, Silica and Arkona-Widder Formations was from the Acadian Orogen, as for equivalent strata in New York. There does not appear to be any other source available in the mid-continent.

The Arkona and Silica Shales are justifiably famous for extraordinary preservation of crinoids and trilobites in multiple obrution deposits (see Kesling & Chilman 1975; Brett 1998), even in western localities where individual shale units are extremely thin. This observation raises very important questions as to how clays were dispersed so widely and deposited as blankets during single events hundreds of kilometres from their source. We suggest that these represent enormous mud plumes that were dispersed following giant storm floods along water mass boundaries and thus extended far west of the foreland basin.

Palaeogeographical implications

As noted, the upper Hamilton units of New York are physically separated from coeval strata in western Ontario and northern Ohio by a north–south trending area in which these units are truncated by an unconformity. Dark-grey shales of the Skaneateles sequence (Skaneateles–Plum Brook formations) are overlain by Upper Devonian Huron black or dark-grey shales at a cryptic contact seen in both northern Ohio and southern Ontario east of London. However, the fact that counterparts of the Skaneateles and Ludlowville formations can be correlated on either side of this region suggests that this area became uplifted and truncated only after deposition of these units.

The Oatka Creek Formation can be traced across this region in the subsurface of southern Ontario and the transition from black, laminated Chittenango Shale facies to grey, fossiliferous mudstones of the Bell Shale can be documented in the series of drill cores from Port Stanley near London to Ipperwash and finally Sarnia, Ontario. This change is accompanied by thinning. For example the Bell Shale thins from 23 to 17.5 m between Ipperwash and Sarnia and exhibits a much richer fauna, largely of species typical of the Hamilton Group in the Appalachian Basin. Likewise, the overlying Arkona Shale shows a general slight westward thinning and increase in fossil content in this region and is presumed to pass westward into the thinner and more fossiliferous, limestone-rich Silica Shale west of Sarnia, Ontario, in Michigan. Fossiliferous Silica Shale is known from Detroit, 80 km to the south (Kesling & Chilman 1975). Overlying Ludlowville units show only minor westward thinning.

The Plum Brook Shale near Sandusky, Ohio, is of similar facies to the Arkona Shale near its type area and represents an approximate extension of the facies belt about 320 km to the south–southwest. The facies shift observed between dark shale strata in eastern Loraine County, resembling the Skaneateles Formation of New York, to strata of Arkona lithology in western Loraine County takes place over a short lateral distance, ~25 km (Fig. 2). The location of this facies change is the same as that at which thickness change takes place in the underlying Oatka Creek sequence, indicating the presence of a long-term basinal feature dividing the siliciclastic-rich shales to the east from more calcareous shales to the west.

The fact that the Plum Brook succession approximately matches thinner but similar units in the Silica Shale just west of the present Findlay Arch indicates that the eastern Findlay Arch and western Findlay Arch regions were not separate depositional basins during deposition of the lower to middle Hamilton Group (Fig. 2). As with the Arkona Shale, this thinning transition from Plum Brook to Silica is accompanied by a greater proportion of tempestitic limestones and increased diversity and abundance of fossils indicating a westward shallowing. Moreover, the gradual thinning of the Silica Shale from its type area into western Ohio and Indiana suggests progressive westward shallowing onto a platform.

Collectively, this evidence indicates a north–northeastward facies strike perpendicular to an east-dipping ramp from the present day western Ohio–Indiana line through eastern Ohio into New York. The Findlay Arch, as expressed in the modern outcrop patterns, was not present, at least in its present location, during deposition of the upper Skaneateles and lower Ludlowville sediments. This further suggests that there was no physical barrier between the Appalachian and Michigan Basins during deposition of the lower to middle Hamilton strata. This inference is further supported by strong faunal similarities between these units (see below). Indeed, it is questionable whether the Michigan Basin, as such, existed during this time as a distinctive tectono-sedimentary feature.

Conversely, strong condensation of the upper Hamilton (upper Ludlowville–Moscow formations) in the Sandusky area of Ohio may further indicate that this region was experiencing gentle upwarping during later Givetian time, which perhaps resulted in the late Middle Devonian truncation of most of the previously deposited Hamilton sediments slightly farther east between Sandusky and the Appalachian Basin. This could be thought of as a precursor of the Findlay Arch, although it lay substantially to the east of the present arch. We suggest that this erosion was related to crustal flexure, possibly a forebulge, that developed in response to renewed thrust loading during the third tectophase of the Acadian Orogeny in late Middle Devonian time (Ettensohn 1987).

Faunal comparisons and biogeography

A few distinct taxa occur in the Hamilton Group of southwestern Ontario. For example, the spiriferid brachiopod '*Mucrospirifer*' *thedfordensis* is exceptionally abundant in the lower Thedford member of the Widder Formation but is rare in the Appalachian Basin; only a few specimens have been found in the Ledyard Shale of New York. This taxon also occurs in the upper portion of the Traverse Group of Michigan (Tillman 1964) and in the Hudson Bay lowland region of northern Ontario/Manitoba (Norris 1993). A few other unique species, particularly of spiriferid brachiopods, have been reported (e.g. see Ehlers & Wright 1955; Wright & Wright 1963), although several of these are closely similar to forms found in New York State. In addition, a few of the coral genera are shared with the Michigan Basin, such as the colonial rugosan *Hexagonaria*. Despite these few anomalous occurrences, however, a considerable majority of the species and most genera found in the Arkona area are shared with the Appalachian Basin fauna, indicating that in spite of its proximity to the Michigan Basin, southwestern Ontario was faunally more closely associated with the Appalachian Basin of New York. This connection is further underscored by the coincidence in timing of several epiboles (e.g. the abundance of *Tropidoleptus* at the base of the Skaneateles sequence and *Eumetabolotoechia* just below the Ludlowville sequence boundary, as well as the pyritic goniatite beds of the upper Arkona Shale, Pompey Member and Plum Brook Shale).

The Plum Brook Shale/Prout Limestone in eastern Ohio show striking faunal similarities with the Skaneateles Formation and Centerfield Member of New York and are clearly developed in the same basin. The Silica Shale is viewed herein as the upramp equivalent of the Arkona–Plum Brook formations and it shares a majority of taxa with these units; most genera are common across the basin, but the Silica shows the lowest similarity with the New York Hamilton fauna. However, a notable proportion of Silica Formation taxa are shared with the Traverse Group in the northern portion of lower peninsular Michigan; these include various taxa, such as the brachiopod *Schizophoria* and the colonial rugosan *Hexagonaria*. These observations suggest that the Silica, although still firmly within the Appalachian Basin Faunal Subprovince, was influenced by the bordering Michigan Basin Subprovince (Koch 1981).

Conclusions

The lower Givetian rocks of Ontario, Ohio and Indiana all show marked similarity in both stratal pattern and faunal composition to the more proximal portions of the Appalachian Basin in New York. Sediments previously identified as lying within separate depositional basins (e.g. Silica Formation of Ohio and Sellersburg Formation of Indiana) can be shown, both faunally and stratigraphically, to be more closely related to the Appalachian Basin than to either the adjacent Michigan or Illinois Basins, indicating that during this time these regions should be viewed as distal, upramp areas within the Appalachian Basin. The existence of the Findlay Arch during this time is challenged, with the above evidence indicating that it was not an active feature at least during the early Givetian. Strata in Ontario, though deposited in a 'transitional' area between the Appalachian and Michigan Basins, show a stronger relationship to the Appalachian Basin during this time. Strata in northwestern Ohio (Silica and Ten Mile Creek formations), though they display the least amount of faunal similarity to the Appalachian Basin, having a large percentage of Michigan Basin faunal elements, still have more in common faunally with the Appohimchi Faunal sub-province than with the Michigan Basin sub-province, and can be shown to have been under similar depositional influences as the more proximal portions of the Appalachian Basin during this time. The ability to correlate high-order (4th-order and above) depositional sequences across the entire study area suggests that the controls on depositional processes were allocyclic in nature and not confined to merely the more proximal portions of the basin where they were initially delineated (Brett & Baird (1996). It is now possible, with the delineation of a fine-scale stratigraphic framework (Fig. 9) to delve more deeply into questions about faunal migration and the interplay between local and regional depositional processes.

The authors wish to acknowledge the assistance of Gordon Baird, Michael DeSantis, Charles Ver Straeten, Patrick McLaughlin, Lindsay Leighton, Cameron Tsujita, and Arnold Miller for both help in the field and in discussing the ideas presented herein. Tim Phillips aided with figure preparation. This paper benefited greatly from reviews by R. Thomas Becker and William Kirchgasser. We would also like to acknowledge Ron Rhea from the

H. R. Collins Core Laboratory of the Ohio Department of Natural Resources. Research by A.J.B. was supported by grants from the Geological Society of America, American Museum of Natural History, and the University Research Council of the University of Cincinnati. This paper is a contribution to IGCP-499.

References

BAIRD, G. C. 1979. Sedimentary relationships of Portland Point and associated Middle Devonian rocks in central and western New York. *New York State Museum Bulletin*, **433**, 24.

BAIRD, G. C. & BRETT, C. E. 1986. Erosion on an anerobic seafloor: significance of reworked pyrite deposits from the Devonian of New York. *Palaeogeography, Palaeoclimatology, Palaeoecology*, **57**, 157–193.

BAIRD, G. C., BRETT, C. E. & VER STRAETEN, C. 1999. The first great Devonian Flooding episodes in western New York: reexamination of Union Springs, Oatka Creek, and Skaneateles formation succession (latest Eifelian–lower Givetian) in the Buffalo–Seneca Lake region. *In*: BAIRD, G. C. & LASH, G. G. (eds) *Field Trip Guidebook New York State Geological Association*, 71st Annual Meeting, Freedonia, New York, A1-Sat. A44.

BAIRD, G. C., BRETT, C. E. & VER STRAETEN, C. 2000. Facies and fossils of the lower Hamilton Group (Middle Devonian) in the Livingston County–Onondaga County region. *In: Field Trip Guidebook New York State Geological Association*, 72nd Annual Meeting, Geneva, NY, pp. 155–175.

BARTHOLOMEW, A. J., BRETT, C. E., DESANTIS, M., BAIRD, G. C. & TSUJITA, C. 2006. Sequence stratigraphy of the Middle Devonian at the border of the Michigan Basin: Implications for sea-level change and paleogeography. *Northeastern Geology*, **28**, 2–33.

BECKER, R. T. 2005. Ammonoids and Substage Subdivisions in the Givetian Open Shelf Facies. *Contributions to the Devonian Terrestrial and Marine Environments: From Continent to Shelf*, IGCP 499/Subcommission on Devonian Stratigraphy joint field meeting, Novosibirsk, Russia, pp. 29–31.

BECKER, R. T. & HOUSE, M. R. 2000. Devonian ammonoid zones and their correlation with established series and stage boundaries. *Courier Forschungsinstitut Senckenberg*, **220**, 113–151.

BEST, E. W. 1953. Pre-Hamilton Devonian stratigraphy of southwestern Ontario, Canada. Unpublished doctoral thesis, University of Wisconsin.

BOUCOT, A. 1990. Silurian and pre-Upper Devonian bio-events. *In*: WALLISER, O. H. (ed.) *Lecture Notes in Earth Sciences*, **30**, Springer-Verlag, Berlin–Heidelberg–New York.

BRETT, C. E. (ed.) 1986. Dynamic Stratigraphy and Depositional Environments of the Hamilton Group (Middle Devonian) in New York State, Part I. *New York State Museum Bulletin*, **457**, 156pp.

BRETT, C. E. 1998. Sequence stratigraphy, palaeoecology, and evolution: biotic clues and responses to sea-level fluctuations. *Palaios*, **13**, 241–262.

BRETT, C. E., BOUCOT, A. J. & JONES, B. 1993. Absolute depths of Silurian benthic assemblages. *Lethaia*, **26**, 25–40.

BRETT, C. E. & BAIRD, G. C. 1994. Depositional sequences, cycles, and foreland basin dynamics in the late Middle Devonian (Givetian) of the Genesee Valley and western Finger Lakes Region. *In*: BRETT, C. E. & SCATTERDAY, J. (eds) *Field Trip Guidebook New York State Geological Association*. 66th Annual Meeting, Rochester, New York, 505–586.

BRETT, C. E. & BAIRD, G. C. 1995. Coordinated stasis and evolutionary ecology of Silurian to Middle Devonian marine biotas in the Appalachian basin. *In*: ERWIN, D. & ANSTEY, R. (eds) *New Approaches to Speciation in the Fossil Record*. Columbia University Press, New York, 285–315.

BRETT, C. E. & BAIRD, G. C. 1996. Middle Devonian sedimentary cycles and sequences in the northern Appalachian Basin. *In*: WITZKE, B. J. & DAY, J. (eds) *Paleozoic Sequence Stratigraphy; View from the North American Craton*. Geological Society of America, Special Paper, **306**, 213–241.

BRETT, C. E., BAIRD, G. C. & BARTHOLOMEW, A. J. 2004. Sequence stratigraphy of highly variable middle Devonian strata in central Kentucky: implications for regional correlations and depositional environments. *Great Lake Section SEPM Annual Field Conference Guidebook*, 35–60.

BULTYNCK, P. 1987. Pelagic and neritic conodont succession from the Givetian of pre-Sahara Morocco and the Ardennes. *Bulletin de l'Institut Royal des Sciences Naturelles de Belgique, Sciences de la Terre*, **57**, 149–181.

CAMPBELL, G. 1942. Middle Devonian stratigraphy of Indiana. *Geological Society of America Bulletin*, **53**, 1055–1072.

CATUNEANU, O. 2002. Sequence stratigraphy of clastic systems: concepts, merits, and pitfalls. *Journal of African Earth Sciences*, **35**, 1–43.

COE, A. L. (ed.) 2003. *The Sedimentary Record of Sea-Level Change*. Cambridge University Press.

CONKIN, J. E. & CONKIN, B. M. 1984. Paleozoic metabentonites of North America, Part I. *University of Louisville Studies in Palaeontology and Stratigraphy*, **16**, 136.

COOPER, G. A. 1930. Stratigraphy of the Hamilton Group of New York. *American Journal of Science*, **19**, 116–134, 214–236.

COOPER, G. A. 1933. Stratigraphy of the Hamilton Group of Eastern New York. *American Journal of Science*, **26**, 537–551; **27**, 1–12.

COOPER, G. A. & WARTHIN, A. S. 1942. New Devonian (Hamilton) correlations. *Geological Society of America Bulletin*, **53**, 873–888.

COOPER, G. A. *ET AL*. 1942. Correlation of Devonian sedimentary rocks of North America. *Geological Society of America Bulletin*, **53**, 1729–1793.

DESANTIS, M. K., BRETT, C. E. & VER STRAETEN, C. A. 2007. Persistent depositional sequences and bioevents in the Eifelian (early Middle Devonian) of eastern Laurentia: North American evidence of the Kačák Events? *In*: BECKER, R. T. & KIRCHGASSER, W. T. (eds) *Devonian Events and Correlations*.

Geological Society, London, Special Publications, **278**, 83–104.

DONATO, S. 2002. Palaeoecology of the Hungry Hollow Formation. Unpublished M.S. Thesis, University of Western Ontario.

EHLERS, G. M. & WRIGHT, J. D. 1955. The type species of *Spinocyrtia* Fredricks and new species of this brachiopod genus from Southwestern Ontario. *Contributions to the Museum of Paleontology, University of Michigan*, **13**, 1–32.

ELDER, W. P., GUSTASON, E. R. & SAGEMAN, B. B. 1994. Correlation of basinal carbonate cycles to nearshore parasequences in the Late Cretaceous Greenhorn Seaway, Western Interior U.S.A. *Geological Society of America Bulletin*, **106**, 892–902.

EMERY, D. & MEYERS, K. J. 1996. *Sequence Stratigraphy*, Blackwell Science Ltd., 297 pp.

ETTENSOHN, F. R. 1987. Rates of relative plate motion during the Acadian Orogeny based on spatial distribution of black shales. *Journal of Geology*, **95**, 572–582.

GRABAU, A. W. 1899. Geology and palaeontology along 18-Mile Creek. *Buffalo Society of Natural Sciences*, Buffalo, New York, 390 pp.

GRASSO, T. X. 1986. Redefinition, stratigraphy, and depositional environments of the Mottville Member (Hamilton Group) in central and eastern New York. *In*: BRETT, C. E. (ed.) Dynamic stratigraphy and depositional environments of the Hamilton Group (Middle Devonian) in New York State, Part I. *New York State Museum Bulletin*, **457**, 5–31.

GRAY, L. M. 1991. The palaeoecology, origin, and significance of a shell-rich bed in the lowermost part of the Ludlowville Formation (Middle Devonian, Central New York). *In*: LANDING, E. & BRETT, C. E. (eds) Dynamic stratigraphy and depositional environments of the Hamilton Group (Middle Devonian) in New York State, Part II. *New York State Museum Bulletin*, **469**, 93–106.

HOLLAND, S. M. & PATZKOWSKY, M. E. 2004. Ecosystem structure and stability: Middle Upper Ordovician of central Kentucky. *Palaios*, **19**, 316–331.

HOUSE, M. R. 1965. A study in the Tornoceratidae: the succession of *Tornoceras* and related genera in the North American Devonian. *Philosophical Transactions of the Royal Society of London B*, **250**, 79–130.

HOUSE, M. R. 1978. Devonian ammonoids for the Appalachians and their bearing on international zonation and correlation. The Palaeontological Association, *Special Papers in Palaeontology*, **21**, 70.

HOUSE, M. R. 1981. Lower and Middle Devonian goniatite biostratigraphy. *In*: OLIVER, W. A., JR. & KLAPPER, G. (eds) Devonian biostratigraphy of New York, Part I. *IUGS-SDS*, 33–38.

HOUSE, M. R. 1991. Devonian sedimentary microrhythms and a Givetian time scale. *Proceedings of the Ussher Society*, **7**, 392–395.

HOUSE, M. R. 1995. Devonian precession and other signatures for establishing a Givetian timescale. *In*: HOUSE, M. R. & GALE, A. S. (eds) Orbital forcing timescales and cyclostratigraphy. *Geological Society Publication*, **85**, 37–49.

KAUFMANN, B. 2006. Calibrating the Devonian time scale: a synthesis of U-Pb ID-TIMS ages and conodont stratigraphy. *Earth Science Reviews*, **76**, 175–190.

KESLING, R. V. & CHILMAN, R. B. 1975. Strata and megafossils of the Middle Devonian Silica Formation. *University of Michigan Papers on Paleontology*, **8**, 408.

KIRCHGASSER, W. T. 2000. Correlation of stage boundaries in the Appalachian Devonian, Eastern United States. *CFS. Courier Forschungsinstitut Senckenberg*, **225**, 271–284.

KLAPPER, G. 1971. Sequence within the conodont genus *Polygnathus* in the lower Middle Devonian. *Geologie et Paleontologie*, **5**, 59–79.

KLAPPER, G. 1981. Review of New York Devonian conodont biostratigraphy. *In*: OLIVER, W. A., JR. & KLAPPER, G. (eds) Devonian biostratigraphy of New York, Part I. *IUGS SDS*, 57–68.

KLAPPER, G. & ZIEGLER, W. 1967. Evolutionary development of the *Icriodus latericrescens* group (Conodonta) in the Devonian of Europe and North America. *Palaeontographica, Abt. A*, **127**, 68–83.

KLOC, G. J. 1983. Stratigraphic distribution of ammonoids from the Middle Devonian Ludlowville Formation in New York. Unpublished M.S. Thesis, University of Rochester, 78.

KLUG, C. R. 1983. Conodonts and biostratigraphy of the Muscatatuck Group (Middle Devonian), South-Central Indiana and North-Central Kentucky. *Wisconsin Academy of Sciences, Arts and Letters*, **71**, 79–112.

KOCH, W. F. 1981. Brachiopod community palaeoecology, paleobiogeography, and depositional topography of the Devonian Onondaga Limestone and correlative strata in Eastern North America. *Lethaia*, **14**, 83–104.

LANDING, E. & BRETT, C. E. 1987. Trace fossils and regional significance of a Middle Devonian (Givetian) Disconformity in Southwestern Ontario. *Journal of Paleontology*, **61**, 205–230.

LANDING, E. & BRETT, C. E. (eds). 1991. Dynamic Stratigraphy and Depositional Environments of the Hamilton Group (Middle Devonian) in New York State, Part II. *New York State Museum Bulletin*, **469**, 177 pp.

MITCHELL, S. W. 1967. Stratigraphy of the Silica Formation of Ohio and the Hungry Hollow Formation of Ontario, with paleogeographic interpretations. *Papers of the Michigan Academy of Sciences, Arts and Letters*, **52**, 175–196.

MILLER, K. G., KOMINZ, M. A. *ET AL.* 2005. The Phanerozoic record of global sea-level change. *Science*, **310**, 1293–1298.

NORRIS, A. W. 1993. Brachiopods from the Lower Shale Member of the Williams Island Formation (Middle Devonian) of the Hudson Platform, Northern Ontario and Southern District of Keewatin. *Geological Survey of Canada Bulletin*, **460**, 113.

ORR, R. W. 1971. Conodonts from the Middle Devonian Strata of the Michigan Basin. *Indiana Dept. of Natural Resources Geological Survey Bulletin*, **45**, 110.

ORR, R. W. & KLAPPER, G. 1968. Two new conodont species from the Middle–Upper Devonian boundary beds of Indiana and New York. *Journal of Paleontology*, **42**, 1066–1075.

ORR, R. W. & POLLOCK, C. A. 1968. Reference sections and correlations of the Beechwood Member (North Vernon Limestone, Middle Devonian) of southern Indiana and northern Kentucky. *American Association of Petroleum Geology Bulletin*, **52**, 2257–2262.

PROSH, E. C. 1990. The Devonian ammonoid *Agoniatites* from Hungry Hollow, southwestern Ontario. *Canadian Journal of Earth Sciences*, **27**, 999–1001.

RICKARD, L. V. 1984. Correlation of the subsurface Lower and Middle Devonian of the Lake Erie region. *Geological Society of America Bulletin*, **95**, 814–828.

SAGEMAN, B. B. & HOLLANDER, D. J. 1999. Cross correlation of paleoecological and geochemical proxies; a holistic approach to the study of past global change. *Geological Society of America Special Paper*, **332**, 365–384.

SPARLING, D. R. 1985. Correlation of the subsurface Lower and Middle Devonian of the Lake Erie region: alternative interpretation and reply. *Geological Society of America Bulletin*, **96**, 1213–1220.

SPARLING, D. R. 1988. Middle Devonian stratigraphy and conodont biostratigraphy, North-Central Ohio. *Ohio Journal of Science*, **88**, 2–18.

SPARLING, D. R. 1992. On the age of the Hungry Hollow Limestone. *Journal of Paleontology*, **66**, 339.

SPARLING, D. R. 1995. Conodonts from the Middle Devonian Plum Brook Shale of North-Central Ohio. *Journal of Paleontology*, **69**, 1123–1139.

SPARLING, D. R. 1999. Conodonts from the Prout Dolomite of North-Central Ohio and Givetian (Upper Middle Devonian) correlation problems. *Journal of Paleontology*, **73**, 892–907.

STAUFFER, C. R. 1909. The Middle Devonian of Ohio. *Geological Survey of Ohio, Fourth Series*, **10**, 204.

STAUFFER, C. R. 1915. The Devonian of southwestern Ontario. *Canadian Dept. of Mines and Geological Survey*, Memoirs, **34**, 341.

STUMM, E. C. 1942. Fauna and stratigraphic relations of the Prout Limestone and the Plum Brook Shale of northern Ohio. *Journal of Paleontology*, **16**, 549–563.

STUMM, E. C. & WRIGHT, J. D. 1958. Check list of fossil invertebrates described from the Middle Devonian rocks of the Thedford–Arkona region of southwestern Ontario. *Contributions to the Museum of Paleontology, University of Michigan*, **14**, 81–132.

TILLMAN, J. R. 1964. Variation in species of *Mucrospirifer* from Middle Devonian rocks of Michigan, Ontario, and Ohio. *Journal of Paleontology*, **38**, 952–964.

TILLMAN, J. R. 1970. The age, stratigraphic relationships, and correlation of the lower part of the Olentangy Shale of central Ohio. *Ohio Journal of Science*, **70**, 202–217.

TUCKER, R. D., BRADLEY, D. C., VER STRAETEN, C. V., HARRIS, A. G., EBERT, J. R. & McCUTCHEON, S. R. 1998. New U–Pb zircon ages and the duration and division of Devonian time. *Earth and Planetary Science Letters*, **158**, 175–186.

UYENO, T. T., TELFORD, P. G. & SANFORD, B. V. 1982. Devonian conodonts and stratigraphy of southwestern Ontario. *Geological Survey of Canada Bulletin*, **332**, 32.

VAIL, P. R., AUDEMARD, F., BOWMAN, S. A., EISNER, P. N. & PÉRÈZ-CRUZ, C. 1991. The stratigraphic signatures of tectonics, eustacy, and sedimentology: An overview. *In*: EINSELE, G., RICKEN, W. & SEILACHER, A. (eds) *Cycles and Events in Stratigraphy*. Berlin, Springer-Verlag, 617–659.

VANUXEM, L. 1842. *Geology of New York*, Part III. Albany, White and Visscher, 307 pp.

VER STRAETEN, C. A., GRIFFING, D. H. & BRETT, C. E. 1994. The lower part of the Middle Devonian Marcellus 'Shale,' central to western New York State. *In*: *Stratigraphy and Depositional History*, New York State Geological Association, 66th Annual Meeting Guidebook, 271–322.

VER STRAETEN, C. A. & BRETT, C. E. 2006. Pragian to Eifelian Strata (mid Lower to lower Middle Devonian), Northern Appalachian Basin–stratigraphic nomenclatural changes. *Northeastern Geology and Environmental Sciences*, **28**, 80–95.

VOGEL, K., BRETT, C. E. & GOLUBIC, S. 1987. Endolith associations and their relation to facies distribution in the Middle Devonian of New York State. *Lethaia*, **20**, 263–290.

WALLISER, O., BULTYNCK, P., WEDDIGE, K., BECKER, R. T. & HOUSE, M. R. 1995. Definition of the Eifelian–Givetian Stage boundary. *Episodes*, **18**, 107–115.

WRIGHT, J. D. & WRIGHT, E. P. 1961. A study of the Middle Devonian Widder Formation of Southwestern Ontario. *Contributions to the Museum of Paleontology, University of Michigan*, **16**, 287–300.

WRIGHT, J. D. & WRIGHT, E. P. 1963. The Middle Devonian Ipperwash Limestone of southwestern Ontario and two new brachiopods therefrom. *Contributions to the Museum of Paleontology, University of Michigan*, **18**, 117–134.

ZIEGLER, W., KLAPPER, G. & JOHNSON, J. G. 1976. Redefinition and subdivision of the *varcus*-zone (conodonts, Middle-? Upper Devonian) in Europe and North America. *Geology et Palaeontology*, **10**, 109–140.

Recognizing the Kačák Event in the Devonian terrestrial environment and its implications for understanding land–sea interactions

J. E. A. MARSHALL[1], T. R. ASTIN[2], J. F. BROWN[3], E. MARK-KURIK[4] & J. LAZAUSKIENE[5]

[1]*School of Ocean and Earth Science, National Oceanography Centre Southampton, University of Southampton, Waterfront Campus, European Way, Southampton SO14 3ZH, UK (e-mail: jeam@noc.soton.ac.uk)*

[2]*School of Human and Environmental Science, The University of Reading, Whiteknights, PO Box 217, Reading RG6 6AH, UK*

[3]*The Park, Hillside Road, Stromness, Orkney, KW16 3AH, Scotland*

[4]*Institute of Geology, Tallinn University of Technology, Ehitajate tee 5, 19086 Tallinn, Estonia*

[5]*Vilnius University, 21 Ciurlionio, LT-2009 Vilnius, Lithuania*

Abstract: The Kačák Event is a late Eifelian (Mid-Devonian) episode of marine dysoxia/anoxia with associated extinctions. It has been widely recognized in the shelf seas that surrounded the Old Red Sandstone continent. It was contemporary with the lacustrine Orcadian Basin in Scotland. This basin contains the distinctive Achanarras lake horizon that contains a rich and diverse fish fauna. The Achanarras lake was wide and deep and would have been filled by rainfall from a monsoon system at an insolation maximum. Faunal elements within the lake are in common with the Kernavė Member in Estonia and this level can be conodont dated as late Eifelian *eiflius* or *ensensis* Zone. Therefore the group of lacustrine flooding climatic events that occur at and above the Achanarras level can be correlated with the marine Kačák Event (*sensu lato*) and both can be regarded as having a common climatic cause and driven by an insolation maximum. A reconstruction of the Orcadian Basin drainage system and a water balance model based on the calcium flux within the lake shows that a very significant volume of water would have been seasonally discharged to the Rheic Ocean and would have caused an additional environmental effect.

In 1985, House named the Kačák Event after a marked environmental change that had been identified close to the Eifelian–Givetian boundary. The event was first documented, in what is now the Czech Republic, by Chlupáč (1960) at the level where the typical Barrandian carbonates gave way to black shales, a change that was accompanied by significant extinctions within the benthic invertebrates. The definition of this and other Palaeozoic events (e.g. House 1985, 2002) has created a hierarchy of temporal scales that includes major episodes such as the Frasnian–Famennian mass extinction, major and minor sea-level perturbations and Milankovitch climate cycles, some with accompanying black shale events and biocrises. In 1991, House further suggested that a potential Kačák correlative was the Achanarras level in the lacustrine Orcadian Basin of Scotland. This was primarily based on matching major perturbations within both systems. At that time, it was difficult to accept such a correlation. First, there were problems with the estimated age of the Achanarras level being older than the Kačák Event. Secondly, the Achanarras level marked the greatest expansion of the Orcadian Basin lake(s). Our then (and now) understanding of the control of eustatic sea-level rise within a non-glacial Earth system was based on climatic warming causing the movement of water from lakes, rivers and groundwater to the sea, which together with the associated thermal expansion of the upper layer of the oceans, was able to increase sea-level by several metres (Jacobs & Sahagian 1995). Hence, a sea-level high would be unlikely to coincide with the greatest lacustrine flooding event seen within the basin.

Our understanding of Devonian events and their chronostratigraphy has advanced very substantially since 1991. The age and nature of the Kačák Event has been clarified and there is now substantive palaeontological evidence that enables us to attempt a more direct correlation between the fish faunas of the Achanarras level and the marine record. As will be shown, this correlation provides us with a powerful tool for understanding how events in the marine realm can be driven by changes in the Earth system.

The Orcadian Basin

In the Scottish Old Red Sandstone (ORS), there is a distinctive northern accumulation (Fig. 1) of lacustrine, fluvial and aeolian sediments that is known as the Orcadian Basin (Trewin & Thirlwall 2002).

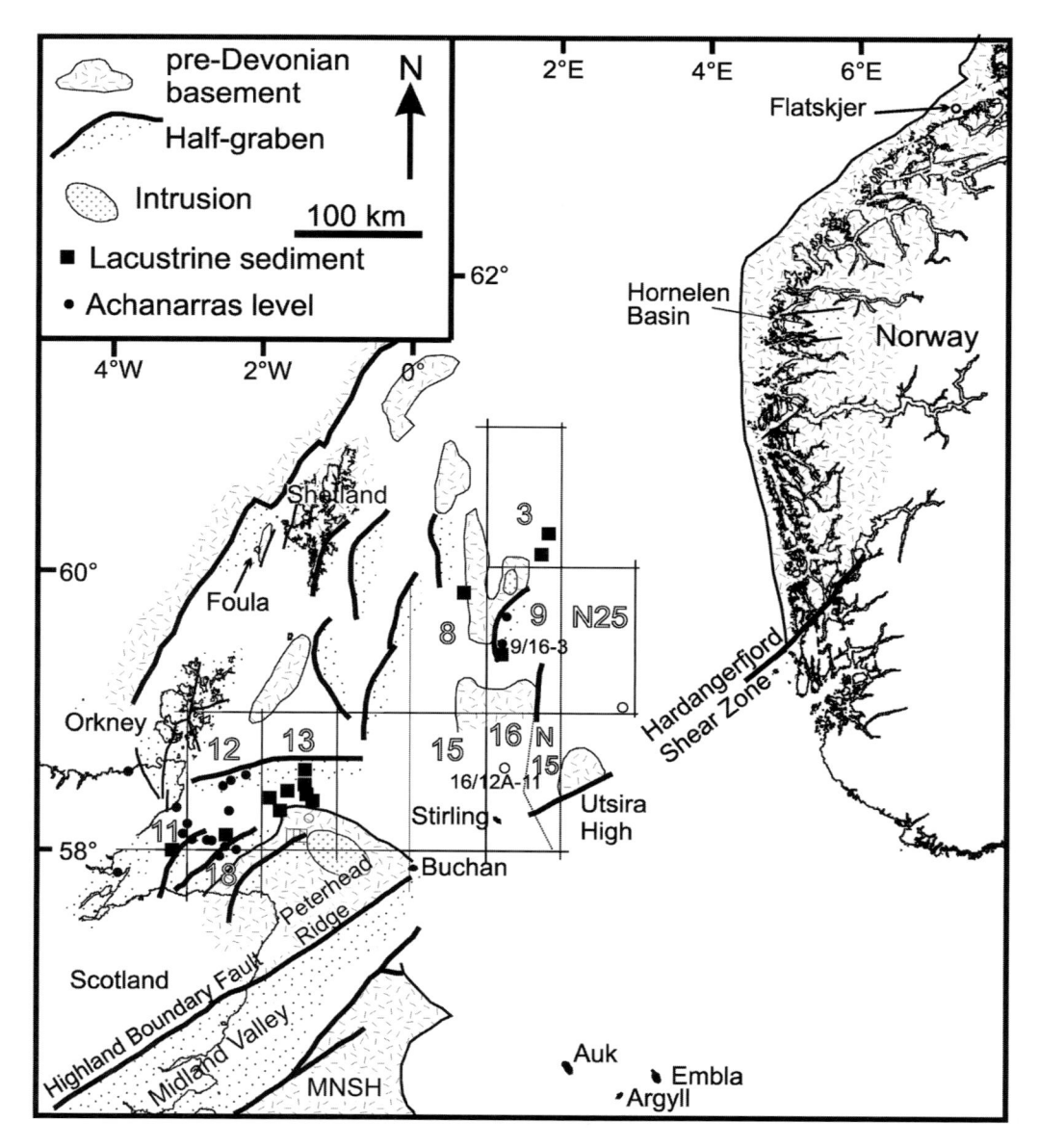

Fig. 1. Map of Orcadian Basin including known extension into the offshore North Sea. Well penetrations of the Achanarras horizon and Middle Devonian lacustrine sediments (both Eifelian Orcadia Formation and Givetian Eday Flagstone Formation) are shown. Proven Achanarras horizon extends as far east as Quads 3 and 9. Identifiable half-graben structures are marked together with basement highs. Devonian oilfields are marked; Auk, Argyll and Embla include Givetian marine limestones. The southern margin of the Orcadian Basin is the Highland Boundary Fault and its continuation into Norway. Devonian sediments occur in Norway as far as 63° N with lacustrine sediments present on Flatskjer. Modified from Marshall & Hewett (2003). MNSH is the Mid North Sea High.

These sediments were deposited in a series of extensional half-graben that are generally aligned North–South, in contrast to the NE–SW basin alignments in the Midland Valley of Scotland. The southern limit to the Orcadian Basin and the boundary between the two systems being the Highland Boundary Fault (Marshall & Hewett 2003). The Orcadian Basin half-graben represent a distinct episode of extension that started in the early Pragian to earliest Emsian interval (Wellman 2004), the process being driven by the gravitational collapse of the Caledonian orogen. The half-graben are present as a number of interlinked structures that occur from the southern shore of the Moray Firth to at least Shetland in the north. They extend eastwards across the North Sea Basin to connect with a number of related structures along the western margin of Norway (Steel et al. 1985). Similar sediments that are, in part, contemporary also occur in East Greenland (Marshall & Astin 1996). The characteristic sedimentary facies within the Orcadian Basin is lacustrine and developed as a series of monotonous cycles that alternate from deep permanent to shallow playa lake. The majority of time within each cycle is represented by the playa lake sediments, that are characteristically thinly bedded laminated sands and silts, with abundant evidence of emergence as demonstrated by the wide variety of desiccation structures that are present, such as mud cracks and evaporite pseudomorphs (Rogers & Astin 1991; Astin & Rogers 1991, 1992, 1993). The cycles are climatic in origin (Donovan 1980) and represent an alternation of wet and dry conditions within the basin. The control being a Milankovitch cyclicity (Hamilton & Trewin 1988; Astin 1990) where the obvious lacustrine alternations was identified (Marshall 1996) as being driven by the 100 ka eccentricity cycle. This identification was based on the then accepted likely duration of the Eifelian of between 5–8 million years. The major development of the lacustrine sediments was in the Eifelian to early Givetian interval as represented by the Caithness Flagstone Group (Caithness), the Stromness Group (Orkney) and the Orcadia Formation (North Sea offshore basins). In addition, there were both earlier (the Pragian to Emsian Struie Group) and later (the Givetian Eday Flagstone Formation) lacustrine developments. However, these were both laterally less extensive and shorter in duration. The characteristic fossils within the lacustrine succession are fish, plants and spores. Although some of these fish were widely distributed around the ORS continent and beyond, and hence clearly inhabited both the marine and lacustrine environment, it is quite clear that, apart from a few brief distinct intervals (Marshall et al. 1996), the sediments are non-marine in origin.

It is during the Eifelian that the lacustrine sediments achieved a maximum in their development. This is the well-known Achanarras Fish Bed level of Caithness with its distinctive and widely distributed fauna. This level represents a time when the deep permanent lake facies is found in all parts of the Orcadian Basin and indeed overlaps onto metamorphic basement (Parnell 1983). The level is recognized by either its distinctive fish fauna or the characteristic sedimentary motif and spore assemblage. It is known from very many localities (Fig. 2) and, as such, has tended to be given many

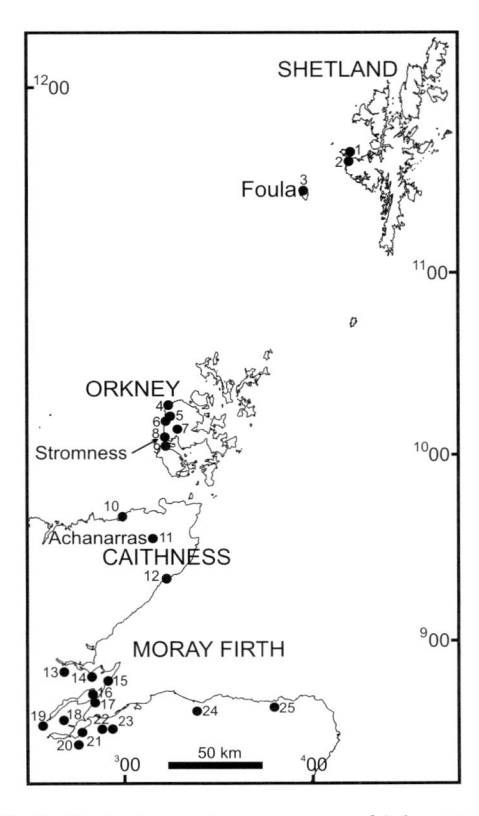

Fig. 2. Map to show onshore occurrences of Achanarras horizon and stratigraphical equivalents. Localities are (1) Papa Stour, (2) Melby, (3) Foula, (4) Birsay, (5) Cruaday Quarry, (6) Bay of Skaill, (7) Brookan Quarry, (8) Noost of Netherton, Stromness, (9) North Hoy, (10) Graemsay, (11) Nirex Dounreay No. 1 borehole, (12) Niandt, (13) Blackpark, (14) Tain No. 1 well, (15) Tarbat coast section, (16) Nigg, (17) Cromarty nodule bed & Navity Shore, (18) Killen Burn, (19) Kinkell, (20) Easter Town Burn, (21) Clava, (22) Knockloam, (23) Lethen Bar, (24) Tynet Burn, (25) Gamrie. The grid used is from the Ordnance Survey. Sources include Marshall 1988, Mykura & Phemister 1976 and Blackbourn & Marshall 1988 (Shetland); Mykura 1976, Black 1978 (Orkney); Crampton & Carruthers 1914 (Caithness); Stephenson 1977, Fletcher et al. 1996 (Easter Ross); Trewin & Davidson 1999 (south side of the Moray Firth) plus Westoll 1977, Barclay et al. 2005 and pers. obs. for all.

different local names. These include the Achanarras level and the Niandt Fish Bed, which is also in Caithness. The correlative level also occurs from a number of localities (Fig. 2) in Orkney, in West Shetland, in the southern part of the Orcadian Basin (Easter Ross) and along the southern shore of the Moray Firth where the lake level generally occurs within marginal facies. There are also, as yet, undated Mid-Devonian lakes (Pollard et al. 1982) found in the Hornelen Basin (62° N) in Western Norway (Fig. 1) plus distinctive playa lake type sediments further north (63°) on the skerry of Flatskjer (Bryhni 1974). In addition, there are contemporary Eifelian lacustrine cycles much further north in East Greenland (Marshall & Astin 1996) on Gauss Halvø (73° N) and in Canning Land (71° N). However, recent tight fit continental reconstructions (Torsvik et al. 2001) of Greenland and the Norwegian Margin place these localities at latitudes (relative to present-day Norway) that are between 64° and 69° N. They are therefore not much further north than the Norwegian localities but importantly contain unequivocal Eifelian lacustrine sediments. These occurrences point to the widespread development of this facies within, at least, an interconnected series of rifts encompassing some 11 degrees of palaeolatitude, a length comparable to the Triassic–Jurassic Newark Rift system from eastern North America (Olsen 1990).

Figure 1 shows the distribution of lacustrine sediments in the offshore Orcadian Basin. This compilation includes wells where the Achanarras horizon has been identified on a combination of log character, palynology, palynofacies and correlation (Marshall & Hewett 2003). In addition, wells are also shown where lacustrine sediments occur (either or both the Orcadia and Eday Flagstone Formations) but the Achanarras level cannot be proven or has not been penetrated. Most of the lacustrine occurrences are concentrated in the Inner Moray Firth in Quads 11, 12, 13 and 18. In addition, there is a further group of isolated lacustrine penetrations in Quads 3, 8 and 9 including an Achanarras correlative in 9/16-3 (Duncan & Buxton 1995). Figure 1 also shows an interpreted distribution of half-graben and basement highs compiled from available well and geophysical data (Marshall & Hewett 2003). The distribution and density of these half-graben indicates the likely area and distribution of lakes. The southern limit of the Orcadian Basin is the offshore extension of the Highland Boundary Fault. Mid-Devonian sediments are generally absent south of this line until the late Givetian when marine limestones deposited during the Taghanic marine transgression are found on post-Devonian graben highs such as Auk, Argyll and Embla (Marshall & Hewett 2003). There are a number of long penetrations (i.e. 700 m) of Devonian sediments in UK Quads 15, 16 and 21 (Buchan Field) and Norway Quads 15 and 25 (Marshall & Hewett 2003, figs 6.5 & 6.20) that are close to the line of the Highland Boundary Fault. However, these sections do not include lacustrine facies and are instead dominated by sand-rich fluvial sediments. Most of these intervals are poorly dated, but where known, are of late Mid- to Late Devonian age. They represent the late post-rift stage of the Orcadian Basin which was dominated by fluvial systems (Upper Eday Sandstone Formation) and indicate a general sediment depocentre and hence dispersal in this direction from more northerly parts of the Orcadian Basin. A core sample from one well section (16/12A-11) is of proven mid–late Frasnian age (pers. obs.) and clearly older than any known Upper ORS section in the Midland Valley of Scotland. This shows a southerly-directed supply of sand that pre-dated the onset of Upper ORS sedimentation in the Midland Valley of Scotland. By implication, this is a continuation of sediment dispersal from the earlier lacustrine Mid-Devonian sedimentary system. The Norwegian Devonian Basins, although containing some lacustrine sediment (Pollard et al. 1982; Steel et al. 1985), are very much dominated by very thick fluvial sequences. This may be a response to their much more proximal location in the Caledonian system where active basin extension promotes a different style of sedimentation (e.g. Braathen et al. 2002). These large fluvial systems will feed into an, as yet unknown, distal complex. Clearly, as shown by the limited well evidence, this may have been a sandy terminal complex (e.g. Kelly & Olsen 1993) that is the counterpart to the Orcadian lacustrine system. At times of high discharge, it would similarly have flowed into the southern North Sea.

The Achanarras level is thus representative of a very large lacustrine system (i.e. a mega-lake). This was clearly a very significant event in the Orcadian Basin and raises questions as to why this lacustrine flooding occurred, and what was the significance both internally and externally of such a large body of water. Although the 'type locality' for the lacustrine event is at Achanarras, the present exposure of the fish bed part of the cycle is now inaccessible through being in a disused and flooded quarry (Trewin 1986). In addition, it is an inland hilltop locality (to promote natural drainage during quarrying) and thus difficult to place within its wider stratigraphic succession (e.g. Miles & Westoll 1963). Therefore, the Sandwick Fish Bed from Orkney will be used as the reference section. Here the level is well exposed (Fig. 2) in a number of very accessible coastal sections along the western seaboard of the mainland of Orkney and therefore can be both easily studied and placed in stratigraphical context. However, note that there is no locality at Sandwick. The fish bed in fact being, named after the several excellent sections which occur within the parish of Sandwick, Orkney (Wilson et al. 1935).

The Sandwick Fish Bed on Orkney

Figure 3 shows a stratigraphical log for the Stromness Group. The group is subdivided at the base of the Sandwick Fish Bed into the Lower and Upper Stromness Flagstone Formations. The Lower Stromness Flagstone Formation has some 49 lacustrine cycles. These were measured from the West Shore section at Stromness where pre-Devonian basement (Strachan 2003) is exposed on a half-graben footwall. Here the cycles are thinner (average 5.25 m) than those from equivalent strata further north on Mainland Orkney. The overlying Upper Stromness Flagstone and 'Rousay Flagstone Formation' cycles were measured from the more northern localities where the cycles are thicker. These formations include a further 38 lake and near-lake cycles with an average thickness of about 12 m (Rogers & Astin 1991). The permanent lake part of each cycle (often referred to as fish beds) has an average thickness of 1.5 m and is characterized by lamination on a sub-millimetre scale with alternations of organic-rich, carbonate and silt laminae (Rayner 1963; Trewin 1986). The lake cycles are symmetrical with a progressive deepening into each cycle followed by shallowing. The shallowing trend is recognized by the change from lacustrine laminites to silts with shallow-water wave-rippled sands to the laminated sands and silts of the playa lake facies with their characteristic mudcracks (Rogers & Astin 1991). Fish fossils are not present in every cycle and are also normally restricted to discrete horizons within each cycle. The fish fauna outside the Sandwick Fish Bed is generally represented by a small number of genera such as *Dipterus*. Other characteristic features of the permanent lake sediments include stromatolites (Fannin 1969) and Magadi-type cherts (Parnell 1986). Within the Stromness Group sequence (Fig. 3), there are distinct intervals where groups of cycles are less well developed and more sand-rich. This permitted the definition by Astin (1990) of a number of larger-scale bundles of lacustrine cycles that represent times when the climate became drier or wetter.

A detailed log through the interval of the Sandwick Fish Bed is shown in Figure 4. This demonstrates how the cycle character changes from the 'typical' 5 m cycles, i.e. cycles 33, 34a and 34b, to the Sandwick Fish Bed cycle at 35a (at 0 m on the log). The sediments then remain mostly as permanent lake facies through two complete cycles (Fig. 4, 35a/1, 0–3 m & 35b/2, 3–22.5 m) until the presence of numerous mudcracks show the return of the playa lake environment and the more typical cyclic alternation. However, it is emphasized that rare incipient crack arrays and desiccation

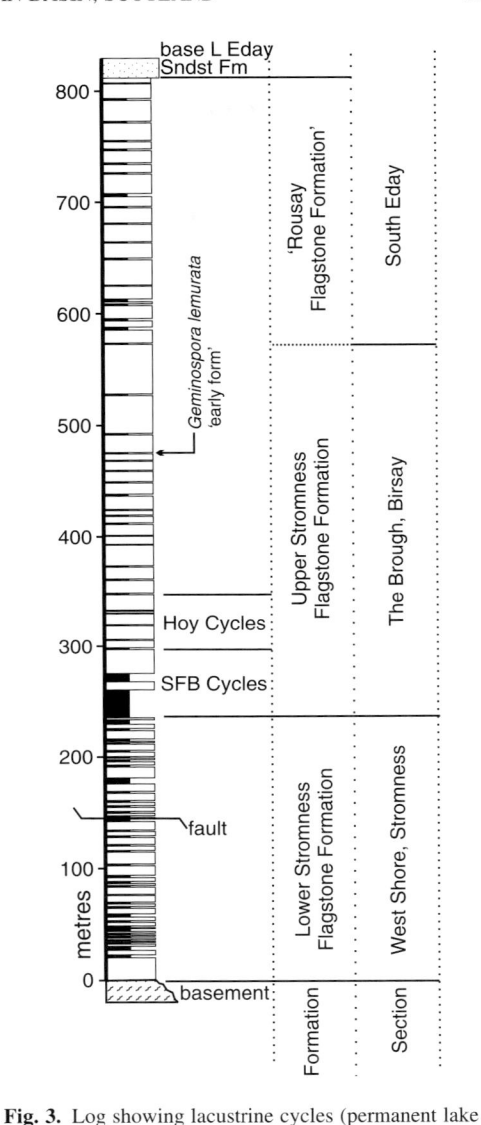

Fig. 3. Log showing lacustrine cycles (permanent lake facies is solid black) in Lower and Upper Stromness Flagstone and Rousay Flagstone Formations. The log is split between the West Shore, Stromness; Birsay, North Mainland and South Eday sections. The Lower and Upper Stromness Flagstone Formation sections are correlated at the Sandwick Fish Bed. The Upper Stromness and 'Rousay Flagstone Formation' logs are from Astin (1990). The interval that includes the Sandwick Fish Bed (*sensu* Fannin in Mykura 1976) and Hoy Cycles is thicker and represent more permanent lakes or a greater fluvial influence. Based on the inception of *Geminospora lemurata*, the base of the Givetian can be placed in the Hoxa section in the upper part of the Upper Stromness Flagstone Formation. This level can then be correlated (Speed 1999) to the Broch of Birsay section.

polygons occur within this interval (Fig. 4, 6 m & 7 m) including immediately above the main lake unit (Astin & Rogers 1992). This demonstrates that the Sandwick Fish Bed lake did dry out and was not a single continuous long-lived lake. Cycle 3 (cycle renumbering restarts at the base of the Sandwick Fish Bed) still shows a somewhat thicker interval of permanent lake facies (as at 35 m on log) and then the next four cycles (the Hoy Cycles, Fig. 3) show a significant fluvial influence (Fannin in Mykura 1976). As the name implies, these are particularly well developed on Hoy in Orkney, where cross-bedded channel sandstones over 10 m thick occur within lacustrine cycles. This fluvial influence is pervasive across Orkney and shows that fluvial processes were active at the lake margins, although with insufficient input of water to maintain a permanent lake. Similar thick fluvial intercalations are also known on the southern side of the Moray Firth at Tynet Burn (Trewin & Davidson 1999). These sand-rich cycles represent somewhat wetter than normal conditions with an active lake-margin fluvial phase between the playa lake and permanent lake part of the cycle.

The actual 'fish bed' part of the Sandwick Fish Bed is thin and restricted to an interval of 0.5 m (Fig. 4). The same situation occurs at Achanarras Quarry (Trewin 1986) where the fish bed occupies less than 2 m of section. The majority of the Sandwick Fish Bed cycle in Orkney is dominated by laminated lacustrine siltstones with thin parallel laminated sandstones. These sands represent lacustrine underflow (e.g. Trewin 1986) and are distinct from the channelled sheet flood sands so common within the playa lake parts of the cycle.

The Achanarras lake

This brief description of the Sandwick Fish Bed establishes that the interval represents a time of sustained lake stability (in Orkney) that lasted for three normal lake cycles, a time of some 300 ka. However, the lake only supported a diverse fish fauna for a much shorter interval of this time. This diverse fish fauna, with its frequent articulated specimens (Trewin 1986), is coincident with high TOCs, well-developed sediment lamination and a high relative abundance of AOM (amorphous organic matter) within the kerogen. It therefore represents a time interval when there was the strong development of lake water stratification and hence relatively deeper water (i.e. maximum flooding).

The lake water depth is more difficult to determine but various estimates have been attempted. Some of these are unrealistic, such as the depth being equivalent to the decompacted thickness of sediment between the deepest lake and first signs

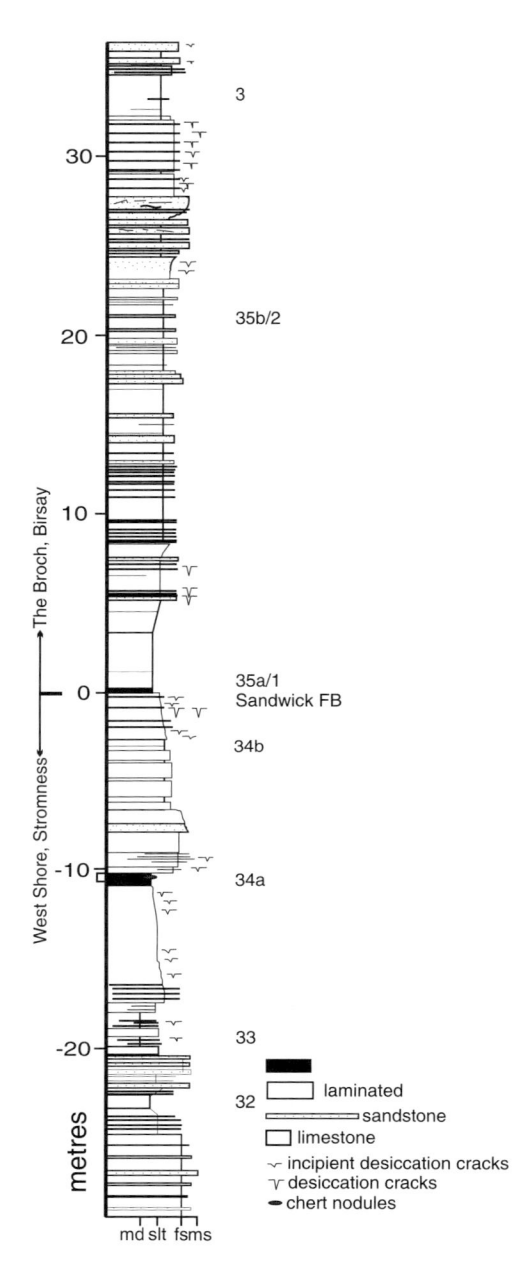

Fig. 4. Log of the Sandwick Fish Bed and adjacent cycles from the West Shore Stromness and Birsay, Orkney. The fish bed is at 0 m on the log. The two lacustrine cycles from 0 to 21 m are generally represented by permanent lake conditions. Above 24 m, playa lake conditions return as represented by the occurrence of abundant mudcracks. But note the occurrence of both incipient mudcracks and deep desiccation crack sets at 5–7 m. These demonstrate that the Sandwick Fish Bed did dry out during the cycle.

of emergence (e.g. Fannin in Mykura 1976; Trewin 1986; Stephenson et al. 2006). All that this change implies is that the basin becomes dry as opposed to the accommodation space becoming filled with sediment up to the previous lake-full base level. Other water depth estimates are based on the relationship between the distance of wind fetch across the lake surface and the depth reached by the resulting wavebase (Olsen 1990). Various values have been estimated, e.g. 75 m or 122 m (Duncan & Hamilton 1988), depending on the fetch distance. However, there is one direct measurement available for the Achanarras correlative from Nairnside where the lake transgresses onto basement topography (Marshall & Fletcher 2002). This gives a minimum value of 25–30 m for this basin margin location. A value of 15–20 m was determined by Astin (1985) for the Givetian Eday Group from the gradient of an alluvial fan that was transgressed by a lake. A somewhat different method based on wave ripples (Allen 1981) provided a depth estimate of <10 m for water depth and a fetch of <20 km for a Givetian lake in SE Shetland. The same Givetian stratigraphical level from East Greenland has also provided a more direct estimate (Marshall & Stephenson 1997) that is again based on the transgression of a single lake unit across basement topography at a lake margin. This gives a value of 100 m. All these values suggest that, at the time of the Achanarras horizon, a lake depth of at least 25 m would be typical for laminite formation at the margins and with the maximum depth probably reaching to well over 100 m within the deeper parts of the basin.

The size of the lake/lake basin is more difficult to determine. The total West–East extent of the ORS depositional system from western Orkney to Norway is some 400 km. However, it is clear that this includes several discrete half-graben (Fig. 1). The North–South dimension is more problematic but an Achanarras correlative can be identified as far north as Shetland, a distance of some 400 km excluding post-Devonian strike-slip movement. In addition, there was the high probability of a series of interconnected basins reaching as far north as East Greenland (Canning Land and Vilddal, Gauss Halvø), a distance of 11°, i.e. about 1500 km when Greenland is placed within a tight fit reconstruction.

The character of the marginal occurrences of the Achanarras equivalents is revealing as to the duration of the maximum extent of the lake. These marginal occurrences can onlap sequences as diverse as conglomeratic alluvial fans, fluvial sandstones and metamorphic basement. However, these correlatives are much attenuated and often characterized by nodular limestone (e.g. Tynet Burn, Trewin & Davidson 1999). This suggests that the duration of the largest lake, which was the time when there would have been the greatest interconnection between the different systems and hence permitted the free movement and homogenization of fish faunas, was probably restricted to the thin interval of the main fish bed. During the remainder of the cycle, the deeper parts of the main basins would have been full and the lakes generally permanent throughout the cycle but not necessarily interlinked. Conversely, in the marginal areas the lake would have been short-lived before reverting to the original sedimentary facies.

The laminations in the fish bed part of the cycle have been used to estimate its time duration based on the assumption that each carbonate/organic matter/silt doublet or triplet represents an annual accumulation. This gave Rayner (1963) an estimate of 4 ka for the 3.3 m of laminated lacustrine sediment approximating to the fish bed. Rayner regarded this as an indication of the order of time occupied by the interval, a value that has generally been accepted by subsequent authors (e.g. Astin 1990) although Trewin (1986) regarded 4 ka as perhaps an underestimate given the lower silt content within the central part of the fish bed. This view is confirmed by measurements of lamination in the Sandwick Fish Bed, Orkney (Fig. 5a) where some 2663 doublets and triplets occur that have individual modal thicknesses of 0.225 mm. This suggests a more realistic duration of approximately 2 ka for the 0.5 m interval of the fish bed and some 80 ka for the 20 m of laminated section. This value is also significant as it is greater than the typical decadal to low millennial recycling time for both water and ions within a large stratified lake (e.g. Yuretich & Cerling 1983; Ojiambo & Lyons 1996).

The type of carbonate present within the laminated permanent lake facies of the Orcadian Basin lacustrine cycles is significant (Trewin 1986). The upper and lower parts of the fish-bearing part of the Achanarras laminate have carbonate laminations that are secondarily altered to dolomite whilst the central section contains coarse polyhedral calcite crystals. This calcite mineralogy and crystal size of the central section, when contrasted to the dolomitic margins, indicate formation from water with the highest Ca/Mg ratio. Therefore, it would appear that the only part of the Achanarras Fish Bed that is an open system is the central calcitic part. This is also the interval with the most diverse fish fauna and indicates that the lake had both a diversity of input plus a direct extra-basinal outlet.

The prevailing climate during the deposition of the Orcadian Basin sequence was arid to semi-arid. There are numerous direct sedimentological indicators for this arid climate that can be found in the playa lake sediments. These include the

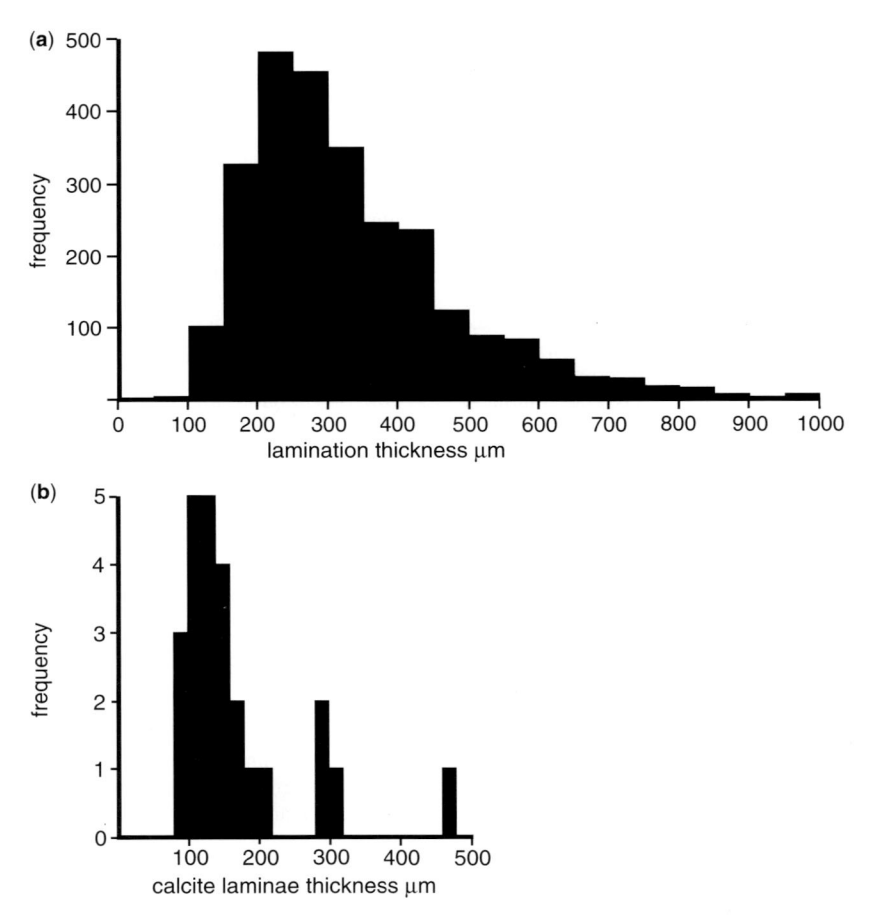

Fig. 5. (**a**) Histogram showing thicknesses of individual lacustrine doublet and triplet laminae measured from the Sandwick Fish Bed, Stromness, Orkney. Blocks of the section were digitally imaged and the laminae then recognized and measured by line scans. (**b**) Histogram showing thicknesses of individual calcite laminae measured from a petrological thin section of Beds 2–3 from Achanarras Quarry, Caithness. Individual laminae were measured with a motorized x–y stage with a 1 μm resolution.

ubiquitous desiccation cracks, gypsum and rarer halite pseudomorphs, calcretes and the aeolian dune systems that developed on the dry lake surfaces (Rogers & Astin 1991). Clearly, the permanent lake part of the cycles represents times when the precipitation/evaporation (P/E) ratio changed substantially and perennially to fill the basins. During the Sandwick Fish Bed cycle, this increased P/E was sufficient to sustain the deeper parts of the system as a standing body of water through much of the normally longer arid parts of the climate cycle. The Devonian palaeolatitudinal location of Orkney was 20° S (Torsvik & Cocks 2004) and hence at approximately the southern margin of the arid climate zone. It remained within this arid zone during the deposition of the entire Orcadian succession.

The fish fauna of the Achanarras level

The fish faunas in the Orcadian Basin are well known, having been studied for over 150 years (Andrews 1982). The faunas have been documented in numerous publications although, as yet, the fish assemblages are typically still described by the formation rather than reported as discrete ranges tied to the more recently established cycle-based lithostratigraphy (Astin 1990). There has also been an understandable emphasis on particular stratigraphic levels such as at Achanarras where fish faunas are diverse and well preserved. This means that it can be difficult to place such assemblages in their lithological and biostratigraphical context (e.g. Miles & Westoll 1963). Importantly, Trewin (1986) gives a very detailed distribution of a very significant collection of fish through the 1.95 m thick fish bed level at Achanarras Quarry. The fish fauna starts (and ends) with *Dipterus*, the lake-colonizing fish which dominates the lower half-metre of the interval. The fauna then diversifies for the next 0.6 m of permanent lake facies into an assemblage dominated by *Pterichythodes*, *Mesoacanthus* and larval *Dipterus* (*Palaeospondylus*, Thomson *et al.* 2003). These fish then decline in abundance and are largely replaced by *Coccosteus cuspidatus* which then becomes the most abundant fish through the next 0.6 m of section before itself declining as the fauna reverts to *Dipterus* to complete the interval. Significantly, *C. cuspidatus* is always represented by mainly adult specimens with no individual found that has an estimated body length less than 150 mm. Trewin (1986) noted this anomaly and attributed it to either *post-mortem* sorting, differential preservation or the ecology of *Coccosteus*. Subsequently, Hamilton & Trewin (1988) explained this population of adult specimens as representing the result of migration from the marine environment.

Amongst the other fish present at Achanarras are rare specimens of *Rhamphodopsis threiplandi*, which occurs sporadically through the upper part of the fish bed. It is only known from the Achanarras horizon at Achanarras itself and Blackpark (Miles 1967) together with a related species (*R. trispinatus*) that occurs at the same level at Cromarty, Tynet Burn and Lethen Bar (Watson 1938; Trewin 1986; Trewin & Davidson 1999).

A climatic versus tectonic control on lake development?

Extensional lacustrine systems such as the Orcadian Basin are the result of an interplay between tectonics and climate. An episode of extension is required to form the half-graben, which then provides accommodation space within the sedimentary system. If the climate is sufficiently wet, these can then fill with water to form lakes, until, and in the absence of continuing extension, they become infilled with sediment. There has always been a debate (e.g. Cohen 1990) as to whether lacustrine events, such as found at the Achanarras level, could be the result of an extensional episode that would generate short-lived accommodation within the system and hence a short-term increase in lake depth before the sedimentary system responded. Such extensional episodes can be clearly identified in the Orcadian Basin, a good example (Astin 1985) being at the base of the Lower Eday Sandstone Formation. However, these events are easily recognized as they are accompanied by uplift as shown by localized sediment reddening, the reworking of lacustrine sediments and the spread of short-lived alluvial fans with the introduction of a characteristic basement clast assemblage. In contrast, the sedimentological context of the Achanarras level is opposite in its character. The fish bed was deposited on, and within, all the contemporaneous environments seen in the Orcadian Basin, including the most proximal such as fans and basement. Therefore the lake floods across all sedimentary environments and hence was not a response to any tectonic-driven change in sediment pattern.

Why the Achanarras mega-lake

The Achanarras level marks a time of sustained water supply to the lake and hence high rainfall within the drainage basin. This rainfall must have been maintained throughout most of the 100 ka cycle to prevent the lake from evaporating away during the normally longer, drier part of each cycle. Normally, this lake permanence is attributed to an increasingly wet climate and certainly the immediate environment of the lake basin would have been more humid. However, analogues with modern lake systems (e.g. Hamilton 1982; Ruddiman 2001) at the equivalent latitude show that the control is a greater strength of the monsoon system, i.e. an increased seasonal insolation rather than a general overall reduction in aridity. The increasing insolation acts to reinforce the continental high pressure cell and therefore eventually draw in more moisture-laden cooler air that condenses to rain. The paradox being that, in this situation, increased solar radiation produces a larger, more sustained lake. There are clear examples of this Milankovitch-driven expansion of arid zone lake systems during the Quaternary insolation maxima (Partridge *et al.* 1998; Gasse 2000). Understanding this mechanism gives us the ability to place the Achanarras lake in the context of other events on the eastern margin of the ORS continent and hence derive a common explanation of contemporaneous changes that occur within both the marine and terrestrial environments. Importantly, the increased water supply that maintains the Achanarras lake may not occur within the immediate area but may result from the cooling air releasing its moisture in the more distant Caledonian highland areas in Greenland and northern Norway. These highland areas still existed at least into the early Devonian, as shown by the isotopic content of contemporary metamorphic fluids that were sourced by rainwater from an altitude of some 5 km (Barker *et al.* 2000).

A water budget for the Achanarras lake

There is now enough information on size, climate, basin configuration and water chemistry to be able to construct a simple model for the Achanarras lake. In constructing this model, we shall consider only the middle calcite-precipitating part of the fish bed (~0.5 m) when the lake was at its maximum extent. At this time, the lake was deep and stratified as shown by the presence of laminated sediments, the high AOM content and preservation of intact fossil fish. To maintain both the lake water column and stratification at this palaeolatitude (i.e. in an arid zone), the supply of lake water must have been continually replenished. This contrasts with the upper and lower parts of the permanent lake facies where evaporation has increased the Mg/Ca ratio and caused dolomitization. It is well known that standing bodies of water at similar latitudes today have very high evaporation rates. For example, the modern Lake Nasser (23° N, Shahin 1985) and Lake Chad (13° N, Eugster & Hardie 1978) both lose a depth equivalent to 2 m of water each year whilst lakes in the southern states in the western USA, somewhat further from the equator (35° N), typically lose about 1.5 m annually (Meyers & Nordenson 1962). Some tropical African lakes (Johnson 1996) are effectively closed systems with very little water leaving the lake as evaporation from the surface balances both inflow and precipitation onto the lake surface. However, these lakes tend to become saline.

Clearly, the Achanarras lake had a significant inflow because if it had been subjected to an annual surface evaporation loss of 2 m it would, at an estimated 100 m depth, have lasted for less than a century. This constant inflow of water is also the mechanism for recharging the calcium supply; otherwise, carbonate precipitation would cease. The alternative scenario of evaporation losses equalling inflow with the chemical balance maintained by calcite precipitation is also unlikely as the middle part of the fish bed does not become dolomitic, i.e. the Ca/Mg ratio remains low with any accumulating Mg^{2+} being flushed from the system by outflow. This contrasts with the upper and lower parts of the fish bed when the carbonate is dolomitic and the system evaporative and accumulating Mg^{2+}, i.e. it is a closed system at these times.

The mechanism for carbonate production within a normally undersaturated lacustrine system (Kelts & Hsü 1978) is seasonal algal growth that removes CO_2 from the lake water and causes precipitation of calcium carbonate. Measurement (Fig. 5b) of compacted thickness of each calcite laminae at Achanarras in Beds 2–3 of Trewin (1986) gives an average value of about 150 μm. The carbonate laminae are generally pure and do not contain appreciable quantities of both quartz silt and organic matter. An average value for the carbonate purity is estimated from petrography at 70%. Therefore, to produce a 150 μm thick lamination requires that each m^2 of lake floor will represent the production from the water column of about 0.41 kg of calcite, equivalent to about 0.16 kg of calcium. Clearly, not all the available Ca^{2+} will be precipitated from the lake water. Comparison with values for annual carbonate flux versus the Ca^{2+} content in the water from a carbonate-precipitating system such as modern Lake Zurich (Kelts & Hsü 1978) shows that the conversion, which is not temperature-dependent, turns 60% of the available Ca^{2+} into carbonate each year. In contrast, estimates from a modern tropical system, such as Lake Turkana (Yuretich & Cerling 1983) show a near complete removal of the carbonate each year. The value adopted in the model is a 75% conversion of Ca^{2+} to calcite. The question of a lateral reduction in carbonate production across a single lacustrine system was also considered. However, it is known that the more marginal Achanarras correlatives (e.g. the Clava Mudstone Member) still have significant carbonate content (Fletcher et al. 1996) with fish bed thicknesses at least equivalent to more central locations. Hence this effect is discounted.

Assuming that the lamination is annual gives us a method of estimating from the Ca^{2+} content of river water the inflow required to maintain this carbonate flux. However, this value is a minimum as it neglects the contribution of any rainfall directly onto the lake surface and which has no Ca^{2+} content. The chemistry of river waters very much reflects the geology of the catchment area. Compilations of the Ca^{2+} content of the world's rivers (e.g. Stumm & Morgan 1995) show that values above 50 mg l^{-1} are unusual. In addition, the larger river systems drain bigger areas and hence will tend to average Ca^{2+} concentrations of around 13 mg l^{-1}. However, the rivers that drain catchment areas consisting of older crust tend to higher values. Accepting that the lake waters of the Achanarras lake produced evaporites both above and below the fish bed suggests that a somewhat higher range of values is more likely. A higher value is also supported by the provenance of the water from areas with exposed limestone bed rock (the Durness Limestone in the NW Highlands of Scotland) as evidenced by clasts of this lithology in Eday Group conglomerates (Astin 1985) plus the more extensive areas of Precambrian and Lower Palaeozoic carbonate rocks in East Greenland that sub-crop beneath Devonian sediments (Cowie & Adams 1957), thus indicating its likely very widespread previous distribution. However, the model was constructed using a range of values of Ca^{2+} concentration rather than selecting any single value. This model has then been inverted to produce a nomogram (Fig. 6) where a range of lake areas can be compared to a range of likely Ca^{2+} concentrations to give the volume of water required to maintain carbonate export to the lake bed. This volume of water,

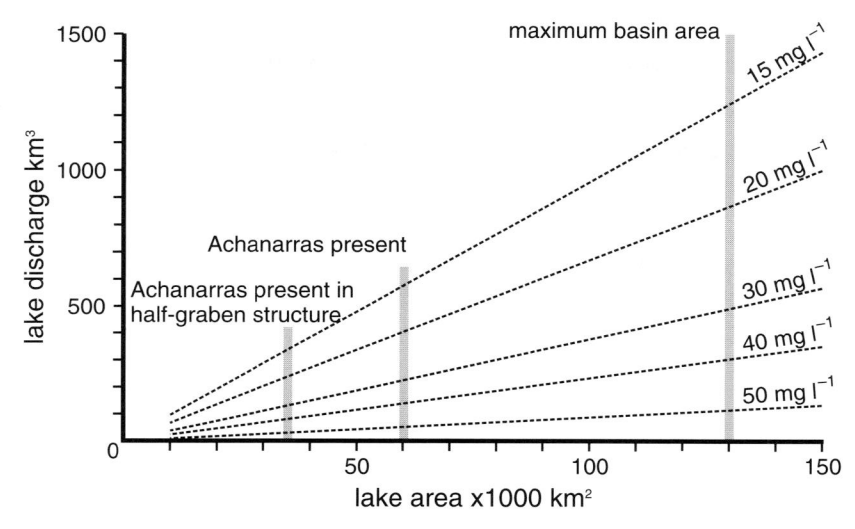

Fig. 6. Nomogram showing Achanarras lake water budget model results. The area of the lake can be compared to a range of lake water Ca^{2+} values (in mg l^{-1}) and hence converted to a likely discharge from the lake after correction for evaporation. Vertical grey bars mark the maximum basin area, the area that is underlain by a known Achanarras level and where these are within discrete half-graben.

corrected for annual evaporation as a standing body of water, gives an estimate of annual discharge from the lake.

The range of lake areas selected was based on the likely area of the Orcadian Basin including its offshore extension. The areas were directly calculated from Figure 1 using IsoCalc.com in CorelDraw©. A maximum estimate of the entire area (e.g. Ziegler 1990) gives values of about 130 000 km², This, when linked to a Ca^{2+} concentration of 30–40 mg l^{-1}, indicates a water discharge of 300–500 km³. A more realistic total basin area that only includes areas where the Achanarras level is known to be present is some 60 000 km², and this reduces to about 35 000 km² when restricted to the actual recognized half-graben structures that are known to contain the Achanarras correlative. These smaller areas indicate seasonal outflows of between 100–200 and 100–150 km³ respectively with the same range of lake water Ca^{2+} concentration. These results indicate both likely limits to the size of the basin and the outflow. A maximum estimate of basin size gives very large outflows consistent with a major continental-sized catchment driven by a sustained flow. The lower values (i.e. about 100 km³ plus any direct lake surface rainfall) are more consistent with a large monsoon-driven ephemeral lake system.

The position of the outflow is unclear but the preponderance of younger sandy facies in Quads 15 and 16 (Figs 1 & 7) would suggest at least a drainage route to the west of the Utsira High. The water would then have to flow in a generally SE direction around the northern part of the Mid North Sea High since south of the Mid North Sea High late Givetian sediments occur directly on basement (Fig. 1, well Q-1; Ziegler 1990). The distance to the marine margin (which is not present within a central graben; compare Marshall et al. 1996 with Coward et al. 2003) is about 500 km. Comparisons with a river system such as the Nile (Shahin 1985), that similarly has a remote lacustrine source and then runs through arid conditions without further tributaries, suggest water volume losses through seepage and evaporation for a river exiting the Orcadian Basin would be of the order of 5%. However, it should be noted that seepage losses may have been higher during early Mid-Devonian times, given the rather limited plant cover and general regolith aspect of Devonian mineral soils (Algeo et al. 2001). These model results suggest that a series of interconnected half-graben could deliver a volume of at least an order of 100 km³ of water each year to the Rheic Ocean. For comparison, the pre-Aswan Dam Nile had an annual discharge of 90 km³.

The age of the Sandwick Fish Bed

Determining an accurate age for the Sandwick Fish Bed against the marine standard has been problematic. Broad fish-based correlations have been made that place the sequence, for reasons that are not always clear (compare Mykura 1976 with

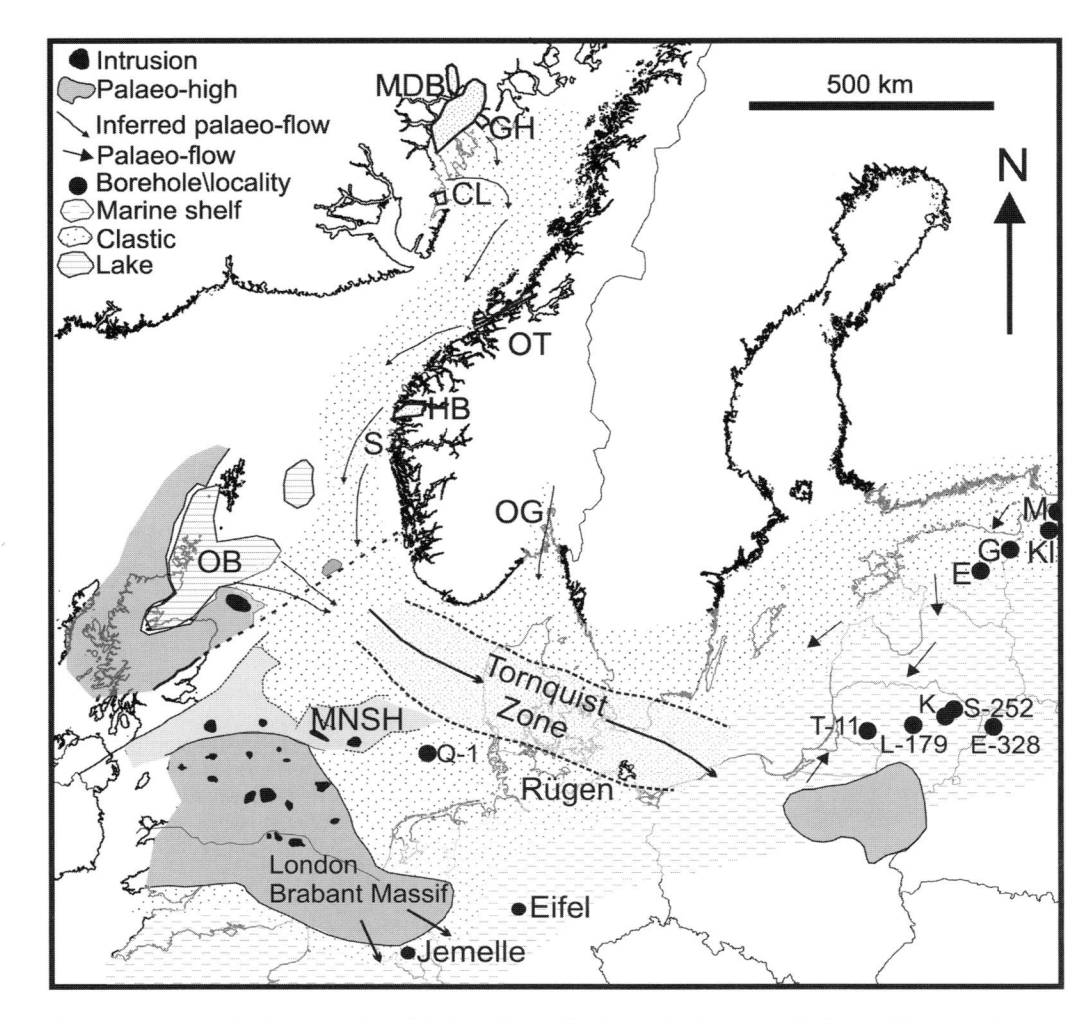

Fig. 7. Reconstructed palaeogeography of the Orcadian Basin (OB) at the Achanarras horizon and its external drainage. The flow from the Orcadian Basin will be diverted southeastwards by the Mid North Sea High (MNSH) and London Brabant Massif. A significant thickness of poorly dated Eifelian sandstone occurs in deep boreholes on Rügen. This lies within the Tornquist Zone and will divert drainage towards the marine margin. The Olso Graben (OG) flow is conjectural but reflects a return to an earlier Siluro-Devonian palaeoflow. The eastern part of the reconstruction is from Kuršs (1992a, b). Greenland is placed in a tight fit reconstruction (Torsvik et al. 2001; Mosar et al. 2002). This brings a major development of Eifelian lacustrine sediments closer to the ORS basins in Norway and the Orcadian Basin. Main Devonian Basin (MDB), Gauss Halvø (GH), Canning Land and Wegener Halvø (CL), Outer Trondheim (OT), Hornelen Basin (HB), Solund (S), Taurage-11 (T-11), Ledai-179 (L-179), Kaniukai well (K), Svedasai-252 (S-252), Eividovichi-328 (E-328), Gorodenka (G), Klenny (Kl), Marino (M), Emajõgi River dredge (E).

Mykura & Phemister 1976), within the late Eifelian or early Givetian. Spores have brought greater precision to age dating in the ORS and particularly significant is the inception of *Geminospora lemurata* which has a conodont-dated first occurrence that is earliest Givetian (Loboziak et al. 1991). Marshall (1996) showed that the first occurrence of unequivocal *Geminospora lemurata* at Windwick, Orkney (Figs 3 & 8) was in the upper part of the 'Rousay Flagstone Formation'. However, it shared its inception with *Chelinospora concinna*, a distinctive spore that is now known (Turnau 1996) to have an approximately mid-Givetian first occurrence (Fig. 8). Subsequent investigation of a more accessible section at Hoxa shows an inception for the 'early form' of *G. lemurata* at least 25 lacustrine

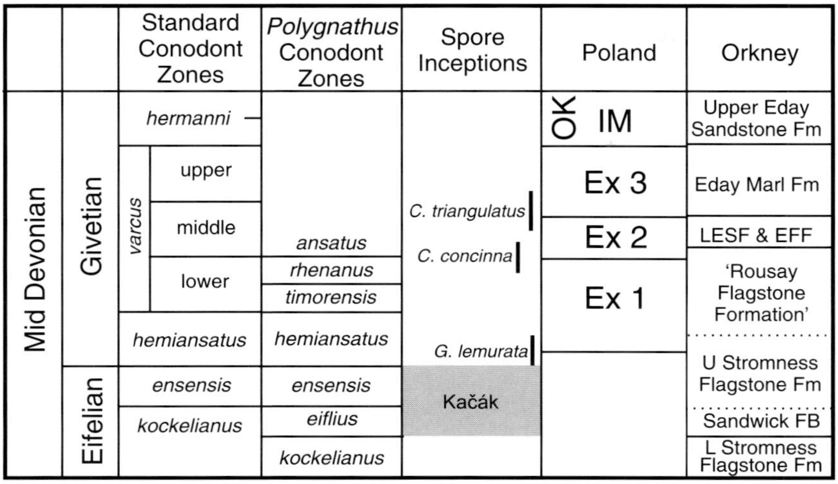

	Standard Conodont Zones		Polygnathus Conodont Zones	Spore Inceptions	Poland	Orkney
Mid Devonian / Givetian	hermanni				OK IM	Upper Eday Sandstone Fm
	varcus	upper		C. triangulatus	Ex 3	Eday Marl Fm
		middle	ansatus		Ex 2	LESF & EFF
		lower	rhenanus	C. concinna	Ex 1	'Rousay Flagstone Formation'
			timorensis			
	hemiansatus		hemiansatus	G. lemurata		U Stromness Flagstone Fm
Eifelian	ensensis		ensensis	Kačák		
	kockelianus		eiflius			Sandwich FB
			kockelianus			L Stromness Flagstone Fm

Fig. 8. The Orkney succession with major spore inceptions and correlation to the northern European spore zones of Turnau (1996) and Turnau & Racki (1999) from Poland. This permits an approximate correlation with the global conodont standard. The range bars on the spore inceptions are the error in their correlation to conodont zones. The 'standard' conodont zones are shown together with revised polygnathid zonation. LESF is Lower Eday Flagstone Formation, EFF is Eday Flagstone Formation.

cycles (and possibly 21) above the base of the Sandwick Fish Bed. Hence, the Sandwick Fish Bed is clearly Eifelian in age and the local base of the sequence in Orkney can be estimated as early but not earliest Eifelian (Marshall 1996). In addition, the Stromness Group lacustrine cycles allow for the estimation of an internal time-scale. There are some 90 lacustrine cycles within the Stromness Group with the base of the Givetian occurring between 10–20 cycles beneath the top of the Group. If these were 100 ka eccentricity cycles, then this gives a duration for the Eifelian part of the sequence of between 7–8 million years. This is an acceptable assumption using the geochronological estimates for the entire Eifelian of 5.7 ma (Gradstein et al. 2004) and 6.5 ma (Williams et al. 2000). However, a revision of the Devonian time-scale (Kaufmann 2006) has proposed a marked reduction in the duration of the Eifelian stage to 4 ma which if accepted may require re-evaluation of the 100 ka cyclicity.

It has become progressively apparent from the 1970s that some faunal elements that occur in the Orcadian Basin were also present in the Baltic region. Many of these identifications resulted from the recognition of synonymies that had stemmed from the fact that the Baltic fishes are from marginal marine sediments and hence generally preserved as isolated plates whereas those from the Orcadian are complete but crushed specimens. Hence, two parallel taxonomic systems have developed. The limited Baltic material, consisting of isolated skeletal elements with some characters of particular individual and unknown interspecific variations, has led to taxa being erected that are different from those in the Orcadian Basin. It is important to note that before the publication of Kuršs (1975), the sediments were regarded as continental in origin and hence it was considered normal that the two terrestrial basins could have different fish faunas.

However, this correlation has not immediately resulted in any increased precision as the Baltic succession is marginal marine (Kuršs 1992a, b), with the fauna generally restricted to fish and rare levels with low diversity shelly faunas and hence equally difficult to correlate to the Devonian standard time-scale. The Baltic sequence is also thin and composed of a number of quite distinct depositional units that conceal significant time gaps within the succession. A significant discovery was the recognition (Mark-Kurik 1991, 2000) that *Coccosteus cuspidatus* was present in the Kernavė Member of Estonia, a level that is well known for its diverse fish fauna (Mark-Kurik 1995). Systematic excavations in the Gorodenka locality, NE Estonia (Fig. 7), yielded a number of *Coccosteus* plates, which were generally larger than those known under the name *C. orvikui* (Obrucheva 1962, 1966) and coming from the same locality. The new material did not reveal taxonomically significant characters to distinguish the two species of *Coccosteus*. Although the specimens of *C. cuspidatus* from Estonia and the Leningrad District still have plates that are not as large as those from

Achanarras localities in Scotland, they can still be considered as skeletal elements of adult fishes but of a smaller size. Obrucheva recorded *C. orvikui* from Marino (Slavyanka River), Klenny (Luga River), both in the Leningrad District, and an unspecified borehole in Belarus. In addition to the Gorodenka locality, Gross (1940) indicated localities with *C. orvikui* as in the Emajõgi River dredge near Tartu, and also at Aruküla, Tamme and Haaslava. However, according to Obrucheva (1962, 1966), the latter three localities are records of another *Coccosteus* species, *C. grossi* O.Obr.

The Kernavė Formation

The Kernavė Formation (part of the Narva Regional Stage in Lithuania, but note it is accorded member status in Estonia) is a well-defined and distinctive interval within the Baltic Devonian (Paškevičius 1997; Narbutas 2004). It has a type section in the Ledai-179 borehole in Lithuania (Fig. 7) where it is 13 m thick. Lithologically, the lower part is characterized by claystones intercalated with marlstones, siltstones and clayey dolomitic limestone. This part of the member has a fauna consisting of brachiopods, bivalves, gastropods, tentaculites, conodonts and trilobites. The upper part of the member is characterized by bright-coloured (red, violet and brown) marlstones and sandstones with a fauna of lingulid brachiopods and vertebrates.

Palaeogeographically, the Kernavė Member has been interpreted (Kleesment 1997; Paškevičius 1997; Narbutas 2004) as representing the final stage of the Narva transgression characterized by the influx of terrigenous clastics and freshwater that restored the area to a normal marine salinity followed by a minor recession at the very end of Kernavė time. However, it is more realistically represented as reflecting the dual influx of increased clastic sediment from the north coupled with a marine transgression from the south that brought in a more diverse marine fauna in contrast to the surrounding intervals. The Kernavė Member has a diverse fish fauna (Mark-Kurik 1995, 2000; Valiukevičius 1995, 2000). This includes the smallish *Coccosteus cuspidatus* formerly referred to as *Coccosteus orvikui* and *Millerosteus*? *orvikui*. Another faunal element in common, as a comparison record, between the two areas is *Rhamphodopsis* cf. *threiplandi* from the Kernavė Formation (Lithuania, Taurage borehole, 659.85–659.95 m) which in the Orcadian Basin occurs only at the Achanarras level.

The Kernavė Formation contains a diverse RL spore zone assemblage (Avkhimovitch *et al.* 1993). The elements of this assemblage are also present at the Achanarras level and hence from within the same rather broad spore zone. However, the significant difference about the Kernavė assemblage is that, unlike most Baltic Devonian palynofloras, it contains abundant but low diversity marine microfossils. These include a low diversity chitinozoan faunule (Kaniukai well, Vaitiekūnienė 1983), rare acanthomorph acritarchs, scolecodonts, plus a previously unreported monospecific bloom of acritarchs. Selected elements are shown on Figure 9 from the Svedasai-252 well in Lithuania.

An Achanarras–Kernavė correlation?

What links the Achanarras level with the Kernavė is the occurrence of two fossil fish (*Coccosteus*; *Rhamphodopsis*), identified on species level and, in addition, other fish are known from both areas at the generic level, e.g. the arthrodire *Homostius*, osteolepiform sarcopterygians (Mark-Kurik 1991, 2000) and acanthodians (Young 1995). Of these, *C. cuspidatus* is particularly characteristic of the Achanarras horizon, being both very widely distributed and particularly abundant (72%) in the upper part of the bed. However, although it is replaced above the Sandwick Fish Bed by *Dickosteus threiplandi*, it does occur sporadically beneath this level to the base of the Lower Stromness Flagstone Formation. Significantly, all the specimens of *C. cuspidatus* at Achanarras are adult specimens. One obvious interpretation (Hamilton & Trewin 1988) is that *C. cuspidatus* migrated into the Orcadian Basin via its external drainage. The specimens now identified as *C. cuspidatus* from Estonia include smaller individuals which raises the possibility that it was in areas such as the Baltic that the fish was reproducing and then able to colonize the Orcadian Basin when linked to the sea. However, it appears that it was unable to breed within the Orcadian Basin. Initially, this link was sporadic, as evidenced by the rare finds of *C. cuspidatus* beneath the Achanarras level. However, it was at times of maximum lake expansion and hence the strongest and most stable outflow that *C. cuspidatus* was able to freely migrate into the Orcadian Basin.

The other fish that the Achanarras level and the Kernavė Member have in common is the comparison record of *Rhamphodopsis threiplandi*. This is a unique occurrence at both these levels but it must always be acknowledged that these are probably occurrences within, an as yet, undefined total range.

What is distinctive about both these levels is that they represent unusual events within their respective successions, the Achanarras being a lake of both exceptional size and duration and containing a diverse fish fauna. The Kernavė Formation similarly has a very distinctive and diverse fish

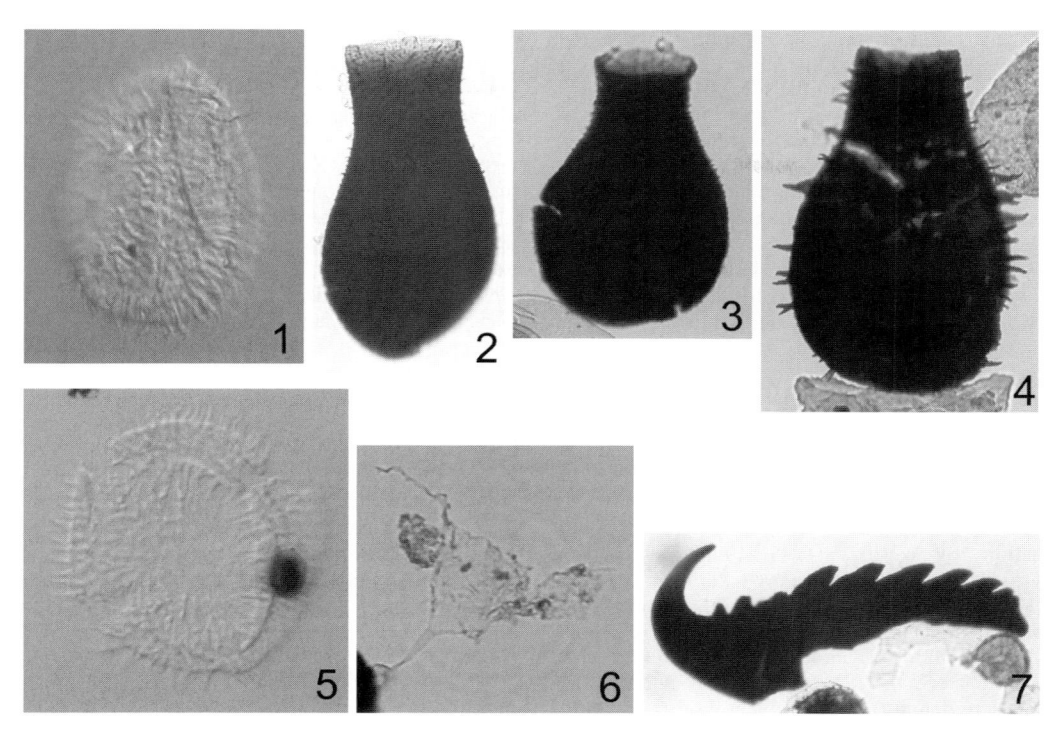

Fig. 9. Significant Middle Devonian palynomorphs from the Kernavė Formation, Lithuania, Svedasai-252 well, 208.8 m. 1 & 5 acritarchs, ×750. 2 & 3 ?*Desmochitina minor* of Pichler 1971 ×250. 4. *Angochitina devonica* of Pichler 1971 ×250. 6. *Veryhachium trispinosum*, ×500. 7. scolecodont ×100.

assemblage and represents a ?marine incursion into the Baltic area that is normally characterized by rather variable and elevated salinities with, in consequence, an impoverished invertebrate fauna. This combination of shared fauna and distinctiveness is accepted as sufficient to maintain the correlation of these two levels.

The age of the Kernavė Member

Conodonts are not known from the *C. cuspidatus*-bearing Kernavė Member localities in Estonia. However, a conodont assemblage (*Icriodus struvei, Polygnathus linguiformis linguiformis, P. l. alveolus, P. parawebbi, P.* cf. *xylus ensensis, P.* cf. *costatus oblongus*) has been recovered in Belarus from a microvertebrate preparation of the Kastyukovichi Horizon in the Eividovichi-328 borehole (Valiukevičius *et al.* 1995, Fig. 2) which is a correlative of the Kernavė Member. To place this discovery in perspective, this assemblage is one of only two known conodont assemblages in the region. This correlation is secure given both the cratonic nature of the succession and the extensive subsurface knowledge of both the Baltic Devonian and its extension into Belarus (Sorokin *et al.* 1981;

Valiukevičius *et al.* 1986). This assemblage has been variously assigned to the *kockelianus* Zone (including the lower *ensensis* subzone, Valiukevičius 1995) or the lower *ensensis* Zone (Valiukevičius *et al.* 1995) and is equivalent to the *Polygnathus parawebbi* Beds of the Eastern European Platform. There has now been a revision of the polygnathid conodont zonation at this level (Fig. 8) with the *kockelianus* Zone now generally subdivided into the *kockelianus, eiflius* and *ensensis* Zones. A conodont range compilation from the more complete Eifelian sections in the Anti-Atlas of Morocco (Belka *et al.* 1997) shows that *P. parawebbi* has an inception at the base of the *eiflius* Zone. *Icriodus struvei* has a last occurrence in the Eifel Mountains (Weddige 1988), where it occurs rarely in the lowest Givetian (basal *hemiansatus* Zone). The presence of *P.* cf. *xylus ensensis*, although a comparison record, supports the *ensensis* Zone assignment. Consequently, the recorded fauna from Belarus could be from the *eiflius* Zone or, given the comparison record of *P.* cf. *xylus ensensis*, the *ensensis* Zone, the latter being the main event level of the Kačák. Therefore, the conodonts show that the Kernavė Formation and the Achanarras horizon occur somewhere within the *eiflius* to *ensensis* interval. Here, the internal cycle-based stratigraphy for the Stromness Group is again useful. There are some 50 cycles present from the base of the succession to the Sandwick Fish Bed followed by an estimated 21 to 25 cycles present up to the base of the Givetian. Using a cycle duration of 100 ka this gives a duration close to 2 ma for the Sandwick Fish Bed to base Givetian interval and hence suggests a position low in the combined zones. A shorter precession cycle duration (e.g. Astin 1990) is less likely, given that lamination counts for the Sandwick Fish Bed suggest that these units have a duration greatly in excess of 20 ka.

The Kačák Event

The Kačák Event has been the subject of significant debate as to its definition, limits, expression, and the intensity or even presence of associated extinctions. These arguments are reviewed in a series of contributions by Walliser (1985, 1990, 1995, 2000), Truyóls-Massoni et al. (1990), Chlupáč & Kukal (1986, 1988), Budil (1995a, b), May (1995), Schöne (1997), Becker & House (1994), House (1996, 2002) and Garcia-Alcalde (1998). Part of this debate is the result of different authors using different definitions of the event. For example, the event has been variously defined (according to Walliser 1995) on the occurrence of the entire black shale interval of the Kačák Member (House 1996) or only the base of the black shale (Walliser 1995) together with the somewhat later first appearance of the tentaculid *Nowakia otomari sensu stricto* (the *otomari* Event of Walliser, 1985). This onset of the black shale, although later than the first appearance of the goniatite *Cabrieroceras crispiforme* (=*rouvillei*), was also previously used to define the base of the Givetian (i.e. the *Maenioceras* Stufe) further adding to the confusion as to the exact level of the Kačák Event. This lack of clarity was compounded by the use of different definitions for the base of the Givetian Stage (which in addition gave different dates for the event) together with an assumption that the biotic changes were directly related to the lithological and environmental change. Much of this confusion has now become clarified during the formal process of the definition of the Eifelian–Givetian boundary (Walliser 2000). What has now become clear is that there is a sequence of marked lithological changes that occur in the late Eifelian and that these occur, with associated extinctions, starting in the uppermost *kockelianus* Zone (Schöne 1997), with the main change (the L'Ei 1 of Walliser 2000) at the base of the *ensensis* Zone. There is then a second later event (L'Ei 2) that is just below the Eifelian–Givetian boundary and which is also accompanied by extinctions. The Kačák Event interval continues into the lowest part of the Givetian. A similar sequence of events within the wider Kačák interval was also detailed in House (1996) and discussed by Garcia-Alcalde (1998). This confusion has often led to the event being referred to as the Kačák-*otomari* Event or KOE (e.g. Truyóls-Massoni et al. 1990).

The association of the lithological expression of the interval with extinctions is also contentious. For example, the Kačák Event in Germany (Schöne 1997) was characterized by a marine transgression and the rapid spread of oxygen-depleted waters onto the outer shelf. This environment had an associated fauna adapted to low O_2 environments. However, the level was not associated with significant extinctions, with the pre-event benthic assemblages being restricted to a more-oxygenated narrow shallow zone on the shelf. Following termination of the Event, these oxic benthic assemblages then reoccupy the deeper-water environment. In contrast, the Kačák Event in the Barrandian Basin was accompanied (Chlupáč & Kukal 1986) by significant faunal changes that included the disappearance of the diverse trilobite population of the Chotěc Limestone (Chlupáč 1994) together with other main benthic groups (brachiopods, crinoids and corals), with most of these extinctions occurring in the uppermost *kockelianus* Zone. There was less impact on the planktonic and nektonic groups but the fauna becomes impoverished. Interestingly, fossils of land plants are recorded in abundance (Budil 1995a, b) for the first time.

There is now an emerging isotope record of the Kačák Event. However, data acquisition is difficult through the event as the underlying lithology changes from limestone to carbonate-poor organic-rich mudstones that generally lack biota. Whole rock sample analysis shows a positive excursion in oxygen and both inorganic and organic carbon in the deeper water but not the shallow-water localities (Hladíková et al. 1997). This distinct positive $\delta^{13}C$ excursion is replicated in the much larger bulk carbonate isotope study of Buggisch & Mann (2004) although with less-detailed sampling across the Kačák interval. A parallel study of palaeotemperatures (Joachimski et al. 2004) determined from both conodont apatite and brachiopod calcite shows a general elevated temperature during the late Eifelian although again with a low sample resolution through the interval.

An interesting study (Hladil et al. 2005) that included the Kačák interval was the combined magnetic susceptibility and gamma-ray logging of platform carbonates in the Czech Republic. When subjected to a moving normalization, these logs show significant anomalies at and above the Kačák Event. These were attributed to an increased flux of atmospheric dust. Although this was, in common with Ellwood *et al.* (2003), attributed to an impacting extraterrestrial bolide, other possibilities included increased aeolian or riverine input.

How the Eifelian terrestrial and marine environments might interact

This investigation of the Achanarras level has revealed that:

1. The interval in the Orcadian Basin is a composite event with a deep permanent lacustrine environment that was sustained for three 100 ka lacustrine cycles plus the four more fluvial-dominated Hoy cycles.

2. For prolonged periods, there was sustained flow out of the lake basin. In addition, there was probably a parallel fluvial flow to the marine margin from the Norwegian part of the system.

3. At these times, the annual seasonal flow would have easily been of the order of 100 km^3.

4. This time of maximum lake expansion was in the *eiflius* or ?more likely *ensensis* conodont Zones and hence contemporaneous with a number of the Kačác events.

Although it is not possible to directly match events around the Achanarras and Kačák intervals, the latter can now be understood as being both a consequence of the climatic events that sustained the lake and the result of the external drainage from the lake. Figure 7 is a palaeogeography of the relevant parts of northern Europe and East Greenland using the tight-fit reconstruction of Torsvik *et al.* (2001). The East Greenland Devonian Basins are shown, including those in Canning Land and Gauss Halvø that are known to contain Eifelian lacustrine sediments. Any external drainage has the potential to link with that from the Norwegian Devonian Basins and ultimately and probably indirectly with that from the Orcadian Basin. Any drainage exiting the system across the Highland Boundary–Hardangerfjord Shear Zone line would be directed eastwards by the presence of 'granite'-cored basement highs that are present in northern and eastern England and under the Mid North Sea High. There is no evidence for any Devonian sediment in this area until the late Givetian. However, an established depocentre is that within the Tornquist Zone. The arguments for this were summarized in Marshall *et al.* (1996). There is also direct evidence for a thickness of at least a kilometre of Eifelian clastic sediments in deep boreholes under Rügen (McCann 1999). Therefore, it appears feasible that any water exiting the Orcadian Basin would be diverted along this line to the Rheic Ocean. Also shown is a compilation of the palaeogeographies for the Kernavė Formation which shows a clastic marine embayment with a westerly-directed current. Clearly, any fish inhabiting this area would be well placed to encounter any flow exiting from the Orcadian Basin.

The Achanarras lake can only have been sustained by a strengthening of the monsoon system, i.e. increased insolation related to orbital forcing. This increased insolation would also have occurred right across the ORS continent with increased run-off in many areas that were not in the direct path of drainage from the Orcadian Basin. This is the explanation for the widespread sand supply into the Baltic Devonian at this level of the Kernavė Member. It also provides an explanation for the increased dust content (by either aeolian input or terrestrial runoff) and abundance of plant fossils within distal carbonate successions such as in the Czech Republic. The late Eifelian was also a time when other short-lived clastic deposits occur peripheral to the ORS continent. These include the Lomme Formation in Belgium (Bultynck & Dejonghe 2001). This formation is defined (type section at Jemelle, Fig. 7) by a clastic intercalation of sandy shales including immature sandstones into an otherwise carbonate-rich sequence and contains conodonts of *kockelianus* and *ensensis* age. It also marks a hiatus within the succession that is part of the 'Great Gap' of Struve (1982). This gap and spread of the clastic facies could have similarly resulted from the increased insolation and precipitation that also filled the Achanarras lake.

This increased insolation could also be responsible for the effects of the Kačák Event as recognized in the marine environment. These are the increased sea temperatures, decreased stratification and hence increased nutrient recycling and higher productivity (e.g. Southam *et al.* 1982) with the spread of dysoxia together with a transgression forced by thermal expansion of the upper layer of the ocean (cf. May 1995). All these effects would have been episodic, i.e. related to the individual climatic cycles. The seven cycles have a combined duration of some 700 ka and hence were capable of giving rise to the series of events as seen in the Kačák. The external drainage from the greater Orcadian Basin would also have had a profound effect on the immediate area. Figure 10 is a reconstruction of the Rheic Ocean after Scotese (2002). Superimposed on this is the Achanarras lake (OB) together with localities where the Kačák Event has been recognized. Although there is a continuing debate as to the extent of the Rheic Ocean in Mid-Devonian times (e.g. Martínez Catalálan *et al.* 2004) and particularly the location of Armorica, most reconstructions now show a width approximating to 2000 km with Euramerica and Gondwana in close proximity although not necessarily in contact. The seasonal discharge of many km^3 of freshwater into a closing and restricted ocean would have had a significant effect on the marine environment with the temporary formation of a low salinity surface layer. For example, a discharge of 100 km^3 could cover an area of 100 000 km^2 with a metre-thick layer. Although two orders of magnitude less than the entire area of the Rheic Ocean shown in Figure 10, this water would have a deleterious environmental effect in the immediate vicinity in areas such as the Eifel (E). The 'modern' analogue of this situation is the discharge from the Nile system where enhanced seasonal outflow during post-glacial insolation maxima are generally accepted as causing stratification, dysoxia and the deposition of sapropels with the temporary disappearance of the fauna to beyond the range of the outflow (Rohling et al. in press). Subsequently, at times of lower discharge, oxic bottom conditions return and the surviving fauna recolonizes the affected area.

It is doubtful whether the discharge from the Orcadian Basin could have had any influence on the more distant Kačák Event localities (Fig. 10), such as in the Barrandian Basin (B), the Montagne Noire (MN), the Cantabrians (C) and Morocco (Mo). However, these areas would still be influenced by the environmental changes within the Rheic Ocean and, as in the case of the Barrandian Basin, do show evidence of increased terrigenous input (i.e. increased runoff) together with plant fossils sourced from the local land areas. Further afield (Fig. 10), the Kačák Event has been recognized in the Hamilton Group on the western side of the ORS continent (Brett & Baird 1997). Although, clearly outside the range of any influence from the Orcadian Basin, this location, which is peripheral to the ORS continent, would have been subjected to the similar dual terrestrial and marine environmental effects from increased insolation. The most distant Kačák Event to be recognized is that in South China at Liujing (Bai et al. 1994). However, this area is still close to the equatorial zone and would have been subjected to a similar set of environmental effects.

In conclusion, this analysis of the age, correlation and significance of the Achanarras lake has demonstrated that it represents one of a distinct sequence of insolation-driven climatic perturbations. Fish-based correlations show it to have a marine correlative level that is part of the sequence of events known as the Kačák. From this correlation, we can conclude that the climatic perturbation that caused the Achanarras lake would also have had a profound effect on the marine environment. In addition, the magnitude of the eventual discharge to the sea from the lake would also have had a significant additional local effect.

We have benefited greatly from the contributions of Michael House to Devonian geology and particularly his efforts towards international correlation which has made

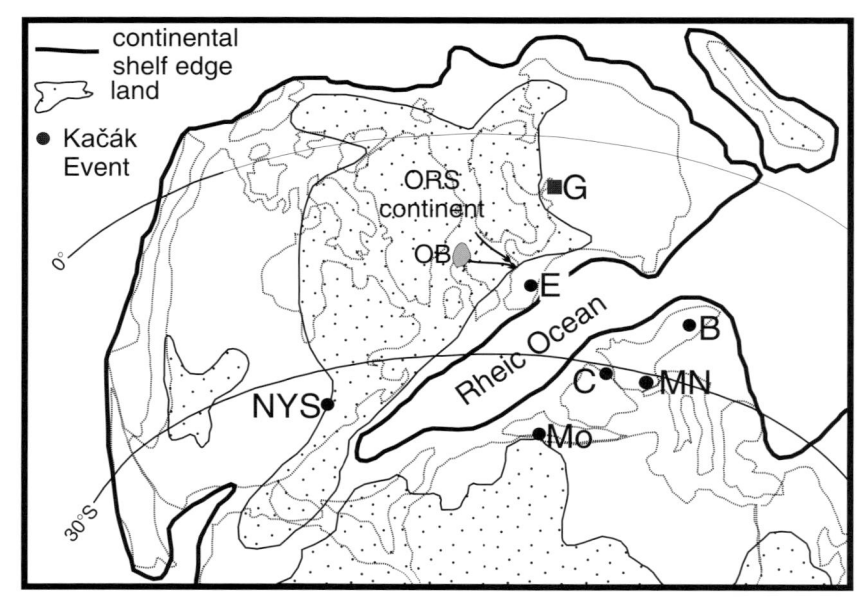

Fig. 10. Reconstruction of the Rheic Ocean from Scotese (2002). The locations of recognized Kačák levels are marked. They mostly occur on the margins of the Rheic Ocean. Orcadian Basin (OB), Eifel (E), Gorodenka (G), NYS (New York State), Mo (Morocco), C (Cantabria), B (Barrandian Basin), MN (Montagne Noire).

this paper possible. One of us in particular (JEAM) gratefully acknowledges Michael House's friendship and advice during his time in Southampton. Shir Akbari is thanked for preparing the palynological samples. Nacho Valenzuela-Ríos and Jau-Chyn Liao are acknowledged for a particularly informative discussion on the conodont assemblage. Thomas Becker & Christoph Hartkopf-Fröder provided important comments.

References

ALGEO, T. J., SCHECKLER, S. E. & MAYNARD, J. B. 2001. Effects of the Middle to Late Devonian spread of vascular land plants on weathering regimes, marine biotas, and global climate. In: GENSEL, P. G. & EDWARDS, D. (eds) *Plants Invade the Land*. Columbia University Press, New York, 213–236.

ALLEN, P. A. 1981. Wave generated structures in the Devonian lacustrine sediments of South-east Shetland and ancient wave conditions. *Sedimentology*, **28**, 369–379.

ANDREWS, S. M. 1982. *The Discovery of Fossil Fishes in Scotland up to 1845*. Royal Scottish Museum Studies, Royal Scottish Museum, Edinburgh.

ASTIN, T. R. 1985. The palaeogeography of the Middle Devonian Lower Eday Sandstone, Orkney. *Scottish Journal of Geology*, **21**, 353–375.

ASTIN, T. R. 1990. The Devonian lacustrine sediments of Orkney, Scotland; implications for climatic cyclicity, basin structure and maturation history. *Journal of the Geological Society, London*, **147**, 141–151.

ASTIN, T. R. & ROGERS, D. A. 1991. 'Subaqueous shrinkage cracks' in the Devonian of Scotland reinterpreted. *Journal of Sedimentary Petrology*, **61**, 850–859.

ASTIN, T. R. & ROGERS, D. A. 1992. 'Subaqueous shrinkage cracks' in the Devonian of Scotland reinterpreted—reply. *Journal of Sedimentary Petrology*, **62**, 923–924.

ASTIN, T. R. & ROGERS, D. A. 1993. 'Subaqueous shrinkage cracks' in the Devonian of Scotland reinterpreted—reply. *Journal of Sedimentary Petrology*, **63**, 566–567.

AVKHIMOVITCH, V. I. & TCHIBRIKOVA, E. V., ET AL. 1993. Middle and Upper Devonian miospore zonation of Eastern Europe. *Bulletin des Centres de Recherches Exploration-Production Elf Aquitaine*, **17**, 79–147.

BAI, S. L., BAI, Z. Q., MA, X. P., WANG, D. R. & SUN, Y. L. 1994. *Devonian Events and Biostratigraphy of South China*, Peking University Press, Beijing, China.

BARCLAY, W. J., BROWNE, M. A. E., MCMILLAN, A. A., PICKETT, E. A., STONE, P. & WILBY, P. R. 2005. *The Old Red Sandstone of Great Britain, Geological Conservation Review Series*, **31**, Joint Nature Conservation Committee, Peterborough.

BARKER, A. J., BENNETT, D. G., BOYCE, A. J. & FALLICK, A. E. 2000. Retrogression by deep infiltration of meteoric fluids into thrust zones during late-orogenic rapid unroofing. *Journal of Metamorphic Geology*, **18**, 307–318.

BECKER, R. T. & HOUSE, M. R. 1994. International Devonian goniatite zonation, Emsian to Givetian, with new records from Morocco. *Courier Forschungsinstitut Senckenberg*, **169**, 79–135.

BELKA, Z., KAUFMANN, B. & BULTYNCK, P. 1997. Conodont-based quantitative biostratigraphy for the Eifelian of the eastern Anti-Atlas, Morocco. *Geological Society of America Bulletin*, **109**, 643–651.

BLACK, G. P. 1978. *Orkney, Localities of Geological and Geomorphological Importance, Geology and Physiography Section*, Nature Conservancy Council, Newburg.

BLACKBOURN, G. A. & MARSHALL, J. E. A. 1985. The Geology of Foula, Shetland. In: BLACKBOURN, G. A. (ed.) *Geological Field Guide to Foula, Shetland*, Britoil, Glasgow, 1–46.

BRAATHEN, A., OSMUNDSEN, P. T., NORDGULEN, Ø., ROBERTS, D. & MEYER, G. B. 2002. Orogen-parallel extension of the Caledonides in northern Central Norway: an overview. *Norwegian Journal of Geology*, **82**, 225–241.

BRETT, C. E. & BAIRD, G. C. 1997. Epiboles, outages, and ecological evolutionary bioevents: taphonomic, ecological and biogeographic factors. In: BRETT, C. E. & BAIRD, G. C. (eds) *Palaeontological Events, Stratigraphic, Ecological and Evolutionary Implications*. Columbia University Press, New York, 249–284.

BRYHNI, I. 1974. Old Red Sandstone of Hustadvika and an occurrence of dolomite at Flatskjer, Nordmøre. *Norges Geologiske Undersøkelse*, **311**, 49–63.

BUDIL, P. 1995a. The Middle Devonian Kačák Event in the Barrandian area. *Geolines*, **3**, 7–8.

BUDIL, P. 1995b. Demonstrations of the Kačák Event (Middle Devonian, uppermost Eifelian) at some Barrandian localities. *Věstník Českého Geologického ústavu*, **70**, 1–24.

BUGGISCH, W. & MANN, U. 2004. Carbon isotope stratigraphy of Lochkovian to Eifelian limestones from the Devonian of central and southern Europe. *International Journal of Earth Science*, **93**, 521–541.

BULTYNCK, P. & DEJONGHE, L. 2001. Devonian lithostratigraphic units (Belgium). *Geologica Belgica*, **4**, 39–69.

CHLUPÁČ, I. 1960. Stratigrafiká studie o vrstvách srbských (givet) ve středočeském devonu. *Sborník Ústředního Ústavu Geologického, Oddil Geologicky*, **26**, 143–185.

CHLUPÁČ, I. 1994. Devonian trilobites—evolution and events. *Geobios*, **27**, 487–505.

CHLUPÁČ, I. & KUKAL, Z. 1986. Reflection of possible global Devonian events in the Barrandian area, C.S.S.R. In: WALLISER, O. H. (ed.) *Global Bio-Events, a Critical Approach. Lecture Notes in Earth Sciences*, **8**, 169–179.

CHLUPÁČ, I. & KUKAL, Z. 1988. Possible global events and the stratigraphy of the Palaeozoic of the Barrandian (Cambrian–Middle Devonian), Czechoslovakia. *Sborník geologických věd—geologie*, **43**, 83–146.

COHEN, A. S. 1990. Tectono-stratigraphic model for sedimentation in Lake Tanganyika, Africa. In: KATZ, B. J. (ed.) *Lacustrine Basin Exploration—Case Studies and Modern Analogues*. AAPG Memoirs, **50**, 137–150.

COWARD, M. P., DEWEY, J., HEMPTON, M. & HOLROYD, J. 2003. Tectonic evolution. In: EVANS, D., GRAHAM, C., ARMOUR, A. & BATHURST, P.

(eds) *The Millennium Atlas: Petroleum Geology of the Central and Northern North Sea*, Geological Society, London, 17–33.

COWIE, J. W. & ADAMS, P. J. 1957. The geology of the Cambro-Ordovician rocks of Central East Greenland. *Meddelelser om Grønland*, **153**, 1–193.

CRAMPTON, C. B. & CARRUTHERS, R. G. 1914. *The Geology of Caithness (sheets 110 and 116 with parts of 109, 115 and 117)*, Memoirs of the Geological Survey, HMSO, Scotland, Edinburgh.

DONOVAN, R. N. 1980. Lacustrine cycles, fish ecology and stratigraphic zonation in the Middle Devonian of Caithness. *Scottish Journal of Geology*, **16**, 35–50.

DUNCAN, A. D. & HAMILTON, R. F. M. 1988. Palaeolimnology and organic geochemistry of the Middle Devonian in the Orcadian Basin. *In*: FLEET, A. J., KELTS, K. & TALBOT, M. R. (eds) *Lacustrine Petroleum Source Rocks*. Geological Society of London, Special Publications, **40**, 173–201.

DUNCAN, W. I. & BUXTON, N. W. K. 1995. New evidence for evaporitic Middle Devonian lacustrine sediments with hydrocarbon source potential on the East Shetland Platform, North Sea. *Journal of the Geological Society, London*, **152**, 251–258.

ELLWOOD, B. B., BENOIST, S. L., EL HASSANI, A., WHEELER, C. & CRICK, R. E. 2003. Impact ejecta layer from the Mid-Devonian: possible connection to global mass extinctions. *Science*, **300**, 1734–1737.

EUGSTER, H. P. & HARDIE, L. A. 1978. Saline Lakes. *In*: LERMAN, A. (ed.) *Lakes, Chemistry, Geology, Physics*. Springer-Verlag, New York, 237–293.

FANNIN, N. G. T. 1969. Stromatolites from the Middle Old Red Sandstone of western Orkney. *Geological Magazine*, **106**, 77–88.

FLETCHER, T. P., AUTON, C. A., HIGHTON, A. J., MERRITT, J. W., ROBERTSON, S. & ROLLIN, K. E. 1996. *Geology of the Fortrose and eastern Inverness district*. Memoirs of the British Geological Survey, Sheet 84W (Scotland).

GARCIA-ALCALDE, J. L. 1998. Devonian events in northern Spain. *Newsletters in Stratigraphy*, **36**, 157–175.

GASSE, F. 2000. Hydrological changes in the African tropics since the Last Glacial Maximum. *Quaternary Science Reviews*, **19**, 189–211.

GRADSTEIN, F. M., OGG, J. G. & SMITH, A. G. 2004. *A Geologic Time Scale 2004*. Cambridge University Press, Cambridge.

GROSS, W. 1940. Acanthodier und Placodermen aus Heterostius-Schichten Estlands und Lettlands. *Annales Societatis Rebus Naturae Investigrandis in Universitate Tartuensi Constitutae*, **46**, 1–89.

HAMILTON, A. C. 1982. *Environmental History of East Africa: A Study of the Quaternary*. Academic Press, London.

HAMILTON, R. F. M. & TREWIN, N. H. 1988. Environmental controls on fish faunas of the Middle Devonian Orcadian Basin. *In*: MCMILLAN, N. J., EMBRY, A. F. & GLASS, D. J. (eds) *Devonian of the World*, Canadian Society of Petroleum Geologists Memoirs, **14**, 589–600.

HLADÍKOVÁ, J., HLADIL, J. & KŘÍBEK, B. 1997. Carbon and oxygen isotope record across Pridoli to Givetian stage boundaries in the Barrandian basin (Czech Republic). *Palaeogeography, Palaeo- climatology, Palaeoecology*, **132**, 225–241.

HLADIL, J., GERSL, M., STRNAD, L., FRANA, J., LANGROVA, A. & SPISIAK, J. 2005. Stratigraphic variation of complex impurities in platform limestones and possible significance of atmospheric dust: a study with emphasis on gamma-ray spectrometry and magnetic susceptibility outcrop logging (Eifelian–Frasnian, Moravia, Czech Republic). *International Journal of Earth Science*, DOI 10.1007/s00531-005-0052-8.

HOUSE, M. R. 1985. Correlation of mid-Palaeozoic ammonoid evolutionary events with global sedimentary perturbations. *Nature*, **313**, 17–22.

HOUSE, M. R. 1991. The Devonian System. *Encyclopædia Britannica*, **19**, 804–814.

HOUSE, M. R. 1996. The Middle Devonian Kačák Event. *Proceedings of the Ussher Society*, **9**, 79–84.

HOUSE, M. R. 2002. Strength, timing, setting and cause of mid-Palaeozoic extinctions. *Palaeogeography, Palaeoclimatology, Palaeoecology*, **181**, 5–25.

JACOBS, D. K. & SAHAGIAN, D. L. 1995. Milankovitch fluctuations in sea level and recent trends in sea-level change: ice may not always be the answer. *In*: HAQ, B. U. (ed.) *Sequence Stratigraphy and Depositional Response to Eustatic, Tectonic and Climatic Forcing*. Kluwer Academic Publishers, Netherlands, 329–366.

JOACHIMSKI, M. M., VAN GELDERN, R., BREISIG, S., BUGGISCH, W. & DAY, J. 2004. Oxygen isotope evolution of biogenic calcite and apatite during the Middle and Late Devonian. *International Journal of Earth Science*, **93**, 542–553.

JOHNSON, T. C. 1996. Sedimentary processes and signals of past climatic change in the large lakes of the East African Rift Valley. *In*: JOHNSON, T. C. & ODADA, E. O. (eds) *The Limnology, Climatology and Palaeoclimatology of the East African Lakes*. Gordon & Breach, 367–412.

KAUFMANN, B. 2006. Calibrating the Devonian time scale: a synthesis of U–Pb ID-TIMS ages and conodont stratigraphy. *Earth-Science Reviews*, **76**, 175–190.

KELLY, S. B. & OLSEN, H. 1993. Terminal fans—a review with reference to Devonian examples. *Sedimentary Geology*, **85**, 339–374.

KELTS, K. & HSÜ, K. J. 1978. Freshwater carbonate sedimentation. *In*: LERMAN, A. (ed.) *Lakes, Chemistry, Geology, Physics*. Springer-Verlag, New York, 295–323.

KLEESMENT, A. 1997. Devonian sedimentation basin. *In*: RAUKAS, A. & TEEDUMÄE, A. (eds) *Geology and Mineral Resources of Estonia*. Estonian Academic Publishers, Tallinn, 205–208.

KURŠS, V. M. 1975. *Lithology and mineral resources of the terrigenous Devonian of the Main Devonian Field*. Zinatne, Riga [in Russian].

KURŠS, V. M. 1992a. *Devonian terrigenous sedimentary deposits of the Main Devonian Field*. Zinatne, Riga [in Russian].

KURŠS, V. M. 1992b. Depositional environments and burial conditions of fish remains in the Baltic Middle Devonian. *In: Fossil Fishes as Living Animals*, Academia, **1**, 251–260. Academy of Sciences of Estonia.

LOBOZIAK, S., STREEL, M. & WEDDIGE, K. 1991. Miospores, the *lemurata* and *triangulatus* levels and their faunal indices near the Eifelian/Givetian boundary in the Eifel (F.R.G.). *Annales de la Société Géologique de Belgique*, **113**, 299–313.

McCann, T. 1999. Middle to Late Devonian basin evolution in the Rügen area, NE Germany. *Geologie en Mijnbouw*, **78**, 57–71.

Mark-Kurik, E. 1991. On the environment of Devonian fishes. *Proceedings of the Estonian Academy of Sciences. Geology*, **40**, 122–125.

Mark-Kurik, E. 1995. Trophic relations of Devonian fishes. *Geobios, M.S.* **19**, 121–123.

Mark-Kurik, E. 2000. The Middle Devonian fishes of the Baltic States (Estonia, Latvia) and Belarus. *Courier Forschungsinstitut Senckenberg*, **223**, 309–324.

Marshall, J. E. A. 1988. Devonian miospores from Papa Stour, Shetland. *Transactions of the Royal Society of Edinburgh: Earth Sciences*, **79**, 13–18.

Marshall, J. E. A. 1996. *Rhabdosporites langii, Geminospora lemurata* and *Contagisporites optivus*: an origin for heterospory within the Progymnosperms. *Review of Palaeobotany and Palynology*, **93**, 159–189.

Marshall, J. E. A. & Astin, T. R. 1996. An ecological control on the distribution of the Devonian fish *Asterolepis*. *Newsletters on Stratigraphy*, **33**, 133–144.

Marshall, J. E. A. & Fletcher, T. P. 2002. Middle Devonian (Eifelian) spores from a fluvial dominated lake margin in the Orcadian Basin, Scotland. *Review of Palaeobotany and Palynology*, **118**, 195–209.

Marshall, J. E. A. & Hewett, A. J. 2003. Devonian. *In*: Evans, D., Graham, C., Armour, A. & Bathurst, P. (eds) *The Millennium Atlas: Petroleum Geology of the Central and Northern North Sea*, Geological Society, London, 65–81.

Marshall, J. E. A., Rogers, D. A. & Whiteley, M. J. 1996. Devonian marine incursions into the Orcadian Basin, Scotland. *Journal of the Geological Society, London*, **153**, 451–466.

Marshall, J. E. A. & Stephenson, B. J. 1997. Sedimentological response to basin initiation in the Devonian of East Greenland. *Sedimentology*, **44**, 407–419.

Martínez Catalálan, J. R., Fernández-Suárez, J., Jenner, G. A., Belousova, E. & Díez Montes, A. 2004. Provenance constraints from detrital zircon U–Pb ages in the NW Iberian Massif: implications for Palaeozoic plate configuration and Variscan evolution. *Journal of the Geological Society, London*, **161**, 463–476.

May, A. 1995. Relationship among sea-level fluctuation, biogeography and bioevents in the Devonian: an attempt to approach a powerful, but simple model for complex but long-range control of biotic crises. *Geolines*, **3**, 38–49.

Meyers, J. S. & Nordenson, T. J. 1962. Evaporation from the 17 Western States, with a section on evaporation rates. *United States Geological Survey, Professional Paper*, **272-D**, 71–100.

Miles, R. S. 1967. Observations on the ptyctodont fish, *Rhamphodopsis* Watson. *Journal of the Linnean Society (Zoology)*, **47**, 99–120.

Miles, R. S. & Westoll, T. S. 1963. Two new genera of coccosteid arthrodire from the Middle Old Red Sandstone of Scotland, and their stratigraphical distribution. *Transactions of the Royal Society of Edinburgh*, **65**, 180–210.

Mosar, J., Eide, E. A., Osmundsen, P. T., Sommaruga, A. & Torsvik, T. H. 2002. Greenland–Norway separation: A geodynamic model for the North Atlantic. *Norwegian Journal of Geology*, **82**, 281–292.

Mykura, W. 1976. *British Regional Geology, Orkney and Shetland*, HMSO, Edinburgh.

Mykura, W. & Phemister, J. 1976. *The Geology of Western Shetland, Sheet 127 and parts of 125, 126 and 128*. Memoirs of the Geological Survey of Great Britain, Scotland.

Narbutas, V. 2004. Devonian and Carboniferous *In*: Baltrunas, A. (ed.) *Evolution of earth underground and forecast of its resources in Lithuania*. National Scientific Programme 'Litosfera', Vilnius. pp. 700 [in Lithuanian with English summary].

Obrucheva, O. P. 1962. *Placoderm Fishes of the Devonian of the USSR (coccosteids and dinichthyids)*. Moscow University Press, Moscow [in Russian].

Obrucheva, O. P. 1966. New facts concerning the coccosteids (Placodermi) from the Baltic Devonian. *In*: Grigelis, A. (ed.) *Palaeontology and Stratigraphy of the Baltic and Byelorussia*, **I (VI)**, Mintis, Vilnius, 151–189 [in Russian].

Ojiambo, B. S. & Lyons, W. B. 1996. Residence times of major ions in Lake Naivasha, Kenya, and their relationship to lake hydrology. *In*: Johnson, T. C. & Odada, E. O. (eds) *The Limnology, Climatology and Palaeoclimatology of the East African Lakes*. Gordon & Breach, 267–278.

Olsen, P. E. 1990. Tectonic, climatic, and biotic modulation of lacustrine ecosystems—Examples from the Newark Supergroup of Eastern North America. *In*: Katz, B. J. (ed.) *Lacustrine Basin Exploration, Case Studies and Modern Analogs*. AAPG Memoirs, **50**, 209–224.

Parnell, J. 1983. Ancient duricrusts and related rocks in perspective: a contribution from the Old Red Sandstone. *In*: Wilson, R. C. L. (ed.) *Residual Deposits: Surface Related Weathering Processes and Materials*. Special Publication of the Geological Society, London, **11**, 197–209.

Parnell, J. 1986. Devonian magadi-type cherts in the Orcadian Basin, Scotland. *Journal of Sedimentary Petrology*, **56**, 495–500.

Partridge, T. C., Demenocal, P. B., Lorentz, S. A., Paiker, M. J. & Vogel, J. C. 1998. Orbital forcing of climate over South Africa: a 200,000-year rainfall record from the Pretoria Saltpan. *Quaternary Science Reviews*, **16**, 1125–1133.

Paškevičius, J. 1997. *The Geology of the Baltic Republics*. Vilnius University, Geological Survey of Lithuania.

Pichler, R. 1971. Mikrofossilien aus dem Devon der südlichen Eifeler Kalkmulden. *Senckenbergiana Lethaea*, **52**, 315–357.

Pollard, J. E., Steel, R. J. & Undersrud, E. 1982. Facies sequences and trace fossils in lacustrine/fan delta deposits, Hornelen Basin (M. Devonian), Western Norway. *Sedimentary Geology*, **32**, 63–87.

RAYNER, D. H. 1963. The Achanarras Limestone of the Middle Old Red Sandstone, Caithness, Scotland. *Proceedings of the Yorkshire Geological Society*, **34**, 117–138.

ROGERS, D. A. & ASTIN, T. R. 1991. Ephemeral lakes, mud pellet dunes and wind-blown sand and silt: reinterpretations of Devonian lacustrine cycles in north Scotland. *In*: ANADON, P., CABRERA, L. & KELTS, K. (eds) *Lacustrine Facies Analysis. Special Publications International Association of Sedimentologists*, **13**, 199–221, Blackwell Scientific Publications, Oxford.

ROHLING, E. J., ABU-ZIED, R., CASFORD, C. S. L., HAYES, A. & HOOGAKKER, B. A. A. (in press). The Mediterranean Sea: Present and Past. *In*: WOODWARD, J. (ed.) *Physical Geography of the Mediterranean Basin*. Oxford University Press.

RUDDIMAN, W. F. 2001. *Earth's Climate, Past and Future*. Freeman, New York.

SCHÖNE, B. R. 1997. Der *otomari*-Event und seine Auswirkungen auf die Fazies des Rhenoherzynischen Schelfs (Devon, Rheinisches Schiefergebirge). *Göttinger Arbeiten zur Geologie und Paläontologie*, **70**, 1–140.

SCOTESE, C. R. 2002. Paleomap Project, http://www.scotese.com.

SHAHIN, M. 1985. *Hydrology of the Nile Basin*. Developments in Water Science, **21**. Elsevier, Amsterdam.

SOROKIN, V. S., LYARSKAYA, L. A. & SAVVAITOVA, L. S. 1981. *Devonian and Carboniferous of the Baltic Regions*. Zinatne, Riga [in Russian].

SOUTHAM, J. R., PETERSON, W. H. & BRASS, G. W. 1982. Dynamics of anoxia. *Palaeogeography, Palaeoclimatology and Palaeoecology*, **40**, 183–198.

SPEED, R. G. 1999. *Kerogen variation in a Devonian half-graben system*. Ph.D. thesis, University of Southampton.

STEEL, R., SIEDLECKA, A. & ROBERTS, D. 1985. The Old Red Sandstone basins of Norway and their deformation: a review. *In*: GEE, D. G. & STURT, B. A. (eds) *The Caledonide Orogeny-Scandinavia and Related Areas*. Wiley, New York, 293–315.

STEPHENSON, D. 1977. Intermontane basin deposits associated with an early Great Glen feature in the Old Red Sandstone of Invernessshire. *In*: GILL, G. (ed.) *The Moray Firth area Geological Studies*, Inverness Field Club, 35–45.

STEPHENSON, M. H., LENG, M. J., MICHIE, U. & VANE, C. H. 2006. Palaeolimnology of Palaeozoic lakes, focusing on a single lake cycle in the Middle Devonian of the Orcadian Basin, Scotland. *Earth-Science Reviews*, **75**, 177–197.

STRACHAN, R. A. 2003. The metamorphic basement geology of Mainland Orkney and Graemsay. *Scottish Journal of Geology*, **39**, 145–149.

STRUVE, W. 1982. The Great Gap in the record of marine Middle Devonian. *Courier Forschungsinstitut Senckenberg*, **55**, 433–448.

STUMM, W. & MORGAN, J. J. 1995. *Aquatic Chemistry, Chemical Equilibria and Rates in Natural Waters*. Wiley, New York.

THOMSON, K. S., SUTTON, M. & THOMAS, B. 2003. A larval Devonian lungfish. *Nature*, **426**, 844–834.

TORSVIK, T. H. & COCKS, L. R. M. 2004. Earth geography from 400 to 250 Ma: a palaeomagnetic, faunal and facies review. *Journal of the Geological Society, London*, **161**, 555–572.

TORSVIK, T. H., VAN DER VOO, R., MEERT, J. G., MOSAR, J. & WALDERHAUG, H. J. 2001. Reconstructions of the continents around the North Atlantic at about the 60th parallel. *Earth and Planetary Science Letters*, **187**, 55–69.

TREWIN, N. H. 1986. Palaeoecology and sedimentology of the Achanarras fish bed of the Middle Old Red Sandstone, Scotland. *Transactions of the Royal Society of Edinburgh*, **77**, 21–46.

TREWIN, N. H. & DAVIDSON, R. G. 1999. Lake-level changes, sedimentation and faunas in a Middle Devonian basin-margin fish bed. *Journal of the Geological Society, London*, **156**, 535–548.

TREWIN, N. H. & THIRLWALL, M. 2002. Old Red Sandstone. *In*: TREWIN, N. H. (ed.) *The Geology of Scotland*. 4th edn., The Geological Society, London, 213–249.

TRUYÓLS-MASSONI, M., MONTESINOS, R., GARCIA-ALCALDE, J. L. & LEYVA, F. 1990. The Kacak-otomari event and its characterization in the Palentine Domain (Cantabrian Zone, NW Spain). *In*: KAUFFMAN, E. G. & WALLISER, O. H. (eds) *Extinction Events in Earth History, Lecture Notes in Earth Sciences*, **30**, Springer-Verlag, Berlin, 133–143.

TURNAU, E. 1996. Miospore stratigraphy of Middle Devonian deposits from Western Pomerania. *Review of Palaeobotany and Palynology*, **93**, 107–125.

TURNAU, E. & RACKI, G. 1999. Givetian palynostratigraphy and palynofacies: new data from the Bodzentyn Syncline (Holy Cross Mountains, central Poland). *Review of Palaeobotany and Palynology*, **106**, 237–271.

VAITIEKŪNIENĖ, G. K. 1983. On the Devonian complex of spores of the Kernavė suite of the south Baltic area. *In*: *Palynologic Researches in Geological Studies of the Baltic Region and the Baltic Sea*. Riga, Zinātne, 27–31 [in Russian].

VALIUKEVIČIUS, J. J. 1995. Acanthodians from marine and non-marine Early and Middle Devonian deposits. *Geobios, MS*, **19**, 393–397.

VALIUKEVIČIUS, J. J. 2000. Acanthodian biostratigraphy and interregional correlations of the Devonian of the Baltic States, Belarus, Ukraine and Russia. *Courier Forschungsinstitut Senckenberg*, **223**, 271–289.

VALIUKEVIČIUS, J. J., KLEESMENT, A. E., KURIK, E. J. & VAITIEKŪNIENĖ, G. K. 1986. Correlation and organic remains of deposits of the Narva Regional Stage. *In*: BRANGULIS, A. P. (ed.) *Biofacies and fauna of the Silurian and Devonian of Baltic*, Zinatne, Riga, 73–122 [in Russian].

VALIUKEVIČIUS, J. J., TALIMAA, V. & KRUCHEK, S. 1995. Complexes of vertebrate microremains and correlation of terrigenous Devonian deposits of Belarus and adjacent territories. *Ichthyolith Issues, Special Publication*, **1**, 53–59.

WALLISER, O. H. 1985. Natural boundaries and commission boundaries in the Devonian. *Courier Forschungsinstitut Senckenberg*, **75**, 401–408.

WALLISER, O. H. 1990. How to define 'global bioevents'. *In*: KAUFFMAN, E. G. & WALLISER, O. H. (eds) *Extinction Events in Earth History, Lecture*

Notes in Earth Sciences, **30**, Springer-Verlag, Berlin, 1–3.

WALLISER, O. H. 1995. Global events in the Devonian and Carboniferous. *In*: WALLISER, O. H. (ed.) *Global Events and Event Stratigraphy*. Springer-Verlag, Berlin, 225–250.

WALLISER, O. H. 2000. The Eifelian–Givetian stage boundary. *Courier Forschungsinstitut Senckenberg*, **225**, 37–47.

WATSON, D. M. S. 1938. On *Rhamphodopsis*, a Ptyctodont from the Middle Old Red Sandstone of Scotland. *Transactions of the Royal Society of Edinburgh*, **59**, 397–410.

WEDDIGE, K. 1988. Conodont distribution within the event interval. *Courier Forschungsinstitut Senckenberg*, **102**, 132–133.

WELLMAN, C. H. 2004. Palaeoecology and palaeophytogeography of the Rhynie Chert plants: evidence from integrated analysis of *in situ* and dispersed spores. *Proceedings of the Royal Society of London*, **B271**, 985–992.

WESTOLL, T. S. 1977. Northern Britain. *In*: HOUSE, M. R., RICHARDSON, J. B., CHALONER, W. G., ALLEN, J. R. L., HOLLAND, C. H. & WESTOLL, T. S. (eds) *A Correlation of the Devonian Rocks in the British Isles*, Geological Society, London, Special Report, **7**, 66–93.

WILLIAMS, E. A., FRIEND, P. F. & WILLIAMS, B. P. J. 2000. A review of Devonian time scales: databases, construction and new data. *In*: FRIEND, P. F. & WILLIAMS, B. P. J. (eds) *New Perspectives on the Old Red Sandstone*. Geological Society, London, Special Publication, **180**, 1–21.

WILSON, G. V., EDWARDS, W., KNOX, J., JONES, R. C. B. & STEPHENS, J. V. 1935. *The Geology of the Orkneys*. Memoirs of the Geological Survey, HMSO, Edinburgh, Scotland.

YOUNG, S. 1995. Micro-remains from Early and Middle Devonian acanthodian fishes from the U.K. and their biostratigraphic possibilities. *Ichthyolith Issues*, Special Publication, 1, 65–68.

YURETICH, R. F. & CERLING, T. E. 1983. Hydrogeochemistry of Lake Turkana, Kenya: Mass balance and mineral equilibria in an alkaline lake: *Geochimica et Cosmochimica Acta*, **47**, 1099–1109.

ZIEGLER, P. A. 1990. *Geological Atlas of Western and Central Europe*, 2nd edn. Shell Internationale Petroleum, Maatschappij B. V., The Hague.

Givetian (Middle Devonian) brachiopod–goniatite–correlation in the Dra Valley (Anti-Atlas, Morocco) and Bergisch Gladbach–Paffrath Syncline (Rhenish Massif, Germany)

V. EBBIGHAUSEN[1], R. T. BECKER[2], J. BOCKWINKEL[3] & Z. S. ABOUSSALAM[2]

[1]*Engstenberger Höhe 12, D-51519 Odenthal, Germany (e-mail: Volker@vxr.de)*
[2]*Geologisch-Paläontologisches Institut, Westfälische Wilhelms-Universität, Corrensstr. 24, D-48149 Münster, Germany (e-mail: rbecker@uni-muenster.de, taghanic@uni-muenster.de)*
[3]*Dechant-Fein-Str. 22, D-51375 Leverkusen, Germany (e-mail: jbockwinkel@t-online.de)*

Abstract: The Givetian pelagic and dysoxic outer shelf facies of the Dra Valley (SW Morocco) yielded as minor benthic faunal elements a number of stringocephalid and uncitid brachiopods that allow a precise correlation of these marker brachiopods with the regional, detailed goniatite zonation. In a reverse situation, the predominant neritic shallow-water succession of the Bergisch Gladbach area (Rhenish Massif, Germany), which is characterized by a detailed succession of stringocephalids and *Uncites*, has yielded rare and new Middle Givetian goniatite species. These findings allow, with some help of conodont data, neritic–pelagic correlations within and between widely separated basins. New species are *Tornoceras* n. sp. from the Büchel Formation (with coloration remains), '*Trevoneites*' *paffrathensis* n. sp. from the Lower Plattenkalk Formation, and *Maenioceras heinorum* n. sp. from the Hornstein Member. New material of stringocephalids and *Uncites* is described from the Dra Valley. The identical, well-defined range of *Uncites (U.) gryphus gryphus* in the lower to middle parts of the Middle Givetian of the Dra Valley and Rhenish Massif underscores the stratigraphical significance of this genus that was widely distributed in Europe, northern Gondwana, the Urals, and Central and Eastern Asia.

The extensive Devonian shelf areas were characterized by strong faunal differences between shallow, photic, neritic inner shelf areas with diverse benthic faunas and pelagic outer shelf areas, dominated by planktonic and nektonic faunas and with a low-diversity, specialized, subphotic benthos, which is not very similar to nearshore bottom-dwelling assemblages. As a result, the precise correlation of the neritic (Rhenish) and pelagic (Hercynian) magnafacies (e.g. Erben 1962) is difficult and has been a major task in Devonian biostratigraphy. Whilst pelagic faunas are characterized by cosmopolitan taxa, neritic faunas tend to comprise mostly endemic forms. Therefore, neritic–pelagic correlations need to be achieved in parallel in many basins, which is far from being resolved in most regions and in large parts of the stratigraphic column. Records of neritic faunal elements in pelagic facies and *vice versa* are in any case remarkable, especially if they concern stratigraphically important marker species.

The Dra Valley in the western part of the Anti-Atlas (southern Morocco) is a key area for Devonian cross-facies correlations because there is almost continuous outcrop over several hundreds of kilometres. Transitions from pelagic to neritic deposition took place roughly from the east to west and, with sea-level cycles, within exposed successions. An overview of regional geology and stratigraphy has been provided by Becker *et al.* (2004*a*). Until recently, most exposures were difficult to access as all are situated close to the politically troubled Moroccan–Algerian border. New detailed fieldwork started in spring 2000 and is continuing. A specimen of *Uncites* Defrance, a supposed typical neritic (Rhenish) brachiopod genus, was discovered in the predominantly pelagic Oued Mzerreb Member of the Ahrerouch Formation in the area south of Tata (Fig. 1), in direct association with a rich pelagic assemblage of goniatites (Ebbighausen *et al.* 2004). During the Dra Valley fieldtrip of the International Subcommission on Devonian Stratigraphy (SDS) in March 2004, additional specimens of *Uncites* as well as important stringocephalids were collected by various SDS members and at various localities. The first joint occurrence of *Uncites*, stringocephalids and goniatites of the Dra Valley, however, was reported much earlier by Drot & Hollard (1965) in a short conference summary. The presence of *Uncites* in the Dra Valley was also noted by H. Hollard (1970) in an excursion guide book dealing with Precambrian correlations in the western and central Anti-Atlas. First records of several species of

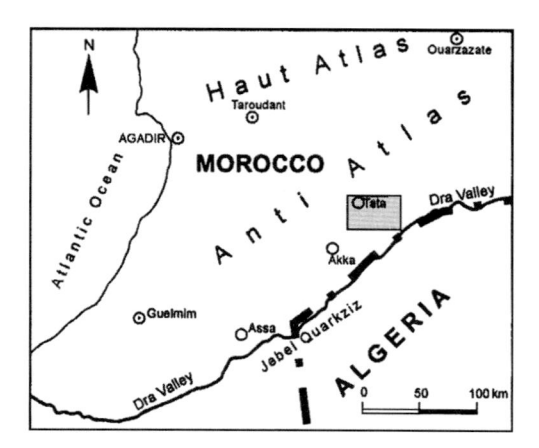

Fig. 1. Map of the western part of the Anti-Atlas and western part of the Dra Valley. The grey rectangle shows the Tata region (details in Fig. 2), where Middle Devonian sections were studied.

correlative value. The 'Unterer Plattenkalk' is long known to contain both neritic (e.g. in storm beds with brachiopod mass accumulations) and pelagic taxa, such as goniatites (e.g. Winterfeld 1894, 1895; Jux & Strauch 1965) and conodonts (Kleinebrinker 1992). These associations contradict its interpretation as a lagoonal facies (Jux & Strauch 1965) and rather suggest deposition on a drowned and storm-influenced carbonate platform with episodic invasions of open-shelf faunal elements. Goniatites and conodonts are completely lacking in the true lagoonal facies of Devonian reef complexes.

The current paper improves the neritic–pelagic correlation in the Middle Givetian (using the proposed definition of Bultynck & Gouwy 2002) by describing for the first time precisely dated new brachiopod material from the deeper-water, argillaceous, goniatite facies of the Dra Valley and, vice versa, by the description of important goniatite occurrences in the reefal to shallow neritic facies of the Rhenish Massif. For a review of global Givetian ammonoid zones, see Becker & House (2000).

Localities and material

Dra Valley, SW Morocco (see Fig. 2)

The first and single *Stringocephalus* sp. of the Anti-Atlas, from near Ksar de Zaïr (northern border of the Tindouf Basin, Hollard & Drot 1958), came from a layer with goniatites then identified as *Agoniatites fulguralis* var. *phillipsi* Wedekind and *Maenioceras terebratum* Sandberger & Sandberger. Hollard & Drot (1958) noted that the stringocephalid showed affinities with juveniles of *S. burtini*. The goniatite identifications need revision but, unfortunately, the location of the specimens is currently unknown. Because their position is unknown in the maenioceratid succession of the Tata area (Becker *et al.* 2004*b*), it is not possible to assign the material to a specific level within the Middle Givetian. Later, Drot (1961) described two additional brachiopods from Foum Medfa south of Assa (as *Stringocephalus* sp. 1) and from Hassi Smeira (as ?*Stringocephalus* sp. 2). Both specimens came from marls at the top of limestones with *A. fulguralis* var. *phillipsi* and *M. excavatum* (Phillips). This may indicate a rather low position in the Middle Givetian since the latter goniatite species now falls in synonymy with *Bensaidites molarius* (Whidborne), representing a genus that last occurs in the basal Middle Givetian in the Dra Valley (Becker *et al.* 2004*b*). In the Assa–Torkoz area, limestones with *Agoniatites* have recently been observed only low in the Givetian, just above the Kačák Event Beds. The *Stringocephalus* sp. 1 possesses a narrow triangular sinus in

stringocephalids, but none identified with certainty, were made by Hollard & Drot (1958) and Drot (1961). All these records confirm that both brachiopod groups belong as a minor benthic element to the faunal inventory of the mostly pelagic Givetian of the Dra Valley.

The genus *Uncites* is best known from the Rhenish Slate Mountains of Germany. It includes a second subgenus, *Uncites* (*Winterfeldia*) Jux & Strauch 1966, and a total of five species/subspecies, one of which (*U. pauciplicatus* Zhang in Zhang & Fu 1983) is endemic to China. Other superficially similar species, such as '*U. Galloisi* Oehlert', have been placed subsequently in the somewhat homoeomorphic genus *Borndhardtina* Schulz. *Uncites* was revised and described in detail in a now-classic study by Jux & Strauch (1966). Both subgenera delivered valuable marker forms (Struve 1982, fig. 12) in the calcareous, neritic Givetian of the Eifel Mountains, in the Bergisch Gladbach area just east of the Rhine, and further on in more eastern parts of the Rhenish Massif. New material extends the known stratigraphic range of some taxa. In Germany, *Uncites* is usually accompanied by stringocephalids ('Strigunc faunas' of Struve 1982). *Stringocephalus burtini* (Defrance in Blainville) was split by Struve (1992) into different genera, species and subspecies, figured after his death by Thormann & Weddige (2001). Well-preserved and abundant stringocephalids and *Uncites* can be collected in the reefal 'Massenkalk facies' of the Büchel Formation and in the Lower Plattenkalk Formation of the Bergisch Gladbach–Paffrath Syncline. In the reefal facies (Büchel Formation), ammonoids are normally lacking but herewith we record the first specimen, which is of

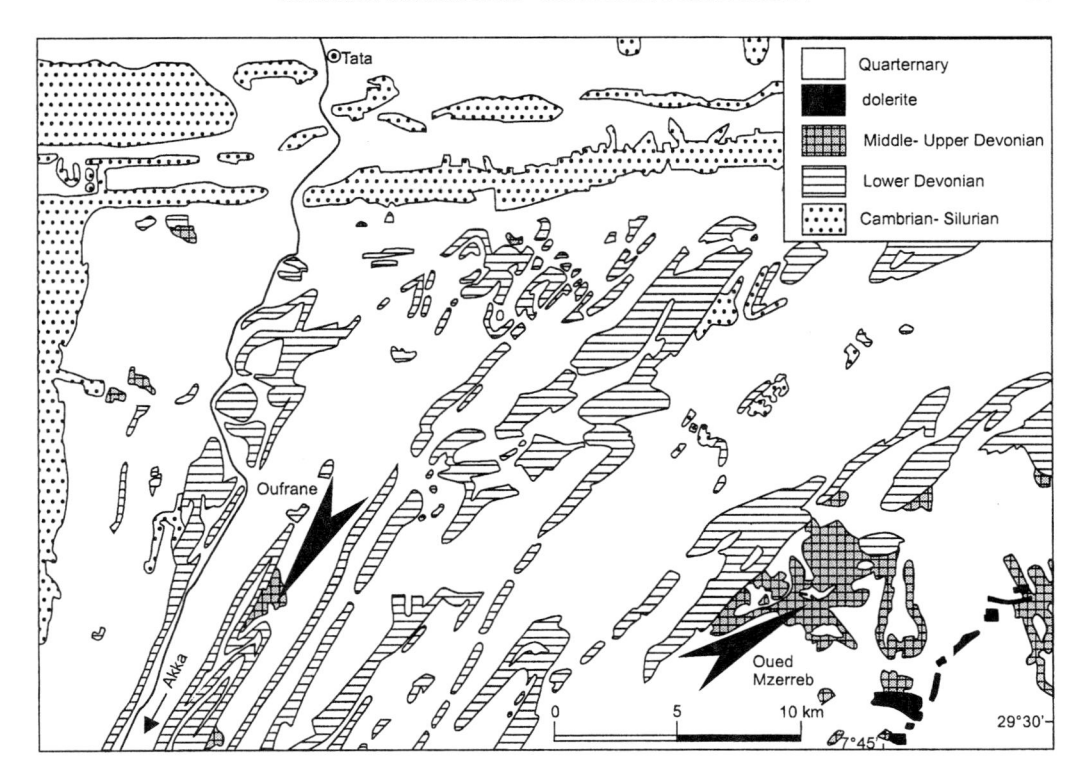

Fig. 2. Geological map of the Tata region (eastern Dra Valley) showing the positions (black arrows) of sampled sections at Oufrane and Oued Mzerreb.

the pedicle valve and shows affinities with *Parastringocephalus dorsalis* (Goldfuss). The badly preserved second specimen was only doubtfully assigned to *Stringocephalus*.

The regionally first specimen of *Uncites gryphus* (Schlotheim) was collected by Drot & Hollard (1965) south of the crest of Mersakhsaï, about 27 km south of Akka, together with two unidentified stringocephalids. Based on the rich associated goniatite fauna, then identified as *M. terebratum*, *A*. sp. ex gr. *fulguralis* (Whidborne) and *A. obliquus* (Whidborne), the *Uncites* specimen was assigned to the Upper Givetian. The Upper Givetian, however, will be formally defined to start much higher (base of *hermanni* Zone, Aboussalam & Becker 2002). The identification of the recorded agoniatitids is now questionable, following the revision of the English species by House (2002). Therefore, the precise position of Drot & Hollard's (1965) specimen within the Middle Givetian is unknown.

In his regional review, Hollard (1967, p. 225) assigned the level of *Stringocephalus* that were found above maenioceratids also to the 'Givetien supérieur' but he did not add precision (see also brief reference in Hollard 1970, p. 177). The new material allows for the first time a precise correlation of subordinate neritic brachiopods with the detailed goniatite succession of the Dra Valley. The new collections are from the Oued Mzerreb Member of the Ahrerouch Formation in the eastern Dra Valley (Tata area) and from the roughly contemporaneous Coral Marl Member of the Ahrerouch Formation in the western Dra Valley (Assa area, Becker *et al.* 2004*a*). The latter unit was named as 'Schistes gris et calcaires noirs á Tentaculites á *Maenioceras* et Stringocephales' in Hollard (1963), suggesting the same level as the material of Drot (1961).

In general, the material was collected together with other fauna (goniatites etc.) from bedding surfaces in very dry desert areas and is thus partly weathered. The detailed re-sampling of beds proved that there was only minor erosive mixing of surface faunas but in cases of gentle slope morphology it cannot be ruled out completely that some of the specimens were slightly transported. However, this has not affected zonal assignments. The following localities yielded brachiopod material:

1. Oufrane West, *ca.* 22 km S of Tata (Fig. 2, GPS N 29° 32,41′ W 07° 59,33′).

References: Aboussalam (2003) and Aboussalam *et al.* (2004).

A single *Stringocephalus* sp. (MB.B.2449) was collected at 20 m distance from the Lower *pumilio* Bed (Bed 1b) and may or may not have been transported into the plain. Bed 3a yielded two *Uncites (Uncites) gryphus gryphus* (MB.B.2450.1-2), *Subsinucephalus* sp. (MB.C.2451, Fig. 5k–m), a questionable *Subsinucephalus* (MB.B.2452), *Stringocephalus* sp. (MB.B.2453.1-3), *Agoniatites costulatus* (d'Archiac & De Verneuil) (MB.C.3885, Fig. 8l–m), *A. meridionalis* Bensaid, *Maenioceras terebratum* (MB.C.3886, Fig. 8j–k), a different new *Maenioceras*, *Wedekindella* n. sp. (MB.C.3887, Fig. 8h–i), '*Trevoneites*' *assessi* Göddertz, and *Sobolewia virginiana* House. Associated fauna consists of subordinate pyritic gastropods, typical deeper-water bivalves (e.g. *Buchiola*, some nuculoids), rare thamnoporids and crinoid stems. A ?*Parastringocephalus* sp. (MB.B.2454, Fig. 5n–o) was collected higher from Bed 6a; a third *U. (U.) gryphus gryphus* (MB.B.2455, leg. M. W. Amler, Marburg) and another *Stringocephalus* sp. (MB.B.2456) were collected below Bed 8. All brachiopods are crushed and distorted. The stringocephalids partly show the median septum but cannot be identified at species level. Rare stromatoporoids, chaetetids, branching tabulate corals (thamnoporids) and colonial rugose corals suggest a minor shallowing and improved seafloor oxygenation in the upper part of the sampled slope (Beds 4 to 6). However, the majority of brachiopods was found in the hypoxic Bed 3a.

2. Oufrane East (GPS N 29° 32,51′ W 7° 59,07′)

References: Aboussalam (2003, p. 105) and Aboussalam *et al.* (2004, p. 56).

The best-preserved *Uncites (Uncites) gryphus gryphus* (MB.B.2457, Fig. 8n–r) was collected from Bed 2a′, which is characterized by mostly poorly preserved but large *Agoniatites costulatus*, *A.* cf. *meridionalis*, *Sobolewia virginiana*, *Maenioceras terebratum*, '*Trevoneites*' *assesi*, rare other tornoceratids, and *Wedekindella* n. sp. (with concave umbilical wall). Bed 2a′ correlates with the rather unfossiliferous upper part of Bed 3a of Oufrane W.

3. Oued Mzerreb, *ca.* 13 km SE of Tata (GPS N 29° 33,12′ W 07° 46,19′, Fig. 2).

References: Detailed section in Aboussalam (2003) and Becker *et al.* (2004*b*).

The oldest and rare neritic brachiopods occur very low in the Givetian, including a squashed *Bornhardtina* cf. *triangularis* Wedekind 1934 (MB.B.2467) from Bed −7a, and an eroded *Chascothyris* cf. *holzapfeli* Torley 1934 (MB.B.2468) from Bed −4c. Four crushed *Stringocephalus* sp., partly showing the median septum of the pedicle valve, were collected in the plain from the surface of the otherwise rather unfossiliferous Bed 4a (MB.B.2458.1-2) and Bed 5a (MB.B.2459, Fig. 5j). There is also one crushed but more or less complete *Uncites (Uncites) gryphus gryphus* (MB.B.2460). These specimens may be older than the Lower *pumilio* level and seem to pre-date *Maenioceras terebratum* that enters in Bed 6. Two additional crushed *Stringocephalus* sp. (MB.B.2461.1-2) were found in the Upper *Maenioceras* Bed (Bed 6), which yielded a very rich goethitic goniatite fauna (two species of *Maenioceras*, *Agoniatites*, '*Trevoneites*', *Wedekindella*, *Sobolewia*, *Tornoceras*), nautiloids, some gastropods, rare hyolitids, tabulate corals, crinoid stems, and the deeper-water bivalve *Buchiola*.

4. Bou Tserfine *ca.* 18.5 km NE of Assa (Fig. 1, GPS N 28° 39,196′ W 09° 16,093′).

For a first but still somewhat preliminary locality description, see Becker *et al.* (2004*c*). More than 100 m south of the prominent Rich 4, Givetian marls with rugose and tabulate corals, the Coral Marl Member of the Ahrerouch Formation, are poorly exposed above a solid dark-grey marker limestone with mass occurrence of '*pumilio*-type' small brachiopods. There are only few brachiopods, including one crushed but almost complete *Uncites (Uncites) gryphus gryphus* (MB.B.2462, leg. E. Schindler, Frankfurt a. M.) and two poorly preserved, incomplete stringocephalids (MB.B.2463.1-2, leg. B. Schröder, Cologne). One of the specimens, an elongated fragment, displays the internal median septum and on the outer side a shallow shell depression as in *Subsinucephalus* or *Parastringocephalus*.

Bergisch Gladbach–Paffrath Syncline, Rhenish Massif (see Fig. 3)

For many centuries, the Bergisch Gladbach–Paffrath Syncline (Fig. 3) has yielded well-preserved neritic faunas from the reef facies of the Büchel Formation and from the overlying Lower Plattenkalk Formation. Most illustrations of *Stringocephalus burtini* in general publications and textbooks were erroneously based on specimens from the area rather on the types from the southern part of the Dinant Syncline in Belgium. As a result, the concept of *S. burtini*, the type-species of the genus, was gradually replaced by incorrect references to *Stringocephalus buchi buchi*, a similar-looking and the most common stringocephalid species of the Bergisch Gladbach–Paffrath Syncline (see Struve 1992).

Goniatites are rare in the Givetian of the Gladbach–Paffrath Syncline. No goniatite has ever been reported from the Büchel Formation. Winterfeld (1895) first mentioned *Tornoceras simplex* (v. Buch), *Anarcestes cancellatus* d'Archiac & De

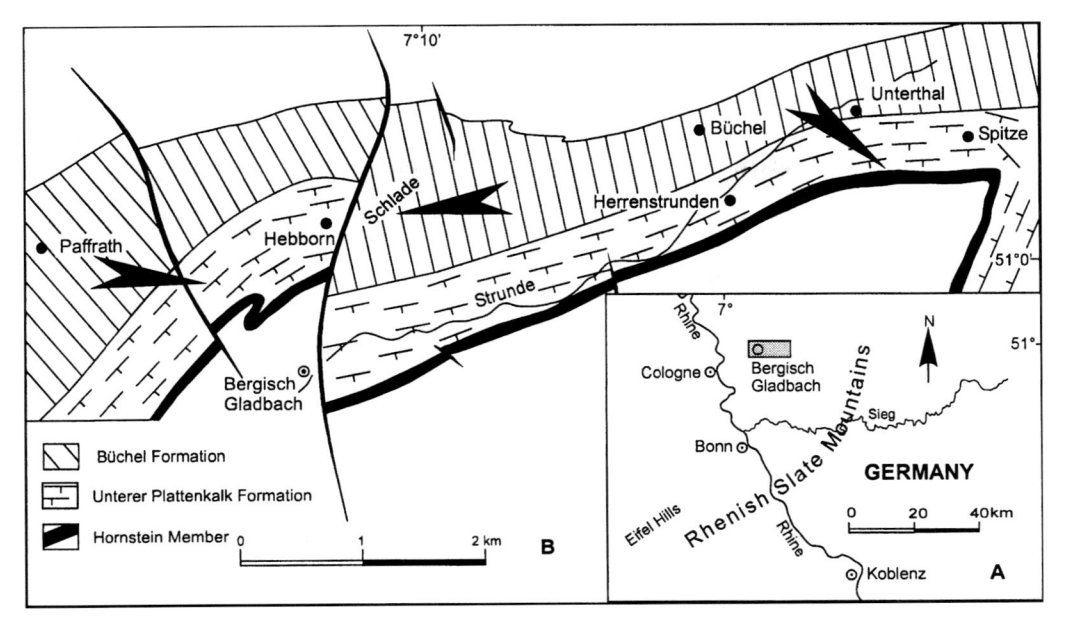

Fig. 3. (**A**) Map of the Rhenish Slate Mountains showing as grey rectangle the position of the Bergisch Gladbach–Paffrath Syncline. (**B**) Simplified geological map of the Bergisch Gladbach–Paffrath Syncline showing outcrop areas of Middle Devonian rocks and the position of sampled localities (black arrows).

Verneuil (now *Sobolewia cancellata*) and *Maenioceras terebratum* from the overlying Lower Plattenkalk Formation. None of these species has ever been figured from the area. *T. simplex* is in fact a *nomen dubium* (Becker 1993), perhaps based on an open umbilicate holzapfeloceratid, and is not a *Tornoceras*. This underlines the importance of taxonomic revisions. Jux & Strauch (1965) added supposed *Tornoceras frechi* Wedekind and *Agoniatites inconstans expansus* (Vanuxem) to the faunal list of the 'Unterer Plattenkalk'. The first species is the type-species of the Frasnian to Lower Famennian genus *Phoenixites* (Becker 1993) and is not yet known from the Givetian; the compressed tornoceratids under question need to be re-studied. The second species is a typical late Eifelian goniatite in North America. The specimen illustrated in Jux & Strauch (1965) may belong to the Middle Givetian species *A. fulguralis* or *A. obliquus* (both revised in House 2002), depending on the whorl width, which was not illustrated. The *M. terebratum* record could refer to any *Maenioceras* species.

The senior author of this publication has intensively collected fossils in the Bergisch Gladbach area for over 25 years. The described brachiopod and goniatite new material comes from the following localities:

1. Unterthal, abandoned quarry in a small village *ca.* 5 km NE of Bergisch Gladbach (GPS N51° 00,553′, E7° 11,700′, Fig. 3).

References: Winterfeld 1898; Kleinebrinker 1992.

The section exposes thin-bedded to platy, dark-grey limestones of the higher part of the Lower Plattenkalk Formation and higher, shaly interbeds, rarely with cherts, that are usually assigned to the subsequent 'Hornstein-Horizont' (e.g. Jux 1956) that forms a distinctive member. The fauna from Unterthal include stromatoporoids, rare solitary and colonial rugose corals, thamnoporids, *Eomegalodon* and other bivalves, scaphopods, orthoconic nautiloids with colour banding, other nautiloids, crinoids, fish remains, and diverse brachiopods, including *Uncites (Winderfeldia) paulinae* Winterfeld (MB.B.2464, Fig. 5g–i), *Stringocephalus* sp., *Parastringocephalus* sp., *Subsinucephalus* sp., *Schizophoria* sp., atrypids, pentamerids, and terebratulids. There are also tornoceratids and a very rare maenioceratid, *Maenioceras heinorum* n. sp. (MB.C.3888, Fig. 8f–g), collected by the Hein brothers.

2. Schladetal, abandoned Zimmermann Quarry in a small valley *ca.* 1.5 km NNE of Bergisch Gladbach (GPS N51° 00,127′, E7° 08,488′, Fig. 3).

References: Jux 1960.

The outcrop is famous for its massive, well-exposed stromatoporoid reefs and well-preserved, diverse reef-dwelling faunas of the Büchel Formation. There are abundant stromatoporoids, including branching amphiporids, some solitary rugose corals, heliolitid tabulates, rare gastropods,

Eomegalodon, as well as *Stringocephalus buchi buchi* and *Uncites gryphus gryphus* (e.g. MB.B.2465, Fig. 5e–f). The facies setting is normally not suitable (too shallow) for ammonoids. A single specimen of a new tornoceratid species (MB.C.3889, Fig. 8c–e, leg. Hein brothers) may have floated into the photic environment.

3. Paffrath area, excavation for a house foundation opposite the Blegge House (GPS N50° 59,920', E7° 06,290').
The ditch exposed the fossiliferous lower part of the Lower Plattenkalk Formation, consisting of yellowish-weathering, poorly bedded, fine-grained, grey marls. The diverse faunal assemblage is composed of stromatoporoids, many gastropods (bellerophontids, *Murchisonia* and others), solitary rugose corals, thamnoporids, scaphopods, rare polyplacophorans, crinoid stems, nautiloids and brachiopods, including lingulids, terebratulids, *Martinia*, rhynchonellids, *Uncites gryphus gryphus* (MB.B.2469, Fig. 5c–d), large *Uncites (Winterfeldia) beuthi* Jux & Strauch (MB.B.2466, Fig. 5a–b), *Stringocephalus* sp., *Subsinucephalus* sp., and *Parastringocephalus* sp. This typical neritic, photic, shallow-water assemblage is completed by six small-sized specimens of the new tornoceratid '*Trevoneites*' *paffrathensis* n. sp. (MB.C.3890.1, Fig. 8a–b, MB.C.3890.2-6).

Brachiopod–goniatite–conodont correlations (see Fig. 6)

Eifel Mountains

Uncites gryphus is a relatively rare form in the Eifel Hills of the western Rhenish Massif (Struve 1982, p. 240, confirmed by our observations). Struve (1986, 1988) placed its entry, later assigned to a smaller subspecies, *U. (U.) gryphus gryphulus* Struve 1992, in the Rossdorf Member of the Kerpen Formation (Fig. 4). In terms of conodont stratigraphy, this level within the Lower Givetian is poorly dated (*Bipennatus bipennatus–Icriodus lilliputensis* assemblage of Weddige 1988) but it seems to pre-date the entry of *I. brevis* within the *timorensis* Zone (lower part of former Lower *varcus* Zone, Fig. 6). *U. (U.) gryphus gryphus* commences low in the Bolsdorf Formation in the Eifel succession, which has been correlated at rather broad scale with the Lower to Middle *varcus* Zone (Weddige 1988, fig. A 14-18/11). There are no ammonoids from the Eifel Mountains, but based on the conodont data, the entry of *U. (U.) gryphus gryphulus* correlates with a level within the geographically restricted *Bensaidites koeneni* Zone (MD II-A) of the goniatite succession (Becker & House 2000), a level that in the Dra Valley has only agoniatitids (Becker *et al.* 2004*b*).

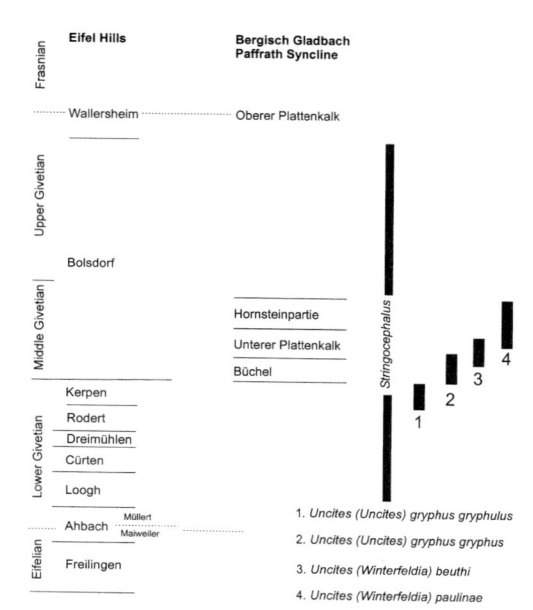

Fig. 4. Lithostratigraphy of the late Eifelian to basal Frasnian of the Eifel Hills and the Bergisch Gladbach Paffrath Syncline, with the stratigraphic range of *Stringocephalus* and species of *Uncites*.

The entry of *U. (U.) gryphus gryphus* is either still high in the Lower Givetian or correlates with a level near the base of the Middle Givetian; precise data are not available. The subsequent Givetian brachiopod record is obscured in the Eifel Mountains by widespread secondary dolomitization. *U. (Winterfeldia) beuthi* is known from a single locality in the Sötenich Syncline (Jux & Strauch 1966, p. 214).

Bergisch Gladbach–Paffrath Syncline (see Fig. 6)

In the Bergisch Gladbach area E of the Rhine, *Uncites (Uncites) gryphus gryphus* is common in the Büchel Formation. *U. (Winterfeldia) beuthi* enters in its upper part (Jux & Strauch 1966, fig. 17; here Fig. 4). The biostromal facies is unsuitable for conodont dating; only the long-ranging *Icriodus brevis* has been recorded by Kleinebrinker (1992). The single new tornoceratid represents a new form and is also of little use for correlation. Taking into account the conodont ages of the under- and overlying formations, the Büchel Formation can be correlated with the lower part of the Middle Givetian (*ca.* MD II-B to lower II-C, within the *rhenanus* Zone = upper part of former Lower *varcus* Zone).

The new collections from the Paffrath Syncline show that *U. (U.) gryphus gryphus* is not restricted to the Büchel Formation (Jux & Strauch 1966) but that

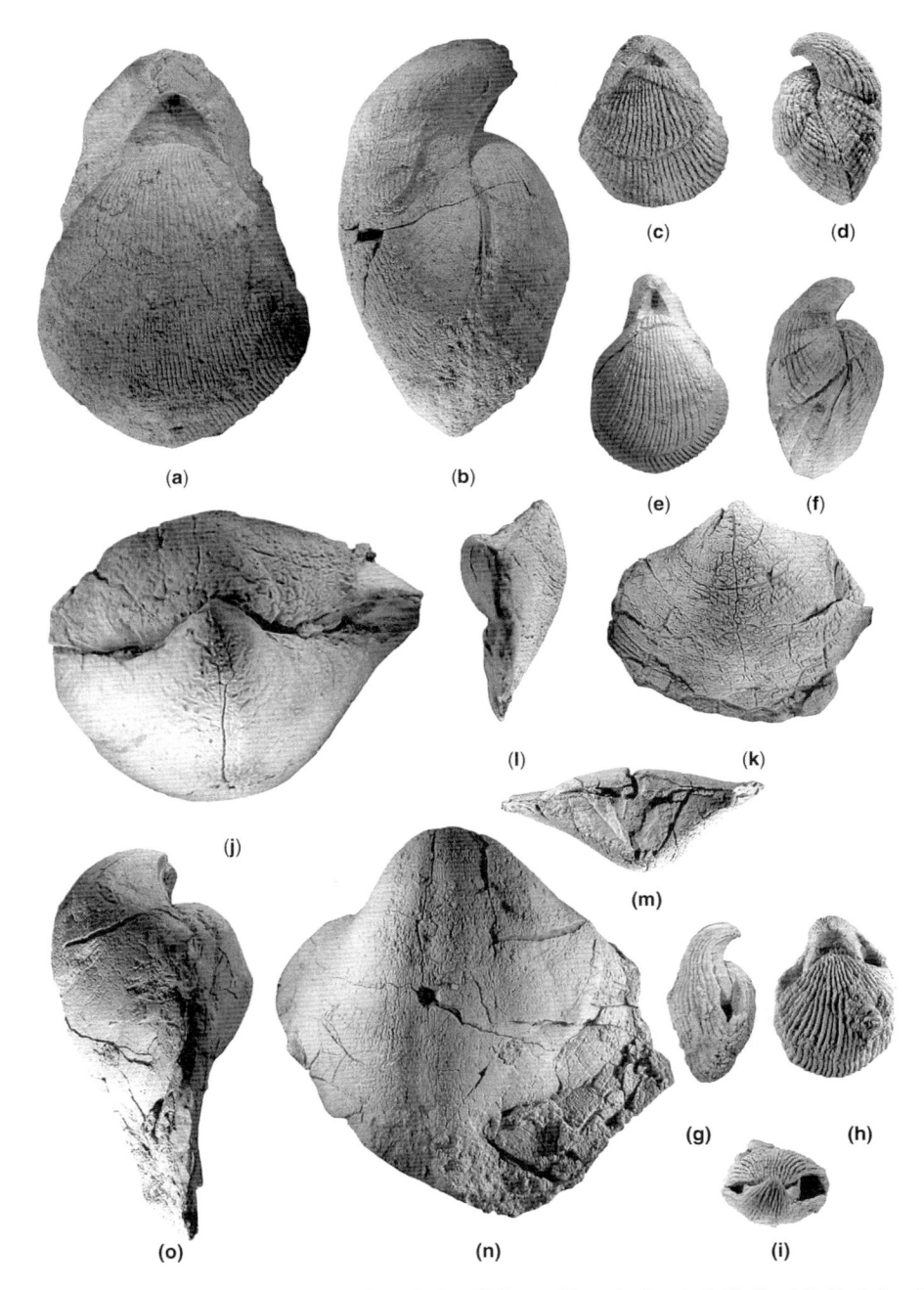

Fig. 5. Species of *Uncites* and *Stringocephalus* from the Dra Valley and from the Bergisch Gladbach Paffrath Syncline, all natural size. **a** & **b**, *Uncites (Winterfeldia) beuthi* Jux & Strauch 1965, MB.B.2466, excavation in Paffrath, Lower Plattenkalk Formation, dorsal and lateral views; **c** & **d**, *U. (U.) gryphus gryphus* (Schlotheim 1820), MB.B.2469, excavation in Paffrath, Lower Plattenkalk Formation, dorsal and lateral views; **e** & **f**, *U. (U.) gryphus gryphus*, MB.B.2465, Schladetal, Büchel Formation, dorsal and lateral views; **g–i**, *U. (W.) paulinae* (Winterfeld 1895), MB.B.2464, Unterthal, Lower Plattenkalk Formation, dorsal, lateral and posterior views; **j**, *Stringocephalus* sp., MB.B.2459.1, Oued Mzerreb, loose from Bed 5, apical view; **k–m**, *Subsinucephalus* sp., MB.B.2451, Oufrane W, Bed 3, ventral, lateral and posterior views; **n** & **o**, ?*Parastringocephalus* sp., MB.B.2454, Oufrane W, Bed 6a, ventral and lateral views.

conodont zones	Dra Valley goniatite zones	key	Bergisch Gladbach goniatite succession	Uncites succession	Bergisch Gladb. lithostratigraphy	
semialternans	Pharciceras aff. amplexum	III-A	Pharciceras sp.		Upper Plattenkalk Fm.	MIDDLE GIVETIAN
ansatus	Afromaenioceras n. sp.	II-D2	Maenioceras heinorum n. sp.	† paulinae	Hornstein Member	
	"Trevoneites" n. sp.	II-D1	"Trevoneites" paffrathensis n. sp.	beuthi	Lower Plattenkalk Formation	
	Maenioceras decheni	II-C2	"Maenioceras terebratum"			
rhenanus/ varcus	Maenioceras terebratum	II-C1	Tornoceras n. sp.	gryphus gryphus	Buechel Formation	
	Maenioceras n. sp. III	II-B2				
	Bensaidites n. sp.	II-B1				
timorensis	Agoniatites aff. vanuxemi	II-A/B		gryphus gryphulus	Torringen Formation	LOWER

Fig. 6. Correlation chart of Givetian chronostratigraphy, global events, Dra Valley goniatite (sub)zones and standard conodont zones with the Bergisch Gladbach lithostratigraphy, goniatite and *Uncites* successions.

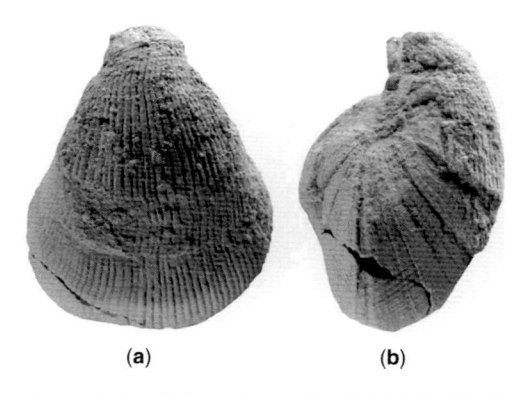

(a) **(b)**

Fig. 7. *Uncites (U.) gryphus gryphus* (Schlotheim 1820) from the mixed neritic–pelagic facies of the volcaniclastic ironstones (MD II-C/D) of the Martenberg, eastern Rhenish Massif.

it ranges with *U. (W.) beuthi* into the lowermost part of the Lower Plattenkalk Formation. The upper part of the range of *U. (W.) beuthi* overlaps with the lower range of *U. (W.) paulinae* (Jux & Strauch 1966, fig. 17; here Fig. 4). The Lower Plattenkalk Formation has yielded (Kleinebrinker 1992) *Polygnathus*

ansatus, the defining species of the *ansatus* Zone (former Middle *varcus* Zone), which enters in the Rhenish Massif and Morocco at the level of the Upper *pumilio* Event (Bultynck 1987; Aboussalam 2003) within the upper part of the *Maenioceras terebratum* Zone (MD II-C, just below the entry of *M. decheni* in the Dra Valley). The goniatite assemblages of the Lower Plattenkalk Formation need further revision and may include both the upper part of MD II-C (*M. terebratum* Zone) and lower II-D (*Afromaenioceras sulcatostriatum* Zone); *Afromaenioceras*, the marker genus of MD II-D (Becker & House 2000), however, has not been found so far. But *'Trevoneites' paffrathensis* n. sp. appears to be related to the still unnamed *'Trevoneites'* n. sp. illustrated by Becker *et al.* (2004b) as a regional zonal index for MD II-D1 (the pre-Taghanic part of the *ansatus* Zone) in the Dra Valley.

U. (W.) paulinae is characteristic of the 'Hornstein-Horizont'. Based on conodont data (Kleinebrinker 1992), this unit has been correlated with the lower/middle parts of the Taghanic Event interval (Aboussalam 2003, p. 33, tab. 4), that is within the upper part of the *ansatus* Zone (former Middle *varcus* Zone, Fig. 6). This correlation is supported by the recognition of the subsequent Upper *varcus*

Zone (=*semialternans* Zone, topmost Middle Givetian, MD III-A) in the overlying basal Upper Plattenkalk Formation. The lower/middle Taghanic Event interval of southern Morocco (MD II-D2) is characterized by advanced, relative tubby species of *Afromaenioceras* and *Maenioceras*. The extremely compressed new *Maenioceras* from Unterthal documents a very different regional evolutionary trend within the Maenioceratidae.

Northern, eastern and southern Rhenish Massif

Uncites (*Uncites*) *gryphus gryphus* that come from roughly contemporaneous units of the Büchel Formation occur widely in the lower part of the Massenkalk of the Wuppertal and Hagen-Balve reef complexes (Holzapfel 1895; Paeckelmann 1922; Jux & Strauch 1966; Koch 1984; Becker 1985, e.g. MB.B.2470 from Letmathe). Higher in the Sauerland succession, *U. (Winterfeldia) beuthi* and *U. (W.) paulinae* are known from brachiopod-rich reefal slope deposits of the Iserlohn area (northern Rhenish Massif, Torley 1908, 1934). At the Schleddenhof (Torley 1908, revised in Jux & Strauch 1966), the neritic beds with *U. (W.) beuthi* yielded rare tornoceratids and two species of *Sobolewia* that are not zonally diagnostic. The outcrop at Schleddenhof no longer exists and thus cannot be sampled for conodonts. In the revised Givetian chronostratigraphy (Aboussalam & Becker 2002), *Uncites* does not range into the Upper Givetian.

An early report of joint goniatite–*Uncites* occurrences dates back to Holzapfel (1895) who reported *U. (U.) gryphus* (identification confirmed by Jux & Strauch 1966) from the haematitic ironstones of the Brilon area, including the famous Martenberg volcanic seamount (MB.B.2471, Fig. 7), where rich mixed neritic–pelagic assemblages lived just below the photic zone. Conodont data (Ziegler 1958; Aboussalam 2003) suggest that the diverse Martenberg fauna at least partly falls in the *ansatus* Zone. This is supported by the presence of *Maenioceras decheni*, the index species of the upper part of the *terebratum* Zone (MD II-C2, starting above the Upper *pumilio* level, Becker *et al.* 2004b), in the ironstones with *Uncites* at several localities. The eastern Rhenish occurrences of *U. (U.) gryphus gryphus*, therefore, confirm the youngest range of the species in the Lower Plattenkalk Formation in the Bergisch Gladbach area (Fig. 4).

Holzapfel (1895) also reported *U. (U.) gryphus* from Villmar in the southern Rhenish Massif (Lahn Dill Syncline), which is the type-locality of *M. terebratum*. The precise local position of both taxa within the *terebratum* Zone (MD II-C) is not known. As noted by Holzapfel (1895, p. 337–338), it is remarkable that *Uncites* is missing in the famous, contemporaneous mixed neritic–pelagic fauna of Fretter (Attendorn–Elspe Syncline, central Rhenish Massif).

Dra Valley

The ammonoid assemblage of the *Uncites*-bearing Bed 3a at Oufrane West falls in the lower part of the *Maenioceras terebratum* Subzone (MD II-C1, Becker *et al.* 2004b; Aboussalam *et al.* 2004). The *Parastringocephalus* from Bed 6a still falls in the same subzone. But the specimens from below Bed 8 may possibly come from the next-younger *M. decheni* Subzone (MD II-C2) since Beds 6b to 7b represent the Upper *pumilio* level, which elsewhere marks the base of the *ansatus* Zone (Bultynck 1987; Aboussalam 2003). *M. decheni* has not been found at Oufrane but is abundant in the same lithological unit just above the Upper *pumilio* Beds at Oued Mzerreb (Becker *et al.* 2004b).

The first new *Uncites* specimen discovered in the Dra Valley (Ebbighausen *et al.* 2004) came at Oufrane East from the same goniatite level (*terebratum* Subzone, MD II-C1) as the lower, main material of the western section.

The oldest neritic brachiopods at Oued Mzerreb fall in the Lower to basal Middle Givetian (MD II-A/B, Becker *et al.* 2004b). The *Stringocephalus* and *U. (U.) gryphus gryphus* collected in the plain below Bed 6 may have been derived from the poorly fossiliferous upper part of the regional *Maenioceras* n. sp. III Zone (MD II-B2), which still lacks constricted wedekindellids. The stringocephalids from Bed 6 fall, as at Oufrane, in the *terebratum* Subzone (MD II-C1). The lack of associated conodonts and goniatites does not allow to give a precise age within the Middle Givetian for the single *Uncites* from Bou Tserfine.

The combined evidence of the new Dra Valley *U. (U.) gryphus gryphus* suggests that they come precisely from the same time-interval as the Bergisch Gladbach material. Specimens from below the Upper *pumilio* Beds at Oufrane and Oued Mzerreb correlate with levels in the Büchel Formation (Fig. 6). The possible record of the species from just above the Upper *pumilio* Bed at Oufrane West is still in accord with its range into the Lower Plattenkalk Formation of the Bergisch Gladbach area (Fig. 4). This good agreement supports the idea that *Uncites* species are of high biostratigraphical significance in widely separated regions. The genus is known widely from a (sub)tropical belt stretching from SW England, to the Boulonnais, Ardennes, Germany, eastern Urals, Central Asia (Uzbekistan), China (Gansu, Yunnan), to Vietnam. This geographical spread shows the potential for future correlations. It is still unknown why the genus is lacking in North America, in northern parts of Russia, Siberia and in Australia.

Unlike the last stringocephalids, but as a range of goniatite genera that have been found in direct association (*Agoniatites, Sobolewia, Wedekindella, 'Trevoneites' assessi* Group), *Uncites* is not known to have survived into the upper, third phase (Aboussalam 2003) of the global Taghanic Event.

Taxonomy

Described and figured specimens are housed in the collection of the Museum für Naturkunde of the Humboldt University, Berlin, with the catalogue numbers MB.B.2449-2471 (brachiopods) and MB.C.3885-3890 (ammonoids). Abbreviations used in the text are: dm, conch diameter; ww, whorl width; wh, whorl height; uw, umbilical width; ah, apertural height; WER, whorl expansion rate; calculated $[dm/(dm-ah)]^2$, IZR, imprint zone rate; calculated $(wh-ah)/wh$.

Family Stringocephalidae King
Subfamily Stringocephalinae King
Stringocephalus sp.
Fig. 5j

Localities and levels. 13 incomplete and mostly crushed specimens from Oufrane, Oued Mzerreb (both Tata region, eastern Dra Valley), and Bou Tserfine (Assa region, western Dra Valley), all from the lower to middle part of the Middle Givetian (MD II-?B2, II-C1, and II-?C2), Oued Mzerreb or Coral Marl Members of the Ahrerouch Formation.

Description. The figured MB.B.2459.1 is a fragment of a medium-sized specimen. Compared with well-preserved material from the Bergisch Gladbach–Paffrath Syncline, the complete size is estimated to have reached *ca.* 8–10 cm. The shell, as far as preserved, is globular, evenly rounded, smooth, without a sinus. The ventral beak is strongly incurved, conspicuous and pushed against the dorsal valve. The palintrope and pedicle foramen are not visible. A similar curvature of the beak can be observed in MB.B.2453.1-3, in MB.B.2458.1 and in MB.B.2461.1. But MB.B.2449 from Oufrane and MB.B.2456 from Oued Mzerreb have less-convex pedicle valves and, at least in the case of the first specimen, this is not caused by secondary shell flattening. The median septum is clearly seen in MB.B.2453.1-2, MB.B.2458.1-2, MB.B.2459.1, MB.B.24493.1 and MB.B.2463.1. A triangular foramen and weathered parts of the deltidial plates are preserved in MB.B.2453.1, MB.B.2453.3 and MB.B.2461.1.

Discussion. Because the material is fragmentary, a specific assignment is not possible. But typical features of the genus, such as shell thickness, median septum, overall size and projecting beak are recognizable. As already suggested by Drot (1961), at least two species of *Stringocephalus* are present in the Dra Valley. The poor preservation allows no meaningful comparison with the diverse species described by Struve (1992) from the Rhenish Massif.

Subsinucephalus sp.
Fig. 5k–m

Locality and level. Two specimens (one questionable) from Oufrane West, Bed 3a, between the Lower and Upper *Pumilio* Beds, middle part of Middle Givetian (MDII-C1), Ahrerouch Formation, Oued Mzerreb Member.

Description. The figured specimen (MB.B.2451) is a medium-sized shell, slightly weathered, with the dorsal valve crushed. The length of the ventral valve is 38 mm, the length of the dorsal valve 33 mm, width 45 mm, and depth *ca.* 17 mm. The shell is widest at midlength and elliptical in outline. The pedicle valve is moderately convex, with a low gentle sinus only visible in oblique light. In the dorsal valve, no sinus is ascertainable. The beak is slightly curved and has an angle of 117°. The interarea is apsacline and slightly curved, the delthyrium is small with conjunct deltidial plates and hypothyrid. The second, questionable specimen (MB.B.2452) is larger and shows the median septa of both valves.

Discussion. The specimen resembles *Subsinucephalus*? *reter* Struve 1992 but differs in its lower depth and higher angle between the beak ridges. For specific determination, more and better-preserved material is needed.

?Parastringocephalus sp.
Fig. 5n–o

Localities and levels. Oufrane West, Bed 6a, below Upper *Pumilio* Bed, leg. G. Plodowski (Frankfurt a. M.), middle part of Middle Givetian (MDII-C1), Ahrerouch Formation, Oued Mzerreb Member.

Description. MB.B.2454 is crushed and fragmentary. The length of the ventral valve is 64 mm, the length of dorsal valve 54 mm, width 65 mm and depth 31 mm. The shell shape is more 'terebratulid' than 'spiriferid' in outline. The pedicle valve is convex with clearly visible sinus starting at the crest of the beak. The dorsal valve is severely crushed and, therefore, no sinus is visible. The interarea is orthocline, incurved, the delthyrium crushed. Despite the weathering, a 'capillate' (fibre-like) shell structure, as described by Struve (1965), is recognizable.

Discussion. The specimen is placed in *Parastringocephalus* with some reservation, and an assignment to *Subsinucephalus* cannot be excluded. A definite Dra Valley record of *Parastringocephalus* requires better-preserved material.

Family Uncitidae Waagen
Uncites (Uncites) gryphus gryphus (Schlotheim 1820)
(Figs 5c–f, 8n–r)

Localities and levels. Five specimens from Oufrane and Oued Mzerreb (Tata region, eastern Dra Valley), one from Bou Tserfine (Assa region, western Dra Valley), all from lower to middle parts of the Middle Givetian (MD II-?B2, II-C1 and II-?C2), Oued Mzerreb and Coral Marl Members, Ahrerouch Formation; one specimen from the Büchel Formation of Bergisch Gladbach (MB.B.2470, leg. VE), one from the Lower Plattenkalk Formation of Paffrath (MB.B.2469, leg. VE), and one from the Massenkalk Formation of Iserlohn-Letmathe (MB.B.2470, leg. RTB) for comparisons.

Description. The specimen from Oufrane-East (MB.B.2457, Fig. 8n–r) has the following parameters: length of pedicle valve 28.5 mm, length of brachial valve 24.5 mm, width 21.3 mm, valve depth 15.7 mm. The shell is ribbed, with 11 more or less even ribs across 10 mm width in the middle part of the pedicle valve. There are irregularly spaced concentric growth lines. The beak is slightly incurved and pointed, with koskinoid perforations. The delthyrium is occupied by a single, concave deltidium (Fig. 8r). All other specimens show the principal characters of the species but are too squashed and distorted for further description. Some reached larger size than the figured specimen.

Discussion. The assignment of the *Uncites* specimens from the Dra Valley to *Uncites (Uncites) gryphus gryphus* is unambiguous. They fall in the field of variation with respect to shell size, form and ribbing (see Jux & Strauch 1966). For comparison, an *U. (Winterfeldia) beuthi* (MB.B.2466, Fig. 5a–b) from the Lower Plattenkalk Formation of the Paffrath type region is illustrated. It shows flanks of the beak which are slightly concave and clearly separated from the back of the beak, unlike in *U. (U.) gryphus gryphus* from the same area (e.g. MB.B.2469, Fig. 5c–d) where the flanks are rounded and fully integrated in the curvature of the beak. *U. (U.) gryphus gryphus* has its greatest width in the middle of the shell, *U. (W.) beuthi* near the slightly curved anterior commissure. *U. (W.) paulinae* (e.g. MB.B.2464, Fig. 5g–i) is generally smaller and distinguished by deep, pouch-like indentations of the shell on both sides of the beak (parathyridia). These are interpreted as 'brood pouches'. *U. (U.) gryphus gryphulus* is smaller than the nominate subspecies; it reaches a length of only 25 mm. The Chinese *U. (U.) pauciplicata* is more coarsely ribbed.

Family Maenioceratidae Bogoslovskiy
Maenioceras heinorum n. sp.
Fig. 8f–g

Type locality and level. Abandoned quarry at Unterthal, small village *ca.* 5 km NE of Bergisch Gladbach (Rhenisch Slate Mountains, Germany), Lower Plattenkalk Formation, Hornstein Member (MD II-D1), *ansatus* Zone.

Derivation of name. After the brothers Hans Peter and Uwe Hein who made their unique specimen available for study.

Material. Only the Holotype, MB.C.3888.

Diagnosis. Maenioceras with strongly compressed (ww/dm = 0.36), thinly discoidal conch, narrow umbilicus and distinctive keel formed by deep ventrolateral furrows at small size. Growth lines probably biconvex, with wide lateral sinus.

Description. The small specimen is not complete but shows details of a very peculiar conch form that is unique amongst known Middle Givetian ammonoids. The flanks are flattened and only very gently rounded, abruptly interrupted towards the venter by deep spiral furrows, which are already typically developed at 4.5 mm wh. The prominent ventral keel is very narrow and slightly rounded in cross-section. The umbilical wall is rounded but towards the end of the preserved whorl there is a faint subumbilical spiral depression. Ornament is only poorly visible under oblique light but a deep lateral sinus, suggesting biconvex growth ornament, can be recognized. Sutures are not preserved. At 12.3 mm dm, ww is 4.5 mm (*ca.* 36.5% dm), wh 6 mm (ww/wh = 0.75), and uw *ca.* 1.5 mm (12.2% dm).

Discussion. Amongst all described Givetian goniatite genera, open umbilicate, strongly compressed and keeled conches are only known in *Maenioceras*. In the Dra Valley, there is a still unnamed keeled tornoceratid but the keel develops long after the umbilicus is completely closed. A comparison with rich populations of *M. terebratum* from Oued Mzerreb and Oufrane shows that the latter species is significantly thicker (*ca.* 46% dm)

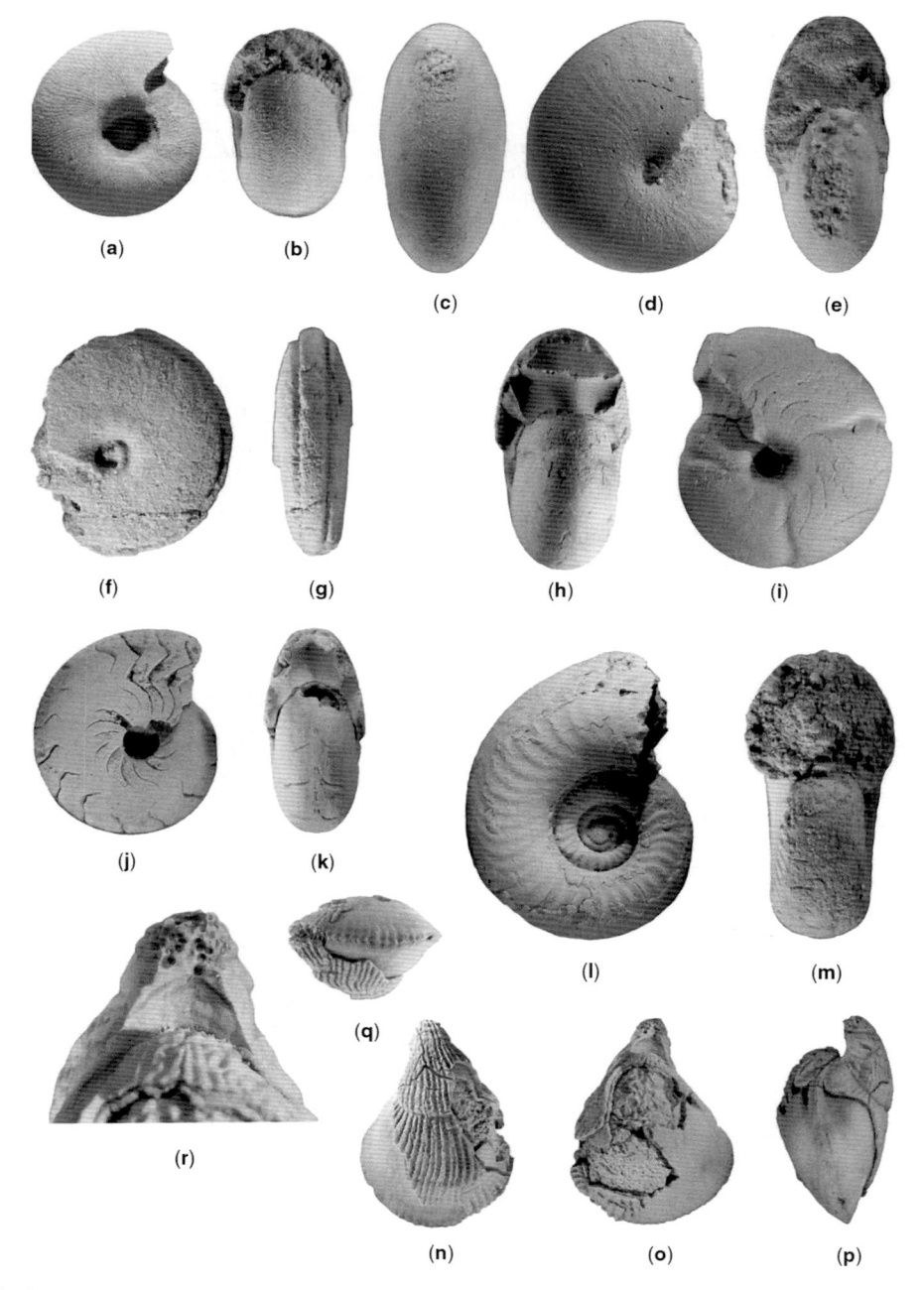

Fig. 8. Species of goniatites from the Dra Valley and from the Bergisch Gladbach–Paffrath Syncline, and *Uncites* from the Dra Valley. a & b, '*Trevoneites*' *paffrathensis* n. sp., holotype, MB.C.3890.1, excavation in Paffarath, Lower Plattenkalk Formation, lateral and ventral views, ×5; c–e, *Tornoceras* n. sp., MB.C.3889, Schladetal, Büchel Formation, dorsal, lateral and ventral views, ×3; f & g, *Maenioceras heinorum* n. sp., holotype, M.B.C.3888, Unterthal, Hornstein Member, lateral and dorsal views, ×3; h & i, *Wedekindella* n. sp., MB.C.3887, Oufrane W, Bed 3a, ventral and lateral views, ×3; j & k, *Maenioceras terebratum* (Sandberger & Sandberger 1851), MB.C.3886, Oufrane W, Bed 3a, lateral and ventral views, ×3; l & m, *Agoniatites costulatus* (D'Archiac & De Verneuil 1842), MB.C.3885, Oufrane W, Bed 3a, lateral and ventral views, ×3; n–r, *Uncites (Uncites) gryphus gryphus* (Schlotheim 1820), MB.B.2457, Oufrane E, Bed 2a′, ventral, dorsal, lateral, and anterior views, ×1, beak, ×4.

at the same size (Fig. 8j–k). The umbilicus is also almost twice as wide (24% at 12.5 mm dm) compared to *M. heinorum* n. sp. and there is no trace of ventrolateral furrows; only ventral edges are present (Fig. 8j–k and Becker *et al.* 2004*b*, pl. 1, figs 17–18). The type material of *M. terebratum* from Villmar is larger-sized but still possesses no keel. Both in German and Moroccan populations of *M. terebratum*, gentle furrows first appear between 20 and 25 mm and even later in ontogeny. They never create such a distinctive keel as in the new species. This is also true for the holotype of *M. tenue* Holzapfel 1895 (see Becker & House 1994, pl. 6, figs 14–15), in which sutures are also unknown and which is most likely a junior synonym of *M. terebratum*. All other described Maenioceratidae have much thicker shells.

Since no similar goniatite has ever been described from the Givetian on a global scale, and with respect to its rareness and the very distinctive shell form, naming it as a new species is warranted.

Family Tornoceratidae Arthaber
Subfamily Tornoceratinae Arthaber
Tornoceras n. sp.
Figs 8c–e, 9a–b

Locality and level. Abandoned quarry in the Schladetal, small valley *ca.* 1.5 km NNE of Bergisch Gladbach (Rhenisch Slate Mountains, Germany), Büchel Formation, Middle Givetian (MD II-B/C1), collected by the Hein brothers.

Material. Only one specimen, MB.C.3889.

Description. The cross-section is compressed tegoid, with the thickest width of the well-rounded flanks above the rounded umbilical wall. The flanks converge to a rounded venter. The shell parameters are as follows: *ca.* 13.6 mm dm, ww 6.9 mm (ww/dm = 0.51), wh 7.7 mm (wh/dm = 0.56, ww/wh = 0.9), ah = 4.6 mm (ah/dm = 0.34), WER = 2.29, IZR = 0.4. The growth lines are fine, bundled and biconvex, with low dorsolateral and ventrolateral salients. The lateral sinus is shallow, the ventral sinus narrow and deep. Coating of shell for photography (Fig. 8c–e) covered a unique feature, preserved colour banding on the flanks and venter (Fig. 9). There are narrow spiral lines symmetrically on both shell sides, crossing the ventrolateral salient of the growth ornament. One side displays a second spiral line towards the mid-flank. On the venter, there is a band of 1–1.5 mm spaced-chevrons.

Discussion. The ventral band was first thought to have been created by internal shell thickenings shining through the shell but the latter is not translucent (no septa visible) and thickenings or mould constrictions are not common in other Givetian

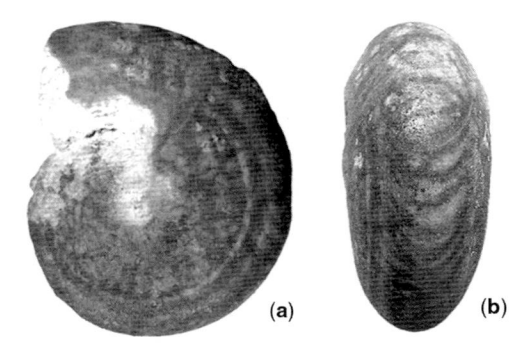

Fig. 9. Colour banding on the flanks (**a**, double spiral line) and venter (**b**, chevrons parallel to growth lines) of *Tornoceras* n. sp. (MB.C.3889) from the Büchel Formation, Bergisch–Gladbach–Paffrath Syncline (Germany).

Tornoceras. The combination with the spiral patterns supports an origin by coloration. Preservation of colour banding is anyway well known from orthoconic nautiloids (zigzag-type patterns), brachiopods and gastropods of the Bergisch Gladbach area (D'Archiac & de Verneuil 1842; Strauch 1985; Weber 2000). However, it has not yet been demonstrated for goniatites and not for any Devonian ammonoid elsewhere. However, spiral and radial colour bands seem to be a common feature in ammonoids (e.g. Mapes & Davis 1996; Keupp 2005) that is just rarely preserved. The observed patterns suggest that colour camouflage was probably widespread in Devonian cephalopods.

The rounded shell form of the single *Tornoceras* from the Büchel Formation does not resemble any other Givetian species. Both the American *T. uniangulare* Group and the European *T. typum* Group (including *T. bertrandi* Frech and *T. hughesi* Whidborne = *T. whidbornei* Foord & Crick) have flat flanks. Juveniles of Givetian *T. uniangulare* subspecies are also thinner (House 1965) but juveniles of European species need further study. MB.C.3889 most likely represents a new species, but similar specimens have been observed in the Lower Plattenkalk Formation and naming of the new taxon will await recognition of the suture in such additional material.

'*Trevoneites*' *paffrathensis* n. sp.
Fig. 8a–b

Type locality and level. Excavation for a house foundation (2004) in Paffrath, part of the village of Bergisch Gladbach (Rhenisch Slate Mountains, Germany), Lower Plattenkalk Formation, level with *Uncites (Winterfeldia) beuthi* and *Uncites (Uncites) gryphus gryphus*, Middle Givetian.

Table 1. Shell parameters of 'Trevoneites' paffrathensis n. sp. (in mm)

No.	dm	ww	wh	uw	ah	WER	ww/dm	ww/wh	uw/dm	IZR
MB.C.3890.2	8.95	5.15	3.89	2.37	1.97	1.64	0.58	1.32	0.26	0.49
MB.C.3890.1	6.32	3.97	2.67	1.68	1.24	1.55	0.63	1.49	0.27	0.54
MB.C.3890.3	5.8	3.42	2.4	1.63	1.25	1.62	0.59	1.43	0.28	0.48
MB.C.3890.4	4.67	3.21	1.84	1.22	1.07	1.68	0.69	1.74	0.26	0.42

Derivation of name. After Paffrath village.

Material. The holotype, MB.C.3890.1, the best-preserved specimen, and five paratypes, MB.C.3890.2-6.

Diagnosis. Small-sized, early stages depressed, slowly expanding (WER 1.55 to 1.68), with well-rounded flanks and venter, umbilicus moderately wide (26–28% dm). Later stages becoming more compressed and, from *ca.* 10 mm dm on, with absolute closure of the umbilicus. Growth lines fine and regular, rursiradiate biconvex, with shallow flank and deeper ventral sinus; flank salients low. Sutures with moderately deep and narrow dorsal lobe, moderately high dorsolateral saddle and wide, rounded adventitious lobe.

Description. The six available specimens show only moderate variability. All have a well-rounded, relative thick conch shape and a moderately wide and deep umbilicus. The umbilical wall is steeply rounded but becomes subangular and straight in the largest paratype (MB.C.3890.2). This specimen also shows dorsal sutures and adorally a septal face, suggesting the presence of a rounded flank lobe and a moderately high dorsolateral saddle, as typical for tornoceratids. The whorl shape starts to become less depressed and a ridge that was part of the now-missing succeeding whorl suggests that the umbilicus became closed at maturity. The growth ornament is visible on all specimens and is slightly bent backwards (rursiradiate). It consists of rather regular fine ribs. Shell measurements are given in Table 1. They show the decreasing ww/wh ratio during ontogeny and relative constant umbilical width and WER below 10 mm dm.

Discussion. Becker & House (1994) introduced the genus *Trevoneites* for open umbilicate Givetian tornoceratids that resemble Upper Devonian protornoceratids. *Protornoceras foxi* House 1963 from North Cornwall was chosen as type-species. The Algerian species *Protornoceras assessi* Göddertz 1987 was included in *Trevoneites* with some reservation, mostly because of its different, regular, somewhat rursiradiate ornament lacking ventrolateral furrows. Rich new collections of the *assessi* Group from the Dra Valley proved that mature specimens completely close their umbilicus whilst true *Trevoneites* remain open umbilicate throughout ontogeny. Until revision of the group based on the Moroccan collections, species of the *assessi* Group are assigned to '*Trevoneites*'. The shell shape and ornament of the new species from Paffrath resemble '*Tr.*' *assessi* but are significantly thicker. They are even more rotund than the still-undescribed '*Trevoneites*' n. sp. figured from Oued Mzerreb in Becker *et al.* (2004*b*). '*Tr.*' *paffrathensis* n. sp. provides the first evidence that the *assessi* Group was not confined to northern Gondwana.

This manuscript benefited from close cooperation between the Institute of Geology and Palaeontology at Münster and the Instutut Scientifique, Rabat, especially with Prof. Dr. A. El Hassani, who enabled access to sections in the Dra Valley. Dr. P. Sartenaer (Brussels) drew our attention to important brachiopod literature. The Hein brothers (Wermelskirchen) made rare goniatite specimens available for study. We thank all members of the 2004 SDS excursion that delivered their Dra Valley brachiopod specimens for this study. Prof. Dr. Ma Xueping (Beijing) and Prof. Dr. W. T. Kirchgasser (Potsdam, N. Y.) kindly reviewed and improved the manuscript.

References

ABOUSSALAM, Z. S. 2003. Das 'Taghanic-Event' im höheren Mittel-Devon von West-Europa und Marokko. *Münstersche Forschungen zur Geologie und Paläontologie*, **97**, 1–332.

ABOUSSALAM, Z. S. & BECKER, R. T. 2002. The Base of the *hermanni* Zone as the Base of an Upper Givetian Substage. *Document submitted to the International Subcommission on Devonian Stratigraphy, Annual Meeting, Toulouse 2002*, 1–10 [also: *SDS Newsletter*, **19**, 25–34].

ABOUSSALAM, Z. S., BECKER, R. T., BOCKWINKEL, J. & EBBIGHAUSEN, V. 2004. Givetian biostratigraphy and facies development at Oufrane (Tata region, eastern Dra Valley, Morocco). *Documents de l'Institut Scientifique*, **19**, 53–59.

BECKER, R. T. 1985. Devonische Ammonoideen aus dem Raum Hohenlimburg-Letmathe (Geologisches Blatt 4611 Hohenlimburg). *Dortmunder Beiträge zur*

Landeskunde, naturwissenschaftliche Mitteilungen, **19**, 19–34.

BECKER, R. T. 1993. Stratigraphische Gliederung und Ammonoideen-Faunen im Nehdenium (Oberdevon II) von Europa und Nord-Afrika. *Courier Forschungsinstitut Senckenberg,* **155**, 1–405.

BECKER, R. T. & HOUSE, M. R. 1994. International Devonian goniatite zonation, Emsian to Givetian, with new records from Morocco. *Courier Forschungsinstitut Senckenberg,* **169**, 79–135.

BECKER, R. T. & HOUSE, M. R. 2000. Devonian ammonoid zones and their correlation with established series and stage boundaries. *Courier Forschungsinstitut Senckenberg,* **220**, 113–151.

BECKER, R. T., JANSEN, U., PLODOWSKI, G., SCHINDLER, E., ABOUSSALAM, S. Z. & WEDDIGE, K. 2004a. Devonian litho- and biostratigraphy of the Dra Valley—an overview. *Documents de l'Institut Scientifique,* **19**, 3–18.

BECKER, R. T., ABOUSSALAM, Z. S., BOCKWINKEL, J., EBBIGHAUSEN, V., EL HASSANI, A. & NÜBEL, H. 2004b. The Givetian and Frasnian at Oued Mzerreb (Tata region, eastern Dra Valley). *Documents de l'Institut Scientifique,* **19**, 29–43.

BECKER, R. T., BOCKWINKEL, J., EBBIGHAUSEN, V., ABOUSSALAM, Z. S., EL HASSANI, A. & NÜBEL, H. 2004c. Lower and Middle Devonian stratigraphy and faunas at Bou Tserfine near Assa (Dra Valley, SW Morocco). *Documents de l'Institut Scientifique,* **19**, 90–100.

BULTYNCK, P. 1987. Pelagic and neritic conodont successions from the Givetian of pre-Sahara Morocco and the Ardennes. *Bulletin de l'Institut Royal des Sciences Naturelles de Belgique,* **57**, 149–181.

BULTYNCK, P. & GOUWY, S. 2002. Towards a standardization of global Givetian substages. Document presented to the International Subcommission on Devonian Stratigraphy, Annual Meeting, Toulouse, **2002**, 1–6.

D'ARCHIAC, V. & DE VERNEUIL, M. E. 1842. On the fossils of the Older Deposits in the Rhenish Provinces; preceded by a general Survey of the Fauna of the Palaeozoic Rocks, and followed by a Tabular ist of the Organic Remains of the Devonian System in Europe. *Transactions of the Geological Society of London,* **6**, 303–410.

DROT, J. 1961. Quelques formes de Brachiopodes Givétiens du Dra (Maroc présaharien) peu communes en Afrique du Nord. *Notes et Mémoires du Service Géologiques,* **20**, 59–68.

DROT, J. & HOLLARD, H. 1965. Présence d'*Uncites gryphus* (Schlotheim 1822) (Brachiopode) dans le givétien du maroc présaharien. *Compte rendu sommaire des séances de la société géologique de France,* **1965**, 21.

EBBIGHAUSEN, V., BECKER, R. T., BOCKWINKEL, J. & ABOUSSALAM, Z. S. 2004. A neritic alien in the pelagic Givetian of the Dra Valley. *In*: EL HASSANI, A. (ed.) *Devonian Neritic-Pelagic Correlation and Events, International Meeting on Stratigraphy, Rabat, Morocco, March 1–10, 2004, Abstracts,* 22–23.

ERBEN, H. K. 1962. Zur Analyse und Interpretation der Rheinischen und Hercynischen Magnafazies des Devons. *2. Internationale Arbeitstagung zur Silur/Devon-Grenze und Stratigraphie des Silurs und Devon, Bonn 1960, Symposium-Band,* 42–61.

GÖDDERTZ, B. 1987. Devonische Goniatiten aus SW-Algerien und ihre stratigraphische Einordnung in die Conodonten-Abfolge. *Palaeontographica, Abteilung A,* **197**(4–6), 127–220.

HOLLARD, H. 1963. Une tableau stratigraphique du Dévonien du Sud de l'Anti-Atlas. *Notes du Service géologique du Maroc,* **23**, 105–109.

HOLLARD, H. 1967. Le Dévonien du Maroc et du Sahara Nord-Occidental. *In:* OSWALD, D. H. (ed.) *International Symposium on the Devonian System,* **1**, 203–244, Calgary.

HOLLARD, H. 1970. Silurien—Dévonien—Carbonifère. Colloque international sur les corrélations du Précambrien. Agadír-Rabat—3-23 mai 1970—Livret-guide de l'excursion: Anti-Atlas occidental et central. *Notes et Memoires du Service géologique du Maroc,* **229**, 161–188.

HOLLARD, H. & DROT, J. 1958. Présence du genre *Stringocephalus* dans le Maroc présaharien. *Compte Rendu sommaire des séances de la Société Géologique de France,* **13**, 313–314.

HOLZAPFEL, E. 1895. Das Obere Mitteldevon (Schichten mit *Stringocephalus Burtini* und *Maeneceras terebratum*) im Rheinischen Gebirge. *Abhandlungen der Königlich Preussischen Geologischen Landesanstalt, Neue Folge,* **16**, 1–459.

HOUSE, M. R. 1963. Devonian ammonoid successions and facies in Devon and Cornwall. *Quarterly Journal of the Geological Society of London,* **119**, 1–27.

HOUSE, M. R. 1965. A study in the Tornoceratidae: The succession of *Tornoceras* and related genera in the North American Devonian. *Philosophical Transactions of the Royal Society of London, Series B. Biological Sciences,* **763**(250), 79–130.

HOUSE, M. R. 2002. Devonian (Givetian) goniatites from Wolborough, Barton and Lummaton, South Devon. *Geosciences in South-West England,* **10**, 281–292.

JUX, U. 1956. Stratigraphie, Faziesentwicklung und Tektonik des jüngeren Devons in der Bergisch Gladbach-Paffrather Mulde. *Neues Jahrbuch für Geologie und Paläontologie, Abhandlungen,* **102**, 295–328.

JUX, U. 1960. Die devonischen Riffe im Rheinischen Schiefergebirge, Teil 1. *Neues Jahrbuch zur Geologie und Paläontologie, Abhandlungen,* **110**, 186–258.

JUX, U. & STRAUCH, F. 1965. Die '*Hians*'-Schille aus dem Mitteldevon der Bergisch Gladbach–Paffrather Mulde. *Fortschritte in der Geologie von Rheinland und Westfalen,* **9**, 51–86.

JUX, U. & STRAUCH, F. 1966. Die Mitteldevonische Brachiopoden-Gattung *Uncites* Defrance 1825. *Palaeontographica, Abteilung A,* **125**, 176–222.

KEUPP, H. 2005. Das Geheimnis der Spiralbänder: Farbmuster auf Ammonitengehäusen. *Fossilien,* **6/05**, 369–375.

KLEINEBRINKER, G. 1992. Conodonten-Stratigraphie, Mikrofazies und Inkohlung im Mittel- und Oberdevon des Bergischen Landes. *Geologisches Institut der Universität zu Koeln, Sonderveröffentlichungen,* **85**, 1–101.

KOCH, L. 1984. *Aus Devon, Karbon und Kreide: Die fossile Welt des nordwestlichen Sauerlandes*. 159pp., Hagen (v. d. Lippe Verlagsgesellschaft).

MAPES, R. H. & DAVIS, R. A. 1996. Color Patterns in Ammonoids. *In*: LANDMAN, N., TANABE, K., & DAVIS, A. (eds) *Ammonoid Paleobiology, Topics in Geobiology*, **13**, 103–127.

PAECKELMANN, W. 1922. Der mitteldevonische Massenkalk des Bergischen Landes. *Abhandlungen der Preußischen Geologischen Landesanstalt, Neue Folge*, **91**, 1–113.

SANDBERGER, G. & SANDBERGER, F. 1849–1856. Die Versteinerungen des rheinischen Schichtensystems in Nassau. Mit einer kurzgefassten Geognosie dieses Gebietes und mit steter Berücksichtigung analoger Schichten anderer Länder. I-IV, 1-564, pls 1-41, Wiesbaden [pp. 73–104 and pls 9–13 published in 1851].

SCHLOTHEIM, E. F. BARON VON, 1820. Die Petrefactenkunde auf ihrem jetzigen Standpunkte durch die Beschwreibung seiner Sammlung versteinerter und fossiler Überreste des Thier- und Pflanzenreiches der Vorwelt erläutert. I-LXII, 1–437, Gother.

STRAUCH, F. 1985. Farberhaltung bei Fossilien. *Arbeitskreis Paläontologie Hannover*, **13**, 16–31.

STRUVE, W. 1965. Über *Geranocephalus (Stringomimus)* n. sg. und Verwandte. *Senckenbergiana lethaea*, **46**, 459–472.

STRUVE, W. 1982. Schaltier-Faunen aus dem Devon des Schwarzbach-Tales bei Ratingen, Rheinland. *Senckenbergiana lethaea*, **63**, 183–283.

STRUVE, W. 1986. Sektion Paläozoologie III. *Courier Forschungsinstitut Senckenberg*, **85**, 257–273.

STRUVE, W. 1988. Stop A 18: Nollenbach Glen–Roßberg–Rodert area, Hillesheim Syncline. *Courier Forschungsinstitut Senckenberg*, **102**, 143–149.

STRUVE, W. 1992. Neues zur Stratigraphie und Fauna des rhenotypen Mittel-Devon. *Senckenbergiana lethaea*, **71**, 503–624.

THORMANN, F. & WEDDIGE, K. 2001. Addendum zu STRUVE W. (1992), Neues zur Stratigraphie und Fauna des rhenotypen Mittel-Devon: Abbildungen der Holotypen. *Senckenbergiana lethaea*, **81**, 307–327.

TORLEY, K. 1908. Die Fauna des Schleddenhofes bei Iserlohn. *Abhandlungen der Königlich Preußischen Geologischen Landesanstalt, Neue Folge*, **53**, 1–56.

TORLEY, K. 1934. Die Brachiopoden des Massenkalkes der Oberen Givet-Stufe von Bilveringsen bei Iserlohn. *Abhandlungen herausgegeben von der senckenbergischen naturforschenden Gesellschaft*, **43**, 67–148.

WEBER, H. M. 2000. Farbmuster-Erhaltung bei mitteldevonischen Mollusken aus der Paffrather Mulde. *Archäologie im Rheinland*, **1999**, 19–21.

WEDDIGE, K. 1988. Eifel Conodonts. *Courier Forschungsinstitut Senckenberg*, **102**, 103–110.

WEDEKIND, R. 1934. Kritische Bemerkungen zur Gliederung des Eifler Mitteldevons. *Zeitschrift der Deutschen Geologischen Gesellschaft*, **86**, 19–28.

WINTERFELD, F. 1894. Über den mitteldevonischen Kalk von Paffrath. *Zeitschrift der deutschen geologischen Gesellschaft*, **46**, 687–696.

WINTERFELD, F. 1895. Ueber eine *Caiqua*-Schicht, das Hangende und Liegende des Paffrather Stringocephalen-Kalkes. *Zeitschrift der deutschen geologischen Gesellschaft*, **46**, 687–696.

WINTERFELD, F. 1898. Der Lenneschiefer, geologische Studien des Bergischen Landes. I. Teil. *Deutsche geologische Zeitschrift*, **1898**, 1–53.

ZHANG, Y. & FU, L. 1983. Brachiopoda (Devonian). *In*: Xían Institute of Geology Mineral Resources (ed.) *Paleontological Atlas of Northwest China: Shaanxi, Gansu and Ningxia Volume, Part 2, Upper Palaeozoic*, 244–425, pls. 88–143, Geological Publishing House, Beijing.

ZIEGLER, W. 1958. Conodontenfeinstratigraphische Untersuchungen an der Grenze Mittel-/Oberdevon und in der Adorf-Stufe. *Notizblätter des Hessischen Landesamtes für Bodenforschung*, **87**, 7–77.

The end-Frasnian mass extinction in the Eifel Mountains, Germany: new insights from organic matter composition and preservation

C. HARTKOPF-FRÖDER[1], M. KLOPPISCH[2], U. MANN[2], P. NEUMANN-MAHLKAU[1], R. G. SCHAEFER[2] & H. WILKES[3]

[1]*Geologischer Dienst NRW, De-Greiff-Str. 195, D-47803 Krefeld, Germany*
(e-mail: hartkopf-froeder@gd.nrw.de)

[2]*Forschungszentrum Jülich, ICG-V: Sedimentäre Systeme, D-52425 Jülich, Germany*

[3]*GeoForschungsZentrum Potsdam, Sektion 4.3 Organische Geochemie, Telegrafenberg, D-14473 Potsdam, Germany*

Abstract: The Büdesheimer Bach borehole in the Prüm Syncline, Eifel Mountains, Germany encountered upper Frasnian and lowermost Famennian sediments including the Upper Kellwasser Horizon (UKW) and a limestone-dominated sequence ('LKW') which can be correlated with the Lower Kellwasser Horizon in other sections. The palynofacies is characterized by a high abundance of amorphous organic matter (AOM), prasinophytes, miospores and acritarchs indicative of a fully marine, rather distal and oxygen-deficient environment. AOM and a low sterane/hopane ratio suggest that cyanobacteria were important primary producers and that bacterial reworking and oxidation influenced the organic matter composition resulting in reduced total organic carbon (TOC) contents and lower hydrogen index (HI) values. The 'LKW' and the UKW can be distinguished from the adjacent units by the abundant prasinophytes and some geochemical parameters (e.g. higher HI, lower pristane/phytane ratio, lower aryl isoprenoid ratio) but these differences are not highly significant. However, the sediments between these two horizons show an increased input of bacteria and terrestrial plant material as a result of the regression that follows after the deposition of the Lower Kellwasser Horizon. Aryl isoprenoids originating from diagenetic transformation of the carotenoid isorenieratene, which are markers for anoxygenic photosynthetic green sulphur bacteria, have been detected in all samples but are more abundant in the 'LKW' and UKW. Hence, photic zone anoxia seems to have been more pronounced during deposition of these horizons, supporting the view that widespread anoxia was an important trigger of the massive end-Frasnian biotic decline.

In the late Frasnian, two major extinction events took place: the Lower Kellwasser Event (=late Frasnian Event) and the Upper Kellwasser Event (= Frasnian–Famennian Boundary Event, F/F Event), the latter being one of the most intense mass extinctions in the Phanerozoic (e.g. Sepkoski 1996; Copper 2002; House 2002) with the shallow-water marine fauna of tropical and subtropical regions being severely decimated by both events. The Lower Kellwasser Event mainly affected trilobites (Feist & Schindler 1994) and ammonoids (e.g. Becker & House 1994). In most areas, just before this event the planktonic styliolinids became extinct (Schindler 1993; You Xing Li 2000). It was around the Lower Kellwasser Event that the metazoan reef ecosystem almost completely vanished and suffered its final extinction in the latest Frasnian (Schindler 1993; Copper 2002). In the Famennian, this ecosystem was replaced by cyanobacterial calcimicrobe–algae reefs representing opportunistic disaster forms.

During the Upper Kellwasser Event, 63% of the trilobite families disappeared (Feist & Schindler 1994), a suborder of the goniatites (Gephuroceratina) was completely wiped out (Becker 1993, pp. 137–140; Becker & House 2000, pp. 114, 132), more than 70% of the benthic ostracodes were exterminated, the planktonic entomozoacean ostracodes first experienced substantial losses and then temporarily vanished (Olempska 2002), and the tentaculitids were at least considerably reduced (You Xing Li 2000; Sandberg et al. 2002, p. 481)—to list only some of the more important groups (for more data, see e.g. Schindler 1993; Walliser 1996; McGhee 1996). Some confusion has arisen concerning the fate of the acritarchs, highly resistant organic-walled microfossils of unknown exact biological affinity, most of which are primary producers and hence of utmost importance for the marine food web. For instance, it has been stated that the phytoplankton/acritarchs were considerably reduced during (1) the

Late Devonian (Frasnian) (Tappan 1986, p. 551); (2) during the end-Frasnian extinction event (e.g. Mendelson 1987, p. 84; Schindler 1993, p. 117; Walliser 1996, p. 234; Schülke 1998, p. 93); or (3) by the end of the Devonian and early Mississippian (e.g. Tappan 1968, p. 193, fig. 1; Loeblich 1969, p. 705; Tappan & Loeblich 1972, p. 207, fig. 1). It is now well established that there was no major decline for the acritarchs at or near the Frasnian/Famennian boundary but rather during the late Famennian to early Mississippian (e.g. Strother 1996; Le Hérissé et al. 2000). Compared to the marine fauna, much less is known as to how and what extent the terrestrial ecosystem was disturbed. Although many freshwater fish species did not survive into the Famennian, the losses seem to be less dramatic, than for their marine counterparts (Long 1993). The diversity of land plants seems to decline during the late Frasnian to mid-Famennian but it is not yet possible to determine whether this is gradual or a sudden collapse. However, land plants seem to have played a crucial role in the massive biotic demise (see below).

It is noteworthy to state that not all groups disappeared simultaneously or went extinct with the onset of close to the Kellwasser Events but that the two events display a staggered succession of extinction phases (Schindler 1993). This is why the term Kellwasser Crisis spanning a time interval from the late Frasnian to the earliest Famennian has been introduced. During the Upper Kellwasser Event alone, Sandberg et al. (2002) recognized four steps that led to the dramatic biodiversity losses at the very end of the Frasnian. Following the far-reaching biotic turnover in the marine ecosystem, survivors in refuges like deep basins subsequently gave rise to the highly diverse Famennian ecosystem (e.g. Sandberg et al. 2002, p. 480).

In some regions, the end-Frasnian mass extinction is coupled with two hypoxic/anoxic events which are manifest in two discrete black bituminous limestone-shale levels of some decimetre thickness, the Lower and Upper Kellwasser Horizons. Typically developed Kellwasser Horizons are easily recognizable in the cephalopod limestone facies deposited on the submarine highs of the pelagic realm but they have also been described from basinal shales (e.g. Piecha 1993) and rarely from very shallow-water settings (e.g. Piecha 2002). Both, the limestone and the accompanying black shales yield a characteristic fauna including mass occurrences of homoctenids and entomozoaceans. The Lower Kellwasser Horizon occurs near the base of the Late *rhenana* conodont Zone, the Upper Kellwasser Event near the top of the *linguiformis* conodont Zone. The age of the Kellwasser Event is approximately 377.2 ± 1.7 Ma as revealed from U–Pb zircon dating of a bentonite layer intercalated between the two Kellwasser Horizons in the famous Steinbruch Schmidt section (Kaufmann et al. 2004).

A considerable number of studies have addressed the biotic changes during the end-Frasnian but the primary causation(s) which led to the demise of so many taxa is still a matter of vigorous debates (e.g. summarized by Racki 2005). Collision with an extraterrestrial bolide (e.g. reviewed by Racki 1999), explosive gas release from cratonic lithosphere (Morgan et al. 2004), plate tectonic reorganization and massive submarine volcano-hydrothermal activity (Racki 1999), eustatic and climatic changes (e.g. Becker & House 1994; Sandberg et al. 2002; Copper 2002; Joachimski et al. 2004), and global marine anoxia (e.g. Joachimski & Buggisch 1993; Bond et al. 2004) have all been taken into consideration. These factors may have acted as single forces or in combination.

Although it was frequently emphasized that increase in primary bioproductivity, phytoplankton blooms and enhanced organic carbon burial rate in organic-rich sediments are crucial to decipher the mechanisms which led to the end-Frasnian extinction, surprisingly few publications have dealt with the palynofacies (Paris et al. 1996; Filipiak 2002) and organic geochemistry (Joachimski et al. 2001) of the Lower and Upper Kellwasser Horizons and the sediments in-between, immediately below and above. This lack of information on the variation in organic matter composition is mainly due to the specific requirements of unweathered sediment samples of low thermal maturity for organic geochemical studies unlike palynological investigations that can even be realized using metamorphosed rocks. Generally, the vitrinite reflectance of Upper Devonian sediments in the Ardennes–Rhenish Massif is between 3 and 5% R_{max} and reaches up to 7.5% R_{max}. Only in two regions, the Paffrath and the Prüm Syncline (Fig. 1), is the thermal alteration exceptionally low (e.g. Teichmüller & Teichmüller 1979; Hartkopf-Fröder et al. 2004) and does not exceed a Conodont Alteration Index (CAI) of 1.5 (Helsen & Königshof 1994) or a vitrinite reflectance of 1.03% R_{max}. Hence, sediments in these synclines are appropriate for detailed organic geochemical studies of the organic matter. In the Paffrath Syncline, equivalents of the Kellwasser Horizon are poorly known only from small outcrops. However, in the Prüm Syncline located in the Eifel Depression (=Eifel North–South Zone), the entirely cored Büdesheimer Bach borehole encountered the Upper Kellwasser Horizon and sediments showing a positive excursion of $\delta^{13}C_{org}$ which can be correlated with the Lower Kellwasser Horizon (Joachimski et al. 2002). Only one vitrinite reflectance measurement ($R_r = 0.56\%$) from Upper Devonian sediments of

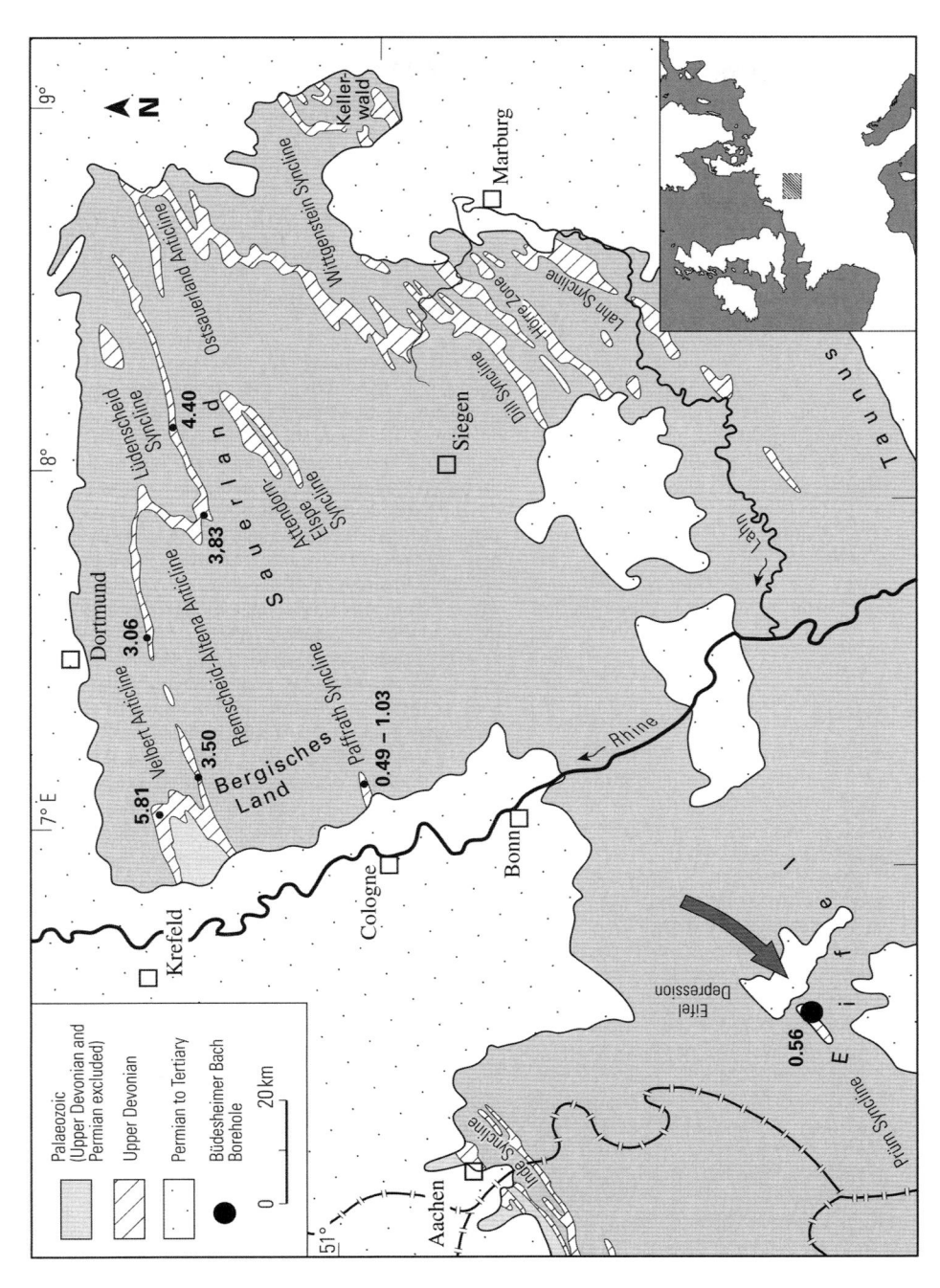

Fig. 1. Outcrop pattern of the Upper Devonian sediments in the Rhenish Massif, western Europe, showing some vitrinite reflectance data (small dots) for sediments of Frasnian age (from Wolf 1972; Teichmüller & Teichmüller 1979; Hartkopf-Fröder et al. 2004, fig. 2) and the location of the Büdesheimer Bach borehole in the Prüm Syncline.

the Prüm Syncline has been published so far but more data from the Middle Devonian succession clearly indicate a considerably lower maturity in the Eifel Depression compared to other regions of the Ardennes–Rhenish Massif (Teichmüller & Teichmüller 1979). Conodonts recovered from sediments of Frasnian and early Famennian age of the Prüm Syncline reveal a CAI of 1.0–1.5, thus emphasizing the very low thermal alteration in this unmetamorphosed area (e.g. Piecha 1994; Königshof & Werner 1994).

In summary, the entirely cored Büdesheimer Bach borehole offers a unique opportunity to study both the palynological and organic geochemical variation of the organic matter because of both the low thermal maturity and the significant amount of information available on the paleoecology and facies of the Late Devonian in the Prüm Syncline. Therefore, the main objectives of this interdisciplinary study are: (1) to investigate which primary producers thrived in the ocean and if there are any fluctuations with regard to their diversity and abundance; (2) to identify short-term transgressive-regressive pulses; (3) to trace variations in oxicity or anoxicity in the water column and/or at the sediment–water interface; and (4) to hence compare the environmental conditions before, during and after the end-Frasnian extinction event.

Geological setting

The Prüm Syncline is located in the Ardennes–Rhenish Massif (Fig. 1), which is a part of the Rhenohercynian fold and thrust belt of the Middle European Variscides (Oncken et al. 1999). In the western Ardennes–Rhenish Massif, the Eifel Depression comprises SW–NE striking synclines of which the Prüm Syncline is the most southwestern one. The Prüm Syncline is the only area in the Eifel Depression where Upper Devonian sediments have been deposited or are still preserved. The very fossiliferous Frasnian and lower Famennian sediments are mainly composed of dolomites, limestones (including reef limestones) and mudstones to marlstones. Some thin bentonites represent volcanic air-fall ash layers. These ashes are important key beds for regional stratigraphical correlations. The Lower and Upper Kellwasser Horizons have both been located in the syncline (for more details on the regional geology and stratigraphy see Meyer 1994 and Grimm 1995). It is noteworthy to state that the typical Büdesheimer goniatite fauna described by Clausen (1969) is older than the Lower Kellwasser Horizon. In composition, this fauna differs considerably from the goniatite assemblage of the Büdesheim Goniatitenschiefer Formation between the two Kellwasser Horizons (Becker & House 1994, text-fig. 17).

Frasnian and Famennian sediments have been deposited on the shelf in a shallow channel ('Eifeler Meeresstraße') of the Rhenish Basin between the Ardennes Shoal in the northwest and the Rhenish Shoal in the southeast. Due to the uplift of these two shoals, the channel both deepened and narrowed until the emergence of the Ardennish–Rhenish Shoal in the late Famennian (Paproth et al. 1986; Paproth 1989).

The Büdesheimer Bach borehole is located at the northeastern rim of the Prüm Syncline in the Eifel Mountains. This borehole reached a total depth of 165.40 m. The cores have been described in detail with regard to lithology and biostratigraphy by Piecha (1994). The fossiliferous sediments consist predominantly of olive grey to dark grey calcareous mudstones, marlstones and brownish grey, grey to black limestones (Fig. 2).

The lower part of the succession (125.44–165.40 m) is assigned to the Oos Plattenkalk Formation (middle and upper Frasnian) and is characterized by banded olive grey to dark grey calcareous mudstones to marlstones, allodapic and autochthonous limestones. The allodapic limestones are up to 70 cm thick. They show grading with coarsest material at the base and a higher clay content at the top, thus developing into marlstones. At the erosive bases, load casts and flame structures occur. These horizons represent turbiditic limestones. From two limestone-dominated sequences (at approx. 125–130 m and 136–140 m), Piecha (1994) isolated a conodont fauna typical for the base of the Late *rhenana* conodont Zone. Although the two sequences do not show the typical lithological characteristics of the Kellwasser Horizon, the $\delta^{13}C_{org}$ is raised between *ca.* 125–140.5 m with two pronounced positive excursions ($\delta^{13}C_{org} > -28‰$) between *ca.* 125–131 m and *ca.* 134.1–140.5 m (Joachimski et al. 2002). These two distinct positive shifts can be correlated with those characteristic for the Kellwasser Horizon when compared with other sections. In order to emphasize that the Lower Kellwasser Horizon is not developed with its typical lithological features, we therefore describe this part of the borehole sequence as 'Lower Kellwasser Horizon' ('LKW'). Based on lithology, conodont and carbon isotope data, a tectonic duplication of the two limestone-dominated sequences is very likely (Piecha 1994, p. 340; Joachimski et al. 2002, p. 97, fig. 4). From the bottom of the borehole, a small conodont fauna may be indicative for the Late *hassi* or *jamieae* conodont Zone.

The Oos Plattenkalk Formation is overlain by the upper Frasnian Büdesheim Goniatitenschiefer Formation (125.44–72.85 m). The latter is defined of the formation and consists of dark grey to black, highly bituminous limestone rich in fossils such as tentaculites, ostracodes and cephalopods. Based on the lithology and some rare and rather long-ranging conodonts (*Polygnathus webbi* and *Icriodus alternatus alternatus*), it is assumed that this horizon represents equivalents of the Upper Kellwasser Horizon (Piecha 1994, p. 337). This presumption is corroborated by a distinct positive excursion of $\delta^{13}C_{org}$ ($\delta^{13}C_{org} > -28‰$) between ca. 71.0 and 73.9 m (Joachimski et al. 2002).

The uppermost part of the cores (72.85–0.30 m) represents the Neu–Oos Cypridinenschiefer Formation (lower Famennian). This succession is characterized by olive grey to dark grey, thin-bedded calcareous mudstones and marlstones, rich in pyrite and ostracodes. Very few limestone nodules and thin horizons are intercalated. Allodapic limestones are lacking. In contrast to the Büdesheim Goniatitenschiefer Formation, goniatites are less abundant. Two bentonite horizons have been localized at 46.70 m and 56.65 m. At 2.20 m, an ostracode assemblage has been assigned to the *serratostriata–nehdensis* ostracode Zone (pers. comm. M. Piecha) indicative for an early Famennian age.

Material and methods

All samples were obtained from cores of the Büdesheimer Bach borehole (R ^{25}40 900 H ^{55}65 690, sheet 5705 Gerolstein of the topographic map 1 : 25 000) which was drilled in 1963. For a lithological description of the full sequence encountered in Büdesheimer Bach borehole, we refer to the preceding chapter.

Palynological samples were processed by following standard techniques including demineralization in HCl–HF–HCl. The resultant residue was then concentrated by sieving with a 10 μm polyester fabric mesh. Because of the low thermal alteration, no oxidation was necessary. Permanent strew mounts were produced with Cellosize as the mounting medium and Elvacite 2044™ epoxy resin as the embedding medium. For quantitative palynofacies analysis, per sample two slides were prepared and 150 palynomorphs counted per slide. They were not differentiated to genus or species level. Incident light fluorescence microscopy was performed with a Zeiss 09 filter block which incorporated a blue light excitation filter (BP 450–490 nm bandwidth), a chromatic beam splitter (FT 510 nm), and a barrier filter (LP 520 nm). All slides are housed in the palynological collection of the Geological Survey of North Rhine–Westphalia.

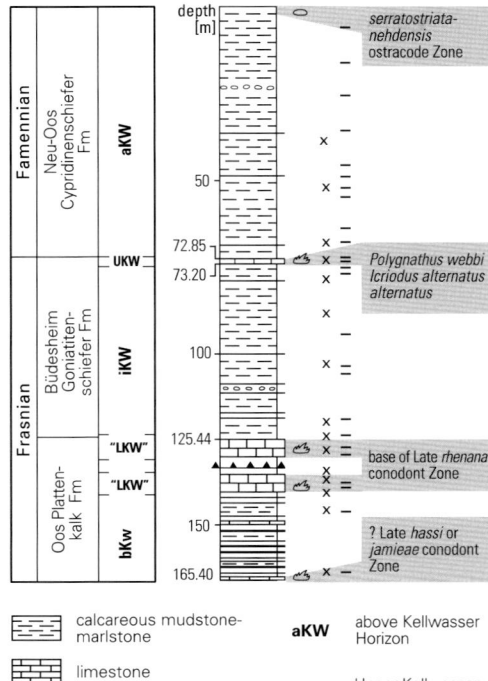

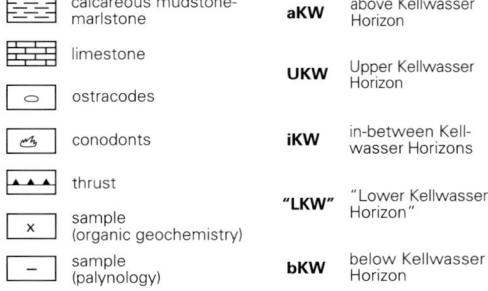

Fig. 2. Büdesheimer Bach borehole section with litho- and chronostratigraphical division, position of samples, definition of borehole units (bKW, 'LKW', iKW, UKW and aKW), and position of conodont/ostracode assemblages (mostly based on data from Piecha 1994 and pers. comm. M. Piecha). The two detrital limestone-dominated sequences between 125.44 and 139.55 m do not show the distinctive lithological features of the Kellwasser Horizon but are characterized by positive excursions in $\delta^{13}C_{org}$ (for more information see text).

by monotonous olive grey to dark grey, thin-bedded calcareous mudstones to marlstones with few thin autochthonous limestone horizons and limestone nodules intercalated. In contrast to the Oos Plattenkalk Formation, allodapic limestones do not occur. The calcareous mudstones and marlstones are very fossiliferous and rich in finely dispersed pyrite. Most of the fossils are pyritized. A distinctive horizon between 72.85–73.20 m marks the top

For the geochemical investigations, the samples were first powdered in a rotary disc mill. Total carbon, total sulphur and, after removal of inorganic carbon by hydrochloric acid, total organic carbon were measured in duplicate using a LECO apparatus (CS 200/CS 225). A Rock-Eval II (Delsi) instrument was used for pyrolysis as described by Espitalié et al. (1977).

For the determination of trace and rare earth elements, rock samples were treated with a lithium borate composite before the fluxing procedure. Concentrations were determined using a quadrupole-based, inductively coupled plasma-mass spectrometer (Elan 6000 Perkin-Elmer/Sciex Corp). For extraction of the C_{15+} soluble organic matter (SOM), samples were treated with dichloromethane and methanol (99:1, v/v) using the modified flow-blending method after Radke et al. (1978). Separation of the soluble organic matter into C_{15+} saturated hydrocarbons (SAT), C_{11+} aromatic hydrocarbons (ARO) and nitrogen, sulphur and oxygen (NSO) compound fractions, was performed by medium-pressure liquid chromatography (MPLC) after Radke et al. (1980). Internal standards (5α-androstane for the SAT fractions and a mixture of 1-phenylhexane, 1-phenylheptane, 1,8-dimethylnaphthalene, 1-phenylnaphthalene, 2-ethylpyrene and 2-butylpyrene for the ARO fractions) for quantification were added to each sample prior to MPLC.

The SAT fractions were analysed by gas chromatography (GC) using an HP 5890 Series II GC equipped with an on-column injector, an HP Ultra 1 column (50 m length, 0.20 mm inner diameter, 0.33 μm film thickness) and a flame ionization detector. The oven temperature was programmed from 90 °C (initial hold time 2 min) to 310 °C (final hold time 63 min) at a rate of 4 °C/min. Helium was used as carrier gas.

Gas chromatography-mass spectrometry (GC-MS) was accomplished using a gas chromatograph (HP 5890 Series II GC) coupled with a Finnigan MAT 95SQ mass spectrometer. The gas chromatograph was equipped with a temperature-programmable injection system (Gerstel KAS) and a SGE BPX5 fused silica capillary column (50 m length, 0.22 mm inner diameter, 0.25 μm film thickness). Helium was used as carrier gas at a flow rate of 1 ml/min. The oven temperature was programmed from 110 to 340 °C (final hold time 10 min) for the SAT or from 120 to 320 °C (initial hold time 2 min and final hold time 21 min) for the ARO fractions, at a rate of 3 °C/min. The MS was operated in electron impact ionization mode at an ionization energy of 70 eV and a source temperature of 260 °C. Full scan mass spectra were recorded over a mass range of 50 to 650 Da at a total scan time of 1.2 s.

Results

Palynofacies

Nearly all samples yielded abundant palynomorphs, but usually the preservation is mediocre. Most samples are dominated by amorphous organic matter (Fig. 3a–f). Pyrite crystals may be incorporated in palynomorphs (Fig. 3g).

Usually, miospores are common to abundant with the content ranging from 14 to 54% (Fig. 4). However, the assemblage is characterized by a low diversity. Miospores only become rare in two samples; one of these with 7% is just below unit UKW (for definition of borehole units see Fig. 2) and the other (2%) is from within unit UKW. The highest abundance of miospores is in unit iKW where they reach nearly 60% of the palynomorph total (Figs 4, 5). Most miospores are small (ca. 40–50 μm), laevigate or have a fine ornament. Large, thick-walled miospores with prominent processes (e.g. *Hystricosporites*) are rare, tetrads (Fig. 3g and h) or broken miospore tetrads even rarer.

The acritarch assemblage is marked by the low diversity and dominance of small simple forms of the *Veryhachium/Micrhystridium* complex (Fig. 3c and d). The total abundance of spiny acritarchs varies between 2% and 24% (Fig. 5). In the units 'LKW' and UKW, the percentage is usually reduced (below 6% except for one sample at 18%) but the more complex morphotypes predominate over *Veryhachium* and *Micrhystridium*. In unit aKW, the abundance of spiny acritarchs is relatively high compared to the other units. To enable a comparison with the data presented by Filipiak (2002), reticulate genera like *Dictyotidium* and *Cymatiosphaera* are not included in Figure 4 under the prasinophytes but grouped with the acritarchs although these two genera are now regarded as cysts or phycomata of prasinophycean algae (e.g. Tyson 1995, pp. 186–187). Only in unit aKW is the percentage of *Dictyotidium* and *Cymatiosphaera* significant (up to 17%).

The prasinophytes (*Leiosphaeridia*, *Maranhites*, *Dictyotidium* and *Cymatiosphaera*) comprise 32% to 94% of the palynomorph assemblage (Figs 4 and 5). In this study, all leiospheres are assigned to the prasinophytes although some may not belong to this group (Tyson 1995, p. 187) but can more properly be classified with the acritarchs (Sphaeromorphitae). Because of the lack of unambiguous features, a taxonomic approach for smooth-walled leiospheres is extremely problematic. Prasinophytes increase in abundance in the units 'LKW' and UKW where one sample with 94% represents a prasinophyte bloom (Fig. 3a and b). The percentage declines in the lower part of unit iKW to less than 35%. In this unit, the number of

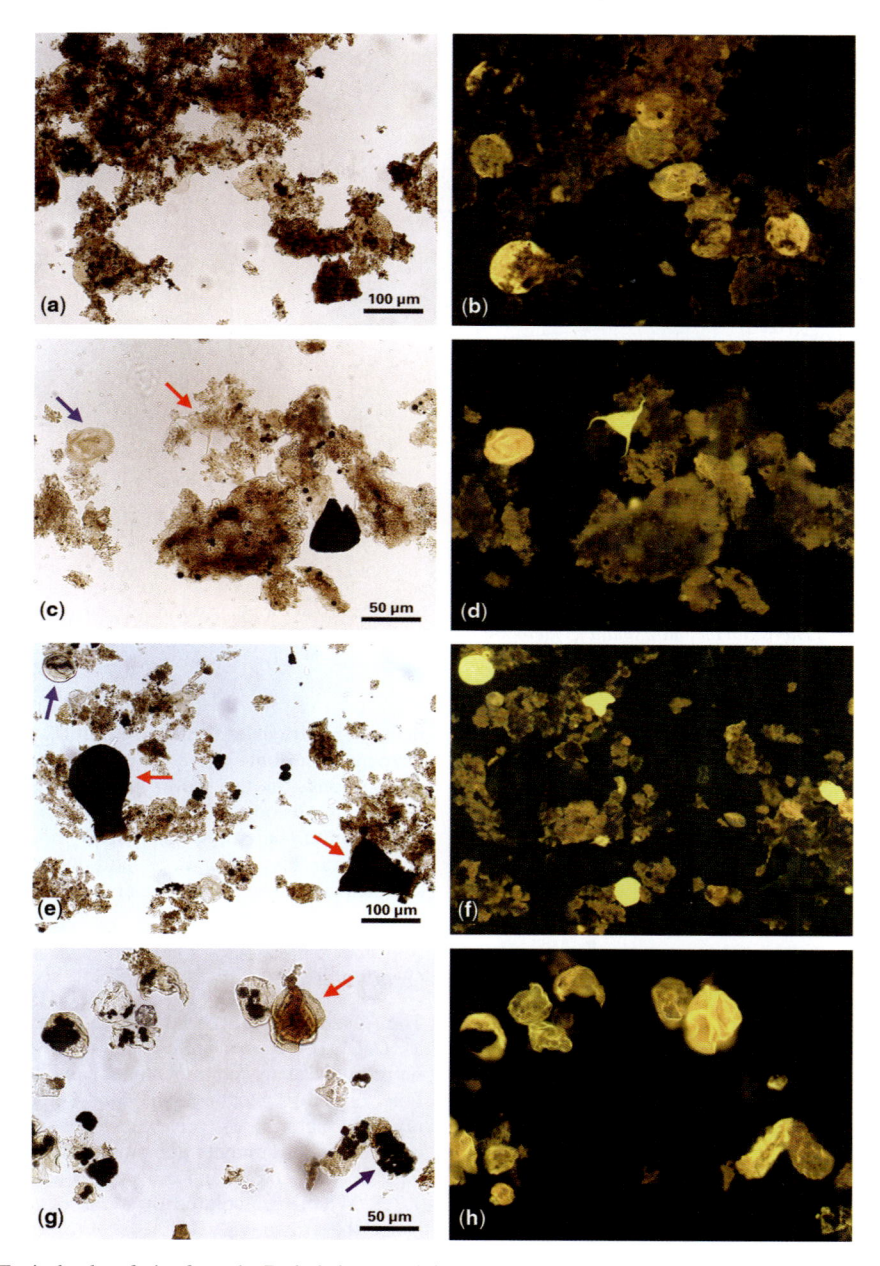

Fig. 3. Typical palynofacies from the Büdesheimer Bach borehole. All unoxidized residues, >10 μm preparation, showing palynomorphs that exhibit bright yellow fluorescence (apart from chitinozoans). (**a**) Abundant amorphous organic matter with prasinophytes. (**b**) Same as in (a), incident blue light fluorescence. Sample from unit UKW, 72.90 m. England Finder L28. (**c**) Abundant amorphous organic matter with prasinophytes (blue arrow) and *Veryhachium* (red arrow) masked by AOM. (**d**) Same as in (c), incident blue light fluorescence. Sample from unit aKW, 64.80 m. England Finder T41/4. (**e**) Prasinophytes (blue arrow) and chitinozoans (red arrows). (**f**) Same as in (e), incident blue light fluorescence. Sample from unit aKW, 45.90 m. England Finder P35/1. (**g**) Prasinophytes, miospore tetrad (red arrow) and strongly degraded palynomorphs (not identified) infilled with pyrite crystals (blue arrow). (**h**) Same as in (g), incident blue light fluorescence. Sample from unit bKW, 145.75 m. England Finder C29/1.

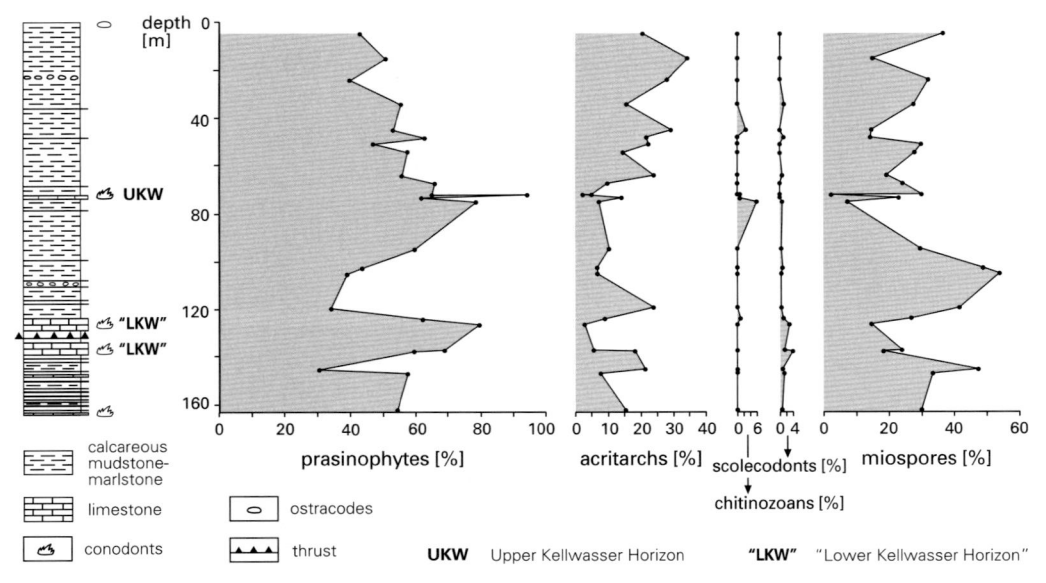

Fig. 4. Palynomorph composition in the Büdesheimer Bach borehole. To allow comparison with the diagrams in Filipiak (2002), *Dictyotidium* and *Cymatiosphaera* are included in the acritarchs.

prasinophytes is reduced except for one sample just above the unit 'LKW' and one sample 2 m below unit UKW.

Chitinozoans and scolecodonts are rare throughout the complete sequence (Figs 3e and f, 4). Only in one sample located 2 m below unit UKW do chitinozoans contribute more than 5% to the palynomorph total. In the lower part of the borehole section, scolecodonts remain between 0.4% and 4% but in the units UKW and aKW the percentage declines to less than 1.3% and more often the samples are barren of scolecodonts.

Geochemistry

The 31 samples yield carbonate contents (calculated as $CaCO_3$) between 8.3% to 48.6%, and thus represent calcareous mudstones and marlstones. For organic geochemical analyses, sufficient organic material is present. The total organic carbon (TOC) contents are shown in Figure 6 and vary between 0.3% and 5.3% with an average of 1.7%. When recalculated on a carbonate-free (cf) basis, the values vary from 0.3% to 9.5% TOC_{cf}. The two TOC maxima correspond stratigraphically to the 'LKW' and UKW. The content of total sulphur (TS) ranges from 0.16% to 1.99% (average of 0.97%). The Rock Eval data (Hydrogen Index (HI) values of 50 to 526 mg HC/g TOC and T_{max} values of 431 ± 5 °C) indicate the presence of variable mixtures of type II (algae and bacteria) and type III (terrestrial plant material) kerogen (Fig. 7). With the exception of the lowermost and uppermost samples in unit iKW, TOC, TS and HI values are conspicuously low throughout this interval.

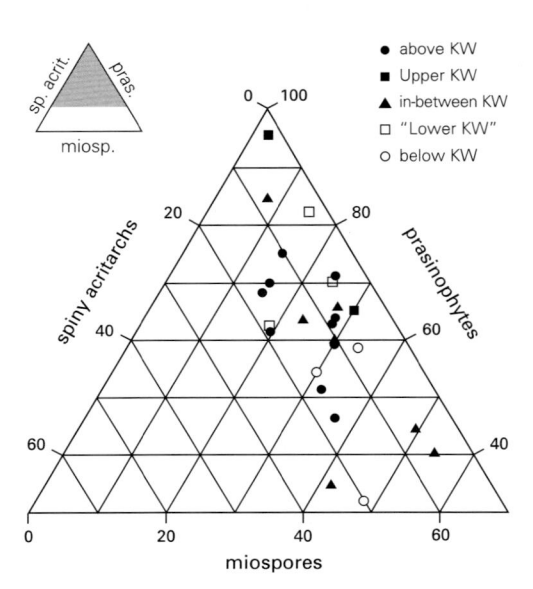

Fig. 5. Ternary diagram showing distribution of spiny acritarchs, prasinophytes (including *Dictyotidium* and *Cymatiosphaera*) and miospores.

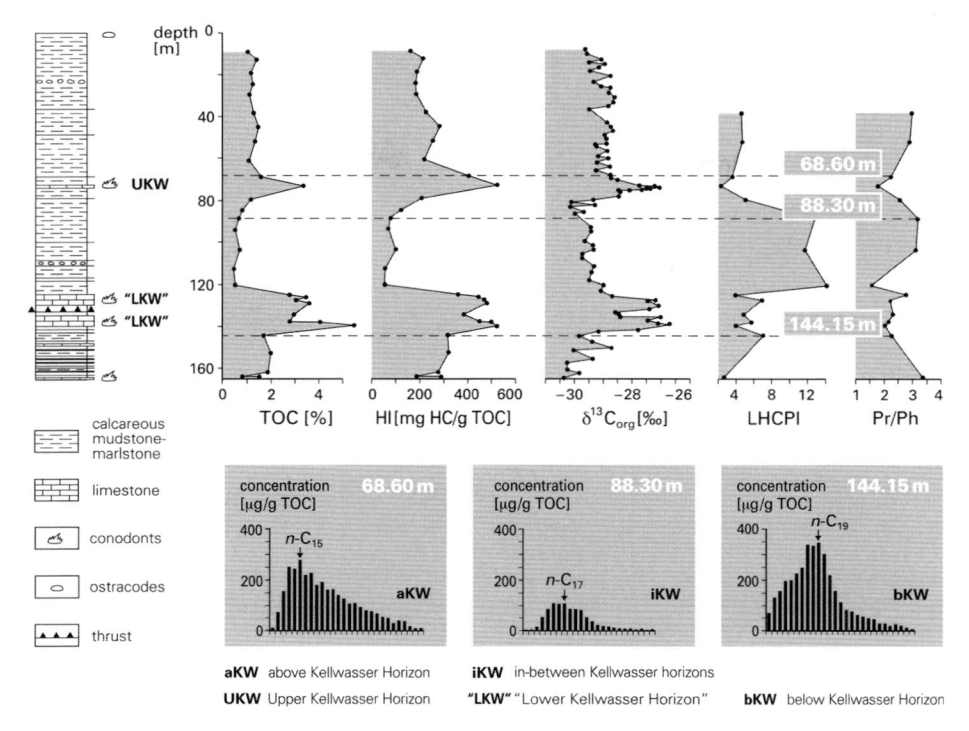

Fig. 6. Total organic carbon (TOC in %), hydrogen index (HI) values, isotopic data ($\delta^{13}C_{org}$ in ‰) (after Joachimski et al. 2002), LHCPI (n-C_{15} + n-C_{17} + n-C_{19}/ n-C_{27} + n-C_{29} + n-C_{31}), pristane/phytane ratio, and n-alkane distributions of selected samples from the Büdesheimer Bach borehole.

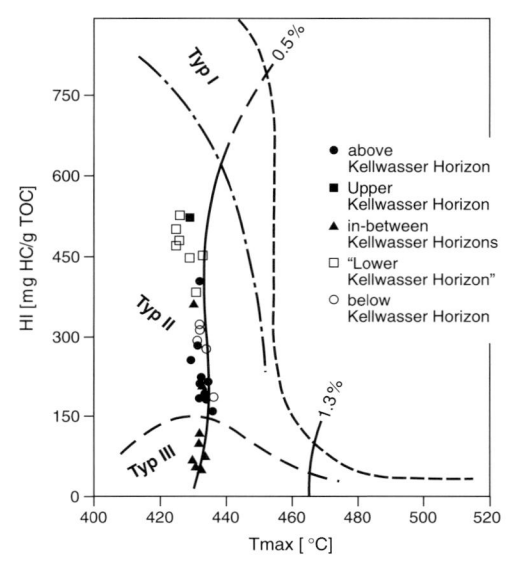

Fig. 7. HI/T_{max} plot after Espitalié & Marquis (1985) gives an overview concerning thermal maturity and variation of organic matter quality. Boundary lines (0.5% and 1.3%) indicate the vitrinite reflectance.

For investigations of the palaeoredox conditions, ten samples representing all units (from the 'LKW' and UKW itself and from the intervals bKW, iKW and aKW) were selected for trace element analyses. Four indices have been used for evaluation of the palaeoredox conditions (Fig. 8, Table 1). The V/Cr concentration ratio in unit bKW is higher (average of 4.91) than in the other intervals (averages between 1.20 and 2.94). The ratios of V/(V + Ni) and Th/U differ only marginally along the sequence with values of 0.52 to 0.80 and 1.36 to 4.10, respectively. The Ni/Co ratio shows higher values in the lower part of the sequence including the lower part of unit iKW below 124.60 m (average of 11.30). Data for the upper part of the borehole section are significantly lower (3.49 on average).

The pristane/phytane concentration ratio, a commonly used organic geochemical palaeoredox indicator, ranges from 1.5 to 3.2 indicating a general predominance of pristane. The pristane/phytane ratio does not show any depth related trend (Fig. 6, Table 1).

Based on the results of the bulk organic geochemical parameters, 15 samples representative of

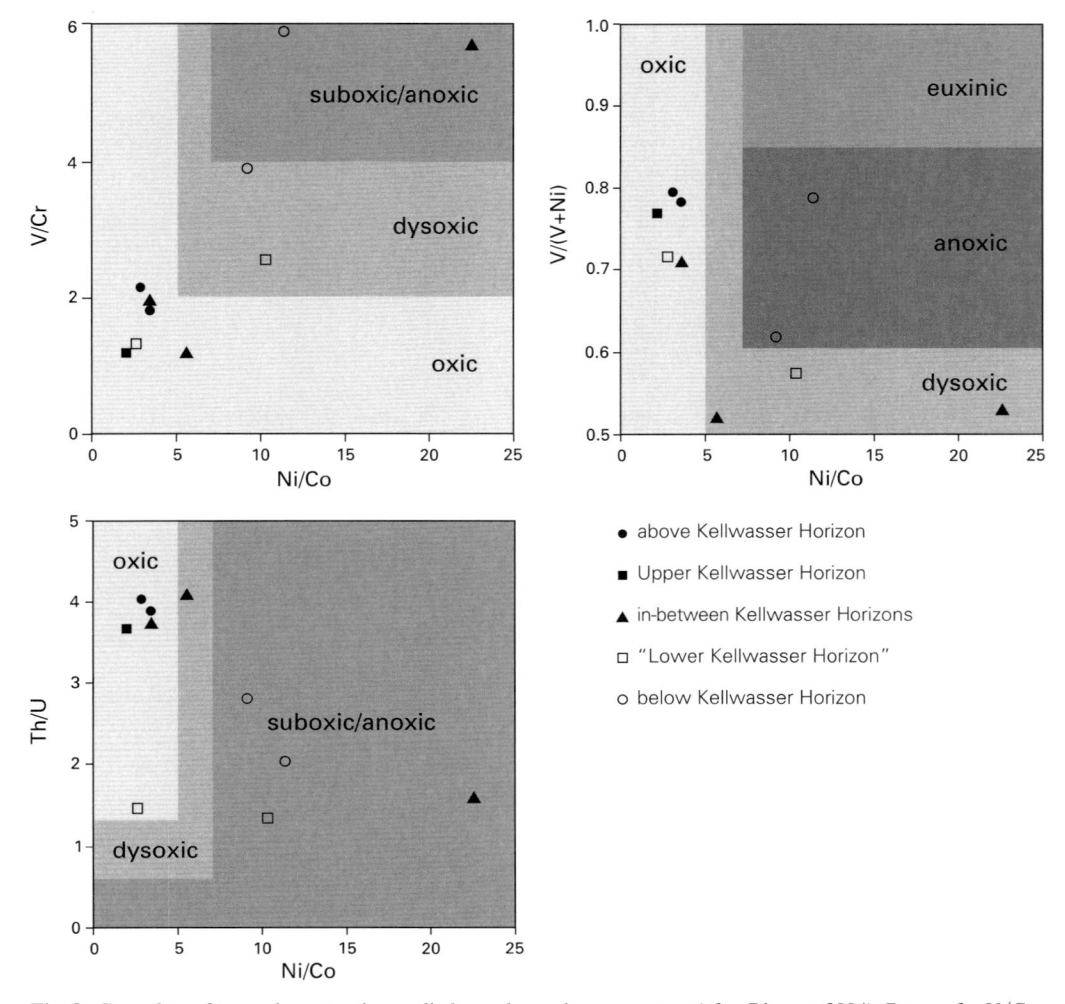

Fig. 8. Crossplots of trace element ratios applied as palaeoredox parameters (after Rimmer 2004). Ranges for V/Cr, Ni/Co and Th/U are adopted from Jones & Manning (1994), for V/(V + Ni) from Hatch & Leventhal (1992).

the individual units, were chosen for evaluation of the molecular composition of the C_{15+} soluble organic matter (SOM). Average values of the extractable material are 1.3 mg SOM/g of rock and 19.9 mg SOM/g TOC. The total yield of the n-alkanes reaches maximally 3928 μg/g TOC. The highest yields of individual n-alkanes are up to 346 μg/g TOC at n-C_{19}. In unit iKW, lower yields can be found (0.1 mg SOM/g sample, 8.9 mg SOM/g TOC; maximal total n-alkane yield 2319 μg/g TOC). As illustrated in Figure 6, the n-alkane distributions are characterized by biased unimodal envelope curves. The bias is most developed at the short-chained n-alkanes (C_{15} to C_{19}) with a slight odd-carbon-number predominance. Higher homologues in the range of n-C_{25} up to n-C_{35} are less abundant. The Light Hydrocarbon Preference Index (LHCPI), as calculated in this study (see Fig. 6 for details), shows high values (average of 8.28) in the unit iKW and low values (averages between 1.86 and 4.24) in the other parts of the borehole section.

With respect to various saturated hydrocarbon biomarkers, only the unit iKW is significantly different from the rest of the section (Fig. 9). The variations are explicit for the parameters (i) regular steranes plus diasteranes over C_{29}–C_{33}-17α(H)-hopanes (steranes/hopanes), (ii) hopane

Table 1. Geochemical bulk and palaeoredox parameters. For definitions of borehole units see Figure 2

Depth [m]	Unit	TOC [%]	S [%]	HI mg HC/g Toc	Pr/Ph	Ts/(Ts+Tm)	Ster/Hop	Hop/(Hop+Mor)	Hopane 22S/(22S+22R)	Sterane 20S/(20S+20R)	Sterane ββ/(ββ+αα)	V/Cr	V/(V+Ni)	Ni/Co	Th/U
37.90	aKW	1.3	1.00	222	2.96	0.42	0.66	0.79	0.56	0.23	0.31	2.14	0.80	2.90	4.04
51.60	aKW	1.2	1.15	253	2.88	0.53	0.66	0.84	0.58	0.34	0.21	1.82	0.78	3.44	3.90
68.60	aKW	1.4	1.17	400	2.23	0.47	0.60	0.83	0.57	0.38	0.33	n.d.	n.d.	n.d.	n.d.
72.90	UKW	3.2	1.73	522	1.76	0.47	0.21	0.82	0.57	0.43	0.37	1.20	0.77	2.02	3.68
79.30	iKW	1.0	1.41	205	2.54	0.43	0.64	0.79	0.56	0.37	0.31	n.d.	n.d.	n.d.	n.d.
88.30	iKW	0.5	0.40	76	3.17	0.12	0.23	0.67	0.58	0.34	0.28	1.18	0.52	5.58	4.10
103.45	iKW	0.5	0.67	101	3.11	0.07	0.28	0.61	0.63	0.36	0.20	1.94	0.71	3.49	3.74
120.15	iKW	0.3	0.24	55	1.53	0.50	0.11	0.66	0.52	0.18	0.22	n.d.	n.d.	n.d.	n.d.
124.60	iKW	2.6	1.99	358	2.79	0.35	0.45	0.78	0.57	0.42	0.27	5.71	0.53	22.60	1.61
127.00	'LKW'	2.9	1.09	469	2.20	0.47	0.44	0.85	0.58	0.44	0.37	2.54	0.57	10.34	1.36
134.10	'LKW'	2.8	1.98	380	2.32	0.42	0.50	0.80	0.57	0.40	0.30	n.d.	n.d.	n.d.	n.d.
137.70	'LKW'	3.9	0.92	500	2.13	0.48	0.55	0.85	0.58	0.41	0.39	n.d.	n.d.	n.d.	n.d.
139.30	'LKW'	5.3	1.65	525	2.01	0.44	0.45	0.83	0.59	0.41	0.29	1.33	0.72	2.65	1.47
144.15	bKW	1.5	1.22	314	2.26	0.54	0.55	0.82	0.55	0.39	0.32	3.91	0.62	9.17	2.82
161.75	bKW	1.7	0.66	275	3.24	0.52	0.60	0.84	0.57	0.40	0.33	5.90	0.79	11.74	2.04

Cut-off values for individual trace element ratios: 1–3 after Jones & Manning (1994); 4 after Hatch & Leventhal (1992).
1. V/Cr: anoxic >4.25; dysoxic between 4.25 and 2.00; oxic <2.00.
2. Ni/Co: anoxic >7.00; dysoxic between 7.00 and 5.00; oxic <5.00.
3. Th/U: anoxic <0.8; dysoxic between 0.8 and 1.33; oxic >1.33.
4. V/(V+Ni): euxinic >0.82; anoxic between 0.82 and 0.54; dysoxic between 0.60 and 0.46; oxic <0.46.

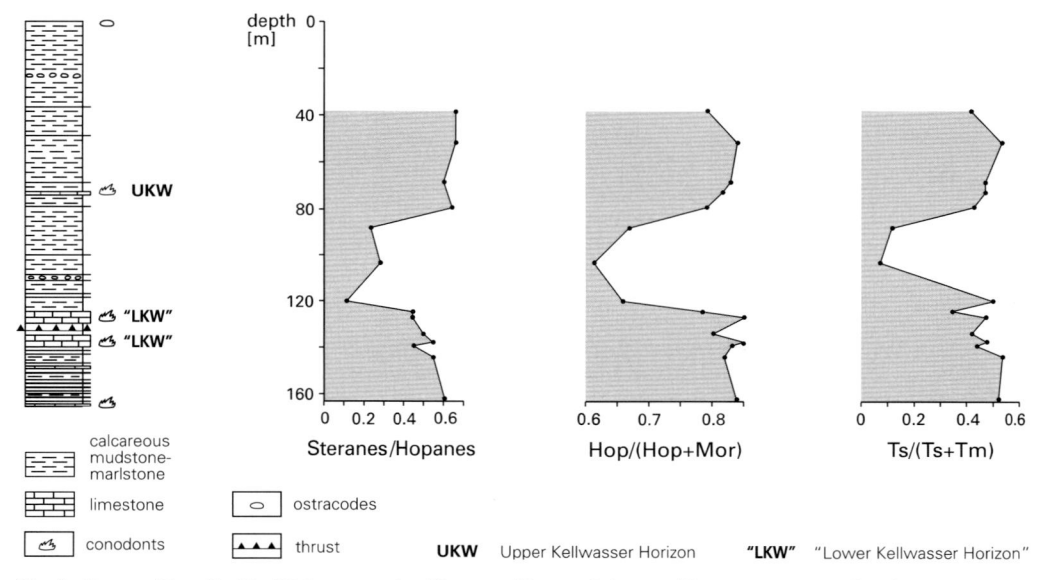

Fig. 9. Sterane/C_{29}–C_{33}17α(H)-hopane ratios [Steranes/Hopanes], hopane/(hopane + moretane) ratios [Hop/(Hop + Mor)] and 18α(H)-22,29,30-Trinorneohopane/(18α(H)-22,29,30-Trinorneohopane + 17α(H)-22,29,30-Trinorhopane) ratios [Ts/(Ts + Tm)] vs. depth.

over hopane plus moretane [Hop/(Hop + Mor)], and (iii) 18a(H)-22,29,30-Trinorneohopane over 18α(H)-22,29,30-Trinorneohopane plus 17α(H)-22,29,30-Trinorhopane [Ts/(Ts + Tm)].

Disregarding the lowest and uppermost sample, very low values for the steranes/hopanes (0.1 to 0.3) and the Hop/(Hop + Mor) ratios (0.6) were detected in the unit iKW. Other units show higher values up to 0.7 for steranes/hopanes and 0.8 for Hop/(Hop + Mor). Only two samples within the unit iKW have significantly lower values for the Ts/(Ts + Tm) ratio (0.07 and 0.12, see Table 1).

The ternary diagram (Fig. 10) displaying the relative proportions of the C_{27}, C_{28} and C_{29} 5α(H),14α(H),17α(H)-steranes points to a rather uniform organic matter composition. There is only one exception in the unit aKW (37.90 m) and one in iKW (120.15 m) with slightly higher proportions of the C_{28} steranes. The rest of the samples are located at a value of about 25 : 25 : 50 ($C_{27} : C_{28} : C_{29}$).

In the aromatic hydrocarbon fractions, a wide range of different compound types including alkylbenzenes, alkylnaphthalenes, alkylphenanthrenes and alkyldibenzothiophenes has been detected. However, only certain aryl isoprenoids will be discussed here in more detail. These compounds have been encountered in all studied samples. Two typical distributions are displayed in Figure 11. The compounds in question show a specific alkyl-2,3,6-trimethyl substitution pattern and are believed to be diagenetic breakdown products of the carotenoid isorenieratene.

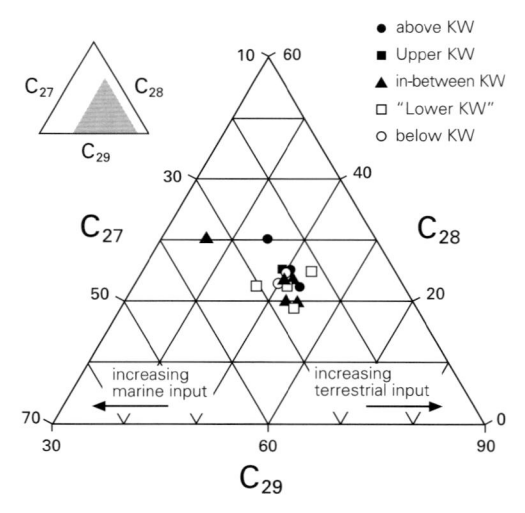

Fig. 10. Ternary diagram showing the C_{27}, C_{28} and C_{29} regular sterane proportion normalized to 100%. Corners of the triangle represent the relative percentage of the corresponding sterane homologues. KW, Kellwasser Horizon.

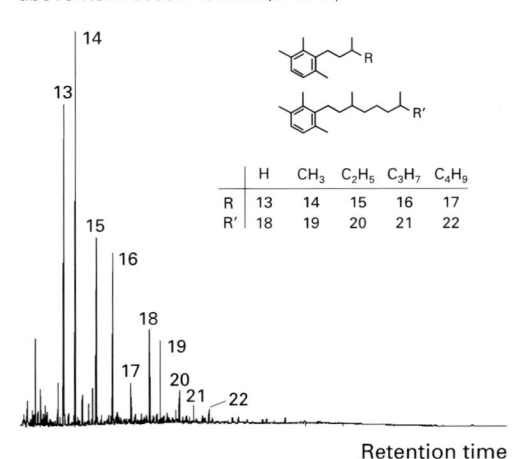

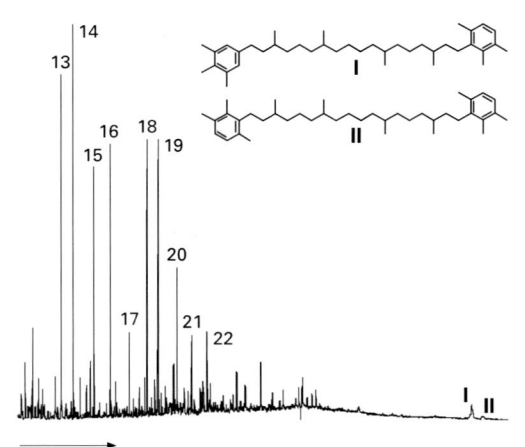

Fig. 11. Extracted ion chromatograms (m/z 133 + 134) displaying the distribution of aryl isoprenoids in two samples from the Büdesheimer Bach borehole section. Numbers and letters on peaks refer to the structures provided in the figure.

Discussion

Maturity

The T_{max} values of 431 ± 5 °C indicate a coalification stage corresponding to a vitrinite reflectance of approximately 0.5% R_r (Espitalié & Marquis 1985), and thus the investigated sediments are considered as marginally mature (Fig. 7). The slight odd–even predominance of the n-alkanes also points to a low maturity. Low maturity is further supported by biomarker maturity parameters (Peters et al. 2005). The hopane $22S/(22S + 22R)$ ratios (0.57 ± 0.02; n = 15) have reached equilibrium values which indicates a minimum maturity equivalent to 0.6% vitrinite reflectance. The Hop/(Hop + Mor) ratio is excluded here for reasons discussed below. However, both the sterane $20S/(20S + 20R)$ ratios (0.38 ± 0.07; n = 15) and the sterane $\beta\beta/(\beta\beta + \alpha\alpha)$ ratios (0.30 ± 0.05; n = 15) are considerably below equilibrium values pointing to a maturity well below 0.9% vitrinite reflectance. Overall, the organic geochemical maturity parameters indicate a uniform maturity throughout the section without any significant increase with depth. Single-walled miospores are yellow to orange in transmitted light and hence have a Thermal Alteration Index of 2 to 2+. Under incident blue light fluorescence, the miospores fluoresce with yellow to yellowish orange colours and the acritarchs/leiospheres with bright yellow colours (Fig. 3), both indicating a low level of thermal alteration. The CAI of conodonts from the Büdesheimer Bach borehole exhibit values of 1.5 as determined by Piecha (1994). Following all maturity parameters, the borehole section is ideally suited for organic geochemical investigations, and a redistribution of the soluble organic matter is thought very unlikely.

Palynofacies

It has been well established for some 40 years that palynofacies analysis can be successfully applied both to interpret ancient depositional settings and to evaluate the potential of the sediments for hydrocarbon generation. Most studies focus on the marine and deltaic realm and provide information about, for example, the oxygenation, salinity, productivity, degradation of organic matter, distance from land, and the terrestrial source vegetation. Primarily, Mesozoic and Cenozoic sediments have been investigated, but Palaeozoic successions have also been dealt with in some examples and which can be used for comparison with the data from the Büdesheimer Bach borehole (e.g. Becker et al. 1974; Filipiak 2002; for more references see Tyson 1995; Batten 1996).

Amorphous organic matter. Amorphous organic matter (AOM) is a ubiquitous constituent in all samples. AOM of aquatic origin is believed to be derived from microbially altered phytoplankton and from bacteria. Another source may be microbial mats constituted by cyanobacteria and thiobacteria, faecal pellets and organic aggregates reprecipitated from dissolved organic matter (Tyson 1995; Batten 1996). AOM is particularly abundant following phytoplankton blooms provided that the organisms were not degraded in a well-oxygenated setting during lateral transport, settling on the sea floor and in the uppermost few centimetres of sediment. Apart from oxidation, the duration of exposure to microbial attack is also critical for the preservation of AOM. Hence, high productivity does not *per se* result in enhanced AOM content but supply of organic matter in excess favours the chance that more AOM accumulates and is preserved. The highest percentage of AOM is usually found in eutrophic oxygen-deficient (dysoxic to anoxic) environments which were protected from pronounced terrestrial phytoclast input, e.g. during transgressions (Tyson 1995, pp. 249–251; Batten 1996, pp. 1012–1016). In the Büdesheimer Bach borehole, the high amounts of AOM indicate that anoxic to dysoxic conditions prevailed during deposition of most parts of the sequence.

In the samples from the Büdesheimer Bach borehole, high amounts of AOM occur together with numerous pyrite crystals. Abundant pyrite is frequently observed in marine algal/AOM formed under low oxygenated conditions (Batten 1996) and has been reported by Filipiak (2002) from the Frasnian/Famennian boundary sediments in the Holy Cross Mountains (Poland).

Miospores. Generally speaking, the miospores are the second most abundant palynomorph group. In the Late Devonian, miospores were produced by vascular land plants. *Rhacophyton* and the progymnosperm *Archaeopteris* are characteristic components of the Late Devonian vegetation. During the late Famennian, *Rhacophyton* dominated 'coal' swamps while *Archaeopteris* trees at least 30 m high thrived in forests on well-drained alluvial plains (Streel 1999).

It is well known that the composition and preservation of miospore assemblages change with the distance from their source and are dependent on the energy level of currents and wave action. On the whole, in a proximal setting one would expect rich and well-preserved assemblages representing a wide size range. By contrast, poorly preserved well-sorted assemblages which consist of smaller specimens, and are less diverse, are inferred to characterize a distal environment. Also, other land plant-derived organic matter like wood and cuticles will be more widespread in a near-source setting.

The preservation of the miospores from the Büdesheimer Bach borehole is mediocre, and the diversity is fairly low. Grains are around 40–50 μm in diameter, and hence in the lower size range of miospore assemblages typical for the latest Frasnian/Famennian (Loboziak *et al.* 2005, p. 81). Trilete laevigate and weakly ornamented miospores are much more abundant than those with prominent processes. Specimens of relatively large, thick-walled *Hystricosporites* which, when well preserved, bear long and slender bifurcating or grapnel-tipped projections are rare, and when encountered, frequently the tips of the processes are broken off. In the Frasnian of the Campine Basin (Belgium), the frequency of the 'robust' hystricospores is controlled by the grain size of the sediment with these miospores being more abundant in coarse grained sediment. The high relative abundance of hystricospores in the Campine Basin may have been the result of river transport from the uplands while the smaller miospores represent the local vegetation (Becker *et al.* 1974, p. 22). Considering the predominantly fine-grained mudstone and marlstone in the Büdesheimer Bach borehole, the rarity of *Hystricosporites* is obvious.

In the Büdesheimer Bach sequence, not only are large miospores with thick exines rare but also tetrads or parts of tetrads. In this respect, the assemblage is comparable to that from Frasnian/Famennian offshore sediments in the Holy Cross Mountains from where Filipiak (2002) even reported a complete absence of tetrads. This stands in sharp contrast to late Famennian miospore assemblages from the Ourthe valley (Belgium) and the Paffrath Syncline (Germany). In the Ourthe valley, Becker *et al.* (1974, encl. II, fig. 5) recognized abundant tetrads in alluvial–lagoonal facies with less-abundant tetrads in the barrier complex and tidal flat environments. The miospore assemblage is associated with marine phytoplankton which is impoverished with regard to diversity when tetrad abundance is rising. This is due to an increasing alluvial influence. In the Paffrath Syncline, tetrads and incomplete tetrads of at least four species (*Diducites plicabilis*, *Endoculeospora setacea*, *Grandispora cornuta* and *Retispora lepidophyta*) are frequently encountered together with a diverse acanthomorph acritarch association. Although the environment was clearly fully marine, as proved by the invertebrate and vertebrate fauna, the depositional setting is assumed not to have been far from land (Hartkopf-Fröder 2004). In the Büdesheimer Bach borehole sequence, the rarity of miospore tetrads and large thick-walled miospores implies a rather distal sedimentary environment where only organic detritus particles hydrodynamically equivalent to clays and fine silts were deposited.

A well-established method to evaluate the relative distance from palaeoshorelines is to compare the miospore versus marine phytoplankton ratio. In Mesozoic and Cenozoic sediments, dinoflagellates are normally used and there are many studies indicating that the miospore vs. dinocyst ratio declines offshore and in areas with freshwater influence (for references see Tyson 1995). In Palaeozoic marine sediments, dominance of miospores over acritarchs and prasinophytes is regarded to signify proximity to the shore or a fluvio-deltaic prevalence (e.g. Streel 1999). However, increased productivity in the ocean will shift the ratio towards smaller numbers, and enhanced terrigenous input due to an increasing morphological gradient on land will result in higher numbers. In both cases, the distance to the palaeoshorelines can remain the same. In the Büdesheimer Bach borehole, the highest percentages are encountered in the lower part of the Büdesheim Goniatitenschiefer (lower part of unit iKW) and in the Oos Plattenkalk Formation (except for the 'Lower Kellwasser Horizon'), i.e. in unit bKW. At least in the Oos Plattenkalk Formation, the rather high amount of miospores may be carried into the basin by turbidites which are intercalated between the calcareous mudstones to marlstones representing the background sedimentation. A similar mechanism has been described from uppermost Devonian sediments at the northeastern rim of the Ardennes–Rhenish Massif (Paproth & Streel 1970). Hence, for the Oos Plattenkalk Formation a high miospore content does not necessarily reflect proximity to the source area. Transportation by turbidites cannot be applied to the rather high amount of miospores in unit iKW, as in the Büdesheim Goniatitenschiefer Formation turbidites are absent. Enhanced miospore percentages which are paralleled by a dominance of type III kerogen in this unit (see below) can rather be explained by a regression following the Lower Kellwasser Horizon. However, the miospore abundance in unit aKW does not reflect the major regression above the Upper Kellwasser Horizon which, as with the Lower Kellwasser Horizon, corresponds to a eustatic highstand (Devleeschouwer et al. 2002). Apart from the few miospore-rich samples in units bKW and iKW, the miospore content in units UKW and 'LKW' does not differ very much in the adjacent intervals, indicating that terrestrial input was not strikingly dissimilar during deposition of these horizons.

The predominance of small miospores over large ones, the low diversity, the rarity of strongly ornamented thick-walled miospores and tetrads, and the prevalence of marine phytoplankton over miospores all indicate a preferential selection of the original assemblage due to long distance of transport and hence a deposition in a low-energy facies rather distal from the terrestrial sources (cf. Tyson 1995, pp. 261–278; Batten 1996, pp. 1027–1028).

Acritarchs. Acritarchs are a polyphyletic, heterogeneous group of organic-walled microfossils of unknown and varied biological affinities (Strother 1996; Le Hérissé et al. 2000). Probably, most acritarchs are resting cysts and phycomata of phytoplankton and unicellular algae but some may be eggs of planktonic animals (e.g. copepods, crustaceans). Most acritarchs are primary producers and characteristically marine although they have occasionally been reported from freshwater environments. The distribution of acritarchs is believed to be controlled by a multitude of factors of which light, turbulence and currents, temperature, nutrition, salinity and oxicity may be the most important ones. In combination with terrestrial debris, acritarchs are useful to recognize inshore–offshore trends. Nearshore assemblages are of low diversity dominated by thin-walled sphaeromorphs (e.g. *Leiosphaeridia*) and simple short-spined acanthomorphs (e.g. *Micrhystridium*, *Baltisphaeridium*, and some species of *Veryhachium*). The dominance by a few species indicates short-lived acritarch blooms in an extreme environment. Diversity increases on the proximal shelf while sphaeromorphs are reduced. Assemblages from the distal shelf are characterized by highest diversity and complex morphotypes. The deep-water/basinal facies is again typified by low diversity and thick-walled prasinophytes (see Tyson 1995, pp. 319–320; Wicander 2002; Montenari & Leppig 2003 for a review).

Compared to late Frasnian/early Famennian acritarch assemblages from the Dinant Basin, Belgium (e.g. Martin 1984) and the Paffrath Syncline, Germany (Jux 1984), the diversity in the Büdesheimer Bach borehole samples is significantly lower [though the number of species recorded by Jux (1984) may be inflated due to splitting variable genera like *Micrhystridium* into too many taxa]. From Frasnian/Famennian boundary sections in the Holy Cross Mountains, Poland, Filipiak (2002) reported an assemblage that is composed of few acritarch species and dominated by leiospheres. This assemblage is comparable to the one from the Büdesheimer Bach borehole and following Filipiak (2002) is interpreted as indicative of an open marine, deep-water environment. However, it has to be taken into account that the similar composition of nearshore and basinal assemblages (see above) presents problems in delimiting very proximal and deep-water deposits. Nevertheless, a low diversity, with only rare complex forms that are numerically exceeded by *Micrhystridium* and *Veryhachium*, plus the predominance of thick-walled over thin-walled prasinophytes, point to a basinal facies rather than to a nearshore environment for most parts of the Büdesheimer Bach borehole sequence.

Interestingly, the generally ubiquitous genera *Micrhystridium* and *Veryhachium* are reduced and outnumbered by more complex morphotypes in most of the samples from units 'LKW' and UKW. With regard to the ecological demands of the complex forms, environmental conditions were therefore not too harsh during deposition of these two units.

Prasinophytes. Prasinophytes are primitive Chlorophyta and have a holoplanktonic life cycle. Mass occurrences of prasinophycean algae are associated with black shale formation and have been reported from many Palaeozoic and Mesozoic sequences. Prasinophyte blooms are thought to preferentially occur in cold water environments, in the brackish surface layer of salinity-stratified basins, under dysoxic to anoxic conditions especially within the transgressive systems tract, in distal facies with low sedimentation rates, or during episodes of eutrophication in the surface water (Tyson 1995, pp. 299–308). Mass accumulations of prasinophytes are definitely paralleled by the relative decline of the normal phytoplankton (acritarchs, dinoflagellates) and hence are regarded as a strong indication for a stressed environment. Prasinophycean algae have therefore been referred to as 'disaster species' (e.g. Tappan 1986). These accumulations are restricted to thin horizons and may contain up to 10 000 specimens per g of sediment (e.g. Paris *et al.* 1996). Concurrent mass occurrences of prasinophytes with a virtual absence of acritarchs have been reported from many upper Palaeozoic black shales and are summarized in Riegel (1993) and Tyson (1995, pp. 487–488).

Prasinophyte percentage is high in almost all samples and compared to the adjacent units is clearly increased in most samples from units 'LKW' and UKW. Compared to the data from Frasnian/Famennian boundary horizons in the Holy Cross Mountains where Filipiak (2002) observed values between 80% to 95%, the prasinophytes are less abundant in the two units of the Büdesheimer Bach borehole. Only in one sample from unit UKW does the palynomorph assemblage consist of nearly 95% of prasinophytes, with very few miospores and acritarchs and hence can be regarded as evidence for an algal bloom. In the Prüm Syncline, mass occurrences of prasinophytes are not restricted to the Kellwasser Horizons but have also been reported from several horizons in the Oos Plattenkalk Formation (Grimm 1991) and therefore are not a distinguishing feature of the unit UKW.

High primary bioproductivity triggered by enhanced terrestrial nutrient supply, giving rise to widespread anoxia, may have been the main cause of the end-Frasnian ecosystem collapse. Wilder (1989) proposed that the spread of vascular plants into previously non-colonized areas increased chemical weathering which resulted in enhanced nutrient supply into the sea, eutrophication, phytoplankton blooms, anoxia, and the termination of reef growth. Subsequently, Algeo *et al.* (1995, 2001) elaborated this hypothesis in more detail for Mid and Late Devonian extinction events by considering changes in pedogenesis, global climate and global geochemical cycles and by emphasizing the importance of evolutionary breakthroughs among vascular land plants such as the development of tall (at least 30 m high) trees and the appearance of the seed habit. Evolution of an arborescent growth habit is correlated with a deeply penetrating root system which accelerates chemical weathering rates. Resistant seeds allowed plants to reproduce under dry conditions and to spread into well-drained habitats. However, temporal coincidence of widespread ecological disruption in the marine realm and the advent of crucial evolutionary innovations in terrestrial plants is difficult to prove. Interestingly, recent modelling of the biogeochemical carbon cycle clearly shows that the two positive $\delta^{13}C$ excursions in the latest Frasnian can only be explained by incorporating continental processes such as increased weathering, rapid expansion of land plants, and additional phosphorous supply into the ocean (Goddéris & Joachimski 2004).

Chitinozoans. The bottle-shaped chitinozoans are rare throughout the complete sequence. The vesicles are composed of a very resistant biogeomacromolecule. The biological affinity is still unclear and they have been assigned to different taxonomic groups such as the protozoans, metazoans, protists and fungi (Miller 1996; Paris & Nõlvak 1999). However, most authors currently attribute chitinozoans to eggs of oviparous soft-bodied metazoans ('chitinozoophorans'). Chitinozoans almost certainly vanished at the end of the Devonian although some questionable findings may indicate that they survived into the Permian. The ecological demands of the chitinozoans are still difficult to evaluate. They exclusively occur in marine environments where most of them are characterized as planktonic and some as benthonic. In general, diversity and abundance increase with decreasing water temperature and with increasing water depth, while in shallow restricted marine environments and in turbulent water few species occur (for examples see Miller 1996). A remarkable mass occurrence of chitinozoans has been described by Paris *et al.* (1996) from the very base of the Famennian just above the Upper Kellwasser Horizon at La Serre, southern France. Such exceptional concentrations are by far not reached in the Büdesheimer Bach succession. Chitinozoans are also rare in the Łagów–Płucki section and absent in the Kowala section (Holy Cross Mountains, Poland, Filipiak 2002). Obviously, living conditions were not too good for them in neither the Prüm Syncline nor in the Holy Cross Mountains localities.

Scolecodonts. As with the chitinozoans, scolecodonts occur only sporadically in the palynomorph assemblage. Scolecodonts are mineralized scleroproteinaceous mouthparts of benthic polychaete annelids of the orders Eunicida and Phyllodocida of which only the former is present during the Palaeozoic (Szaniawski 1996). The fossilization potential of scolecodonts depends on their composition, i.e. their degree of mineralization and sclerotization. Modern polychaete jaws of Glyceridae, Arabellidae and Dorvilleidae are heavily sclerotized and hence more likely to become fossilized while the weakly sclerotized eunicid and onuphid jaws which are mineralized with relatively unstable aragonite have a very low fossilization potential (Colbath 1988). As the chemical composition of the scolecodonts acts as an important taphonomic filter, the fossilized assemblage may not reflect the original abundance and diversity. Modern Eunicida show a preference to shallow water in a nearshore environment and only some have a pelagic mode of life (Szaniawski 1996). The same seems to be true for Ordovician Eunicida which are more abundant and more diverse onshore than in the deeper parts of the basin even though some species are well adapted to offshore conditions (e.g. Hints 1999). At least in the Mesozoic, some scolecodont families were able to inhabit deep and even suboxic depositional settings (Courtinat 1998). Certainly, the polychaete annelids which produced the scolecodonts were well adapted to oxygen-depleted marine water. As some of the scolecodont specimens from the Büdesheimer Bach borehole are smaller fragments, it must be considered that they are allochthonous and that they have been transported *post mortem* to the site of deposition by weak currents or predatory fish. In addition, the occurrence of benthic polychaete annelids does not preclude a dysaerobic to temporarily quasi-anaerobic biofacies.

The palynofacies from two Frasnian/Famennian boundary sections in the Holy Cross Mountains (Poland) described by Filipiak (2002) is similar to the one in the Büdesheimer Bach borehole sequence but the latter contains both more miospores and a lower relative abundance of leiospheres. A palynofacies type rich in AOM and characterized by abundant leiospheres, reduced acritarch diversity and small miospores is indicative for a rather distal and dysoxic–anoxic environment. The Frasnian/Famennian palynomorph assemblages from the Hony and Sinsin sections (Belgium; summarized by Streel *et al.* 2000) are distinguished from those from the Prüm Syncline in being dominated by miospores and acritarchs. They have been deposited in a much more neritic setting.

Geochemistry

The trace element ratios used for this study are commonly documented in the literature (e.g. Jones & Manning 1994; Rimmer 2004). Cut-off values were adopted from Jones & Manning (1994) for the V/Cr, the Ni/Co and the Th/U concentration ratios and from Hatch & Leventhal (1992) for the V/(V + Ni) ratio. The terminology for the description of the bottom-water oxygenation follows Jones & Manning (1994). It defines the palaeoredox conditions as oxic at an oxygen concentration of 860–2.0 ml/l, as dysoxic at 2.0–0.2 ml/l, as suboxic at 0.2–0 ml/l, and as anoxic at 0 ml/l and the occurrence of sulphide.

The lowest part of the sequence shows a clear dysoxic/anoxic environment. The same conditions can also be inferred for the 'Lower Kellwasser Horizon', with one exception (see Table 1). The values of the overlying units can be only interpreted as more or less oxic. This is the first evidence for the excellent preservation of the organic matter in the 'LKW' and in the unit beneath it. This is confirmed by the high HI values in this part of the section. In contrast, the V/(V + Ni) ratio points to prevailing anoxic conditions over the entire section. This is also confirmed at other locations such as the Kowala section (Holy Cross Mountains/Poland; Joachimski *et al.* 2001) and the La Serre section (France; Tribovillard *et al.* 2004).

The use of the pristane/phytane ratio is well established in geochemical investigations (Peters *et al.* 2005) and, despite possible interferences from source and maturity, is often applied as a palaeoenvironmental redox parameter (Powell & McKirdy 1973). Values higher than 1.5 in this study indicate depositional conditions in an oxygenated environment according to standard geochemical interpretation although it is well known that pristane/phytane ratios may be well above 1.0 for anoxic depositional settings (Frimmel *et al.* 2004). The slightly lower values in the units 'LKW' and UKW (Fig. 6) may indicate less oxic conditions, compared to the other parts of the sequence. The plot of pristane/phytane vs. $(Pr + Ph)/(n - C_{17} + n - C_{18})$ provides information about an increase in oxicity or anoxicity (Fig. 12). Similarly to the pristane/phytane ratio, the $(Pr + Ph)/(n - C_{17} + n - C_{18})$ ratio shows

Fig. 12. Cross correlation of the pristane/phytane ratio vs. $(Pr + Ph)/(n\text{-}C_{17} + n\text{-}C_{18})$ (after Frimmel et al. 2004). Shadowed ellipse illustrates the samples within the 'Lower' and Upper Kellwasser Horizon.

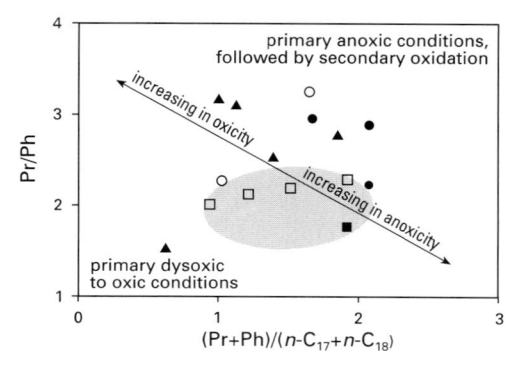

- above Kellwasser Horizon
- Upper Kellwasser Horizon
- in-between Kellwasser horizons
- "Lower Kellwasser Horizon"
- below Kellwasser Horizon

minor variation. For the samples within the 'LKW' and UKW, there is a tendency to more-reducing/less-oxic conditions.

According to the pristane/phytane ratios (Fig. 6) and the trace element ratios (Fig. 8), the Oos Plattenkalk Formation (units bKW, 'LKW') in the Büdesheimer Bach borehole has been deposited under anoxic to dysoxic conditions. High HI values demonstrate the extraordinarily good preservation of the organic material and support this hypothesis. The trace element ratios, analysed for the UKW, would point to oxic conditions. In contrast, the two pristane/phytane ratios established for this unit suggest increasing anoxicity, which is also confirmed by the higher HI values. More oxic conditions prevailed only in units iKW and aKW. The lower TOC and HI values would be in agreement with an enhanced oxidation of the organic matter, possibly resulting from aerobic bacterial reworking.

The small variation displayed in the regular steranes ternary diagram (Fig. 10) reveals that the organic matter is rather uniform in composition. According to Huang & Meinschein (1979), the relative proportions of the C_{27}, C_{28} and C_{29} $5\alpha(H),14\alpha(H),17\alpha(H)$-steranes provide information about the origin of the organic matter. Samples enriched in C_{29} steranes (stigmastanes) suggest a high input from vascular plants. A predominance of steranes with a cholestane skeleton (C_{27}) originates from marine plankton. Ergostanes (C_{28}) are generated from both marine and terrestrial sources. The predominance of the stigmastanes (C_{29}) in the ternary diagram of the C_{27}–C_{29} regular steranes therefore points to the deposition of terrestrial plant material during the late Frasnian and early Famennian.

The sterane/hopane ratio reflects the input of eukaryotic (mainly algae and/or higher plants) vs. prokaryotic biomass into the sediment (Peters et al. 2005). The low values (<1) throughout the borehole section (Fig. 9) indicate a very significant contribution of organic matter from prokaryotes. Therefore it is very likely that cyanobacteria were important primary producers. Particularly low values in unit iKW together with the low TOC and HI values might indicate a strong reworking of the organic matter by heterotrophic and probably aerobic bacteria.

An enhanced sterane/hopane ratio has been reported by Rodrigues et al. (1995, p. 323) for the Frasnian anoxic event in the Parnaíba Basin (Brazil). Based on this ratio and some other biogeochemical indications (e.g. higher relative proportion of C_{27}), these authors assume a dominant algal input during the anoxic event compared to the more significant contribution of land plants before and after the event. Interestingly, in the Büdesheimer Bach borehole the lower Hop/(Hop + Mor) ratios in unit iKW may indicate an increasing input of higher land plants. Hop/(Hop + Mor) is normally used as a biomarker maturity indicator (Peters et al. 2005); however, considering the above assumptions on maturity it is very unlikely that the observed variations of Hop/(Hop + Mor) which are restricted to a specific part of the borehole section, i.e. unit iKW, are due to maturity changes. Changes therefore have to be explained by differences in the specific origin of moretane related to the occurrence of a different facies. Rullkötter & Marzi (1988) interpreted higher Hop/(Hop + Mor) ratios as indicating a lower relative proportion of terrestrial organic matter or a lower direct deposition of moretanes from higher plants. Isaksen & Bohacs (1995) observed increasing C_{30} moretane ratios from transgressive to highstand systems tracts in Lower–Middle Triassic mudrocks from the Barents Sea, corresponding to increased higher land plant input. Applying this to the sequence in the Büdesheimer Bach borehole, it can be assumed that more organic material of higher terrestrial plants was deposited after the 'LKW'. This and the sterane/hopane ratios suggest a regression starting after the 'LKW'. The sample at 79.30 m (below UKW) points to a decrease in bacterial activity. Probably, this latter observation indicates a transgressive sea-level.

The Ts/(Ts + Tm) ratio is a commonly used maturity parameter with a strong dependence on source characteristics (Peters et al. 2005). Assuming again that maturity variations can be neglected higher values point to more oxic conditions and/or a higher clastic input. The two samples in unit iKW, with lowest values suggest dysoxic to anoxic conditions. However, other palaeoredox parameters do not support this interpretation. The low values could be explained by the input of different organic material. We suspect, according to Moldowan et al. (1986), that the source is related to the bacterial and/or terrestrial input.

An early diagenetic product of isorenieratene, isorenieratane, was detected in some of the investigated samples (Fig. 11). Isorenieratane and other aryl isoprenoids identified in this study have previously been reported as constituents of Devonian sediments (e.g. Behrens et al. 1998; Joachimski et al. 2001). Isorenieratane is specifically produced by green sulphur bacteria (Chlorobiaceae) which perform photosynthesis under strictly anoxic conditions in sulphidogenic water columns. Hence, the presence of isorenieratene or its transformation products in sediments is assumed to indicate photic zone anoxia at the time of deposition. Considering the presence of aryl isoprenoids in all investigated samples, we conclude that photic zone anoxia played a role during the entire deposition of the investigated core section.

Schwark & Frimmel (2004) recently suggested a crossplot of the pristane/phytane ratio (Pr/Ph) vs. the newly defined aryl isoprenoid ratio (AIR) in order to better assess the extent and persistence of photic zone anoxia during deposition of the Jurassic Posidonia Shale. An adapted version of this crossplot suitable for our sample set is shown in Figure 13. In general, a positive correlation of pristane/phytane and AIR is observed for the samples from Büdesheimer Bach borehole. All samples from the 'LKW' and UKW plot in the grey area representing the original plot of Schwark & Frimmel (2004). These results indicate that photic zone anoxia may have played a more prominent role during deposition of the 'LKW' and UKW as compared to the adjacent units. Figure 13 further suggests that the organic matter signature in the samples directly below and above the 'LKW' is similar to that within the 'LKW', i.e. environmental conditions leading to this specific organic matter signature may have changed earlier and lasted longer than suggested by the excursions of TOC and $\delta^{13}C_{org}$. Photic zone anoxia has also been documented for the Permian–Triassic extinction event (Grice et al. 2005). This anoxic event coincides with facies changes, which reflect rapid transgression. Grice et al. (2005) suggest that sulphide toxicity was a driving factor of the extinction and in the protracted recovery. Considering the sterane/hopane ratio, we also assume transgressive trends for unit UKW.

Conclusions

1. Based on various parameters, a low thermal maturity corresponding to a vitrinite reflectance of approximately 0.5% R_r has been documented for the sediments encountered in the Büdesheimer Bach borehole. At such a low thermal maturity, significant hydrocarbon generation did not take place. The biomarker maturity parameters revealed a consistently low maturity that excludes misinterpretations due to migrated bitumen.

2. Generally, the palynofacies is characterized by a high amount of amorphous organic matter (AOM) and prasinophytes, an impoverished acritarch assemblage and small miospores. Large thick-walled miospores and tetrads are rare. Such a palynofacies is characteristic for a fully marine, rather distal environment and oxygen-deficient conditions. Higher abundances of miospores in some levels of the Oos Plattenkalk Formation may have been transported into the basin by turbidites. The kerogen dominated by AOM and a low sterane/hopane ratio imply a significant contribution of bacteria, presumably cyanobacteria, to the overall organic matter and, in addition, an influence of bacterial reworking/oxidation on the organic matter composition. Increased hopane abundances representing enhanced bacterial input have been reported by Joachimski et al. (2001) in two of five samples from the Frasnian/Famennian boundary beds in the Kowala section (Holy Cross Mountains, Poland) while in the Büdesheimer Bach borehole hopanes are abundant in the entire sequence and in particular in the interval between the 'Lower' and Upper Kellwasser Horizon.

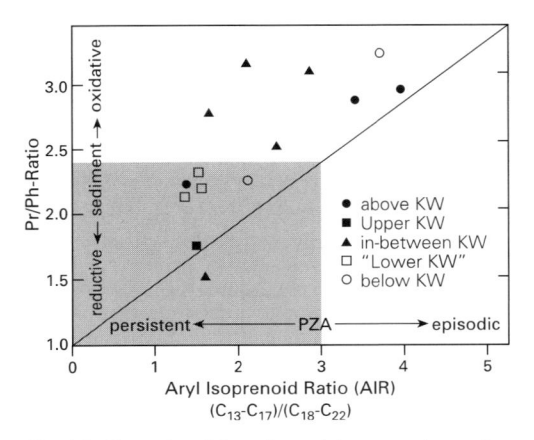

Fig. 13. Crossplot of the pristane/phytane ratio vs. the aryl isoprenoid ratio (AIR, see text for details); the shaded area corresponds to the original plot by Schwark & Frimmel (2004). KW, Kellwasser Horizon.

3. In the Büdesheimer Bach borehole, the Lower Kellwasser Horizon does not show the typical lithological features whereas the Upper Kellwasser Horizon is represented by a characteristic dark grey to black, highly bituminous and fossiliferous limestone 0.35 m in thickness. Despite the fact that the Lower Kellwasser Horizon is developed as a detrital limestone-dominated sequence, the abundance of prasinophytes, TOC and HI values all increase. The same is true for $\delta^{13}C_{org}$ which shows a pronounced positive excursion ($\delta^{13}C_{org} > -28‰$) (Joachimski et al. 2002). Likewise, samples from the Upper Kellwasser Horizon are dominated by prasinophytes including a layer representing a prasinophyte bloom. With respect to the organic geochemical parameters mentioned above, this horizon depicts the same patterns as in the 'Lower Kellwasser Horizon' of the Büdesheimer Bach borehole. It is interesting to note that samples from directly below and above the Upper Kellwasser Horizon are more similar to those from this horizon than to the adjacent units. Also, the $\delta^{13}C_{org}$ excursion covers a longer time interval than the Upper Kellwasser Horizon itself. Obviously, changes in the molecular signature and to some extant changes in the palynomorph composition started before the onset and persisted beyond the end of the deposition of the Upper Kellwasser Horizon. Hence, the environmental conditions resulting in this unique horizon preceded and continued to prevail after deposition of the Upper Kellwasser Horizon, indicating that the Upper Kellwasser Event evolved and gradually came to an end. This is corroborated by the stepwise extinction of some faunal groups (e.g. Schindler 1993; Sandberg et al. 2002).

4. Although the 'Lower' and Upper Kellwasser Horizons are distinguished from the sediments below and above with respect to the geochemical parameters, the differences are not outstandingly significant taking into account that the Kellwasser Horizons are characteristic global marker beds and that the end-Frasnian mass extinction was one of the most devastating biocrises in the Phanerozoic. Changes in the palynomorph composition are even less significant. The palynomorph composition did not change drastically during the latest Frasnian/earliest Famennian as already reported before from sections in Belgium and Poland (e.g. Streel et al. 2000; Filipiak 2002). Changes in the organic matter composition during deposition of the 'Lower' and Upper Kellwasser Horizon were either rather negligible or could not be identified by means of the palynological and geochemical methods applied in this study.

5. The unit in-between the 'Lower' and Upper Kellwasser Horizon differs from all other units by the rather high abundance of miospores in the palynomorph assemblage, the dominance of type III kerogen, and a decrease in the steranes/hopanes and hopane/(hopane + moretane) ratios. These parameters all indicate an increased input of bacteria and terrestrial plant material into the organic matter which is due to the regression that followed the deposition of the Lower Kellwasser Horizon. Just below the Upper Kellwasser Horizon, the decreasing sterane/hopane ratio reflects the start of a transgression. Using microfacies evolution in the Steinbruch Schmidt section (Kellerwald, Germany), Devleeschouwer et al. (2002) showed that the transgressions started below the Kellwasser Horizons which therefore correspond to a eustatic highstand immediately followed by a regressive trend. The sea-level fall above the Upper Kellwasser Horizon (e.g. Devleeschouwer et al. 2002) or that starting within the upper part of this horizon (e.g. Sandberg et al. 2002) is much more severe than the one following the Lower Kellwasser Horizon. For instance, only 130 km north of the Prüm Syncline, Piecha (2002) recognized a rapid regression on the Rhenish shelf carbonate platform which resulted in a hiatus spanning the complete *triangularis* conodont Zone. However, in the Büdesheimer Bach borehole, the sequence above the Upper Kellwasser Horizon does not show an increased miospore content or a change in geochemical parameters indicative for a regression. Therefore, the major eustatic sea-level fall at the base of the Famennian cannot be proved by palynofacies or geochemistry in the Büdesheimer Bach borehole.

6. Based on geochemical evidence and palaeontological data, eutrophication and an increase of bioproductivity have been postulated for the Kellwasser Horizons (e.g. Becker & House 1994; Paris et al. 1996; Joachimski et al. 2001). Prasinophyte blooms, stimulated by increased nutrient fluxes, can be indicative for eutrophication. Hence, one would expect that prasinophytes are extremely abundant in the Kellwasser Horizons. Undoubtedly, Frasnian/Famennian boundary beds at La Serre (Montagne Noire, France) and in the Holy Cross Mountains (Poland) represent episodes of enhanced bioproductivity as evidenced by exceptionally high abundances of tasmanites at La Serre (Paris et al. 1996) and leiospheres in the Holy Cross Mountains (Filipiak 2002). However, in the Holy Cross Mountains the changes in palynofacies are not very drastic in the Frasnian/Famennian boundary beds. In the Büdesheimer Bach borehole, the 'Lower' and Upper Kellwasser Horizon are distinguished from the units below and above with regard to the increased concentration of prasinophytes, but here again the differences are not very pronounced and the overall concentration of prasinophytes is lower than in the Holy Cross Mountains. As for the

prasinophytes, biomarkers which are indicative for algae, such as the C_{27} steranes, did not significantly increase in the 'Lower' and Upper Kellwasser Horizon of the Büdesheimer Bach borehole. This may be due to a more nearshore position as the area of the present Prüm Syncline was located in a narrow channel less than 50 km wide which separated the Ardennes Shoal from the Rhenish Shoal. Therefore, input of terrestrial debris diluting the marine component of the organic matter was higher in the Prüm Syncline than in the Holy Cross Mountains and in La Serre, which were both situated much farther away from land.

7. As summarized above, the widespread anoxic event correlating with the Lower and Upper Kellwasser Horizon seems to have been crucial for the end-Frasnian faunal demise. In the sequence encountered in the Büdesheimer Bach borehole, aryl isoprenoids are present in all samples. These specific compounds are markers for anoxygenic photosynthetic green sulphur bacteria and hence indicate that photic zone anoxia has played a role at least episodically. This is corroborated by sedimentological and biofacies analyses which suggest that the olive grey to dark grey calcareous mudstones to marlstones of the Oos Plattenkalk and Neu–Oos Cyprideinschiefer Formations have been deposited under dysoxic to anoxic conditions (Clausen 1968; Grimm 1995). Although it is difficult to assess the extent and persistence of anoxic conditions, there is evidence that they were more pronounced during deposition of the 'Lower' and Upper Kellwasser Horizon in comparison to the rest of the borehole section. Increased concentration of prasinophytes in these two horizons supports this view. In agreement with findings for another section recording the Frasnian/Famennian boundary (Joachimski et al. 2001) and sections recording the Permian/Triassic boundary (Grice et al. 2005), we therefore conclude that anoxic conditions were an important trigger of the end-Frasnian mass extinction.

U. Amend, J. S. Becker, W. Benders, U. Disko, F. Leistner, W. Lüdtke, U. Lux, U. Mittler, U. Niesel-Tirtey, K. Nogai, C. Pickhardt, A. Richter, J. Schardinel, H. Willsch (all Forschungszentrum Jülich and Geologischer Dienst NRW) are thanked for technical assistance. E. Schindler (Forschungsinstitut Senckenberg) provided access to the cores of the Büdesheimer Bach borehole. The manuscript benefited from reviews by Kliti Grice (Curtin University of Technology) and Hans Kerp (Westfälische Wilhelms-Universität Münster). John E. A. Marshall was kind enough to improve the English. The investigations have been carried out with grants from the Deutsche Forschungsgemeinschaft (NE 303/4-2 and MA 1861/3-2) as part of the Priority Programme 1054.

References

ALGEO, T. J., BERNER, R. A., MAYNARD, J. B. & SCHECKLER, S. E. 1995. Late Devonian oceanic anoxic events and biotic crises: 'rooted' in the evolution of vascular land plants? *Geological Society of America Today*, **5**, 63–66.

ALGEO, T. J., SCHECKLER, S. E. & MAYNARD, J. B. 2001. Effects of the Middle to Late Devonian spread of vascular land plants on weathering regimes, marine biotas, and global climate. *In*: GENSEL, P. G. & EDWARDS, D. (eds) *Plants Invade the Land: Evolutionary and Environmental Perspectives*. Columbia University Press, New York, 213–236.

BATTEN, D. J. 1996. Palynofacies and palaeoenvironmental interpretation. *In*: JANSONIUS, J. & MCGREGOR, D. C. (eds) *Palynology: Principles and Applications*, **3**. AASP Foundation, Dallas, TX, 1011–1064.

BECKER, G., BLESS, M. J. M., STREEL, M. & THOREZ, J. 1974. Palynology and ostracode distribution in the Upper Devonian and basal Dinantian of Belgium and their dependence on sedimentary facies. *Mededelingen Rijks geologische Dienst, Nieuwe Serie*, **25**, 9–99.

BECKER, R. T. 1993. Anoxia, eustatic changes, and Upper Devonian to lowermost Carboniferous global ammonoid diversity. *In*: HOUSE, M. R. (ed.) The Ammonoidea: environment, ecology, and evolutionary change. *Systematics Association Special Volume*, **47**, 115–164.

BECKER, R. T. & HOUSE, M. R. 1994. Kellwasser Events and goniatite successions in the Devonian of the Montagne Noire with comments on possible causations. *Courier Forschungsinstitut Senckenberg*, **169**, 45–77.

BECKER, R. T. & HOUSE, M. R. 2000. Devonian ammonoid zones and their correlation with established series and stage boundaries. *Courier Forschungsinstitut Senckenberg*, **220**, 113–151.

BEHRENS, A., WILKES, H., SCHAEFFER, P., CLEGG, H. & ALBRECHT, P. 1998. Molecular characterization of organic matter in sediments from the Keg River formation (Elk Point group), western Canada sedimentary basin. *Organic Geochemistry*, **29**, 1905–1920.

BOND, D., WIGNALL, P. B. & RACKI, G. 2004. Extent and duration of marine anoxia during the Frasnian–Famennian (Late Devonian) mass extinction in Poland, Germany, Austria and France. *Geological Magazine*, **141**, 173–193.

CLAUSEN, C.-D. 1968. Das Nehden in der Büdesheimer Teilmulde (Prümer Mulde/Eifel). *Fortschritte in der Geologie von Rheinland und Westfalen*, **16**, 205–232.

CLAUSEN, C.-D. 1969. Oberdevonische Cephalopoden aus dem Rheinischen Schiefergebirge. II. Gephuroceratidae, Beloceratidae. *Palaeontographica*, **A132**, 95–178.

COLBATH, G. K. 1988. Taphonomy of recent polychaete jaws from Florida and Belize. *Micropaleontology*, **34**, 83–89.

COPPER, P. 2002. Reef development at the Frasnian/Famennian mass extinction boundary. *Palaeogeography, Palaeoclimatology, Palaeoecology*, **181**, 27–65.

COURTINAT, B. 1998. New genera and new species of scolecodonts (fossil annelids) with paleoenvironmental and evolutionary considerations. *Micropaleontology*, **44**, 435–440.

DEVLEESCHOUWER, X., HERBOSCH, A. & PRÉAT, A. 2002. Microfacies, sequence stratigraphy and clay mineralogy of a condensed deep-water section around the Frasnian/Famennian boundary (Steinbruch Schmidt, Germany). *Palaeogeography, Palaeoclimatology, Palaeoecology*, **181**, 171–193.

ESPITALIÉ, J. & MARQUIS, F. 1985. La pyrolyse rockeval et ses applications, deuxième partie. *Revue de l'Institut français du Pétrole*, **40**, 755–784.

ESPITALIÉ, J., LAPORTE, J. L., MADEC, M., MARQUIS, F., LEPLAT, P., PAULET, J. & BOUTEFEU, A. 1977. Méthode rapide de caractérisation des roches mères, de leur potentiel pétrolier et de leur degré d'évolution. *Revue de l'Institut français du Pétrole*, **32**, 23–42.

FEIST, R. & SCHINDLER, E. 1994. Trilobites during the Frasnian Kellwasser Crisis in European Late Devonian cephalopod limestones. *Courier Forschungsinstitut Senckenberg*, **169**, 195–223.

FILIPIAK, P. 2002. Palynofacies around the Frasnian/Famennian boundary in the Holy Cross Mountains, southern Poland. *Palaeogeography, Palaeoclimatology, Palaeoecology*, **181**, 313–324.

FRIMMEL, A., OSCHMANN, W. & SCHWARK, L. 2004. Chemostratigraphy of the Posidonia Black Shale, SW Germany: I. Influence of sea-level variation on organic facies evolution. *Chemical Geology*, **206**, 199–230.

GODDÉRIS, Y. & JOACHIMSKI, M. M. 2004. Global change in the Late Devonian: modelling the Frasnian–Famennian short-term carbon isotope excursions. *Palaeogeography, Palaeoclimatology, Palaeoecology*, **202**, 309–329.

GRICE, K., CAO, C. ET AL. 2005. Photic zone euxinia during the Permian–Triassic superanoxic event. *Science*, **307**, 706–709.

GRIMM, M. C. 1991. Sedimentologie, Geochemie, Paläontologie, Biostratigraphie und Biofazies des Oos-Plattenkalks (Oberdevon/Büdesheimer Mulde/Eifel). Diplomarbeit Universität Mainz.

GRIMM, M. C. 1995. Sedimentologie und Fazies des Oos-Plattenkalks (Frasnium, Büdesheimer Mulde, Eifel, Deutschland). *Courier Forschungsinstitut Senckenberg*, **188**, 59–97.

HARTKOPF-FRÖDER, C. 2004. Palynostratigraphy of upper Famennian sediments from the Refrath 1 Borehole (Bergisch Gladbach–Paffrath Syncline; Ardennes–Rhenish Massif, Germany). *Courier Forschungsinstitut Senckenberg*, **251**, 77–87.

HARTKOPF-FRÖDER, C., MARSHALL, J. E. A. & VIETH, A. 2004. Organic maturity (vitrinite reflectance, quantitative spore colour) of upper Famennian sediments from the Refrath 1 Borehole (Bergisch Gladbach–Paffrath Syncline; Ardennes–Rhenish Massif, Germany). *Courier Forschungsinstitut Senckenberg*, **251**, 63–75.

HATCH, J. R. & LEVENTHAL, J. S. 1992. Relationship between inferred redox potential of the depositional environment and geochemistry of the Upper Pennsylvanian (Missourian) Stark Shale Member of the Dennis Limestone, Wabaunsee County, Kansas, USA. *Chemical Geology*, **99**, 65–82.

HELSEN, S. & KÖNIGSHOF, P. 1994. Conodont thermal alteration patterns in Palaeozoic rocks from Belgium, northern France and western Germany. *Geological Magazine*, **131**, 369–386.

HINTS, O. 1999. Ordovician scolecodonts of the East Baltic and surrounding areas—an overview. *Acta Universitatis carolinae, Geologica*, **43**, 317–320.

HOUSE, M. R. 2002. Strength, timing, setting and cause of mid-Palaeozoic extinctions. *Palaeogeography, Palaeoclimatology, Palaeoecology*, **181**, 5–25.

HUANG, W. Y. & MEINSCHEIN, W. G. 1979. Sterols as ecological indicators. *Geochimica et Cosmochimica Acta*, **43**, 739–745.

ISAKSEN, G. H. & BOHACS, K. M. 1995. Geological controls of source rock geochemistry through relative sea level; Triassic, Barents Sea. *In*: KATZ, B. J. (ed.) *Petroleum Source Rocks*. Springer, New York, 25–50.

JOACHIMSKI, M. M. & BUGGISCH, W. 1993. Anoxic events in the late Frasnian—causes of the Frasnian-Famennian faunal crisis? *Geology*, **21**, 675–678.

JOACHIMSKI, M. M., OSTERTAG-HENNING, C. ET AL. 2001. Water column anoxia, enhanced productivity and concomitant changes in $\delta^{13}C$ and $\delta^{34}S$ across the Frasnian–Famennian boundary (Kowala—Holy Cross Mountains/Poland). *Chemical Geology*, **175**, 109–131.

JOACHIMSKI, M. M., PANCOST, R. D., FREEMAN, K. H., OSTERTAG-HENNING, C. & BUGGISCH, W. 2002. Carbon isotope geochemistry of the Frasnian–Famennian transition. *Palaeogeography, Palaeoclimatology, Palaeoecology*, **181**, 91–109.

JOACHIMSKI, M. M., VAN GELDERN, R., BREISIG, S., BUGGISCH, W. & DAY, J. 2004. Oxygen isotope evolution of biogenic calcite and apatite during the Middle and Late Devonian. *International Journal of Earth Sciences*, **93**, 542–553.

JONES, B. & MANNING, D. A. C. 1994. Comparison of geochemical indices used for the interpretation of palaeoredox conditions in ancient mudstones. *Chemical Geology*, **111**, 111–129.

JUX, U. 1984. Observations on Upper Devonian acritarch communities from the southern Bergisches Land (Rhenish Massive). *Journal of Micropalaeontology*, **3**, 35–40.

KAUFMANN, B., TRAPP, E. & MEZGER, K. 2004. The numerical age of the Upper Frasnian (Upper Devonian) Kellwasser Horizons: a new U–Pb zircon date from Steinbruch Schmidt (Kellerwald, Germany). *Journal of Geology*, **112**, 495–501.

KÖNIGSHOF, P. & WERNER, R. 1994. Zur Bestimmung der Versenkungstemperaturen im Devon der Eifeler Kalkmulden-Zone mit Hilfe der Conodontenfarbe. *Courier Forschungsinstitut Senckenberg*, **168**, 255–265.

LE HÉRISSÉ, A., SERVAIS, T. & WICANDER, R. 2000. Devonian acritarchs and related forms. *Courier Forschungsinstitut Senckenberg*, **220**, 195–205.

LOBOZIAK, S., MELO, J. H. G. & STREEL, M. 2005. Devonian palynostratigraphy in Western Gondwana. *In*: KOUTSOUKOS, E. A. M. (ed.) *Applied Stratigraphy*. Springer, Dordrecht, 73–99.

LOEBLICH, JR., A. R. 1969. Morphology, ultrastructure and distribution of Paleozoic acritarchs. *Proceedings of the North American Paleontological Convention*, Part G, 705–788.

LONG, J. A. 1993. Early–Middle Palaeozoic vertebrate extinction events. *In*: LONG, J. A. (ed.) *Palaeozoic vertebrate Biostratigraphy and Biogeography*. Belhaven Press, London, 54–63.

MARTIN, F. 1984. Acritarches du Frasnien Supérieur et du Famennien Inférieur du bord méridional du Bassin de Dinant (Ardenne belge). *Bulletin de l'Institut royal des Sciences naturelles de Belgique, Sciences de la Terre*, **55**, 1–57.

MCGHEE, JR., G. R. 1996. *The Late Devonian Mass Extinction: The Frasnian/Famennian Crisis*. Columbia University Press, New York.

MENDELSON, C. V. 1987. Acritarchs. *In*: LIPPS, J. H. (ed.) Fossil prokaryotes and protists. University of Tennessee, *Studies in Geology*, **18**, 62–86.

MEYER, W. 1994. Geologie der Eifel. Schweizerbart, Stuttgart.

MILLER, M. A. 1996. Chitinozoa. *In*: JANSONIUS, J. & MCGREGOR, D. C. (eds) *Palynology: Principles and Applications*, **1**. AASP Foundation, Dallas, TX, 307–336.

MOLDOWAN, J. M., SUNDARARAMAN, P. & SCHOELL, M. 1986. Sensitivity of biomarker properties to depositional environment and/or source input in the Lower Toarcian of S.W. Germany. *Organic Geochemistry*, **10**, 915–926.

MONTENARI, M. & LEPPIG, U. 2003. Die Acritarcha: ihre Klassifikation, Morphologie, Ultrastruktur und paläoökologische/paläogeographische Verbreitung. *Paläontologische Zeitschrift*, **77**, 173–194.

MORGAN, J. P., RESTON, T. J. & RANERO, C. R. 2004. Contemporaneous mass extinctions, continental flood basalts, and 'impact signals': are mantle plume-induced lithospheric gas explosions the causal link? *Earth and Planetary Science Letters*, **217**, 263–284.

OLEMPSKA, E. 2002. The Late Devonian Upper Kellwasser Event and entomozoacean ostracods in the Holy Cross Mountains, Poland. *Acta Palaeontologica Polonica*, **47**, 247–266.

ONCKEN, O., VON WINTERFELD, C. & DITTMAR, U. 1999. Accretion of a rifted passive margin: the Late Paleozoic Rhenohercynian fold and thrust belt (Middle European Variscides). *Tectonics*, **18**, 75–91.

PAPROTH, E. 1989. Devonian palaeogeographic development in Germany and adjacent areas. *DGMK-Berichte, Forschungsbericht*, **468**, 33–52.

PAPROTH, E. & STREEL, M. 1970. Corrélations biostratigraphiques près de la limite Dévonien/Carbonifère entre les faciès littoraux ardennais et les faciès bathyaux rhénans. *In*: STREEL, M. & WAGNER, R. H. (eds) *Colloque sur la stratigraphie du Carbonifère*. Université de Liège, Liège, 365–398.

PAPROTH, E., DREESEN, R. & THOREZ, J. 1986. Famennian paleogeography and event stratigraphy of northwestern Europe. *Annales de la Société géologique de Belgique*, **109**, 175–186.

PARIS, F. & NÕLVAK, J. 1999. Biological interpretation and paleobiodiversity of a cryptic fossil group: the 'chitinozoan animal'. *Geobios*, **32**, 315–324.

PARIS, F., GIRARD, C., FEIST, R. & WINCHESTER-SEETO, T. 1996. Chitinozoan bio-event in the Frasnian-Famennian boundary beds at La Serre (Montagne Noire, Southern France). *Palaeogeography, Palaeoclimatology, Palaeoecology*, **121**, 131–145.

PETERS, K. E., WALTERS, C. C. & MOLDOWAN, J. M. 2005. *The Biomarker Guide*. **Vol. 1** and **2**. Cambridge University Press, Cambridge.

PIECHA, M. 1993. Stratigraphie, Fazies und Sedimentpetrographie der rhythmisch und zyklisch abgelagerten, tiefoberdevonischen Beckensedimente im Rechtsrheinischen Schiefergebirge (Adorf-Bänderschiefer). *Courier Forschungsinstitut Senckenberg*, **163**, 1–151.

PIECHA, M. 1994, with a contribution by Braun, A. Untersuchungen am Kernmaterial der Bohrung 'Büdesheimer Bach' (Kernbohrung 1) aus dem Oberdevon der Prümer Mulde (Eifel)–Stratigraphische, sedimentpetrographische und mikrofazielle Zusammenhänge. *Courier Forschungsinstitut Senckenberg*, **169**, 329–349.

PIECHA, M. 2002. A considerable hiatus at the Frasnian/Famennian boundary in the Rhenish shelf region of northwest Germany. *Palaeogeography, Palaeoclimatology, Palaeoecology*, **181**, 195–211.

POWELL, T. G. & MCKIRDY, D. M. 1973. Relationship between ratio of pristane to phytane, crude oil composition and geological environments in Australia. *Nature Physical Science*, **243**, 37–39.

RACKI, G. 1999. The Frasnian-Famennian biotic crisis: How many (if any) bolide impacts? *Geologische Rundschau*, **87**, 617–632.

RACKI, G. 2005. Toward understanding Late Devonian global events: few answers, many questions. *In*: OVER, D. J., MORROW, J. R. & WIGNALL, P. B. (eds) *Understanding Late Devonian and Permian–Triassic Biotic and Climatic Events: Towards an Integrated Approach*. Elsevier, Amsterdam, 5–36.

RADKE, M., SITTARDT, H. G. & WELTE, D. H. 1978. Removal of soluble organic matter from rock samples with a flow-through extraction cell. *Analytical Chemistry*, **50**, 663–665.

RADKE, M., WILLSCH, H. & WELTE, D. H. 1980. Preparative hydrocarbon group type determination by automated medium pressure liquid chromatography. *Analytical Chemistry*, **52**, 406–411.

RIEGEL, W. 1993. Die geologische Bedeutung der Prasinophyten im Paläozoikum. *Göttinger Arbeiten zur Geologie und Paläontologie*, **58**, 39–50.

RIMMER, S. M. 2004. Geochemical paleoredox indicators in Devonian–Mississippian black shales, Central Appalachian Basin (USA). *Chemical Geology*, **206**, 373–391.

RODRIGUES, R., LOBOZIAK, S., DE MELO, J. H. G. & ALVES, D. B. 1995. Geochemical characterization and miospore biochronostratigraphy of the Frasnian anoxic event in the Parnaíba Basin, Northeast Brazil. *Bulletin des Centres de Recherches Exploration-Production Elf Aquitaine*, **19**, 319–327.

RULLKÖTTER, J. & MARZI, R. 1988. Natural and artificial maturation of biological markers in a Toarcian shale from northern Germany. *Organic Geochemistry*, **13**, 639–645.

SANDBERG, C. A., MORROW, J. R. & ZIEGLER, W. 2002. Late Devonian sea-level changes, catastrophic events, and mass extinctions. *Geological Society of America Special Paper*, **356**, 473–487.

SCHINDLER, E. 1993. Event-stratigraphic markers within the Kellwasser Crisis near the Frasnian/Famennian boundary (Upper Devonian) in Germany. *Palaeogeography, Palaeoclimatology, Palaeoecology*, **104**, 115–125.

SCHÜLKE, I. 1998. Conodont community structure around the 'Kellwasser mass extinction event' (Frasnian/Famennian boundary interval). *Senckenbergiana lethaea*, **77**, 87–99.

SCHWARK, L. & FRIMMEL, A. 2004. Chemostratigraphy of the Posidonia Black Shale, SW-Germany: II. Assessment of extent and persistence of photic-zone anoxia using aryl isoprenoid distributions. *Chemical Geology*, **206**, 231–248.

SEPKOSKI JR, J. J. 1996. Patterns of Phanerozoic extinction: a perspective from global data bases. *In*: WALLISER, O. H. (ed.) *Global Events and Event Stratigraphy in the Phanerozoic*. Springer, Berlin, 35–51.

STREEL, M. 1999. Quantitative palynology of Famennian events in the Ardenne–Rhine regions. *Abhandlungen der Geologischen Bundesanstalt*, **54**, 201–212.

STREEL, M., CAPUTO, M. V., LOBOZIAK, S. & MELO, J. H. G. 2000. Late Frasnian–Famennian climates based on palynomorph analyses and the question of the Late Devonian glaciations. *Earth-Science Reviews*, **52**, 121–173.

STROTHER, P. K. 1996. Acritarchs. *In*: JANSONIUS, J. & MCGREGOR, D. C. (eds) *Palynology: Principles and Applications*, **1**. AASP Foundation, Dallas, TX, 81–106.

SZANIAWSKI, H. 1996. Scolecodonts. *In*: JANSONIUS, J. & MCGREGOR, D. C. (eds) *Palynology: Principles and Applications*, **1**, AASP Foundation, Dallas, TX, 337–354.

TAPPAN, H. 1968. Primary production, isotopes, extinctions and the atmosphere. *Palaeogeography, Palaeoclimatology, Palaeoecology*, **4**, 187–210.

TAPPAN, H. 1986. Phytoplankton: Below the salt at the global table. *Journal of Paleontology*, **60**, 545–554.

TAPPAN, H. & LOEBLICH JR, A. R. 1972. Fluctuating rates of protistan evolution, diversification and extinction. 24. International Geological Congress, Section 7, Paleontology, 205–213.

TEICHMÜLLER, M. & TEICHMÜLLER, R. 1979. Ein Inkohlungsprofil entlang der linksrheinischen Geotraverse von Schleiden nach Aachen und die Inkohlung in der Nord–Süd-Zone der Eifel. *Fortschritte in der Geologie von Rheinland und Westfalen*, **27**, 323–355.

TRIBOVILLARD, N., AVERBUCH, O., DEVLEESCHOUWER, X., RACKI, G. & RIBOULLEAU, A. 2004. Deep-water anoxia over the Frasnian–Famennian boundary (La Serre, France): a tectonically induced oceanic anoxic event? *Terra Nova*, **16**, 288–295.

TYSON, R. V. 1995. *Sedimentary organic matter: organic facies and palynofacies*. Chapman & Hall, London.

YOU XING LI 2000. Famennian tentaculitids of China. *Journal of Paleontology*, **74**, 969–975.

WALLISER, O. H. 1996. Global events in the Devonian and Carboniferous. *In*: WALLISER, O. H. (ed.) *Global Events and Event Stratigraphy in the Phanerozoic*. Springer, Berlin, 225–250.

WICANDER, R. 2002. Acritarchs: Proterozoic and Paleozoic enigmatic organic-walled microfossils. *In*: HOOVER, R. B., LEVIN, G. V., PAEPE, R. R. & ROZANOV, A. Y. (eds) Instruments, methods, and missions for astrobiology, **4**. *Proceedings of the Society of Photo-optical Instrumentation Engineers*, **4495**, 331–340.

WILDER, H. 1989. Neue Ergebnisse zum oberdevonischen Riffsterben am Nordrand des mitteleuropäischen Variscikums. *Fortschritte in der Geologie von Rheinland und Westfalen*, **35**, 57–74.

WOLF, M. 1972. Beziehungen zwischen Inkohlung und Geotektonik im nördlichen Rheinischen Schiefergebirge. *Neues Jahrbuch für Geologie und Paläontologie Abhandlungen*, **141**, 222–257.

Environmental changes at the Frasnian–Famennian boundary in Central Morocco (Northern Gondwana): integrated rock-magnetic and geochemical studies

L. RIQUIER[1], O. AVERBUCH[1], N. TRIBOVILLARD[1], A. EL ALBANI[2], N. LAZREQ[3] & S. CHAKIRI[4]

[1]*UMR 8110, Processus et Bilans en Domaines Sédimentaires, Bâtiment SN5, Université de Lille 1, F-59655 Villeneuve d'Ascq cedex, France (e-mail: Laurent.Riquier@univ-lille1.fr)*

[2]*UMR 6532, Hydrogéologie, Argiles, Sols et Altérations, Bâtiment Sciences Naturelles, Université de Poitiers, 40 Avenue du recteur Pineau, F-86088 Poitiers cedex, France*

[3]*Université Cadi Ayyad de Marrakech, Faculté des Sciences Semlalia, Département de Géologie, Avenue du Prince Moulay Abdellah, BP 2390, 40001 Marrakech, Morocco*

[4]*Université Ibn Tofail, Faculté des Sciences, Laboratoire de Géologie dynamique et appliquée, BP 133, 14000 Kénitra, Morocco*

Abstract: Rock magnetic (magnetic susceptibility and hysteresis parameters) and geochemical analyses (major and trace elements) were carried out on whole rock samples of two Frasnian–Famennian boundary sections, Anajdam and Bou-Ounebdou in the Central Morocco (Western Meseta). During the Frasnian, the decreasing trend of the magnetic susceptibility signal, mainly carried by low-coercivity magnetite grains, indicates a gradual reduction of detrital influx. This decrease in detrital input parallels a Frasnian long-term sea-level rise. In the Late Frasnian Kellwasser Horizons, that are classically considered to represent highstand deposits, the magnetic signal exhibits the lowest intensities in connection with maximum diamagnetic contribution of the carbonate fraction. With respect to geochemical data, the two black carbonate-rich Kellwasser Horizons are characterized by noticeable positive anomalies of bottom-water dysoxic proxies and of marine primary productivity markers. Our data thus suggest that in Central Morocco, the Late Frasnian marine environments were marked by a relatively important biogenic productivity favouring the onset of oxygen-depleted conditions during periods of maximum transgression on the continental platforms.

The Frasnian–Famennian (F–F) boundary (Late Devonian, 374.5 Ma ago; Gradstein *et al.* 2004); was a major period of biodiversity loss and environmental changes. This boundary corresponds to one of the five biggest biological mass extinctions of the Phanerozoic, in which many species of marine organisms died out (e.g. corals, stromatoporoids, trilobites, conodonts, cephalopods) (Sepkoski 1982, 1986; Copper 1986; Sandberg *et al.* 1988; Schindler 1990; Becker & House 1994; McGhee 1996; Walliser 1996; Hallam & Wignall 1999; House 2002; Ma & Bai 2002; Racki *et al.* 2002). This period has been also suggested to be the time of significant climatic variations, attested by the $\delta^{18}O$ signature of marine conodonts apatite (Joachimski & Buggisch 2002; Joachimski *et al.* 2004) and the evolution of the miospore diversity on continents (Streel *et al.* 2000). Such climatic effects are likely to have induced global sea-level fluctuations that, combined with the changes in the erosional processes on land, potentially resulted in varying the intensity of the basinal detrital supply (e.g. Ellwood *et al.* 2000).

On the other hand, the Upper Frasnian interval is generally associated with the deposition of one or two organic-rich units in outer shelf and epicontinental basin settings, i.e. the Kellwasser (KW) Horizons (e.g. Buggisch & Clausen 1972; Sandberg *et al.* 1988; Schindler 1990; Buggisch 1991; Wendt & Belka 1991; Becker & House 1994). These beds have been recognized in many sections, located on the borders of Laurassian (N. America, N. Europe), Gondwanan (S. Europe, Africa), South China and on the southern border of the Siberian continents. Numerous factors controlling the KW organic-rich sediment accumulations have been proposed, such as increased primary productivity or bottom-water oxygen-depleted

From: BECKER, R. T. & KIRCHGASSER, W. T. (eds) *Devonian Events and Correlations.*
Geological Society, London, Special Publications, **278**, 197–217. DOI: 10.1144/SP278.9
0305-8719/07/$15 © The Geological Society of London 2007.

conditions (e.g. Buggisch 1991; Joachimski & Buggisch 1993; Becker & House 1994; Murphy et al. 2000a, b; Bond et al. 2004; Tribovillard et al. 2004a; Riquier et al. 2005). These factors have been connected to different driving mechanisms acting in isolation or combined, such as sea-level fluctuations, climatic variations, land plants spreading, volcanism or mountain building (e.g. Becker & House 1994; Algeo & Scheckler 1998; Racki 1998, 1999; Joachimski & Buggisch 2002; Godderis & Joachimski 2004; Averbuch et al. 2005).

The aims of this study are to present the sedimentary record of the Late Devonian events from two Moroccan sections, where the KW horizons are particularly well exposed. Rock magnetic data, combined with clay mineralogical studies and geochemical analyses of major compounds, will be discussed with respect to the nature and the intensity of the detrital supply versus carbonate productivity during the F–F boundary events. An additional extensive geochemical data set of inorganic elemental ratios is presented here to document marine oxygen (O_2) levels and productivity conditions during deposition of the KW horizons.

Geological setting

The Central Moroccan Meseta belongs to the Variscan orogenic belt of Northern Africa (Fig. 1a). In Devonian times, it represents a transitional zone between platform domains to the west and southwest (Western Meseta) and a deep turbiditic basin to the east and north (Eastern Meseta, Rif; e.g. El Hassani & Tahiri 2000a; Chakiri 2002). This strongly segmented continental margin, which developed at subtropical palaeolatitudes along the northern border of Gondwana, was sequentially dismembered by thrust movements from Late Devonian (the internal metamorphic zones of the Eastern Meseta) to Upper Carboniferous times (the external non-metamorphic Central and Western Meseta; Bouabdelli 1989; Piqué et al. 1993; Bouabdelli & Piqué 1996).

In more detail, the two investigated sections, Bou-Ounebdou and Anajdam, were deposited in the Azrou–Kenifra Basin, which forms the easternmost termination of the Moroccan Central Massif. It developed as a transtensional basin in Lower Devonian times and was inverted not earlier than in Upper Viséan times as the thrust front propagated towards the southwest (Bouabdelli 1989). Both Upper Devonian sections, located in the M'rirt area (Fig. 1b), were thus deposited in a relatively tectonically quiet passive margin-type environment.

The sections of Bou-Ounebdou and Anajdam were well-studied Late Devonian sections: during the last decade, detailed investigations on sedimentary sequences and stratigraphy have been done, particularly within the IGCP projects 421 on 'North Gondwanan Mid-Palaeozoic bioevent/ biogeography patterns in relation to crustal dynamics' (El Hassani & Tahiri 2000a) and by the Subcommission on Devonian Stratigraphy. The Anajdam and Bou-Ounebdou sections were described in detail by Lazreq (1992, 1999), Becker and House (2000), Walliser et al. (2000) and Chakiri (2002). The Bou-Ounebdou section has also been studied by Girard and Albarède (1996), Feist (2002) and Joachimski et al. (2002).

Bou-Ounebdou

The Bou-Ounebdou section, corresponding to the section M'rirt II in Walliser et al. (2000), is located in the M'rirt Nappe, about 5 km southeast of M'rirt (Fig. 1b). The sampled interval records condensed carbonate sedimentation from the Frasnian Lower *hassi* Zone to the Famennian Upper *triangularis* Zone (Fig. 2). The Late Devonian sequence is about 5.5 m thick and exposes a succession of cephalopod limestone beds, mudstones to wackestones in texture, with intercalated calcareous shale levels. The occurrence of pelagic and nekto-benthic faunas argues for a moderately deep outer platform setting (Becker 1993; Becker & House 2000; Chakiri 2002). Two black limestone levels, which are globally time-equivalent to the KW horizons of the Rhenohercynian Massif (Germany), have been identified (Lazreq 1992; Becker 1993). The Lower Kellwasser (LKW) horizon is about 25 cm thick and is observed at the base of the Upper *rhenana* Zone. The 20 cm thick Upper Kellwasser Horizon (UKW) consists of lenticular lenses of black limestones, alternating with black shales. The two KW horizons contain abundant nektonic and planktonic fauna, including orthoconic nautiloids and ammonoids, conodonts and tentaculitids. Previously, the base of the UKW horizon was assigned to the uppermost part of the Upper *rhenana* Zone whereas the end of the UKW horizon was regarded as in the *linguiformis* Zone (Lazreq 1992, 1999). New biostratigraphical data from Becker & House (2000) and Girard et al. (2005) refined this stratigraphical framework showing that the UKW horizon was only restricted to the *linguiformis* Zone (Fig. 2). On top of the UKW horizon, the F–F boundary is represented by a surface of discontinuity (Lazreq 1999; Becker & House 2000; Chakiri 2002). Based on micro- and biofacies analysis, this discontinuity has been interpreted as the result of a significant sea-level fall (Chakiri 2002). However, the

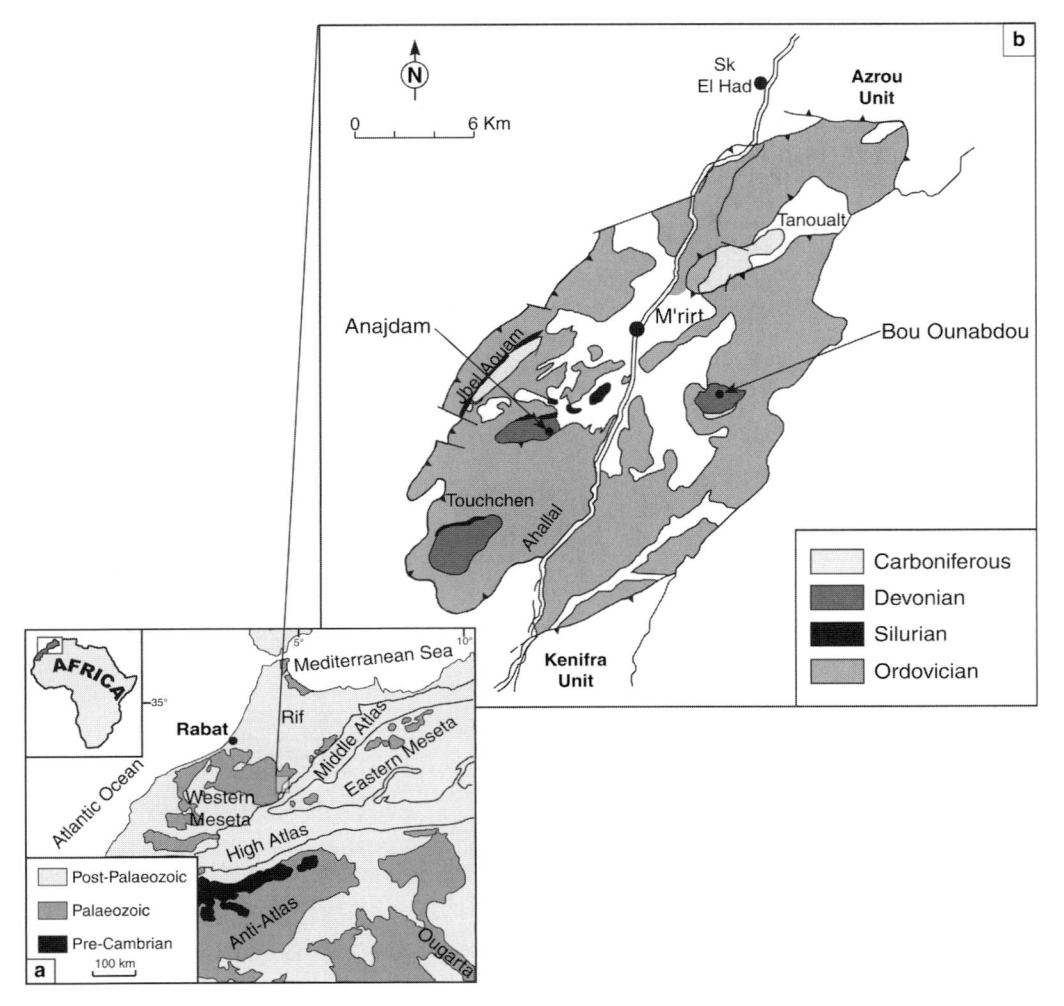

Fig. 1. (a) Main structural subdivision of middle and northern Morocco and (b) location of the studied sections in the M'rirt unit (modified from Bouabdelli 1989).

regression does not appear to have caused a zonal gap at the base of the Famennian (i.e. *triangularis* Zone), as illustrated by the correlation of the *Phoenixites* limestone (Lower *triangularis* Zone) observed between the M'rirt area and the Mont Peyroux Nappe (Montagne Noire) (Becker & House 2000). The sampled Lower Famennian deposits are characterized by 2 m thick, grey griotte limestones.

Anajdam

This section is located in a thrust-sheet developed in the footwall of the M'rirt nappe (Fig. 1b; Bouabdelli 1989). The exposed Upper Devonian succession is about 4.5 m thick and consists of bioclastic limestones with rare marly beds (Fig. 2). Limestone microfacies are represented mostly by packstones and wackestones. The investigated sequence begins within the *falsiovalis* Zone, at the Givetian–Frasnian boundary, and ranges up to the Lower *triangularis* Zone in basal Famennian times (Lazreq 1999). The faunal associations (conodonts, ammonoids, ostracodes and gastropods) characterize an external platform setting. In the Late Frasnian, two 20–45 cm thick dark-grey and black limestone horizons were biostratigraphically identified as the LKW horizon at the base of the Upper *rhenana* Zone, and the UKW horizon, within the *linguiformis* Zone (Fig. 2).

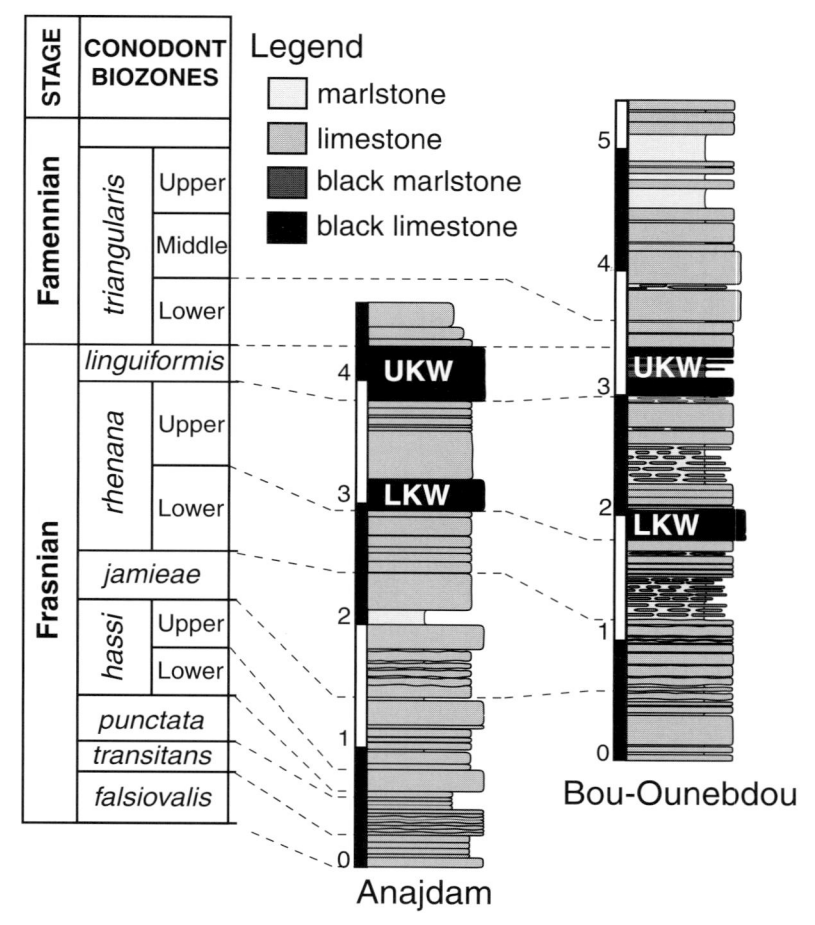

Fig. 2. Stratigraphical columns of the Frasnian–Famennian boundary beds in Anajdam and Bou-Ounebdou (biostratigraphical correlation based on data from Lazreq 1999; Becker & House 2000; Girard *et al.* 2005).

Material and analytical methods

Clay mineralogy

Clay mineral associations have been investigated by X-ray diffraction (XRD) using oriented mounts of the <2 μm size fraction containing non-calcareous particles. Deflocculation of clay aggregates was performed by successive washing with distilled water after decarbonatation of the crushed rock samples. The clay-sized fraction (<2 μm) was separated by settling and centrifugation and placed onto two glass slides and left dry in order to prepare oriented specimens. One of the air-dried, oriented clay-aggregate mounts was saturated in an ethylene glycol atmosphere at 20 °C overnight and another one was heated for two hours at 490 °C (Holtzapffel 1985). So, a series of three X-ray diffractograms was performed for each sample after air-drying, ethylene-glycol solvation and heating. X-ray diffractograms were obtained using a Philips PW 1729 diffractometer, with CuKα radiations and Ni filter, operating at a voltage of 40 KV and a current of 25 mA at a scanning rate of 1°2θ/min.

The identification of clay minerals was made according to the position of the (001) series of basal reflections on the three X-ray diffractograms (Brindley & Brown 1980). Semi-quantitative estimations of major clay minerals, as well as the illite crystallinity index, have been calculated using the Macdiff software (Petschick *et al.* 1996). Measured illite crystallinity index values were converted into the international calibrated scale CIS (Warr & Rice 1994) using a routine correlation process exposed in Robion *et al.* (1999).

The reproducibility of technical works and measurements was tested: the relative error is 5%.

Rock magnetism

The rock magnetic study integrated along-section magnetic susceptibility measurements coupled to representative measurements of magnetic hysteresis loops. The magnetic susceptibility quantifies the ability of the sedimentary rocks under study to be magnetized in a weak magnetic field, according to the respective concentrations of dia- (mainly calcite and quartz), para- (mainly illite, smectite, pyrite) and ferromagnetic *sensu lato* minerals (e.g. magnetite, goethite, haematite, pyrrhotite; see for example Walden *et al.* 1999 or Maher & Thompson 1999 for a more detailed description). In ancient sedimentary rocks, in the absence of intense diagenetic alteration, the magnetic susceptibility has been widely shown to provide a record of the bulk mineralogical changes induced by detrital inputs (magnetite, clays) and biogenic pelagic dilution (diamagnetic calcite; e.g. Crick *et al.* 1997; Devleeschouwer 1999; Ellwood *et al.* 2000; Tribovillard *et al.* 2002). The magnetic susceptibility data set includes 130 samples (70 measurements in Anajdam and 60 in Bou-Ounebdou) collected at 5 to 10 cm intervals, allowing high-resolution study of the magnetic susceptibility fluctuations. Low-field magnetic susceptibility was measured on small rock samples using a Kappabridge KLY-2 susceptibility bridge operating at an alternating-magnetic field of 300 A/m. Each sample was measured three times and a mean of these measurements is reported as the magnetic susceptibility value for that sample. Magnetic susceptibility values are normalized with respect to sample mass. Samples were weighted with a precision of 0.01 g from rock chips. So, results are reported here in mass-normalized magnetic susceptibility, hereafter χ ($m^3\,kg^{-1}$).

Hysteresis parameters were obtained from selected samples (15 measures for each section), using 10–30 mg chips, analysed at room temperature with an Alternating Gradient Force Magnetometer (AGFM 2900) at the Laboratoire des Sciences du Climat et de l'Environnement (Gif-sur-Yvette, France). By definition, hysteresis loops show the evolution of magnetization (M) of magnetic elements as a function of the applied magnetic field (H) (Fig. 3) (e.g. Borradaile *et al.* 1993; Walden *et al.* 1999). The sample is vibrated at room temperature in a magnetic field that is swept from 0 to +1 to −1 T then back to 0 T. From hystereris loop, four parameters are extracted: the saturation (isothermal) magnetization M_s ($\mu A.m^2$), the saturation (isothermal) remanent magnetization M_r ($\mu A.m^2$), the coercive force H_c (mT), and the coercivity of the remanence H_{cr} (mT) (Fig. 3). Values of M_s, M_r and H_c were determined for each hysteresis loop, after corrections for the high-field slope (SC: Slope Correction), which records the combined effects of paramagnetic and diamagnetic contributions. As for χ, magnetic parameters M_s, M_r and SC are normalized with respect to sample mass. H_{cr} was determined by stepwise application of a back-field isothermal remanence to remove the saturation remanence. The hysteresis parameters, described above, will be used here to discuss the origin of the magnetic susceptibility variations by characterizing (1) the relative contributions of the ferromagnetic versus para- or diamagnetic components and (2) the nature and amount of ferromagnetic minerals and especially of magnetite type, which can be generally used as a proxy of the detrital input (e.g. Borradaile *et al.* 1993; Vanderaveroet *et al.* 1999; Walden *et al.* 1999; Ellwood *et al.* 2000).

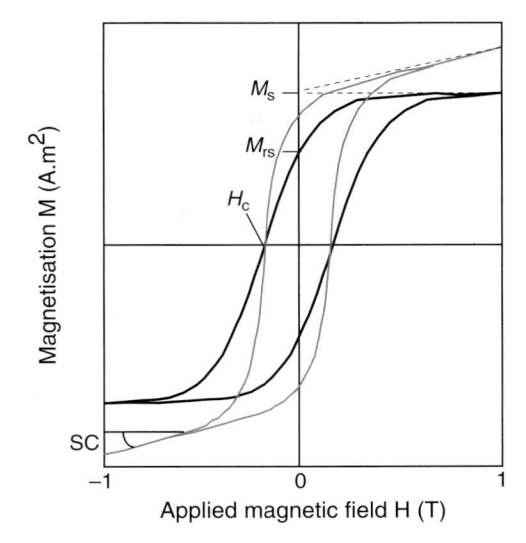

Fig. 3. Schematic hysteresis loop for a magnetite sample. The light grey loop is the original measurement, the black loop represents the same curve after corrections. The parameters saturation magnetization M_s, saturation remanent magnetization M_r, coercive force H_c, coercivity of remanence H_{cr} and slope correction SC are indicated.

Inorganic geochemistry

For each section, about 20 whole-rock geochemical analyses were made allowing the determination of major (Al, K, Si, Ti) and trace (Ba, Co, Cr, Cu, Mo, Ni, Th, U, V) elementary concentrations. Geochemical analyses were performed by ICP-AES (major or minor elements) and ICP-MS (trace elements), at the spectrochemical laboratory of the Service d'Analyse des Roches et des Minéraux of the Centre National de la Recherche Scientifique (Vandœuvre-les-Nancy, France). The samples were prepared by fusion with $LiBO_2$ and HNO_3 dissolution. Precision and accuracy were both found to be better than 1% (mean 0.5%) for major-minor elements, 5% for Co, Cr, Mo, U and V and 10% for Cu, as checked by international standards and analysis of replicate samples, respectively. Carbonate content was determined using a Bernard calcimeter (acid digestion).

The trace-element concentrations were normalized to aluminium to avoid dilution effects by the carbonate fraction and to compare different environmental settings with regard to variations in trace-element contents. Higher element/Al ratios will indicate synsedimentary additions, if a constant composition of the background is assumed. Variations of the Al content in relation to diagenetic formation of clay minerals is unlikely, because in the present study, Al is positively correlated with Ti ($r^2 = 0.99$ for both sections) that is not incorporated during authigenic clay-mineral formation (Calvert & Pedersen 1993).

Results

Clay mineral analysis

X-ray diffraction analyses show that the clay assemblages in both sections are very close and dominantly composed of illite (about 80%) and chlorite (about 10%) (Fig. 4). Diagenesis-sensitive minerals, such as kaolinite or smectite, are either absent or present in traces suggesting a significant diagenetic imprint on the primary detrital clay assemblages. The data of Crystallinity Index Standard (CIS) of illite oscillate globally from 0.25 to 0.50 with an average value of 0.37 for the Bou-Ounebdou section and 0.41 for the Anajdam section (Fig. 4). According to the reference data from Warr & Rice (1994), these analyses indicate that both sections have been affected by a pronounced diagenesis, in the limit of the anchimetamorphic conditions. As a whole, these results are, however, in agreement with recent studies of colour alteration index (CAI) of conodont, suggesting that the maximum temperatures during thrust-related burial in the Central Moroccan Meseta did not exceed about 250 °C (Dopieralska 2003).

Rock-magnetic measurements

Magnetic susceptibility signal. Figure 5 shows the stratigraphical variations of the magnetic susceptibility and the $CaCO_3$ content for the two sedimentary sequences studied. As usually recorded in carbonate formation, the χ signal is broadly low ($<20.0 \times 10^{-8}$ m^3 kg^{-1}) but displays significant evolutions along sections, with short- and long-term variations.

Along the Anajdam section, the χ values mostly vary from 2.6 to 15.3×10^{-8} m^3 kg^{-1} with the exception of the two KW horizons. In these peculiar horizons, the χ values never exceed 1.5×10^{-8} m^3 kg^{-1}. During the Frasnian, two successive slight and gradual decreasing trends, separated by a positive shift to values around 8.9×10^{-8} m^3 kg^{-1}, are observed in the χ signal (Fig. 5). The first decreasing trend occurs from the base of the section (*falsovialis* Zone) up to the early part of the *jamieae* Zone, the second from the middle part of the *jamieae* Zone up to the base of the LKW horizon (Upper *rhenana* Zone). Within the LKW horizon, the χ record shows a sharp decrease to average values around 1.3×10^{-8} m^3 kg^{-1}. The lowest χ value (0.4×10^{-8} m^3 kg^{-1}) is observed just above the LKW horizon. The inter-KW timespan is marked by a well-marked peak (up to the maximum χ value of 15.3×10^{-8} m^3 kg^{-1}). The base of the UKW horizon displays a strong decrease of χ that remains of low values (around 1.3×10^{-8} m^3 kg^{-1}) all along the UKW horizon. An increase of χ is recorded at the base of the Famennian times but its amplitude is difficult to resolve as sampling was not extended higher than the Lower *triangularis* Zone in basal Famennian times.

In the Bou-Ounebdou section, the χ values are about twice lower than those observed in the Anajdam section. They vary from 1.5 to 6.7×10^{-8} m^3 kg^{-1} with an interval of lowest values of 1.0×10^{-8} m^3 kg^{-1}, measured in the LKW horizon (Fig. 5). From the base of the section up to the LKW level, the data are noisy and no clear trend is observed in the χ signal. Above the base of the LKW horizon, a comparable trend to Anajdam is observed with, however, a significant decrease in the amplitude of variations. The UKW horizon is marked by a slight decrease of χ values, following by a slight increasing trend toward average χ values of 3.0×10^{-8} m^3 kg^{-1} during the early Famennian.

The susceptibility data thus allow evidencing a contrast in the sedimentary record of both Moroccan sections. Unlike the more distal Bou-Ounebdou section, the samples of Anajdam exhibit well-defined variations of χ signal with particularly well-marked KW horizons. The trends point to a general long-term decrease during the Frasnian with minimum values localised within the two KW horizons. It is worth noting that these horizons display some significant increases in the carbonate content (up to 95–98%) but, by contrast, the

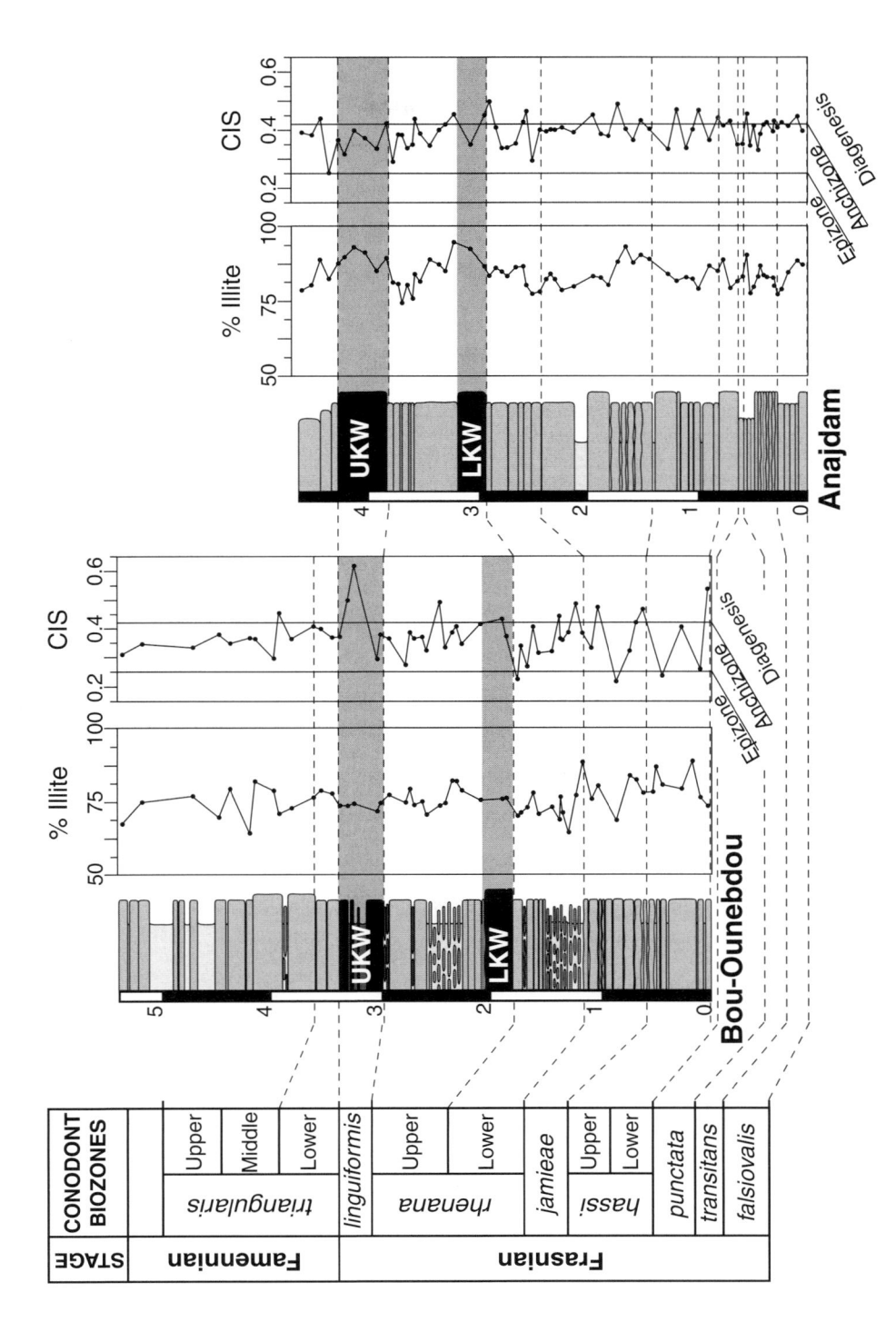

Fig. 4. Stratigraphical variation of percentage of illite and Crystallinity Index Standard of the sedimentary sequence from Bou-Ounebdou and Anajdam. Ranges for CIS are from Warr & Rice (1994).

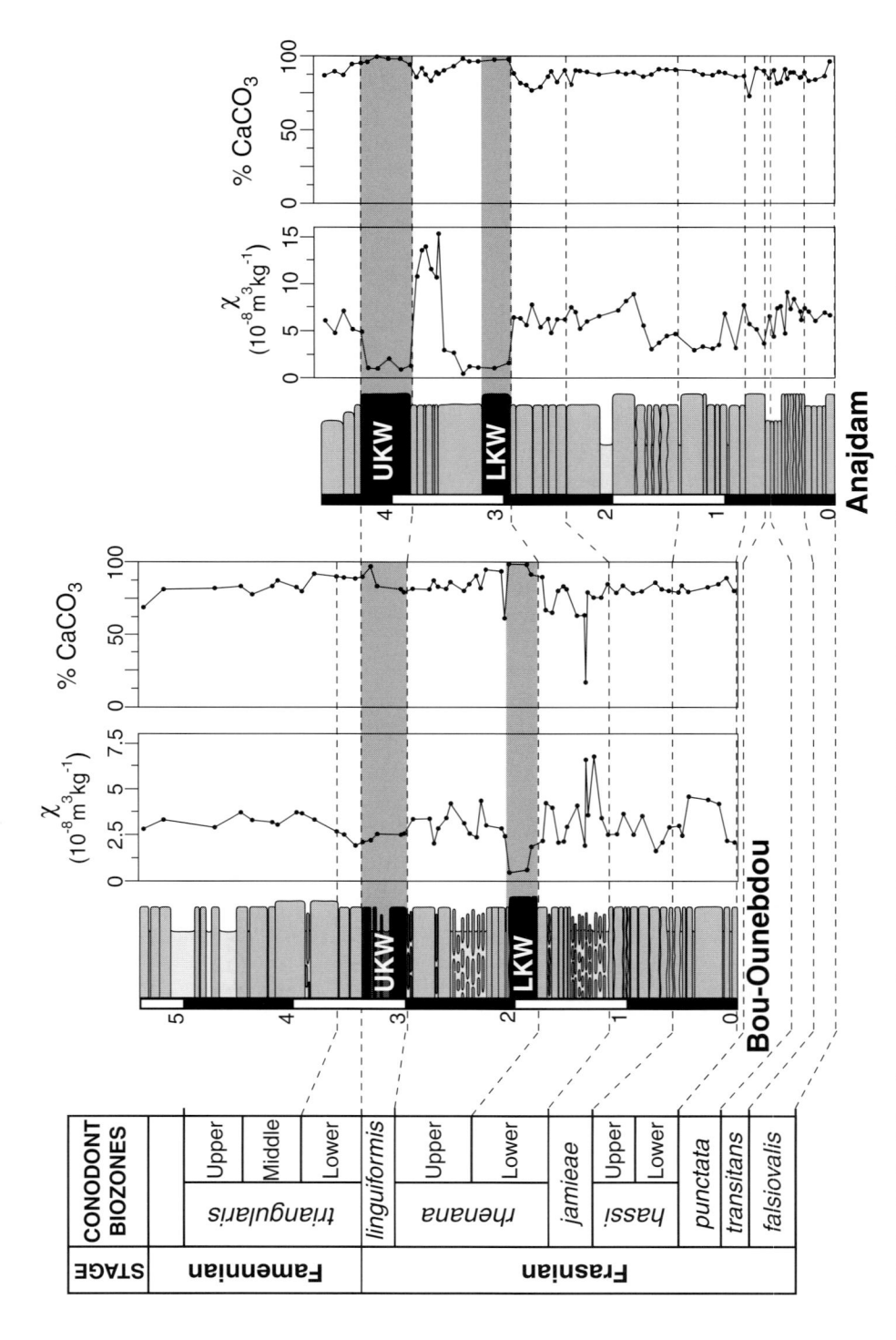

Fig. 5. Stratigraphical variation of mass-normalized magnetic susceptibility and calcium carbonate content of the sedimentary sequence from Bou-Ounebdou and Anajdam.

long-term χ decrease is not accompanied by noticeable variations of the carbonate content (Fig. 5). In both sections, the F-F boundary is marked by an increase in χ values of moderate amplitude.

Magnetic hysteresis properties. Magnetic hysteresis experiments have been carried out for representative samples of both sections (a total of 30 measurements) to provide information about the origin of the χ variations. Measured hysteresis parameters are plotted versus the magnetic susceptibility in Figure 6 to test the possible correlations existing between χ and the different hysteresis parameters.

As expected in carbonate rocks (e.g. Borradaile *et al.* 1993), values of M_s and M_r are globally low. In the Bou-Ounebdou section, as it was observed for χ, the amplitude of variation of M_s is also very low so that any convincing relationship can be drawn between these two parameters. Conversely, in Anajdam, an excellent correlation is observed between χ and M_s (Fig. 6a) thus indicating that the ferromagnetic *sensu lato* contribution controls the χ signal. The H_{cr} values are globally low in Anajdam (around 50 mT) regardless of the χ values (Fig. 6b) but they are significantly higher for much of the samples of Bou-Ounebdou ($60 \leq H_{cr} \leq 250$ mT). These data suggest that, in Anajdam, the χ variations are related to the varying concentration of a low coercivity mineral of magnetite type (Fe_3O_4) whereas, in Bou-Ounebdou, a mixture of magnetite and a higher coercivity phase is put forward. As usually observed in non-red limestone sequences, the latter is probably composed of low susceptibility goethite (FeO(OH); e.g. Heller 1978; Borradaile *et al.* 1993). This suggests an increase of the weathering intensity in the Bou-Ounebdou section compared to that of Anajdam, which is of crucial importance to recognize and understand the difference in the behaviour of iron-bearing minerals in both sections.

Samples with magnetite as the main carrier of the magnetic susceptibility ($H_{cr} < 70$ mT) have been plotted in a classical M_r/M_s versus H_{cr}/H_c graph (Fig. 7) in order to evaluate the bulk grain-size of the ferromagnetic fraction (Day *et al.* 1977). For both sections, the M_r/M_s ratio ranges from 0.1 to 0.25 and the H_{cr}/H_c ratio from 2 to 10. Following boundary values established by Dunlop (1986) and recommended by Borradaile *et al.* (1993), the diagram points to a dominant coarse-grain magnetic fraction with pseudo-single domain (PSD) to multi-domain (MD) behaviours although some points display a slight shift compared to the reference boundaries. Such shift has been observed in other magnetic hysteresis studies of carbonate rocks (e.g. Borradaile *et al.* 1993;

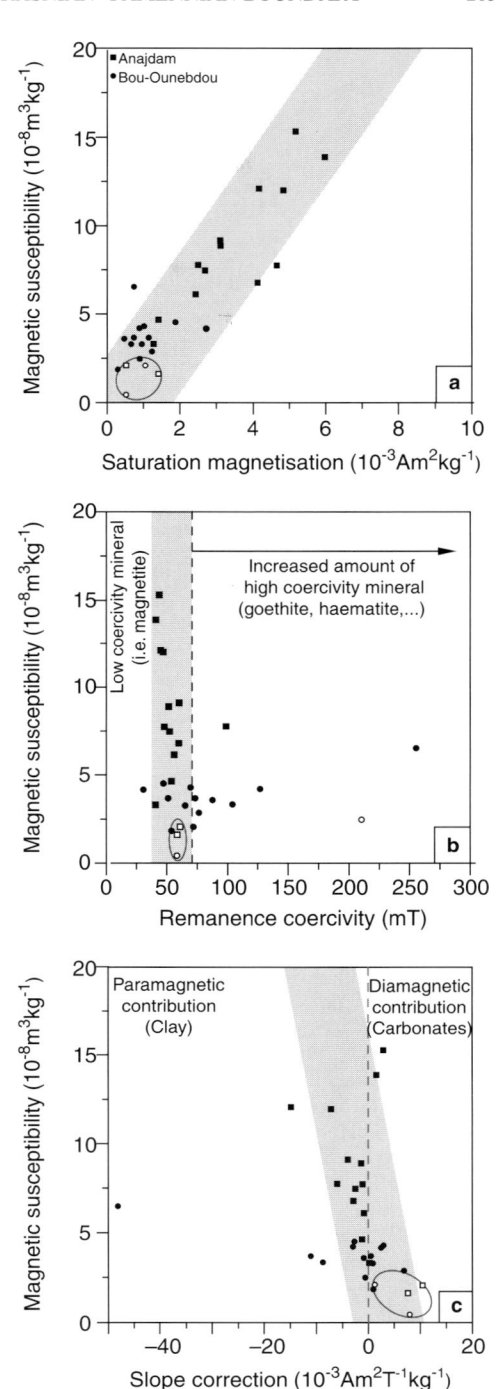

Fig. 6. Hysteresis parameters plotted against low-field mass susceptibility (χ) for representative samples from Anajdam and Bou-Ounebdou. (**a**) Saturation magnetization (M_s) *vs.* χ (**b**), coercivity (H_c) *vs.* χ and (**c**) Slope Correction *vs.* χ.

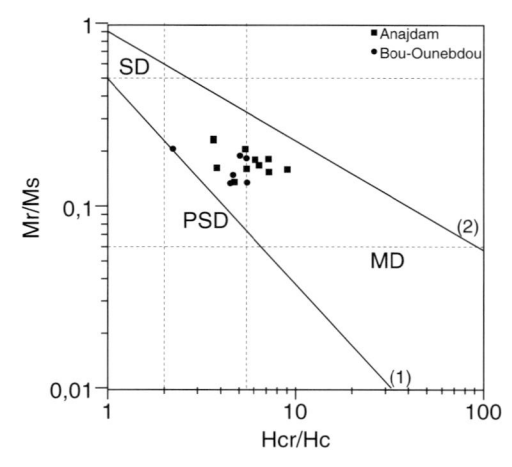

Fig. 7. Hysteresis parameters of samples from Anajdam and Bou-Ounebdou plotted according to Day *et al.* (1977). Ranges for *SD* (single domain), *PSD* (pseudo-single domain) and *MD* (multi-domain) are from Dunlop (1986). Empirical trends are shown for (1) unremagnetized and (2) remagnetized limestones (from Channell & McCabe 1994).

processes which reduce the intensity and amplitude of variations of the χ signal by transforming high susceptibility magnetite into low susceptibility goethite. Therefore, the Bou-Ounebdou section cannot be considered as a reference sequence regarding the analysis of detrital input evolution in the F–F times in the M'rirt area. In the following, the discussion of the magnetic data will thus focus on the Anajdam section.

Inorganic geochemistry analysis

Figure 8 shows the stratigraphical evolutions of some Al-normalized elemental ratios, usually considered as reliable geochemical proxies to estimate the fluctuation of clastic fluxes as well as oxygenation and primary productivity levels in marine environments.

Clastic influx proxies. Some elements, including Ti, Si and K are considered to be indicators of detrital influx (Tribovillard *et al.* 1994; Murphy *et al.* 2000b). Titanium is usually associated with clay minerals and heavy minerals (ilmenite, rutile). Potassium occurs dominantly in clays, whereas Si occurs in both siliciclastic and biogenic fractions (e.g. quartz *vs.* opal). Normalized to Al, which only occurs in aluminosilicates, the Si, Ti and K ratios provide useful information about change in the contribution of elemental concentrations from detrital non-aluminosilicate sources.

In regard to clastic influx, Al-normalized Si, Ti and K concentrations do not show clear trends at the F–F boundary (Fig. 8). The KW horizons are not marked by noticeable increase or decrease of these elements. For both sections, the Ti/Al profile remains almost constant. The values are nearly around the average shale values of 0.053 (Wedepohl 1991) in the major part of the section whereas most of the samples from the KW horizons have a concentration of Ti below the detection limit. Concerning the evolution of the Si/Al ratio in the Anajdam section, the KW horizons are characterized by relatively lower values compared to the concentrations in the overlying and underlying strata. In Bou-Ounebdou, there is a relative stability of the Si/Al ratio during the Late Frasnian, followed by a gradual decrease in the Early Famennian. To some extent, the Si/Al curves show comparable trends for both studied sections. Finally, K/Al ratio is relatively stable; it records the highest values at the end of the two KW horizons (inputs of K-carrying minerals, such as illite-mica or K-feldspar) and the lowest values during the inter-KW timespan for both sections.

Katz *et al.* 2000) and, to our knowledge, has not yet received any satisfactory explanation. Furthermore, M_r/M_s and H_{cr}/H_c values cluster in between the lines established by Channell and McCabe (1994) for diagenesis-induced remagnetized limestones and non-remagnetized limestones (Fig. 7). This, at least, suggests that the primary depositional ferromagnetic signal has not been completely overprinted by diagenetic processes.

The paramagnetic and diamagnetic contributions, quantified by the high-field slope correction SC (Fig. 6c), display noticeable variations showing the varying respective amounts of iron-bearing clays (illites, chlorites) and carbonates. It clearly appears that the matrix of the limestones is diamagnetic in the KW horizons, whereas it is dominantly paramagnetic for the rest of the sedimentary record in both sections. A general anti-correlation of SC and χ is noticed, illustrating the relative increase of clay concentration in levels with high χ (Fig. 6c). Therefore, almost pure carbonates such as the KW levels display the highest diamagnetic contributions and the lowest χ values.

To summarize, hysteresis data thus demonstrate that the χ signal is controlled by the combined contributions of ferromagnetic *sensu lato* magnetite and paramagnetic clays. As shown in Figure 4, the diamagnetic contribution of calcite only exerts a noticeable control on magnetic susceptibility in the almost pure KW limestones. In Bou-Ounebdou, this pattern is perturbed by weathering

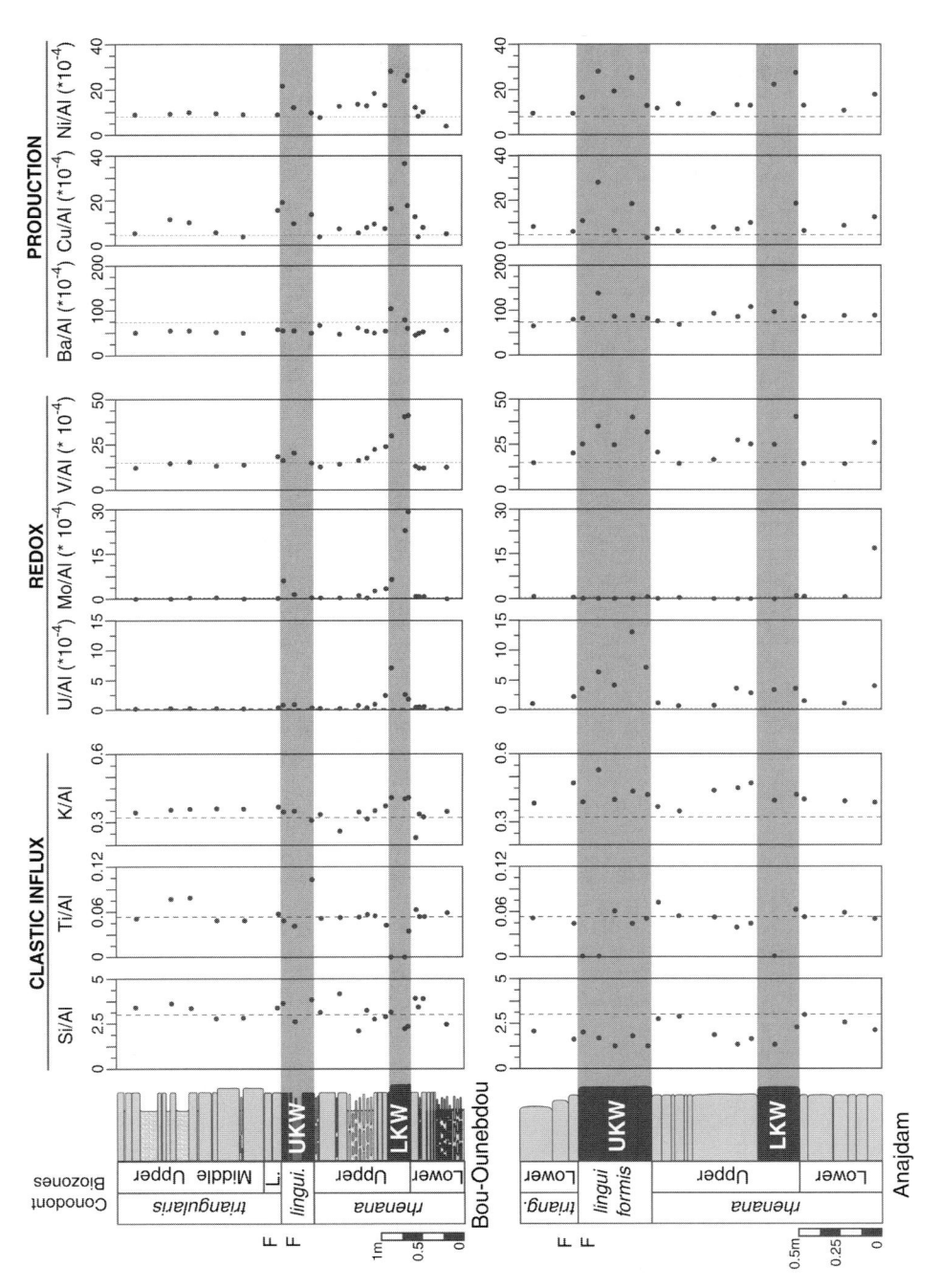

ENVIRONMENTAL CHANGES AT THE FRASNIAN–FAMENNIAN BOUNDARY 207

Fig. 8. Stratigraphical plots of elemental concentrations including proxies for clastic input, redox condition and production for the Anajdam and Bou-Ounebdou sections. Dashed lines indicate the metal/aluminium concentration ratios for average shales (Wedepohl 1971, 1991).

Palaeoredox proxies. In each section studied, a significant feature of the F–F boundary sequence is a remarkable enrichment in Mo, U and V in the KW horizons compared to the concentrations in the overlying and underlying strata (Fig. 8). These elements are thought to be redox-sensitive (Calvert & Pedersen 1993; Crusius *et al.* 1996; Dean *et al.* 1997; Algeo & Maynard 2004). The high ratios of these elements in the KW horizons thus suggest that the sediments were probably O_2-deficient. To confirm these observations, V/Cr, V/(V + Ni), U/Th and Ni/Co ratios, which are considered as reliable redox indexes (Hatch & Leventhal 1992; Jones & Manning 1994), were calculated in order to show the contrasted values between the KW horizons and the other parts of the section and to estimate the level of oxygenation (Fig. 9a–d). For each of the indexes used, two clusters of values can be identified, one corresponding to the samples from the KW horizons, the other corresponding to the samples from the overlying and underlying strata. The samples from the KW horizons show the highest values. They are broadly characterized by values of V/Cr > 1.5; U/Th > 1.5; Ni/Co > 4.25 (Fig. 9a, b and d). The values of the V/(V + Ni) ratios are mainly comprised between 0.5 and 0.7 and do not allow clearly distinguishing the KW horizons from the rest of the section (Fig. 9c).

Palaeoproductivity proxies. Vertical concentration profiles of Ba, Cu and Ni to Al ratios indicate slight enrichment in the KW horizons (Fig. 8). In

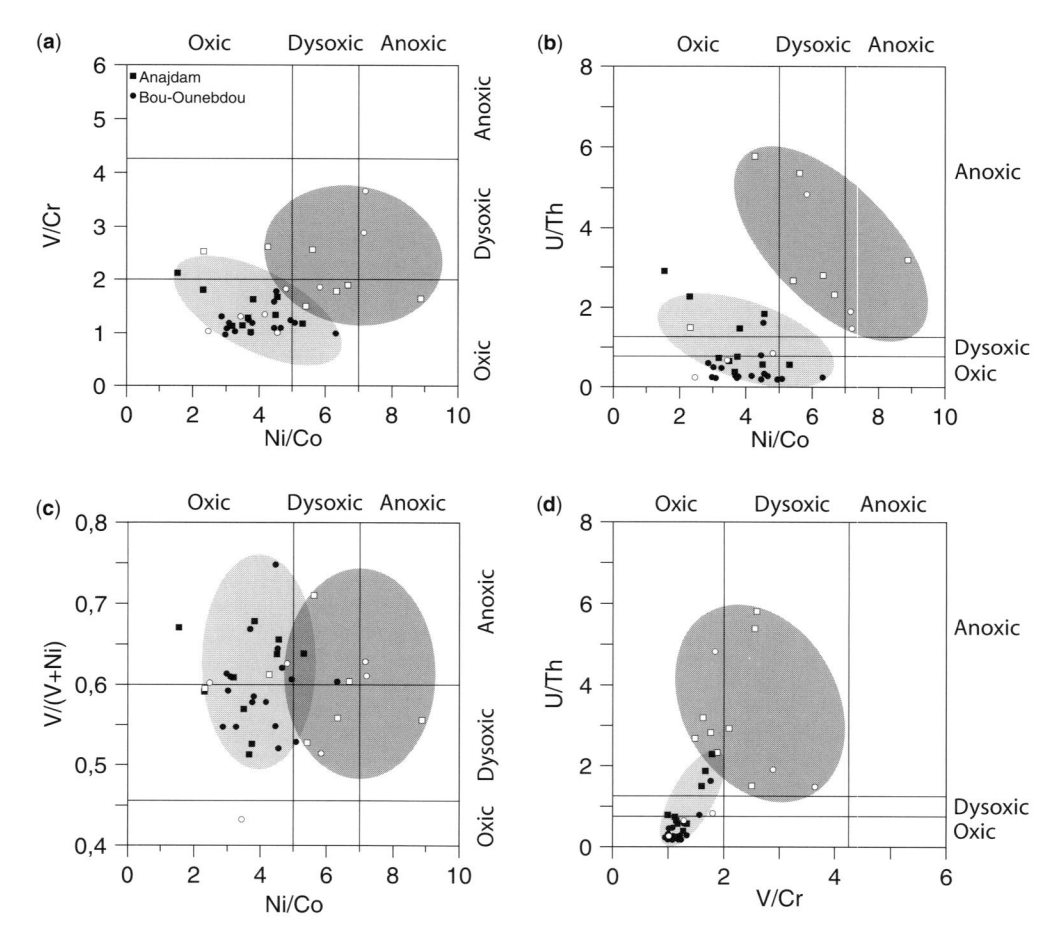

Fig. 9. Crossplots of trace-elements ratios used as palaeoredox proxies. (**a**) V/Cr *vs.* Ni/Co, (**b**) V/(V + Ni) *vs.* Ni/Co, (**c**) U/Th *vs.* Ni/Co and (**d**) U/Th *vs.* V/Cr. Ranges for V/Cr, U/Th and Ni/Co are from Jones & Manning (1994); ranges for V/(V + Ni) are from Hatch & Leventhal (1992). Open symbols: Kellwasser samples.

both sections, most of the KW samples show values of $Cu/Al > 10$ and $Ni/Al > 15$. Ba/Al ratios are higher in Anajdam ($Ba/Al > 90$) than in Bou-Ounebdou ($Ba/Al > 60$) in the KW horizons. Ba abundance is commonly used as a palaeoproductivity marker even if the interpretation is not always straightforward because barite can be dissolved in case of intense sulphate-reduction (McManus et al. 1998). In that case, Ba can be released to pore waters, migrate with them, and then reprecipitate when less-reducing conditions are met. Thus barite migration fronts may form (e.g. McManus et al. 1998). In the present case, the redox proxies indicate that depositional conditions were probably not very reducing (except for the LKW at Bou-Ounebdou). Consequently, Ba distribution must not have been deeply altered by diagenetical remobilization. The distribution of dissolved Ni and Cu in the ocean appears to be related to settling organic debris, since Ni and Cu behave as micronutrients (Calvert & Pedersen 1993; Algeo & Maynard 2004). These elements are readily adsorbed onto organic particles settling through the water column. So, decaying organic matter (OM) easily forms organometal complexes that can be incorporated into the sediment (Algeo & Maynard 2004; Tribovillard et al. 2004b). Thus, these elements are usually trapped with organic particles and are incorporated into iron sulphides during diagenesis. Consequently, they are preserved in the sediment even in the case of complete OM remineralization as well as in the case of marked thermal evolution (Mongenot et al. 1996). Thus, Ni and Cu may serve as indicators of organic inputs (Tribovillard et al. 2000, 2004b).

Summary. The distribution of the geochemical data during the Late Devonian suggests that: (1) the KW horizons coincide with two periods of relative decline of detrital influx, marked by a decrease of Ti and Si concentration; (2) the moderate and limited increase of U, Mo and V, as well as Ba, Cu and Ni concentrations are indicators of periods of dysoxic to anoxic conditions and enhanced productivity, that probably caused an increase of OM deposition.

Discussion

Detrital input versus carbonate production in the Frasnian–Famennian times in central Morocco

As previously mentioned, the χ measurements can be used as a good indicator of detrital input in ancient sedimentary rocks if not strongly altered by diagenetic or weathering processes (e.g. Ellwood et al. 2000). In the Bou-Ounebdou section, the χ signal is thought to have been modified by weathering processes and goethite formation and, therefore, cannot be considered to decipher detrital supply variations. By contrast, in Anajdam, magnetite and clays are the main minerals controlling the χ signal. This section is thus potentially adequate for relating the detrital input evolution through the F–F times. The possible impact of diagenesis on the magnetic signal still remains to be discussed. As shown by clay mineralogical assemblages, the two sections have been affected by a pronounced diagenesis. In such conditions, the transformation of clays of smectite type into illite has been suggested to induce the formation of authigenic magnetite that can potentially overprint or at least alter the primary detrital magnetic signal (e.g. Katz et al. 2000; Zegers et al. 2003). As observed in the Day-Fuller plot (Fig. 6c), the distribution of samples from Anajdam cluster below the line established by Channell & McCabe (1994) for diagenesis-induced remagnetized limestones (Fig. 6c).

Therefore, the primary depositional ferromagnetic signal is likely to have not been completely overprinted by diagenetic processes. Furthermore, M_s, that provides a first-order quantitative estimate of the magnetite concentration, does not correlate with either the illite crystallinity index standard or with the illite percentage (Fig. 10a, b). A slight negative correlation can be roughly observed between M_s and the illite percentage, suggesting that the diagenesis-induced illitization is not accompanied by an increase in the magnetite concentration. On the other hand, early diagenetic reductive dissolution of magnetite in the OM-bearing black KW carbonate levels cannot be proposed as an efficient process in the alteration of the primary magnetic signal (e.g. Machel 1995; Robinson et al. 2000; Tribovillard et al. 2002, 2004b), because the high field SC is strongly diamagnetic (Fig. 6c) and thus does not record any effects of authigenic paramagnetic pyrite. Such a poor control of diagenetic processes upon the magnetic signal is corroborated by the correlation of Ti and Th contents ($r^2 = 0.47$ and 0.66, respectively), classically considered as detrital indicators independent of diagenesis-induced remobilization, with the magnetic susceptibility from the Anajdam section (Fig. 11a, b). These whole data thus demonstrate that, although submitted to significant burial, the χ signal can be confidently used as an indicator of the detrital input evolution in the Anajdam section.

The long-term decrease of χ observed during most of the Frasnian times can thus be interpreted

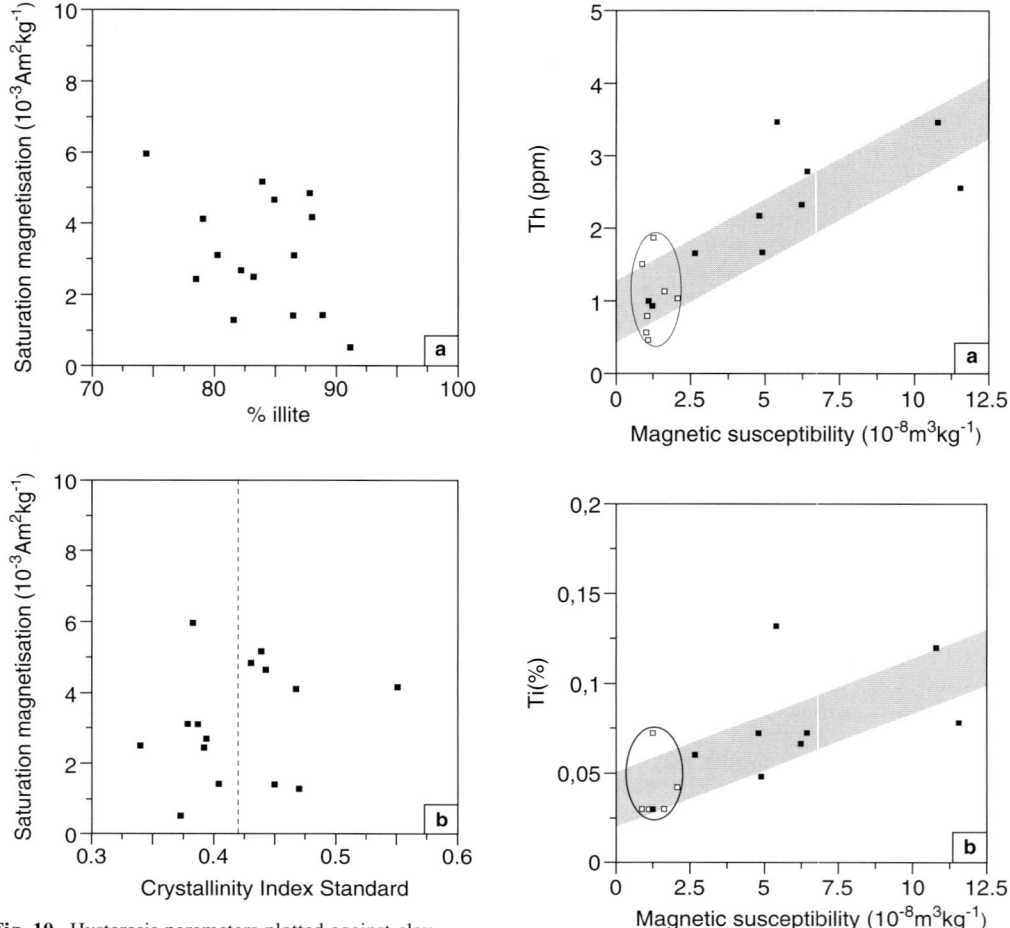

Fig. 10. Hysteresis parameters plotted against clay parameters for representative samples from Anajdam and Bou-Ounebdou. (**a**) Saturation magnetization (M_s) vs. CIS. (**b**) Saturation magnetization (M_s) vs. % of illite.

Fig. 11. Detrital-derived element concentrations plotted against low-field magnetic susceptibility (χ) for representative samples from Anajdam. (**a**) Th (ppm) vs. χ and (**b**) Ti (%) vs. χ.

as a general diminution of the detrital input in the Azrou–Kenifra Basin through Late Frasnian times. This evolution can be traced up to the KW levels that record the lowest χ values, in relation with a minimum detrital input and an enhanced carbonate productivity, as shown by the maximum intensity of the diamagnetic contribution. This is also confirmed by the vertical distribution of Ti/Al and Si/Al ratios related to clastic input (Fig. 8). The F–F boundary represents a significant break in this long-term evolution with a significant increase of the χ, thus suggesting a noticeable increase of the basinal detrital supply. It is worth noting that these variations can be paralleled to the global sea-level fluctuations as synthesized by Sandberg et al. (2002) and indicating a general Frasnian sea-level rise, a Late Frasnian highstand, and a F–F boundary sea-level fall (Fig. 12). These sea-level changes have been well documented in several areas, such as Germany (Piecha 2002; Devleeschouwer et al. 2002), Poland (Racki et al. 2002), NW Australia (Becker & House 1997), China (Chen & Tucker 2003, 2004), and Canada (Mountjoy & Becker 2000).

In a more general perspective, the magnetic susceptibility data from Anajdam are consistent with previous magnetic results obtained from other sections of the northern Gondwana

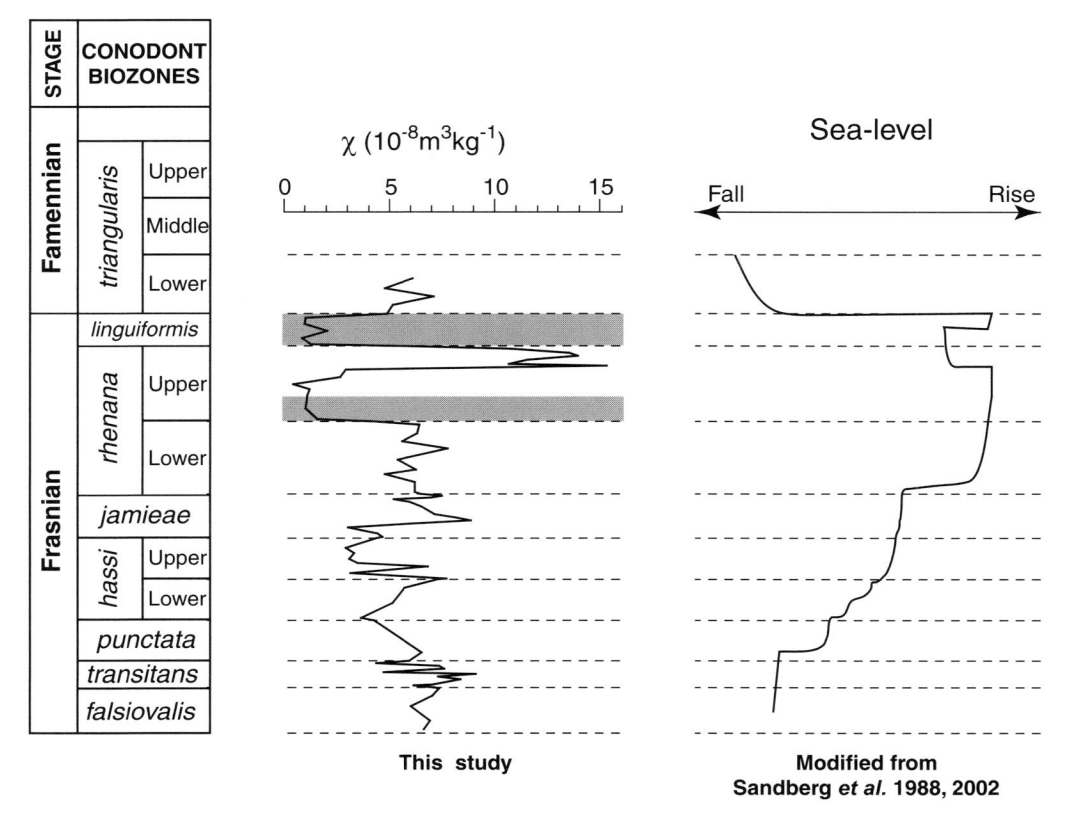

Fig. 12. Synthetic diagram showing the variations of the χ signal obtained from the Anajdam section compared with the Late Devonian sea-level curve (modified from Sandberg *et al.* 1988, 2002).

margin, such as Coumiac (Montagne Noire, S. France), but also from domains of the southern Laurassian margin, such as Steinbruch Schmidt and Beringhausen Tunnel (Rhenish Massif, Germany), Kowala (Holly Cross Mountains, Poland) or Sinsin (Ardennes, Belgium) (Devleeschouwer 1999; Crick *et al.* 2002; Racki *et al.* 2002; Averbuch *et al.* 2005). The mechanisms controlling the first-order observed χ variations have thus to be found in relatively large-scale processes affecting at least the Gondwanan and Laurassian margins. Global climatic warming during Frasnian times with maximum temperatures in the Late Frasnian (Kellwasser events) and global cooling at the F–F boundary are thus the more convincing parameters to account for the observed variations as also suggested by the $\delta^{18}O$ signatures of marine conodont apatite (Joachimski & Buggisch 2002; Joachimski *et al.* 2004) and the miospore diversity evolution on continents (Streel *et al.* 2000).

Marine bottom-water redox-conditions and primary productivity evolution

As mentioned in the 'Results' section, the black limestones of the KW horizons show noticeable variations for nearly all trace element ratios compared to the underlying and overlying limestone beds. Broadly, redox-sensitive elements as well as redox indexes show enrichments in the LKW horizon in Bou-Ounebdou and in the UKW horizon in Anajdam (Fig. 8). Molybdenum concentration has been suggested to be enhanced in the presence of sulphides (Helz *et al.* 1996; Adelson *et al.* 2001; Zheng *et al.* 2000; Vorlicek & Helz 2002; Lyons *et al.* 2003; Algeo & Maynard 2004; Rimmer *et al.* 2004; Tribovillard *et al.* 2004c). The marked Mo enrichment observed in the LKW of the Bou-Ounebdou section argues for sulphidic conditions, whereas in all the other parts of the two sections studied, dysoxic–anoxic conditions are put forward (no Mo-enrichment but relative enrichment in U and V for the KW horizons).

The depletion in O_2 is confirmed by the stratigraphical variations of reliable palaeo-oxygenation indexes such as V/Cr, V/(V + Ni), U/Th, Ni/Co (Fig. 9a–d). In this study, we adopted thresholds proposed by Jones & Manning (1994) and Hatch & Leventhal (1992). For each of the indices used, values indicate anoxic to dysoxic conditions for the two KW horizons. The observed enrichments in redox-sensitive trace elements confirm that the oxic–anoxic interface was located at a shallower depth in the sediment during the deposition of the KW horizons, perhaps even in the water column for the LKW at Bou-Ounebdou. However, the intensity of O_2 depletion is interpreted differently from one ratio to the other (Fig. 9a–d). According to V/Cr and V/(V + Ni) ratios, Late Frasnian conditions were broadly dysoxic to anoxic in outer platform settings, whereas O_2-depletion would be only limited to the KW horizons according to U/Th and Ni/Co. The consistency of the chemical data thus clearly indicates two pulses of oxygen depletion near the base of the Upper *rhenana* Zone and in the *linguiformis* Zone, i.e. within the two KW horizons.

These geochemical results are in agreement with recent biofacies studies (Lazreq 1999; Feist 2002), which demonstrated that in the M'rirt area, the fauna of KW horizons are characterized by a lack of endobenthos and O_2-sensitive epibenthos, such as trilobites and brachiopods. In a more global pattern, our interpretations are in agreement with recent geochemical studies concerning some F–F sections from Germany, France, Poland, Siberia and China (e.g. Racki *et al.* 2002; Ma & Bai 2002; Yudina *et al.* 2002; Bond *et al.* 2004; Tribovillard *et al.* 2004a; Pujol *et al.* in press). In many areas, the F–F boundary records some elemental anomalies related to redox conditions (e.g. U, V and Mo). Thus, both geochemical and palaeontological analyses suggest a noticeable decrease of O_2 concentration in bottom waters during the two KW episodes. Primary productivity tracers such as Ba/Al and OM-related elemental ratios, such as Ni/Al and Cu/Al, display a consistent slight increase through the KW levels. A similar trend is also observed in the carbonate production as shown both by the $CaCO_3$ content and the optimal diamagnetic contribution to the magnetic signal. Trends similar to those observed for the Moroccan sections have been found in the time-equivalent formations of Aeketal (Harz, Germany), along the southern Laurassian outer shelf (Riquier *et al.* 2005). Compared to geochemical results obtained for deeper sections such as La Serre (Montagne Noire, France; Tribovillard *et al.* 2004a), the Moroccan sections display a relatively low enrichment in Ba. According to Bishop (1988), precipitation of barite is favoured in microenvironments, where OM decays from surface water to sea floor. Therefore, barite may not be enriched in shallow, highly productive environments, whereas it is enhanced in deeper sites with similar productivity. The relatively poor record of primary productivity in outer shelves can thus be explained by local conditions. In shallow-water settings, even if conditions become eutrophic, favouring the onset of O_2-deprived conditions, generally intense seawater mixing does not allow a good preservation of OM. Organic elements are remineralized before reaching the water/sediment interface. The observed enrichments in productivity indicators are coeval with two positive excursions of the carbon isotopic signal ($\delta^{13}C_{carb} = +3‰$ V-PDB), recorded in Bou-Ounebdou and in many other F–F sections for inorganic carbon (Joachimski *et al.* 2002). These positive shifts in $\delta^{13}C$ are explained by an increase in the global-scale organic carbon burial rate in the Late Frasnian seas (Joachimski *et al.* 2001, 2002; Murphy *et al.* 2000a, b). In outer shelf settings such as the Moroccan sections, the resultant depletion in marine bottom-water oxygenation and associated OM preservation were probably moderate due to efficient platform ventilation.

Conclusion

In this study, rock magnetic measurements (magnetic susceptibility and hysteresis), combined with clay mineral and geochemical analyses allow to constrain the environmental features of two sections from Central Morocco (Anajdam and Bou-Ounebdou). Although both sections were affected by a pronounced diagenesis (limit of anchizonal conditions), we demonstrate that the magnetic susceptibility represents a semi-quantitative proxy for detrital supply variations. The magnetic signal in the Anajdam section is better defined than that in Bou-Ounebdou, owing to increasing weathering intensity in the latter and transformation of high-susceptibility magnetite into low-susceptibility goethite. The Anajdam section unravels a Frasnian long-term decrease of the basinal detrital input that parallels a general sea-level rise with a major highstand in the Late Frasnian Kellwasser-type levels. The Kellwasser Horizons display the lowest values of magnetic susceptibility and magnetization at saturation, in relation with a minimum detrital input and a maximum carbonate productivity, as shown by the significant diamagnetic contribution to the magnetic signal. The high marine primary productivity in the Late-Frasnian Central Moroccan basin is attested by an increase of the Ba/Al ratio in the Kellwasser levels of both sections. Reliable redox indicators such as V/Cr or U/Th furthermore evidence the general dysoxic conditions prevailing in marine bottom-waters at that time. On top of the Upper Kellwasser Horizon, the Frasnian–Famennian boundary in central Morocco is marked by a break in the sedimentation with possible depositional hiatuses associated with a noticeable sea-level fall (Lazreq 1999; Chakiri 2002). Comparisons with other sections worldwide suggest that the variations in the detrital supply are climatically driven, thus arguing for a long-term Frasnian warming and a significant basal Famennian global cooling event. The latter is subsequent to a Late Frasnian climatic optima leading to a period of particularly increased organic carbon burial (e.g. Joachimski et al. 2002). As previously proposed by different authors (e.g. Becker & House 1994; Godderis & Joachimski 2004; Averbuch et al. 2005), this mechanism would have produced an important drawdown in the atmospheric CO_2 content, thus resulting in a strong reduction of greenhouse effects in Famennian times. Such rapid variations in seawater oxygenation conditions and temperatures are likely to have had drastic repercussions on marine fauna, potentially leading to the Frasnian–Famennian biological crisis.

This study has been financially supported by the Eclipse program of the C.N.R.S. (project Frasnian–Famennian boundary crisis, O. Averbuch coordinator) and is a contribution of the UMR PBDS 8110. We greatly thank C. Kissel for providing access to the Micromag equipment at the Palaeomagnetic Laboratory in LSCE (Gif sur Yvette), J. Morel for the ICP analyses and M. Frere, P. Recourt and D. Malengros for assistance with X-ray analysis. A. Herbosch, I. Berra and C. Crônier are acknowledged for their participation to the sampling fieldtrip in Central Morocco. We also thank Michael Joachimski, Grzegorz Racki and Walter Dean for scientific discussions at earlier stages of this study. Finally, we are grateful to Eberhard Schindler and Thomas Becker; the manuscript benefited from their helpful and constructive comments.

References

ADELSON, J. M., HELZ, G. R. & MILLER, C. V. 2001. Reconstructing the rise of recent coastal anoxia; molybdenum in Chesapeake Bay sediments. *Geochimica Cosmochimica Acta*, **65**, 237–252.

ALGEO, T. J. & SCHECKLER, S. E. 1998. Terrestrial–marine teleconnections in the Devonian: links between the evolution of land plants, weathering processes, and marine anoxic events. *Philosophical Transactions of the Royal Society of London*, **B353**, 113–130.

ALGEO, T. J. & MAYNARD, J. B. 2004. Trace element behavior and redox facies analysis of core shales of Upper Pennsylvanian Kansas-type cyclotherms. *Chemical Geology*, **206**, 289–318.

AVERBUCH, O., TRIBOVILLARD, N., DEVLEESCHOUWER, X., RIQUIER, L., MISTIAEN, B. & VAN VLIET-LANOE, B. 2005. Mountain building—enhanced continental weathering and organic carbon burial as major causes for climatic cooling at the Frasnian–Famennian boundary (ca 376 Ma BP). *Terra Nova*, **17**, 25–34.

BECKER, R. T. 1993. Stratigraphische Gliederung und Ammonoideen-Faunen im Nehdenium (Oberdevon II) von Europa und Nord-Afrika. *Courier Forschungsinstitut Senckenberg*, **155**, 1–405.

BECKER, R. T. & HOUSE, M. R. 1994. Kellwasser Events and goniatite successions in the Devonian of the Montagne Noire with comments on possible causations. *Courier Forschunginstitut Senckenberg*, **169**, 45–77.

BECKER, R. T. & HOUSE, M. R. 1997. Sea-level changes in the Upper Devonian of the Canning Basin, Western Australia. *Courier Forschunginstitut Senckenberg*, **199**, 129–146.

BECKER, R. T. & HOUSE, M. R. 2000. Sedimentary and faunal succession of the allochtonous Upper Devonian at Gara d'Mrirt (Eastern Moroccan Meseta). *Notes et Mémoires, Service Géologique du Maroc*, **399**, 109–114.

BISHOP, J. K. 1988. The barite-opal-organic carbon association in oceanic particulate matter. *Nature*, **332**, 341–343.

BOND, D., WIGNALL, P. B. & RACKI, G. 2004. Extent and duration of marine anoxia during the Frasnian-Famennian (Late Devonian) mass extinction in Poland, Germany, Austria and France. *Geological Magazine*, **141**, 173–193.

BORRADAILE, G. J., CHOW, N. & WERNET, T. 1993. Magnetic hysteresis of limestones: facies control? *Physics of the Earth and Planetary Interior*, **76**, 241–252.

BOUABDELLI, M. 1989. Tectonique et sédimentation dans un bassin orgénique: le sillon Viséen d'Azrou-Khénifra (Est du Massif Hercynien Central du Maroc). Unpublished Ph.D. thesis, Louis Pasteur University, Strasbourg.

BOUABDELLI, M. & PIQUÉ, A. 1996. Du bassin sur décrochement au basin d'avant-pays: dynamique du bassin d'Azrou-Khénifra (Maroc Hercynien central). *Journal of African Earth Science*, **23**, 213–224.

BRINDLEY, G. W. & BROWN, G. 1980. Crystal structures of clay minerals and their identification. *Mineralogical Society Monograph No. 5*. Mineralogical Society, London.

BUGGISCH, W. 1991. The global Frasnian/Famennian "Kellwasser Event". *Geologische Rundschau*, **80**, 49–72.

BUGGISCH, W. & CLAUSEN, C.-D. 1972. Conodonten- und Goniatiten-Faunen aus dem oberen Frasnium und unteren Famennium Marokkos (Tafilalt, Antiatlas). *Neues Jahrbuch für Geologie und Paläontogie Abhandlungen*, **141**, 137–167.

CALVERT, S. E. & PEDERSEN, T. F. 1993. Geochemistry of Recent oxic and anoxic marine sediments:

Implications for the geological record. *Marine Geology*, **113**, 67–88.

CHAKIRI, S. 2002. Sédimentologie et géodynamique du Maroc Central Hercynien pendant le Dévonien. Unpublished Ph.D. thesis, Ibn Tofail University, Kenitra.

CHANNELL, J. E. T. & MCCABE, C. 1994. Comparison of magnetic hysteresis parameters of unremagnetized and remagnetized limestones. *Journal of Geophysical Research*, **99**, 4613–4623.

CHEN, D. & TUCKER, M. E. 2003. The Frasnian/Famennian mass extinction: insights from high-resolution sequence stratigraphy and cyclostratigraphy in South China. *Palaeogeography, Palaeoclimatology, Palaeoecology*, **193**, 87–111.

CHEN, D. & TUCKER, M. E. 2004. Paleokarst and its implication for the extinction event at the Frasnian-Famennian boundary (Guilin, South China). *Journal of the Geological Society of London*, **161**, 895–898.

COPPER, P. 1986. Frasnian/Famennian mass extinction and cold-water oceans. *Geology*, **14**, 835–839.

CRICK, R. E., ELLWOOD, B. B., HASSANI, A., FIEST, R. & HLADIL, J. 1997. Magnetosusceptibility event and cyclostratigraphy (MSEC) of the Eifelian-Givetian GSSP and associated boundary sequences in North Africa and Europe. *Episodes*, **20**, 167–175.

CRICK, R. E., ELLWOOD, B. B., ET AL. 2002. Magnetostratigraphy susceptibility of the Frasnian/Famennian boundary. *Palaeo-geography, Palaeoclimatology, Palaeoecology*, **181**, 67–90.

CRUSIUS, J., CALVERT, S., PEDERSEN, T. & SAGE, D. 1996. Rhenium and molybdenum enrichments in sediments as indicator of oxic, suboxic and sulfidic conditions of deposition. *Earth and Planetary Sciences Letters*, **145**, 65–78.

DAY, R., FULLER, M. D. & SCHMIDT, V. A. 1977. Hysteresis properties of titanomagnetites. Grain size and composition dependence. *Physics of the Earth and Planetary Interior*, **13**, 260–267.

DEAN, W. E., GARDNER, J. V. & PIPER, D. Z. 1997. Inorganic geochemical indicators of glacial-interglacial changes in productivity and anoxia on the California continental margin. *Geochimica et Cosmochimica Acta*, **61**, 4507–4518.

DEVLEESCHOUWER, X. 1999. La transition Frasnien/Famennien (Dévonien sup.) en Europe: sédimentologie, stratigraphie séquentielle et susceptibilité magnétique. Unpublished Ph.D. thesis, Lille and Brussels Universities.

DEVLEESCHOUWER, X., HERBOSCH, A. & PRÉAT, A. 2002. Microfacies, sequence stratigraphy and clay mineralogy of a condensed deep-water section around the Frasnian–Famennian boundary (Steinbruch Schmidt, Germany). *Palaeoecology, Palaeclimalology, Palaeoecology*, **181**, 171–193.

DOPIERALSKA, J. 2003. Neodymium isotopic composition of conodonts as a palaeoceanographic proxy in the Variscan oceanic system. Unpublished Ph.D. thesis, University of Giessen.

DUNLOP, D. J. 1986. Hysteresis properties of magnetite and their dependence on particle size: a test of pseudo-single domain remanence models. *Journal of Geophysical Research*, **91**, 9569–9584.

EL HASSANI, A. & TAHIRI, A. 2000a. The eastern part of central Morocco (western Meseta). *Notes & Mémoires du Service Géologique du Maroc*, **399**, 89–92.

EL HASSANI, A. & TAHIRI, A. (eds) 2000b. Moroccan Meeting of the Subcommission on Devonian Stratigraphy (SDS) – IGCP 421 (April 24th–May 1st 1999): Excursion guidebook. *Notes & Mémoires du Service Géologique du Maroc*, **399**, 1–150.

ELLWOOD, B. B., CRICK, R. E., EL HASSANI, A., BENOIST, S. L. & YOUNG, R. H. 2000. Magnetosusceptibility event and cyclostratigraphy method applied to marine rocks: detrital input versus carbonate productivity. *Geology*, **28**, 1135–1138.

FEIST, R. 2002. Trilobites from the latest Frasnian Kellwasser Crisis in North Africa (Mrirt, central Moroccan Meseta). *Acta Palaeontologica Polonica*, **47**, 203–210.

GIRARD, C. & ALBARÈDE, F. 1996. Trace elements in conodont phosphate from the Frasnian/Famennian boundary. *Palaeogeography, Palaeoclimatology, Palaeoecology*, **126**, 195–209.

GIRARD, C., KLAPPER, G. & FEIST, R. 2005. Subdivision of the terminal Frasnian *linguiformis* conodont Zone, revision of the correlative interval of Montagne Noire Zone 13, and discussion of stratigraphically significant associated trilobites. *In*: OVER, J., MORROW, J. R. & WIGNALL, P. (eds) *Understanding Late Devonian Triassic Biotic and Climatic Events; Towards an Integrated Approach*, 181–198.

GODDERIS, Y. & JOACHIMSKI, M. M. 2004. Global change in the Late Devonian: Modelling the Frasnian-Famennian short-term carbone isotope excursions. *Palaeogeography, Palaeoclimatology Palaeoecology*, **202**, 309–329.

GRADSTEIN, F. M., OGG, J. G. ET AL. 2004. *A Geologic Time Scale 2004*. Cambridge University Press.

HALLAM, A. & WIGNALL, P. B. 1999. Mass extinctions and sea-level changes. *Earth Science Reviews*, **48**, 217–250.

HATCH, J. R. & LEVENTHAL, J. S. 1992. Relationship between inferred redox potential of the depositional environment and geochemistry of the Upper Pennsylvanian (Missourian) Stark Shale Member of the Dennis Limestone, Wabaunsee County, Kansas, U.S.A. *Chemical Geology*, **99**, 65–82.

HELLER, F. 1978. Rock magnetic studies of Upper Jurassic limestones from Southern Germany. *Zeitschrift für Geophysik*, **44**, 525–543.

HELZ, G. R., MILLER, C. V., CHARNOCK, J. M., MOSSELMANS, J. L. W., PATTRICK, R. A. D., GARNER, C. D. & VAUGHAN, D. J. 1996. Mechanisms of molybdenum removal from the sea and its concentration in black shales: EXAFS evidences. *Geochimica Cosmochimica Acta*, **60**, 3631–3642.

HOLTZAPFFEL, T. 1985. Les minéraux argileux. Préparation, analyse diffractométrique et détermination. *Mémoire Société Géologique du Nord*, **12**, 136pp.

HOUSE, M. R. 2002. Strength, timing, setting and cause of mid-Palaeozoic extinctions. *Palaeogeography, Palaeoclimatology, Palaeoecology*, **181**, 5–25.

JOACHIMSKI, M. M. & BUGGISCH, W. 1993. Anoxic events in the Late Frasnian—causes of the Frasnian–Famennian faunal crisis? *Geology*, **21**, 675–678.

JOACHIMSKI, M. M. & BUGGISCH, W. 2002. Conodont apatite $\delta^{18}O$ signatures indicate climatic cooling as a trigger of the Late Devonian mass extinction. *Geology*, **30**, 711–714.

JOACHIMSKI, M. M., OSTERTAG-HENNING, C., ET AL. 2001. Water column anoxia, enhanced productivity and concomitant changes in $\delta^{13}C$ and $\delta^{34}S$ across the Frasnian–Famennian boundary (Kowala—Holy Cross Mountains/Poland). *Chemical Geology*, **175**, 109–131.

JOACHIMSKI, M. M., PANCOST, R. D., FREEMAN, K. H., OSTERTAG-HENNING, C. & BUGGISCH, W. 2002. Carbon isotope geochemistry of the Frasnian–Famennian transition. *Palaeogeography, Palaeoclimatology, Palaeoecology*, **181**, 91–109.

JOACHIMSKI, M. M., VAN GELDERN, R., BREISIG, S., BUGGISCH, W. & DAY, J. 2004. Oxygen isotope evolution of biogenic calcite and apatite during the Middle and Late Devonian. *International Journal of Earth Science*, **93**, 542–553.

JONES, B. & MANNING, D. A. C. 1994. Comparison of geochemical indices used for the interpretation of palaeoredox conditions in ancient mudstones. *Chemical Geology*, **111**, 111–129.

KATZ, B., ELMORE, D., COGOINI, M., ENGEL, M. H. & FERRY, S. 2000. Association between burial diagenesis of smectite, chemical remagnetization, and magnetite authigenesis in the Vocontian trough, SE France. *Journal of Geophysical Research*, **105**, 851–868.

LAZREQ, N. 1992. The Upper Devonian of M'rirt (Morocco). *Courier Forschungsinstitut Senckenberg*, **154**, 107–123.

LAZREQ, N. 1999. Biostratigraphie des conodontes du Givétien au Famennien du Maroc Central. Biofaciès et événement Kellwasser. *Courier Forschungsinstitut Senckenberg*, **214**, 1–111.

LYONS, T. W., WERNE, J. P., HOLLANDER, D. J. & MURRAY, R. W. 2003. Contrasting sulfur geochemistry and Fe/Al and Mo/Al ratios across the last oxic-to-anoxic transition in the Cariaco Basin, Venezuela. *Chemical Geology*, **195**, 131–157.

MA, X. P. & BAI, S. L. 2002. Biological, depositional, microspherule, and geochemical records of the Frasnian/Famennian boundary beds, South China. *Palaeogeography, Palaeoclimatology, Palaeo-ecology*, **181**, 325–346.

MCGHEE, G. R. 1996. The Late Devonian Mass Extinction. The Frasnian-Famennian Crisis. Columbia University Press, New York.

MCMANUS, J., BERELSON, W. M., ET AL. 1998. Geochemistry of barium in marine sediments: Implications for its use as a paleoproxy. *Geochimica Cosmochimica Acta*, **62**, 3453–3473.

MACHEL, H. 1995. Magnetic mineral assemblages and magnetic contrasts in diagenetic environments—with implications for studies of paleomagnetism, hydrocarbon migration and exploration. *In*: TURNER, P. & TURNER, A. (eds) Paleomagnetic application in hydrocarbon exploration and production. *Geological Society Special Publication*, **98**, 9–29.

MAHER, B. A. & THOMPSON, R. 1999. *Quaternary climates, environments and magnetism*. Cambridge University Press, New York.

MONGENOT, T., TRIBOVILLARD, N.-P., DESPRAIRIES, A., LALLIER-VERGÈS, E. & LAGGOUN-DEFARGE, F. 1996. Trace elements as palaeoenvironmental markers in strongly mature hydrocarbon source rocks: the Cretaceous La Luna Formation of Venezuela. *Sedimentary Geology*, **103**, 23–37.

MOUNTJOY, E. W. & BECKER, S. 2000. Frasnian to Famennian sea-level changes and the Sassenach Formation, Jasper Basin, Alberta Rocky Mountains. *In*: HOMEWOOD, P. W. & EBERLI, G.P. (eds). Genetic stratigraphy on exploration and production shales—Case studies from the Pennsylvanian of the Paradox Basin and the Upper Devonian of Alberta. *Bulletin des Centres de Recherches Elf Exploration Production Memoire*, **24**, 181–201.

MURPHY, A. E., SAGEMAN, B. B. & HOLLANDER, D. J. 2000a. Eutrophication by decoupling of the marine geochemical cycles of C, N and P: a mechanism for the Late Devonian mass extinction. *Geology*, **28**, 427–430.

MURPHY, A. E., SAGEMAN, B. B., HOLLANDER, D. J., LYONS, T. L. & BRETT, C. E. 2000b. Black shale deposition and faunal overturn in the Devonian Appalachian Basin: clastic starvation, seasonal water-column mixing, and efficient biolimiting nutrient recycling. *Paleoceanography*, **15**, 280–291.

PETSCHICK, R., KUHN, G. & GINGELE, F. 1996. Clay mineral distribution in surface sediments of the South Atlantic: sources, transport, and relation to oceanography. *Marine Geology*, **130**, 203–229.

PIECHA, M. 2002. A considerable hiatus at the Frasnian/Famennian boundary in the Rhenish shelf region of northwest Germany. *Palaeogeography, Palaeoclimatology, Palaeoecology*, **181**, 195–211.

PIQUÉ, A., BOSSIÈRE, G., BOUILLIN, J.-P., CHALOUAN, A. & HOEPFFNER, C. 1993. Southern margin of the Variscan belt: the north-western Gondwana mobile zone (eastern Morocco and northern Algeria). *Geologische Rundschau*, **82**, 432–439.

PUJOL, F., BERNER, Z. & STÜBEN, D. Chemostratigraphy of some european Frasnian-Famennian boundary key section. *Palaeogeography, Palaeoclimatology, Palaeoecology*, **240**, 120–145.

RACKI, G. 1998. Frasnian-Famennian biotic crisis: undervalued tectonic control? *Palaeogeography, Palaeoclimatology, Palaeoecology*, **141**, 177–198.

RACKI, G. 1999. The Frasnian-Famennian biotic crisis: how many (if any) bolide impacts? *Geologische Rundschau*, **87**, 617–632.

RACKI, G., RACKA, M., MATYJA, H. & DEVLEESCOUWER, X. 2002. The Frasnian/Famennian boundary interval in the South Polish–Moravian shelf basins: integrated event-stratigraphical approach. *Palaeogeography, Palaeoclimatology, Palaeoecology*, **181**, 251–297.

RIMMER, S. M., THOMPSON, J. A., GOODNIGHT, S. A. & ROBL, T. L. 2004. Multiple controls on the preservation of organic matter in Devonian–Mississippian marine black shales: geochemical and petrographic evidence. *Palaeogeography, Palaeoclimatology, Palaeoecology*, **215**, 125–154.

RIQUIER, L., TRIBOVILLARD, N., AVERBUCH, O., JOACHIMSKI, M. M., RACKI, G., DEVLEESCHOUWER, X., EL ALBANI, A. & RIBOULLEAU, A. 2005. Bottom water redox conditions at the Frasnian-Famennian boundary on both sides of the Eovariscan Belt constraints from trace element geochemistry. *In*: OVER, J., MORROW, J. R. & WIGNALL, P. (eds) *Understanding Late Devonian and Permian-Triassic Biotic and Climatic Events; Towards an Integrated Approach*, 199–225.

ROBINSON, S. G., SAHOTA, J. T. S. & OLDFIELD, F. 2000. Early diagenesis in North atlantic abyssal plain sediments characterized by rock-magnetic and geochemical indices. *Marine Geology*, **163**, 77–107.

ROBION, P., AVERBUCH, O. & SINTUBIN, M. 1999. Fabric development and metamorphic evolution of Lower Paleozoic slaty rocks from the Rocroi massif (French-Belgian Ardennes): new constraints from magnetic fabrics, phyllosilicate preferred orientation and illite crystallinity data. *Tectonophysics*, **309**, 257–273.

SANDBERG, C. A., ZIEGLER, W., DREESEN, R. & BUTLER, J. L. 1988. Late Frasnian mass extinction: conodont event stratigraphy, global changes, and possible causes. *Courier Forschunginstitut Senckenberg*, **102**, 263–307.

SANDBERG, C. A., MORROW, J. R. & ZIEGLER, W. 2002. Late Devonian sea-level changes, catastrophic events, and mass extinctions. *In*: KOEBERL, C. & MACLEOD, K. G. (eds) Catastrophic Events and Mass Extinctions: Impacts and Beyond: *Geological Society of America Special Paper*, **356**, 473–487.

SCHINDLER, E. 1990. Die Kellwasser Krise (hohe Frasne-Stufe, Ober Devon). *Göttinger Arbeiten zur Geologie und Paläontologie*, **46**, 1–115.

SEPKOSKI, JR, J. J. 1982. Mass extinctions in the Phanerozoic oceans: A review. *Geological Society of America, Special Papers*, **190**, 283–289.

SEPKOSKI, JR, J. J. 1986. Phanerozoic overview of mass extinctions. *In*: RAUP, D. M. & JABLONSKI, D. (eds) *Patterns and Processes in the History of Life*. Springer-Verlag, Berlin, 277–295.

STREEL, M., CAPUTO, M. V., LOBOZIAK, S. & MELO, J. H. G. 2000. Late Frasnian-Famennian climates based on palynomorph analyses and the question of the Late Devonian glaciations. *Earth-Science Reviews*, **52**, 121–173.

TRIBOVILLARD, N., DUPUIS, C. & ROBIN, E. 2000. Sedimentological and diagenetical conditions of the impact level of the Cretaceous/Tertiary boundary in Tunisia: no anoxia required. *Bulletin de la Société Géologique de France*, **171**, 629–636.

TRIBOVILLARD, N. P., DESPRAIRIES, A., LALLIER-VERGES, E., BERTRAND, P., MOUREAU, N., RAMDANI, A. & RAMANAMPOSOA, L. 1994. Geochemical study of organic-matter rich cycles from the Kimmeridge clay formation of Yorkshire (UK): productivity versus anoxia. *Palaeogeography, Palaeoclimatology, Palaeoecology*, **108**, 165–181.

TRIBOVILLARD, N., AVERBUCH, O., BIALKOWSKI, A. & DECONINCK, J.-F. 2002. The influence of early diagenesis of marine organic matter on the magnetic susceptibility signal of sedimentary rocks. *Bulletin de la Société Géologique de France*, **172**, 295–306.

TRIBOVILLARD, N., AVERBUCH, O., DEVLEESCHOUWER, X., RACKI, G. & RIBOULLEAU, A. 2004a. Deep-water anoxia over the Frasnian-Famennian boundary (La Serre, France): a tectonically induced oceanic anoxic event? *Terra Nova*, **16**, 288–295.

TRIBOVILLARD, N., AVERBUCH, O. & RIBOULLEAU, A. 2004b. Influence of marine organic-matter diagenesis on magnetic susceptibility of sedimentary rocks: the sulphide pathway. *Annales de la Société Géologique du Nord*, **11**, 57–67.

TRIBOVILLARD, N., RIBOULLEAU, A., LYONS, T. & BAUDIN, F. 2004c. Enhanced trapping of molybdenum by sulfurized marine organic matter of origin in Mesozoic limestones and shales. *Chemical Geology*, **213**, 385–401.

VANDERAVEROET, P., AVERBUCH, O., DECONINCK, J.-F. & CHAMLEY, H. 1999. Glacial/interglacial cycles in Pleistocene sediments of New Jersey expressed by clay minerals, grain-size and magnetic susceptibility data. *Marine Geology*, **159**, 79–92.

VORLICEK, T. P. & HELZ, G. R. 2002. Catalysis by mineral surfaces: implications for Mo geochemistry in anoxic environments. *Geochimica Cosmochimica Acta*, **66**, 3679–3692.

WALDEN, J., OLDFIELD, F. & SMITH, J. 1999. Environmental magnetism: a practical guide. *Technical Guide No. 6*, Quaternary Research Association, London, 243pp.

WALLISER, O. H. 1996. *Global events and event stratigraphy in the Phanerozoic: Results of international interdisciplinary cooperation in the IGCP Project 216 "Global Biological Events in Earth History"*. Springer-Verlag, Heidelberg.

WALLISER, O. H., EL HASSANI, H. & TAHIRI, A. 2000. M'rirt: a key area for the Variscan Meseta of Morocco. *Notes et Mémoires, Service Géologique du Maroc*, **399**, 93–108.

WARR, L. N. & RICE, A. H. 1994. Interlaboratory standardization and calibration of clay mineral crystallinity and crystallite size data. *Journal of Metamorphic Geology*, **12**, 141–152.

WEDEPOHL, K. H. 1971. Environmental influences on the chemical composition of shales and clays. *In*: AHRENS, L. H., PRESS, F., RUNCORN, S. K. & UREY, H. C. (eds) *Physics and Chemistry of the Earth*, Pergamon, Oxford, pp. 305–333.

WEDEPOHL, K. H. 1991. The composition of the upper Earth's crust and the natural cycles of selected metals. *In*: MERIAN, E. (eds) *Metals and their compounds in the Environment*, VCH-Verlagsgesellschaft, Weinheim, pp. 3–17.

WENDT, J. & BELKA, Z. 1991. Age and depositional environment of Upper Devonian (Early Frasnian to Early Famennian) black shales and limestones (Kellwasser Facies) in the Eastern Anti-Atlas, Morocco. *Facies*, **25**, 51–90.

YUDINA, A. B., RACKI, G., SAVAGE, N. S., RACKA, M. & MALKOWSKI, K. 2002. The Frasnian-Famennian events in deep-shelf succession, Subpolar Urals: biotic, depositional and geochemical records. *Acta Palaeontologica Polonica*, **47**, 355–372.

ZEGERS, T. E., DEKKERS, M. J. & BAILLY, S. 2003. Late Carboniferous to Permian remagnetization of Devonian limestones in the Ardennes: role of temperature, fluids, and deformation. *Journal of Geophysical Research*, **108**, 2357.

ZHENG, Y., ANDERSON, R. F., VAN GEEN, A. & KUWABARA, J. 2000. Authigenic molybdenum formation in marine sediments: a link to pore water sulfide in the Santa Barbara Basin. *Geochimica Cosmochimica Acta*, **64**, 4165–4178.

The biostratigraphical and palaeogeographical framework of the earliest diversification of tetrapods (Late Devonian)

A. BLIECK[1], G. CLEMENT[2], H. BLOM[2], H. LELIEVRE[3], E. LUKSEVICS[4], M. STREEL[5], J. THOREZ[6] & G. C. YOUNG[7]

[1]*Université de Lille 1, Sciences de la Terre, UMR 8014 du CNRS: Laboratoire de Paléontologie et Paléogéographie du Paléozoïque, F-59655 Villeneuve d'Ascq cedex, France (e-mail: Alain.Blieck@univ-lille1.fr)*

[2]*Subdepartment of Evolutionary Organismal Biology, Dept. of Physiology and Developmental Biology, Evolutionary Biology Centre, Uppsala University, Norbyvägen 18A, SE-75236 Uppsala, Sweden*

[3]*Muséum national d'Histoire naturelle, Département Histoire de la Terre, UMR 5143 du CNRS, Case Postale 38, 57 rue Cuvier, F-75231 Paris cedex 05, France*

[4]*University of Latvia, Institute of Geology, Rainis blvd. 19, LV-1586 Riga, Latvia*

[5]*Université de Liège, Département de Géologie, Unité de recherche Paléobotanique-Paléopalynologie-Micropaléontologie, Sart Tilman, B18, B-4000 Liège 1, Belgium*

[6]*Université de Liège, Département de Géologie, Sart Tilman, B18, B-4000 Liège 1, Belgium*

[7]*Australian National University, Department of Earth and Marine Sciences, GPO Box 4, Canberra, ACT 0200, Australia*

Abstract: The earliest diversification of tetrapods is dated as Late Devonian based on 10 localities worldwide that have yielded bone remains. At least 18 different species are known from these localities. Their ages span the 'middle'–late Frasnian to latest Famennian time interval, with three localities in the Frasnian, one at the F/F transition (though this one is not securely dated) and six in the Famennian. These localities encompass a wide variety of environments, from true marine conditions of the nearshore neritic province, to fluvial or lacustrine conditions. However, it does not seem possible to characterize a freshwater assemblage in the Upper Old Red Sandstone based upon vertebrates. Most of the tetrapod-bearing localities (8 of 10) were situated in the eastern part of Laurussia (=Euramerica), one in North China and one in eastern Gondwana (Australia), on a pre-Pangean configuration of the Earth, when most oceanic domains, except Palaeotethys and Panthalassa, had closed.

The earliest record of tetrapods (four-legged vertebrates) in the Late Devonian is one of the key events in the evolution of vertebrates. It had a very important and lasting impact on the terrestrial ecosystem. These vertebrates are the oldest representatives of tetrapods, a major group of animals that today numbers some 24 000 living species.

Our understanding of the origin of tetrapods, better known popularly as the 'fish–tetrapod transition', has progressed greatly thanks to recent fossil discoveries. Today, eight Devonian tetrapod genera out of a total of 12 (and perhaps even 17 when counting the as-yet undescribed taxa; see below the taxonomic section) have been found in the last 15 years. These new finds have made it possible to reconstruct sequences of character change leading to tetrapod morphologies, eventually to improve phylogenetic analyses, and tentatively identify the genetic basis for some of these changes (e.g. Daeschler & Shubin 1998). However, these discoveries have raised new questions about the evolutionary context of the origin of tetrapods, questions which need to be answered if the event is to be properly understood (see, e.g. Schultze 1997; Clack 2002*b*; Long & Gordon 2004).

In this paper, we address the geological context of this event, that is, mainly the biostratigraphically based dating of the Late Devonian tetrapod-bearing localities, and their palaeogeographical context in the frame of a recently proposed global reconstruction (Averbuch *et al.* 2005). We also comment on

From: BECKER, R. T. & KIRCHGASSER, W. T. (eds) *Devonian Events and Correlations.*
Geological Society, London, Special Publications, **278**, 219–235. DOI: 10.1144/SP278.10
0305-8719/07/$15 © The Geological Society of London 2007.

the palaeoenvironmental interpretation of the various localities known to date, based upon the most recent discoveries of fossils (e.g. as in Belgium), and a cluster analysis of Famennian vertebrate localities worldwide (Lelièvre 2002).

Taxonomic overview

In this section, we briefly review the diversity of Late Devonian tetrapods, taking into account the most recent published data which are summarized in Table 1. We focus on the localities that have yielded tetrapod bone remains (either fragmentary remains or articulated skeletons), and we do not review the trace and trackway localities (for this topic, we refer the reader to Clack 1997 and 2002*b*, pp. 92–95).

Pennsylvania (late Famennian)

The limited Red Hill outcrop in Clinton County, Pennsylvania, USA, has provided more than one tetrapod genus. *Hynerpeton bassetti* is known from most of the left cleithrum with its scapulocoracoid (Daeschler *et al.* 1994), and a second left cleithrum. The posterior part of a right lower jaw has provisionally been assigned to the same genus (Daeschler 2000). *Densignathus rowei* (Daeschler 2000) is defined on very distinguishable lower jaws (a left lower jaw and the posterior portion of a right lower jaw, both supposed to be from a single individual). Another tetrapod remnant has been discovered recently from the same locality, i.e. an isolated left humerus (Shubin *et al.* 2004) and as yet undescribed cranial and post-cranial material (Blom & Clément pers. obs.).

East Greenland (late Famennian)

The tetrapods from East Greenland are the most complete and best known of all the Devonian taxa. To date, three genera and five species are known from more than 550 block specimens. The well-known *Ichthyostega* and *Acanthostega* are represented by a large number of well-preserved skulls and lower jaws, as well as almost complete post-cranial material (see, e.g. Clack 2002*b*). In a recent revision, Blom (2005) shows a morphological variation within the cranial material, which justifies the recognition of three species of *Ichthyostega*: *I. stensioei* Säve-Söderbergh 1932 *I. eigili* Säve-Söderbergh 1932 and *I. watsoni* Säve-Söderbergh 1932. Originally, Säve-Söderbergh (1932) defined the genus *Ichthyostegopsis* on the basis of a small skull with proportions and suture boundaries different from those of *Ichthyostega*. *Ichthyostegopsis* is no longer considered valid since the differences are regarded as an expression of juvenile characters (Blom 2005). However, a third tetrapod genus has recently been recognized based on jaw and tooth morphology from specimens collected on the south side of Celsius Bjerg (Blom pers. obs.; Blom *et al.* 2003; Clack *et al.* 2004).

Scotland (mid- to late Frasnian)

The tetrapod material from Scaat Craig near Elgin, Scotland, includes both cranial and post-cranial material (Ahlberg 1991, 1995, 1998; Ahlberg & Clack 1998; Ahlberg *et al.* 2005). Skull bones and lower jaw fragments, comprising the premaxilla plus median rostral and all of the lower jaw except the articular and medial wall of the adductor fossa, form the type material of the stem tetrapod *Elginerpeton pancheni* (Ahlberg 1995; Ahlberg & Clack 1998; Ahlberg *et al.* 2005). The postcranial material provisionally referred to *Elginerpeton* (Ahlberg 1998) comprises the dorsal part of a scapulocoracoid plus the ventral part of a cleithrum, an ilium and one specimen each of humerus, femur, tibia and neural arch (Ahlberg 1998). However, the identification of the humerus has recently been challenged (Shubin *et al.* 2004; see Ahlberg 2004 for a contrasting view).

Belgium (mid- or late Famennian)

Lohest (1888, pl. VIII: 2, 5) misinterpreted a Devonian tetrapod lower jaw as a large fish remain and assigned it to a new species of *Dendrodus*, *D. Traquairi* [*sic*]. This lower jaw was found in the Famennian of Strud, Namur Province, Belgium, from the Evieux Formation extending apparently from the Middle Famennian in this area. However, this taxon appears to be a representative of *Ichthyostega* ('*Ichthyostega*-like form' in Clément *et al.* 2004). The Devonian tetrapod occurrence in Belgium is today only based on this isolated right lower jaw, but current investigations at the rediscovered Strud locality lead us to expect the discovery of new tetrapod material.

Latvia and western Russia (Frasnian)

Obruchevichthys was originally described as a sarcopterygian fish (Vorobyeva 1977). Known only from lower jaw fragments, it may not have had true limbs and digits, although its phylogenetic relationships are likely to be close to tetrapods (Ahlberg 1991; Clack 2002*b*, p. 91; tetrapods in the sense used by Clack 2006: 'a vertebrate with limbs and digits, as this is the sense in which it is most readily understood by the non-specialist').

Table 1. *Devonian tetrapod localities of the world with their formation, age and environment*

Country	Locality	Tetrapod(s)	Formation	Age	Environment	References
Pennsylvania (USA)	Red Hill, Clinton County	*Hynerpeton bassetti*, *Densignathus rowei*, 3rd taxon (?)	uppermost Catskill Formation, Duncannon Member	upper Famennian, VH Miospore Zone	coastal alluvial plain	Daeschler *et al.* 1994; Daeschler 2000; Traverse 2003
East Greenland	Gauss Halvø and Ymer Ø	*Acanthostega gunnari*, *Ichthyostega watsoni*, *I. eigili*, *I. stensioei*, Tetrapoda gen. et sp. nov.	Aina Dal Formation and Britta Dal Formation	upper Famennian (higher part of GF Miospore Zone to upper or uppermost Famennian)	fluvial	Jarvik 1996; Marshall *et al.* 1999; Clack 2002*a*, 2002*b*; Clack *et al.* 2003, 2004; Blom 2005
Scotland	Scat [Scaat] Craig, near Elgin	*Elginerpeton pancheni*	Scat [Scaat] Craig beds	'middle' or upper Frasnian	fluvial	Ahlberg 1991, 1995; Trewin 2002
Belgium	Strud, Namur Province	*Ichthyostega*-like tetrapod	Evieux Formation (?)	upper middle Famennian (lower part of the GF Miospore Zone)	coastal alluvial plain	Thorez *et al.* 1977; Bultynck & Dejonghe 2002; Clément *et al.* 2004 and this paper
Latvia & Russia	Velna-Ala, Abava riv., Latvia, and ?Novgorod district, Russia (exact location unknown)	*Obruchevichthys gracilis*	Ogre Formation, Latvia; unknown in Russia	upper Frasnian, equiv. *rhenana* Conodont Zone	shallow, near-shore marine	Vorobyeva 1977; Sorokin 1978; Ahlberg 1991, 1995
Latvia	Pavari and Ketleri	*Ventastega curonica*, 'second tetrapod genus'?	Ketleri Formation	upper Famennian, equiv. lower *expansa* Conodont Zone (or younger)	low-tidal, near-shore, marine	Ahlberg *et al.* 1994; Esin *et al.* 2000; Luksevics & Zupins 2003, 2004
Russia	Andreyevka-2, Tula region	*Tulerpeton curtum*, Tetrapoda indet.?	Khovanshchina Formation, Zavolzhsky Horizon	uppermost Famennian (Strunian) equiv. *praesulcata* Conodont Zone (or older)	marine (epicontinental sea)	Lebedev & Clack 1993; Alekseev *et al.* 1994; Lebedev & Coates 1995
Russia	Gornostayevka, Oryol region	*Jakubsonia livnensis*	Zadonskian Regional Stage	lower Famennian, equiv. *crepida* Conodont Zone	deltaic near-shore, marine	Esin *et al.* 2000; Lebedev 2003, 2004
China	Ningxia Hui region	*Sinostega pani*	Zhongning Formation	Frasnian, better than 'upper' Famennian	non-marine	Pan *et al.* 1987; Ritchie *et al.* 1992; Zhu *et al.* 2002
Australia	Jemalong Gap, SW of Forbes, N.S.W.	*Metaxygnathus denticulus*	Nangar Subgroup, Cloghnan Shale	lower Famennian or upper Frasnian	fluvial	Campbell & Bell 1977; Young 1993, 1996, 1999; Young *et al.* 2000*a*

Latvia (Famennian: Pavari and Ketleri)

Ventastega is represented by virtually the whole skull and lower jaw, together with most of the shoulder girdle, part of the pelvis, and fragments of the axial skeleton (ribs and tail fin rays), that makes it the most complete Devonian tetrapod besides *Ichthyostega* and *Acanthostega*. Ahlberg *et al.* (1994, p. 322 and fig. 14) suggest the occurrence of a 'second tetrapod genus' in Ketleri, together with *Ventastega*, on the basis of a mandibular fragment. We propose that this conclusion must be considered with caution until more material is collected.

Russia (Famennian, Andreyevka-2)

The state of preservation of the material of *Tulerpeton* is 'perfect and three-dimensional' according to Alekseev *et al.* (1994, p. 44). In this locality, Lelièvre (2002, p. 151; also Lebedev 2004, table I) cites a Tetrapoda indeterminate, in addition to *Tulerpeton*, after Lebedev & Clack (1993) who have cautiously differentiated the holotype of *Tulerpeton* from all the other sarcopterygian remains in the locality. Lebedev & Clack (1993) believe that two different tetrapods occur at Andreyevka-2, on the basis of two types of tabular bones. These bones are variable among sarcopterygians, which could equally be the case among tetrapods. Moreover, one of the bones figured by Lebedev & Clack (1993, fig. 2 H–I) might not be a tabular. So, the supposed second tetrapod of Andreyevka-2 should be considered with caution, although Lebedev has been advocating two different taxa for 10 years. However, this hypothesis is not in disagreement with the fact that two tetrapods do occur together in the same locality of Pennsylvania, and that more than two tetrapods are known from East Greenland (Table 1).

Russia (Famennian, Gornostayevka)

The recently described material of *Jakubsonia livnensis* (Lebedev 2003, 2004) includes disarticulated cranial and postcranial (pectoral girdle) elements, and the posterior part of a skull roof questionably attributed to *Jakubsonia*.

China (Frasnian)

The record of a tetrapod from Asia, *Sinostega pani*, is restricted to an incomplete left mandible from the Zhongning Formation of the Ningxia Hui autonomous region, northwestern China (Zhu *et al.* 2002). Only the medial view is exposed, comprising most of the prearticular, together with the angular and postsplenial.

Australia (Frasnian/Famennian)

Metaxygnathus denticulus from the Cloghnan Shale, near Forbes, New South Wales, Australia was the first Devonian tetrapod to have been described outside Greenland (Campbell & Bell 1977). It is represented by a complete but poorly preserved right lower jaw.

Taxonomic conclusion

Twelve different genera of Late Devonian tetrapods are now known (Fig. 1) with one species each, except *Ichthyostega* with three species. This makes 14 different species. Nevertheless, other taxa have been mentioned or are still imperfectly known: a third taxon in Pennsylvania (?), a third genus in Greenland, a second genus in Ketleri, Latvia (?), and a second taxon in Andreyevka-2, Russia (?). This would give a total of 15 (sure) to 18 (sure and possible) different species, corresponding to 12 (sure) to 17 (sure and possible) different genera. Additionally, an undescribed tetrapod taxon has been recently announced by Clack *et al.* (2004) from the Upper Devonian of Timan, Russia (work in progress after Ervins Luksevics). This would settle the number of separate Late Devonian tetrapod species to 19.

Critical biostratigraphical review

Important associated flora and fauna are sometimes found with tetrapod remains. The vertebrate faunal composition (placoderms, chondrichthyans, acanthodians, actinopterygians, sarcopterygians) is now increasingly used to aid studies in biostratigraphical approach. Correlations between biostratigraphical subdivisions of the different tetrapod-bearing localities of the world (USA, Greenland, Europe, China, Australia) are currently being attempted by an international early tetrapod working group that assembled in Riga, Latvia, during the Gross Symposium 2 (Sept. 8–14, 2003; Schultze *et al.* 2003). Thus, comparisons with contemporaneous tetrapod-bearing localities will improve our understanding of palaeoenvironmental conditions in which the oldest tetrapods were living (see below) as well as their biogeographical distribution. Nevertheless, to be accurate, it is absolutely necessary for these studies to be based on a very consistent and internationally accepted biochronological framework.

Work in progress by the Subcommission on Devonian Stratigraphy has stabilized in favour of a three-fold subdivision of the Frasnian, and a four-fold subdivision of the Famennian (e.g. Bultynck 2004; Streel 2005). Four Famennian substages (lower, middle, upper, and uppermost Famennian or Strunian) could most probably be accepted by most Devonian vertebrate palaeontologists. The most important thing is that the different faunal levels are precisely defined and dated (see Clack 2006).

Pennsylvania (late Famennian)

All Devonian tetrapod remains come from a single limited locality known as Red Hill, Clinton County. These specimens were found in the Duncannon Member, the uppermost subdivision of the Catskill Formation. Palynomorph samples were collected from several levels at the Red Hill locality. According to Traverse (2003), the spore taxa assemblages would place these samples in the upper sixth of the Famennian Stage. Traverse noticed that Maziane et al. (1999), providing a revision based on sections in Belgium with faunal control, proposed that the VH Spore Zone (for *Apiculiretusispora verrucosa–Vallatisporites hystricosus*) becomes the upper part of the previous VCo Zone (with *Apiculiretusispora verrucosa* and *Vallatisporites hystricosus*, but without *Retispora lepidophyta*). However, the range of the VH Spore Zone is more extended in Pennsylvania (wet area) than in western Europe (dry area). The Red Hill outcrop is thus referable to the interval *trachytera* to middle *expansa* Conodont Zones of the upper Famennian Substage (Fig. 1; see Streel & Loboziak 1996).

East Greenland (late Famennian)

Palynological dating has recently unambiguously resolved the stratigraphical age of the tetrapod-yielding parts of the sequence (Marshall et al. 1999). Spore samples bracketing the upper and lower occurrences of tetrapods place them securely between the upper GF (upper Famennian) and LL to LN (upper to uppermost Famennian) Spore Zones. These data contradict the previously controversial study suggesting a Carboniferous age (Hartz et al. 1997, 1998). In fact, these authors did not suggest a Carboniferous age directly, only that the absolute age was much younger than previously expected, an absolute age that would normally put it in the Carboniferous—but all the other absolute dates could also have been wrong! They put an end to a long dispute on the age of the East Greenland tetrapods (Jarvik 1996; Stemmerik & Bendix-Almgreen 1998). More precisely, the lowest occurrence of *Ichthyostega* and *Acanthostega* is higher than, but close to the base of the latest *marginifera* Conodont Zone. Indeed, the GF Miospore Zone in East Greenland contains *Retispora macrotuberculata*, a marker for the base of the middle part of the GF Zone, i.e. the biostratigraphical level 15 of Streel & Loboziak (1996, fig. 4) (Fig. 1).

Scotland (mid- to late Frasnian)

In his original papers on *Elginerpeton*, Ahlberg (1991, 1995, 1998) gives the age of the Scaat Craig Beds, a possibly partial lateral equivalent of the Alves Beds in the South Moray Firth area, as 'upper Frasnian'. This late Frasnian age is indeed classically advocated for the Scaat Craig Beds (e.g. Friend & Williams 1978, fig. 13; Trewin 2002, fig. 8.28). However, it does not seem to be based upon firm data, independently of fishes. Miles (1968, table 2) correlates the Scaat Craig Beds with the lower part of the '*Phyllolepis* Series' of East Greenland (also Mykura 1991, table 9.3), which is now named the Kap Graah Group, and considered as lower Famennian (Jarvik 1996, fig. 9; Clack & Neininger 2000, fig. 2). The Kap Graah Group lies below the Agda Dal and Elsa Dal formations. The topmost part of the latter has been dated as GF Miospore Zone by Marshall et al. (1999), but corresponds (see East Greenland) to the middle part of that zone spanning the middle–upper Famennian boundary, i.e. the upper to uppermost *marginifera* Conodont Zone *sensu* Streel & Loboziak (2000). So, the Kap Graah Group might be at least pre-GF in age, that is lower to lower middle Famennian (*sensu* Streel & Loboziak 2000, in a four-fold subdivision of the Famennian), or older, i.e. upper Frasnian. This means that the Scaat Craig Beds may be or may not be late Frasnian in age. This late Frasnian age is based upon their fish assemblage (Miles 1968), and there is seemingly no independent biostratigraphical marker such as conodonts, miospores or other fossils. Nevertheless, two other arguments may be given for the Frasnian age of the Scaat Craig Beds:

1. The occurrence of the psammosteid ostracoderm genera *Psammosteus* and *Traquairosteus* in the Scaat Craig Beds (Miles 1968, p. 8) is an argument for a pre-Famennian age (see Ahlberg 1998, p. 102, and section China, here below).

2. Miles (1968, table 2) correlated the Scaat Craig Beds with the interval between the 'c-d Shelon-Ilmen' and the 'e-Stage' of the East Baltic area. The Ilmen and overlying beds of Latvia and the Main Devonian Field of NW Russia are correlated with the 'middle' Frasnian, Daugava Regional Stage (Paskevicius 1997, fig. 66), considered as equivalent to the 'middle' Frasnian *punctata* to *jamieae* Conodont Zones (Esin et al. 2000, fig. 1), and the 'e-Stage' is equivalent to the

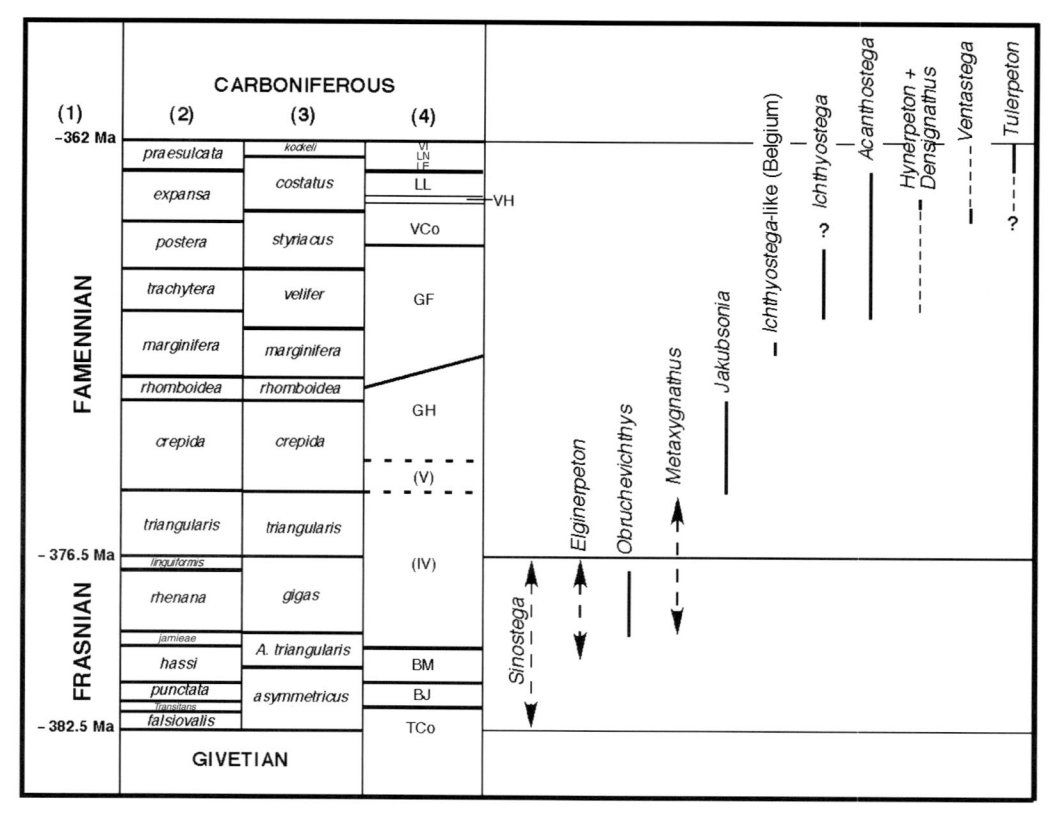

Fig. 1. Biostratigraphical distribution of Devonian tetrapods after data of Table 1. Note that, as stated by Marshall *et al.* (1999, fig. 4), this figure illustrates the age error bar of each taxon. These ages are not ranges, but age durations of conodont/miospore zones in which the taxa have been collected. Dashed lines with arrows indicate uncertainties in datations. *Notes*: (1) radiochronological scale of Williams *et al.* (2000, fig. 8); (2) standard conodont zones of Ziegler & Sandberg (1990, fig. 1); (3) older conodont zones after Ziegler & Sandberg (1990); (4) miospore zones *sensu* Streel in Bultynck & Dejonghe (2002, table 1).

Snezha and Pamusis regional stages, which correspond to the late Frasnian *rhenana* Conodont Zone. Ahlberg *et al.* (1999) correlated the Scaat Craig Beds with 'some part of the Pamusis–Snezha interval' judging from distribution of *Psammosteus*. We thus retain a 'middle' to late Frasnian age for the Scaat Craig Beds (Table 1 and Fig. 1).

The Scaat Craig fauna also comprises placoderms and acanthodians, and appears to be similar to that of two other localities of Scotland: Poolymore and Whitemire (Newman 2005). 'Poolymore was considered to be lower down in the Edenkillie Beds (Miles 1968)' (Newman 2005), that is, in a level equivalent to the 'middle Frasnian' Whitemire Beds (Mykura 1991; Trewin 2002). So, for Newman (2005), the three faunas (Scaat Craig, Poolymore and Whitemire) 'might be stratigraphically closer than considered hitherto' and *Elginerpeton* 'is even older than previously thought'.

Belgium (mid- or late Famennian)

The composition of the sandstone surrounding the tetrapod lower jaw is very unusual (fluviatile deposits generated by flood events). Such a lithofacies is usually known in the Evieux Formation in the northern part of the Dinant synclinorium, Ardenne Allochthon, and the Namur Synclinorium, Brabant Parauthochthon (Thorez *et al.* 1977; Bultynck & Dejonghe 2002) to which it was first attributed (Clément *et al.* 2004). The rediscovery of the one-century old abandoned quarry of Strud, and, more importantly, the finding in November 2005 of the stratum which most probably yielded the tetrapod material in this quarry, have allowed more accurate dating. The palynological study is still in progress, but already it conclusively shows that the Strud locality is older than previously thought (Clément *et al.* 2004). It is now considered to be upper middle Famennian in age (*sensu* Streel in Bultynck

& Dejonghe 2002; Streel 2005; but note that the subdivisions of the Famennian are still under discussion among the Subcommission on Devonian Stratigraphy), i.e. containing the lower part of the GF Miospore Zone and so close to but older than the East Greenland material (Fig. 1). This also means that the Strud quarry Evieux lithofacies might have the same age as the Souverain-Pré Formation elsewhere (Bultynck & Dejonghe 2002, fig. 7). However, a more precise correlation between the different Famennian sections of the Condroz area has still to be processed.

Latvia and western Russia (Frasnian)

The holotype of *Obruchevichthys* comes from the upper Frasnian Ogre Formation of Latvia, which correlates with the *rhenana* Conodont Zone (Esin *et al.* 2000). However, Vorobyeva (1977, pl. XIV: 4, and p. 204) inadequately attributed the sandstones of the type-locality of *Obruchevichthys* to the Nadsnezha Beds, a lithostratigraphical unit which is distributed in the Novgorod district of Russia, along the Lovat river. The type-locality of *Obruchevichthys* is Velna Ala, within the lower, Lielvarde Member of the Ogre Formation. This member consists mainly of fine-grained calcareous sandstones, with a gypsum cement in its lower part, and clay, silt and dolomitic marl in its upper part. Sorokin (1978) supposed that these deposits were formed in shallow waters of a narrow gulf of the Baltic palaeobasin, under conditions of variable salinity. In western Russia, the locality of *Obruchevichthys* is unfortunately not precisely known. Vorobyeva (1977, p. 204 and fig. 46) mentions that it originates perhaps from the Novgorod District. Judging by its state of preservation and the matrix, it is likely that it was collected somewhere along the Lovat river where a rich collection of *Bothriolepis maxima* and other fish remains was gathered. The *Bothriolepis maxima* placoderm 'zone' is biostratigraphically correlated to the *rhenana* Conodont Zone (Esin *et al.* 2000, fig. 1; Luksevics 2001, fig. 2).

Latvia (Famennian: Pavari and Ketleri)

The holotype and main portion of the *Ventastega* material comes from the Pavari locality where the fine-grained sandstone and sand of the Pavari Member of the Ketleri Formation crops out. Other material has been collected from the overlying Varkali Member of the Ketleri Formation represented also by weakly cemented sandstones. No spores or conodonts have been found in the Ketleri Formation, therefore the age of the formation (possibly corresponding to the *expansa* Conodont Zone; Esin *et al.* 2000) can be judged only from its position above the Zagare Formation; besides that, one needs to take into consideration the significant break corresponding to the erosional surface between the Pavari Member and lowermost Nigrande Member of the Ketleri Formation. Therefore, the possibility that the Ketleri Formation could be even younger, corresponding to the latest Famennian, cannot be excluded. Dolomite from the Zagare Formation in Lithuania yields conodonts allowing correlation of this formation with the interval from the *marginifera* to *postera* Zones (Esin *et al.* 2000). Furthermore, the underlying Svete Formation in Lithuania contains conodonts of the *postera* Zone (or Middle–Lower *styriacus* Zone of the previous conodont zonation, as it was reported by Zeiba & Valiukevicius 1972). So, an *expansa* Conodont Zone or younger age is retained for *Ventastega* (Fig. 1).

Russia (Famennian, Andreyevka-2)

The material of *Tulerpeton* comes from a section (ANE-1 or Andreyevka-2 in Alekseev *et al.* 1994) on the right bank of the Tresna river, south of the village of Andreyevka, in the Tula region, south of Moscow. It is located in the lower part of the Khovanshchina Formation, where a limestone-clay sequence contains stromatolites, ostracods, serpulids and charophytes. The skeleton of *Tulerpeton* was found in a layer with abundant sarcopterygian and other fish material. The Khovanshchina Formation is generally dated as equivalent to the *praesulcata* Conodont Zone, even though no conodonts have been found at Andreyevka-2 itself. However, ostracods do occur that are characteristic of the Khovanshchina Formation; they belong to the *Maternella hemisphaerica–Carboprimitia turgenevi* Zone, correlated to the 'Strunian' of the Franco-Belgian basin by Alekseev *et al.* (1994). Nevertheless, the *Maternella hemisphaerica–M. dichotoma* Zone has a much lower range in the Rhenish Massif, down to the upper *postera* Conodont Zone (Groos-Uffenorde *et al.* 2000, fig. 3). This gives the possibility of an older, late Famennian age for *Tulerpeton* (Fig. 1).

Russia (Famennian, Gornostayevka)

The material comes from the Gornostayevka quarry, SW of the town of Livny, Oryol Region of Central Russia, and is dated as equivalent to the ?*triangularis–crepida* Conodont Zone (Lebedev 2004). In this locality, Lebedev (2003) cites *Bothriolepis* cf. *leptocheira*, an antiarchan placoderm. *B. leptocheira* is classically known from the Eleja Regional Stage (RS) of the Russian Platform (Main Devonian Field, including the East Baltic area; Luksevics 2001), a formation which is usually dated as basal Famennian (Esin et al. 2000, fig. 1). The Zadonskian RS of the Oryol region (Central Devonian Field), where the tetrapod comes from, is correlated by Esin et al. (2000, fig. 1) to the Joniskis RS of the Main Devonian Field, just above the Eleja RS (whose latest Frasnian or earliest Famennian age is, however, not solved by the most recent study of its conodont and fish content in Lithuania: Valiukevicius & Ovnatanova 2005). The Zadonskian would thus be equivalent to the upper *curonica* placoderm 'zone' of the Main Devonian Field, and to the *crepida* Conodont Zone (Esin et al. 2000), that is, lower Famennian.

China (Frasnian)

For this locality, Zhu et al. (2002) propose a 'late Famennian' age 'about 355 million years BP', a datum which may be understood as 'latest Famennian' or better 'Famennian/Tournaisian boundary' when using a classical radiochronological scale such as Odin's (1994). However, the most recent revisions of the Devonian scale give −362 to −359 Ma for the Devonian/Carboniferous boundary (Williams et al. 2000; Gradstein & Ogg 2004). So, an age of −355 Ma would fit the earliest Carboniferous better than the Famennian. In fact, Zhu et al. (2002) base their dating of this tetrapod locality upon Pan et al.'s (1987) book for the Famennian age of the Zhongning Formation (Pan et al. 1987, pp. 184–185). However, when considering the miospore assemblage of the Zhongning Formation as listed and figured by Gao in Pan et al. (1987, pp. 120–131, 184–185 and pl. 35–36), it is probably better considered older than the Famennian (G. Playford in litt. to Ritchie et al. 1992, p. 364; S. Loboziak pers. comm. to AB, 12.05.1989—now deceased). For Playford, it '... is certainly older than the latest Devonian *Retispora lepidophyta* Assemblage and could even be pre-Late Devonian ... (and) datable within the interval mid-Givetian to Frasnian' (Ritchie et al. 1992, p. 364). For Loboziak (unpublished), it is likely to be Frasnian. We will thus provisionally consider the *Sinostega* locality as Frasnian in age (Table 1 and Fig. 1).

Incidentally, if the Frasnian age of this locality is confirmed, it reinforces the generally accepted stratigraphical distribution of ostracoderms (*sensu* Janvier 1996). The Zhongning Formation has indeed yielded fragmentary remains of a Galeaspida gen. et sp. indet. (Pan et al. 1987, fig. 17, and pl. 1, 2, 3:1). As all other galeaspids are known only from the Lower to Lower Middle Devonian (Macrovertebrate Assemblages I to VI of China: Zhu 2000, pp. 375–376; Zhu et al. 2000, fig. 2), and as no other ostracoderm is known after the Frasnian (Blieck 1991), the galeaspid from the Zhongning Formation, if Famennian in age, would be the youngest record of galeaspids (Zhu 2000, p. 376), and the youngest record of ostracoderms. But if the Zhongning Formation is considered as Frasnian in age, it is consistent with no ostracoderm being younger than the Frasnian, and with the Frasnian/Famennian biological event being not a simple artefact for agnathans.

Australia (Frasnian/Famennian)

The first evidence of Devonian tetrapods from Gondwana was provided by trackways discovered in Upper Devonian strata of Victoria (Warren & Wakefield 1972). These evidently were made by several unknown tetrapod taxa (Clack 1997). Associated plant remains indicate a Late Devonian age, and Lewis et al. (1994) equated the relevant strata with the Merrimbula Group of southeastern New South Wales, where several fish faunas occur well beneath a Frasnian marine incursion (Young 1993, p. 215). An older trackway described from western Victoria by Warren et al. (1986) was probably not made by a tetrapod according to Clack (1997). The only tetrapod body fossil known from Gondwana is the lower jaw of *Metaxygnathus denticulus* Campbell & Bell (1977), found at Jemalong Quarry, SW of Forbes, New South Wales. The locality is in the Cloghnan Shale, with an associated fish fauna including lungfish (Campbell & Bell 1982; Ahlberg et al. 2001), and placoderms (Young 1993, 1999). Campbell & Bell (1977, p. 369) suggested a late Frasnian or early Famennian age, more likely at the younger end of this time interval, based on consideration of stratigraphical correlations with the Upper Devonian Hervey Group to the east. The tetrapod–fish assemblage was assigned to the 'Jemalong–Canowindra fauna' *sensu* Young (1993: Macrovertebrate Fauna 13), originally dated as early–middle Famennian (Young 1993, fig. 9.2; Young 1996, chart 14). However, a slightly older (late Frasnian) age for the Canowindra fish fauna was suggested by Young (1999, p. 145), indicating approximate alignment to the *rhenana/triangularis* Conodont Zone (MAV 13, Young & Turner 2000, fig. 2). The Canowindra fish fauna occurs near the base of the Hervey Group sequence, associated with evidence of a marine incursion. The tetrapod locality in the Cloghnan Shale is also near the base of the Upper Devonian sequence in the Jemalong Range. Detailed remapping and revision of Hervey Group stratigraphy in central NSW (Young et al. 2000a) supports the correlations initially proposed by Campbell & Bell (1977, pp. 374, 375; but note the nomenclatural change that the 'Pipe Formation' in the vicinity of the Canowindra fish fauna has been renamed the 'Mount Cole Formation').

In their discussion of correlations with the presumed marine/estuarine interval to the east, Campbell & Bell (1977, p. 375) recorded that 'marine rocks at Parkes are at the base of the Mandagery Formation', thus supporting a younger (Famennian) age for the Jemalong occurrence. However, Young (1999) noted that both the lingulid facies and the Canowindra fish fauna to the east of Parkes occur stratigraphically within transition beds between the Mandagery Formation (sandstone), and the overlying finer-grained Mount Cole Formation ('Pipe Formation'). Rather than 'at the base', this ?marine/estuarine interval occurs at the top of the Mandagery Formation, negating the earlier argument for a younger age. The basal marine/estuarine interval is succeeded by a stratigraphical thickness of the Hervey Group estimated in excess of 2.5 km. Both lycopod plants (*Leptophloeum*) and placoderm fishes near the top of the sequence indicate a Late Devonian rather than Carboniferous age (Young *et al.* 2000*a*). The only Late Devonian conodonts from this area (Jones & Turner 2000) come from the Catombal Group at Gap Creek near Orange, 100 km E of Jemalong, and 40 km NE of the Canowindra fish locality. Mawson & Talent (2003) have recently assessed these as indicating an early Famennian age (*crepida* Conodont Zone or younger). The horizon is at least 280 m above the lowest exposed Upper Devonian strata, where corals and bryozoans have been found (base of the sequence obscured by Tertiary basalt; R. K. Jones pers. comm. to GCY, 9.08.2004). A middle level in this sequence was reported to contain spores suggesting an age perhaps as old as Givetian (Webby 1972, p. 119). This is consistent with a SHRIMP zircon U/Pb isotopic age of 376 ± 4 Ma reported by Raymond (1998, p. 220) for the Dulladerry Volcanics, of which the Merriganowry Shale Member is a conformable sequence lying beneath the basal sandstones of the Hervey Group near Cowra (between Canowindra and Forbes; see Young 1999, fig. 1). Thus, the conodonts, bryozoans and brachiopods of the higher '*Lingula* limestone' horizon (Jones & Turner 2000, fig. 2) may represent a separate younger marine incursion (assigned to the Early *marginifera* Zone by Talent *et al.* 2000, p. 253), compared to the single incursion of Frasnian age previously assumed by Webby (1972) and Young (1993).

The similarity analysis of Lelièvre (2002, fig. 21, p. 172) groups the Jemalong–Canowindra fauna with various Northern Hemisphere Famennian tetrapod-bearing localities, including the Britta Dal and Aina Dal formations (East Greenland), the Catskill Formation (Pennsylvania), Andreyevka-2 (Russia), and Pavari (Latvia), but this is influenced by associated placoderms (phyllolepids, *Bothriolepis*, *Remigolepis*). This association is only recorded from Famennian strata in the Northern Hemisphere, but is clearly older (Givetian–Frasnian) in East Gondwana (based upon the whole palaeontological data: see reviews in Young 2003, 2005*a*–*b*). Thus, the problem of the age of the Jemalong fauna must also take account of such biogeographical considerations (Young *et al.* 2000*b*).

Biostratigraphical conclusion

After biostratigraphical information, it appears that most Devonian tetrapods are not exclusively late Famennian in age as thought previously (e.g. Jarvik 1996), but span at least the late Frasnian to late Famennian (Fig. 1; to be compared to Schultze 1997, fig. 1; Clack 2002*b*, fig. 3.2; Long & Gordon 2004, fig. 1). It is certainly difficult to say that tetrapods themselves can be used for correlation until a better fossil record is achieved. However, we need a good biostratigraphical framework which, together with a revised cladistic analysis of the various taxa now known, should lead to a renewed view of the early spreading of tetrapods in Devonian time (see earlier reviews by Schultze 1997; Clack 2002*b*; Long & Gordon 2004).

Palaeoenvironmental considerations

Devonian tetrapods are found in the Frasnian and the Famennian. They had a worldwide, nearly Pangaean palaeogeographical distribution (Laurussia, Gondwana, North China) and, although classically considered as being from terrestrial environments, are found in sedimentary rocks whose original environments are interpreted either as freshwater, brackish or marine (references on Table 1; also Schultze 1997; Lebedev 2004).

The question of the original environment of early tetrapods has been reviewed by Clack (2002*b*, pp. 99–104). She points out the influence of this question upon another one: 'why did tetrapods evolve?' Leaving the older scenarios aside, the question of the environment of Frasnian–Famennian tetrapods has been addressed through comparisons between faunal assemblages of fossiliferous localities by means of cluster analysis, and through analysis of co-occurring aquatic forms with tetrapods. Among others, the different trials by Schultze & Maples (1992), Schultze *et al.* (1994)

and Schultze & Cloutier (1996) (see a summary in Schultze 1997) proposed different analyses bearing on vertebrates, and vertebrates associated with invertebrates of different localities of Devonian and Carboniferous age, mainly from North America. All these analyses share the same phenetic method using general observed similarities of the fauna and flora to cluster the compared localities, where both absence and presence of taxa are considered, an assessment that is inherent to the methodology but that can be discussed depending on the use or not of hierarchical classifications.

More recently, one of us (Lelièvre 2002) has proposed a cluster analysis of 39 Famennian localities bearing early vertebrates. Different methods have been used, viz. classical hierarchical clustering and neighbour-joining (NJ). The latter method is advantageous in that it can use distances with metric and additive properties of the data in order to get a tree of minimal length. Fossil assemblages are treated at a taxonomic resolution to genus and family as most of the species described from those Famennian localities are monotypic. The result of running the NJ method on a localities/taxa matrix gives a single tree, an artefact due to the phenetic method. This tree shows that no assemblage defines a cluster of localities supposed to be freshwater (also Schultze & Cloutier 1996; Schultze 1997; Lelièvre 2002). Some of the localities, i.e. Greenland (where the Aina Dal and Britta Dal Formations are distinguished), Dura Den (Scotland), and Pavari (Latvia), cluster with marine localities such as Andreyevka-2 (Russia), or with coastal estuarine localities such as those from Belgium (Strud, Modave, Esneux, Evieux). This study concludes that vertebrate fossils cannot be used to define freshwater environment in the Upper Old Red Sandstone. The question of whether Late Devonian tetrapods were living in freshwater or nearshore marine environments relies upon other methods such as used in sedimentological and sequence-stratigraphical analyses (references in Friend & Williams 2000). Presently, biological and sedimentological data do not fully agree with each other concerning with this problem.

The recent discovery of a Famennian *Ichthyostega*-like tetrapod in Belgium emphasizes the problem (Clément *et al.* 2004). The bed that originally yielded the tetrapod fossil has been found and identified from its facies (see above). The microconglomerate-sandstone surrounding the lower jaw is composed of fine shale clasts and of palaeosoil clasts. These clasts were generated by flood erosion of the river banks, upstream from a deltaic flood plain. The tetrapod-bearing bed of Strud is thus considered as non-marine, according to one of us (Jacques Thorez): 'The sandstone surrounding the jaw is indicative of fluviatile conditions... This tetrapod therefore lived in rivers and estuaries, but the shoreline at the time of the Evieux Formation was oscillating south to north of Strud' (in Clément *et al.* 2004).

It means that the conclusion of Lelièvre (2002, pp. 175–179) has to be placed in the context of these new data. We can certainly no longer maintain that all Upper Devonian tetrapod-bearing localities were continental. Some were evidently marine such as Velna-Ala, Pavari, Ketleri, Andreyevka-2 and Gornostayevka in Latvia and Russia (references in Table 1; but Long & Gordon [2004, p. 704] propose that the carcass of *Tulerpeton* may have been floated to and deposited in the marine sediments of Andreyevka-2, and that *Tulerpeton* could be non-marine). On the contrary, we can probably not maintain that they all correspond to marine tidal zone deposits (Schultze 1997, 1999). The general idea is expressed by Clack (2002b, p. 99): 'The shallow, swampy waters of marine lagoons, newly populated by emergent plants, might have been the breeding ground for the earliest tetrapods...' [Note, however, that the author did not mean that emergent plants had arisen in Late Devonian time only, as they are known as early as the Silurian; J. Clack pers. comm.]. The Givetian–Frasnian was the time of appearance of substantial trees (Algeo & Scheckler 1998), but certainly not in swampy waters of marine lagoons.

If we add to this corpus of facts and hypotheses the newly developed idea that Late Devonian tetrapods were most probably all aquatic inhabitants, the only conclusion that can be put forward is that any theory of the origin of tetrapods (i.e. origin of limbs with digits, origin of walking, and origin of terrestriality; Clack 2002b, 2006) must be valid for the wide range of animals having occupied habitats ranging from proximal, nearshore marine localities to continental, freshwater lakes and/or rivers. This idea is consistent with the conclusions of Lebedev (2004) who ran a comparison of most Upper Devonian tetrapod sites. Lebedev recognizes that 'these animals dwelled within a wide range of aquatic environments' and that 'The presence of more than one tetrapod in the communities indicates many more diverse tetrapod trophic adaptations than previously considered'. This subject is the topic of a group of scientists from Latvia, Russia, Sweden, the United Kingdom, France, the United States and Australia who are hoping to co-ordinate their activities and produce a more complete theory at a global scale (Clack 2006).

Global palaeobiogeographical context

Because Late Devonian tetrapods are now known in a range of localities worldwide, it is important to consider the global palaeogeographical context of their origin and radiation. A recent palaeogeographical reconstruction is proposed in Figure 2 (Averbuch et al. 2005) with the location of tetrapod-bearing sites. Based primarily on Golonka et al.'s (1994) map, it shows Laurussia (=Euramerica) in a rather southern location, with in particular the palaeoequator running high across Greenland (see also Golonka 2000, time slices 10 and 11; Scotese 2002, maps Devonian). This is compatible with the miospore data, and converges with the position already published by Streel et al. (1990), and emphasized by Streel & Marshall (2006). A consequence is that most of Greenland, among other areas, was in the arid belt on the southern side of a very narrow equatorial belt.

One old question considering palaeogeographical considerations is 'where did tetrapods evolve first?' (if this question can be answered), and how did they disperse to gain a wide geographical distribution from Pennsylvania in the west to Australia in the east? However, we must take care not to fall into a circular reasoning. As Frasnian–Famennian tetrapods are all endemic, and restricted to the locality or region where they have been collected (with the possible exception of *Ichthyostega*, now known both in Greenland and Belgium), we cannot use the occurrence of shared taxa to establish biogeographical relationships among the continental masses with which we are concerned.

We cannot even use the phylogenetic relationships of those taxa compared with their palaeocontinental context because of the lack of consensus over their phylogeny (see discussions and references in Schultze 1997; Clack 2002*b*; Ruta & Coates 2003; Ruta et al. 2003). [The use that the latter authors make of the concepts of total-, crown- and stem-groups, may not be accepted by all of us, but a thorough discussion of this point is beyond the scope of the present paper.] The simplest thing that we can say is that, most probably, as based on current data, tetrapods may have originated in a generalized area including Euramerica, North China, and easternmost Gondwana. There is a higher probability that this occurred in the western, Euramerican part because the sister-group of tetrapods, the tetrapodomorph sarcopterygian taxa *Panderichthys*, *Elpistotege* and a new elpistostegid from Nunavut, Canada (Daeschler et al. 2004) occurred in Euramerica (Fig. 2).

The occurrence of tetrapods on those three landmasses suggests that the latter were closely related in the Late Devonian (e.g. see Scotese & McKerrow 1990; Streel et al. 1990; Golonka 2000; Scotese 2002; and the critical analyses of Young 1981–2003). Given the pre-Pangean disposition of the Late Devonian continents, it is not possible to distinguish between a hypothesis of a northern (through central and southern Asian continental blocks) or a southern (along the northern margin of Gondwana) migration route for tetrapods. Correspondingly, it is not possible to use the distribution of tetrapods to infer the position of the continents without a danger of circular reasoning, and the 'best fit' global palaeogeographical reconstruction should be drawn up on other grounds (for a critical review of this nomenclature, see Cecca 2002). On present knowledge, it is not clear whether major

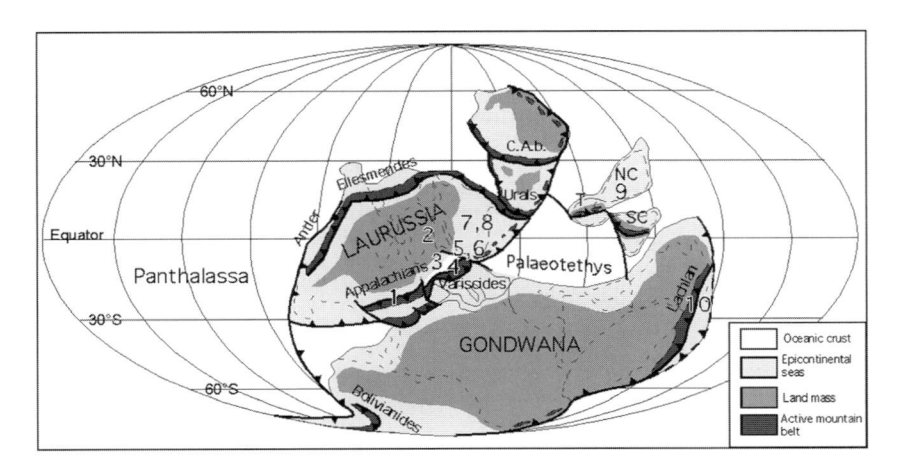

Fig. 2. Late Devonian tetrapod localities plotted on Averbuch's (in Averbuch et al. 2005) palaeocontinental reconstruction, showing geometry of active orogenic systems at the Frasnian/Famennian boundary. 1–Pennsylvania, 2–E. Greenland, 3–Scotland, 4–Belgium, 5–6–Latvia and nearby Russia, 7–Russia: Andreyevka-2, 8–Russia: Gornostayevka, 9–China, 10–Australia. Abbreviations: C.A.b., Central Asian belt; NC, North China; SC, South China; T, Tarim. [Palaeogeographical scheme kindly provided by O. Averbuch, University of Lille 1.]

marine barriers (e.g. between Gondwana and Northern Hemisphere blocks), which evidently separated the main continental areas in Early and Middle Devonian times, were significant in the early dispersal of tetrapods.

As emphasized both by the biotic dispersal of fishes (phyllolepid placoderms: Young *et al.* 2000*b*), land plants (*Callixylon* archaeopterids: Meyer-Berthaud *et al.* 1997), and their miospores (Streel *et al.* 2000) between Gondwana and Laurussia, palaeontological data point to a drastic continental re-organization around the Frasnian–Famennian boundary with, at best, a very narrow residual oceanic domain between these continents. Tectono-metamorphic data in the worldwide Upper Palaeozoic mountain belts corroborate this view, showing that the Late Devonian was characterized by intense tectonic activity with the incipient collision of major continental crustal blocks, *viz.* Laurussia, Gondwana, Kazakhstan and Siberia (Averbuch *et al.* 2005). This may not be evident in long-lasting carbonate platform sequences such as the Frasnian–Famennian of the Tafilalt-Maider and Anti-Atlas in Morocco, but is well known in regions such as the Ardenne Massif in France–Belgium, where the Frasnian is limy when the Famennian is siliciclastic (Bultynck & Dejonghe 2002). That collisional process of continental masses led to closure of oceanic domains and deformation and uplift of wide continental areas (Appalachian belt, European Variscides, Northern African Variscides, Arctic Ellesmerian–Svalbardian belt, Central Asian belt, South Uralian belt); contemporaneous Frasnian–Famennian oceanic subduction led to terrane accretion (western American Antler belt, South American Bolivianides, eastern Australian Lachlan fold belt). These events have certainly modified strongly several marine environments, and seem to have contributed to a significant global cooling event in earliest Famennian time (references in Averbuch *et al.* 2005). Such a cooling event is attested independently both by the miospore distribution on continental areas (Streel *et al.* 2000), and by the $\delta^{18}O$ signature of marine carbonates (Joachimski & Buggisch 2002). The origin and spreading of the first tetrapods is contemporaneous with this global changing context.

Conclusion

Do we know where, when, how and why tetrapod vertebrates appeared and radiated in Late Palaeozoic time? For the time being, we can schematically propose the following provisional answers:

- Where? Somewhere in the area delimited by Pennsylvania in the west and Australia in the east, with a higher probability for its western part because the closest sister taxa of tetrapods, *Elpistostege*, *Panderichthys* and a new elpistostegid from Nunavut were found in Euramerica. The original environments of those very first tetrapods are diverse, from true marine environments of the proximal neritic province to probably true continental environments (fluviatile and/or lacustrine), with apparently a predominance of shallow swampy marine lagoons.
- When? At least by the middle Frasnian, as both the oldest-known tetrapods and their sister-group (see above) are middle Frasnian in age (but see the more detailed argumentation of Ruta & Coates 2003; and the suggestion by Clack 2006, for a period between the mid-Givetian and the mid-Frasnian).
- How? This question is out of the scope of the present paper, and we refer the reader to the papers of Schultze (1997), Clack (2002*b*) and Long & Gordon (2004).
- Why? Various scenarios have been proposed, but none seems convincing enough to be uncritically accepted (see a review in Clack 2002*b*). We just note a coincidence between this biological event and a series of physical-chemical features of the Earth in Late Devonian time, linked to the building of a pre-Pangean configuration of continents.

We dedicate this paper to Michael R. House, and to two other SDS former officers: W. Ziegler and I. Chlupac. All three have been highly influential in the SDS past scientific activities, and in science in general.

This is a contribution to IGCP 491 'Middle Palaeozoic Vertebrate-Biogeography, Palaeogeography, and Climate', and SDS working groups 'Frasnian' and 'Famennian'.

We want to thank Pr. R. Thomas Becker (Münster University, FRG) for having invited us to contribute to this volume in honour of Pr. Michael House. J. A. Clack (University Museum of Zoology, Cambridge, UK) made a review of the manuscript prior to its submission for publication. H.-P. Schultze (University of Kansas, USA) and R. T. Becker did it after submission. All three are sincerely acknowledged. E. B. Daeschler (Academy of Natural Sciences of Philadelphia, PA, USA) made a comment on the material from Pennsylvania. O. Averbuch (University of Lille 1, France) provided the palaeogeographical scheme of Figure 2.

References

AHLBERG, P. E. 1991. Tetrapod or near-tetrapod fossils from the Upper Devonian of Scotland. *Nature*, **354**, 298–301.

AHLBERG, P. E. 1995. *Elginerpeton pancheni* and the earliest tetrapod clade. *Nature*, **373**, 420–425.

AHLBERG, P. E. 1998. Postcranial stem tetrapod remains from the Devonian of Scat Craig, Morayshire,

Scotland. In: NORMAN, D. B., MILNER, A. R. & MILNER, A. C. (eds) A study of fossil vertebrates. *Zoological Journal of the Linnean Society, London*, **122**, 99–141.

AHLBERG, P. E. 2004. Comment on 'The Early Evolution of the Tetrapod Humerus'. *Science*, **305**, 1715.

AHLBERG, P. E. & CLACK, J. A. 1998. Lower jaws, lower tetrapods—a review based on the Devonian genus *Acanthostega*. *Transactions of the Royal Society of Edinburgh*, **89**, 11–46.

AHLBERG, P. E., LUKSEVICS, E. & LEBEDEV, O. A. 1994. The first tetrapod finds from the Devonian (Upper Famennian) of Latvia. *Philosophical Transactions of the Royal Society, London*, **B343**, 303–328.

AHLBERG, P. E., IVANOV, A., LUKSEVICS, E. & MARK-KURIK, E. 1999. Middle and Upper Devonian correlation of the Baltic area and Scotland based on fossil fishes. In: LUKSEVICS, E., STINKULIS, G. & KALNINA, L. (eds) *The Fourth Baltic Stratigraphical Conference: Problems and Methods of Modern Regional Stratigraphy* (Jurmala, 27–30 Sept. 1999). Abstracts: 6–8; Riga.

AHLBERG, P. E., JOHANSON, Z. & DAESCHLER, E. B. 2001. The Late Devonian lungfish *Soederberghia* (Sarcopterygii, Dipnoi) from Australia and North America, and its biogeographical implications. *Journal of Vertebrate Paleontology*, **21**, 1–12.

AHLBERG, P. E., FRIEDMAN, M. & BLOM, H. 2005. New light on the earliest known tetrapod jaw. *Journal of Vertebrate Palaeontology*, **25**, 720–724.

ALEKSEEV, A. S., LEBEDEV, O. A., BARSKOV, I. S., BARSKOVA, M. I., KONONOVA, L. I. & CHIZHOVA, V. A. 1994. On the stratigraphic position of the Famennian and Tournaisian fossil vertebrate beds in Andreyevka, Tula Region, Central Russia. *Proceedings of the Geological Association*, **105**, 41–52.

ALGEO, T. J. & SCHECKLER, S. E. 1998. Terrestrial-marine teleconnections in the Devonian: links between the evolution of land plants, weathering processes, and marine anoxic events. *Philosophical Transactions of the Royal Society, London*, **B353**, 113–130.

AVERBUCH, O., TRIBOVILLARD, N., DEVLEESCHOUWER, X., RIQUIER, L., MISTIAEN, B. & VAN VLIET-LANOË, B. 2005. Mountain building-enhanced continental weathering and organic carbon burial as major causes for climatic cooling at the Frasnian–Famennian boundary (c. 376 Ma)? *Terra Nova*, **17**, 33–42.

BLIECK, A. 1991. Reappraisal of the heterostracans (agnathan vertebrates) of northern Ireland. *Irish Journal of Earth Sciences*, **11**, 65–69.

BLOM, H. 2005. Taxonomic revision of the Late Devonian tetrapod *Ichthyostega* from East Greenland. *Palaeontology*, **48**, 111–134.

BLOM, H., CLACK, J. A., AHLBERG, P. E. & FRIEDMAN, M. 2003. Devonian vertebrates from East Greenland: a review of faunal composition and distribution. In: SCHULTZE, H.-P., LUKSEVICS, E. & UNWIN, D. (eds) *The Gross Symposium 2: Advances in Palaeoichthyology & IGCP 491 meeting* (Riga, Latvia, 8–14 Sept. 2003). *Ichthyolith Issues, Special Publications*, **7**, 13.

BULTYNCK, P. 2004. Message from the chairman. *Subcommission on Devonian Stratigraphy Newsletter*, **20**, 1.

BULTYNCK, P. & DEJONGHE, L. 2002. Devonian lithostratigraphic units (Belgium). In: BULTYNCK, P. & DEJONGHE, L. (eds) Guide to a revised lithostratigraphic scale of Belgium. *Geologica Belgica*, **4** (2001), 39–68.

CAMPBELL, K. S. W. & BELL, M. W. 1977. A primitive amphibian from the Late Devonian of New South Wales. *Alcheringa*, **1**, 369–381.

CAMPBELL, K. S. W. & BELL, M. W. 1982. *Soederberghia* (Dipnoi) from the Late Devonian of New South Wales. *Alcheringa*, **6**, 143–149.

CECCA, F. 2002. *Palaeobiogeography of Marine Fossil Invertebrates: Concepts and Methods*. Taylor & Francis, London & New York.

CLACK, J. A. 1997. Devonian tetrapod trackways and trackmakers; a review of the fossils and footprints. *Palaeogeography, Palaeoclimatology, Palaeoecology*, **130**, 227–250.

CLACK, J. A. 2002a. The dermal skull roof of *Acanthostega gunnari*, an early tetrapod from the Late Devonian. *Transactions of the Royal Society of Edinburgh: Earth Sciences*, **93**, 17–33.

CLACK, J. A. 2002b. *Gaining Ground: The Origin and Evolution of Tetrapods*. Indiana University Press, Bloomington & Indianapolis.

CLACK, J. A. 2006. The emergence of early tetrapods. *Palaeogeography, Palaeoclimatology, Palaeoecology*, **232**, 167–189.

CLACK, J. A. & NEININGER, L. 2000. Fossils from the Celsius Bjerg Group, Late Devonian sequences, East Greenland; significance and sedimentological distribution. In: FRIEND, P. F. & WILLIAMS, B. P. J. (eds) *New Perspectives on the Old Red Sandstone*. Geological Society, London, Special Publications, **180**, 557–566.

CLACK, J. A., AHLBERG, P. E., FINNEY, S. M., DOMINGUEZ ALONSO, P., ROBINSON, J. & KETCHAM, R. A. 2003. A uniquely specialized ear in a very early tetrapod. *Nature*, **425**, 65–69.

CLACK, J. A., AHLBERG, P. E. & BLOM, H. 2004. A new genus of tetrapod from the Devonian of East Greenland. In: The Palaeontological Association, 48th Annual Meeting (Lille & Villeneuve d'Ascq, 17–20 December 2004). Abstracts with programme. *The Palaeontological Association Newsletter*, **57**, 116–117.

CLÉMENT, G., AHLBERG, P. E., BLIECK, A., BLOM, H., CLACK, J. A., POTY, E., THOREZ, J. & JANVIER, P. 2004. Devonian tetrapod from western Europe. *Nature*, **427**(6973), 412–413.

DAESCHLER, E. B. 2000. Early tetrapod jaws from the Late Devonian of Pennsylvania, USA. *Journal of Paleontology*, **74**, 301–308.

DAESCHLER, E. B. & SHUBIN, N. 1998. Fish with fingers? *Nature*, **391**, 133.

DAESCHLER, E. B., SHUBIN, N. H., THOMSON, K. S. & AMARAL, W. W. 1994. A Devonian tetrapod from North America. *Science*, **265**, 639–642.

DAESCHLER, E. B., SHUBIN, N. H. & JENKINS, F. 2004. A new member of the sister group of Tetrapoda: an elpistostegid fish (Sarcopterygii, Elpistostegalia) from the Fram Formation, Ellesmere Island, Nunavut territory, Canada. *Journal of Vertebrate Paleontology, Abstract Volume*, **24**, 50A.

ESIN, D., GINTER, M., IVANOV, A., LEBEDEV, O., LUKSEVICS, E., AVKHIMOVICH, V., GOLUBTSOV, V. &

PETUKHOVA, L. 2000. Vertebrate correlation of the Upper Devonian and Lower Carboniferous on the East European Platform. *In*: BLIECK, A. & TURNER, S. (eds) Palaeozoic Vertebrate Biochronology and Global Marine/Non-Marine Correlation—Final Report of IGCP 328 (1991–1996). *Courier Forschungsinstitut Senckenberg*, **223**, 341–359.

FRIEND, P. F. & WILLIAMS, B. P. J. (eds) 1978. *Devonian of Scotland, the Welsh Borderland and South Wales. In*: International Symposium on the Devonian System (PADS, Bristol, September 1978). Field guide, Palaeontological Association Publication.

FRIEND, P. F. & WILLIAMS, B. P. J. (eds) 2000. *New Perspectives on the Old Red Sandstone*. Geological Society, London, Special Publications, **180**, 623pp.

GOLONKA, J. 2000. Earth history maps: Cambrian-Neogene plate tectonic maps. World Wide Web address: http://www.dinodata.net/Golonka/Golonka.htm; text (1024 Ko), 3 tables, 37 fig.; Kraków b Wydawn, Uniwersytetu Jagiellonskiego.

GOLONKA, J., ROSS, M. I. & SCOTESE, C. R. 1994. Phanerozoic paleogeographic and paleoclimatic modeling maps. *In*: EMBRY, A. F., BEAUCHAMP, B. & GLASS, D. J. (eds) *Pangea: Global Environments and Resources*. Canadian Society of Petroleum Geologists, Memoirs, **17**, 1–47.

GRADSTEIN, F. M. & OGG, J. G. 2004. Geologic Time Scale 2004—Why, how, and where next! *In*: Status of the International Geological Time Scale. *Lethaia*, **37**(2), 175–181 [also World Wide Web address: http://www.stratigraphy.org/scale04.pdf International Commission of Stratigraphy].

GROOS-UFFENORDE, H., LETHIERS, F. & BLUMENSTENGEL, H. 2000. Ostracodes and Devonian Stratigraphy. *In*: BULTYNCK, P. (ed.) Subcommission on Devonian Stratigraphy: Fossil groups important for boundary definition. *Courier Forschungsinstitut Senckenberg*, **220**, 99–111.

HARTZ, E. H., TORSVIK, T. H. & ANDRESEN, A. 1997. Carboniferous age for the east greenland 'Devonian' basin: Paleomagnetic and isotopic constraints on age, stratigraphy, and plate reconstructions. *Geology*, **25**, 675–678.

HARTZ, E. H., TORSVIK, T. H. & ANDRESEN, A. 1998. Carboniferous age for the east greenland 'Devonian' basin: Paleomagnetic and isotopic constraints on age, stratigraphy, and plate reconstructions: Reply. *Geology*, **26**, 285–286.

JANVIER, P. 1996. Palaeontological Association 1995 Annual Address—The dawn of the vertebrates: characters versus common ascent in the rise of current vertebrate phylogenies. *Palaeontology*, **39**, 259–287.

JARVIK, E. 1996. The Devonian tetrapod *Ichthyostega*. *Fossils and Strata*, **40**, 1–213.

JOACHIMSKI, M. & BUGGISCH, W. 2002. Conodont apatite $\delta^{18}O$ signatures indicate climatic cooling as a trigger of the Late Devonian mass extinction. *Geology*, **30**, 711–714.

JONES, R. K. & TURNER, S. 2000. Late Devonian fauna from the Columbine Sandstone (Coffee Hill Member), Gap Creek, central New South Wales. *In*: BLIECK, A. & TURNER, S. (eds) Palaeozoic Vertebrate Biochronology and Global Marine/Non-Marine Correlation—Final Report of IGCP 328 (1991–1996). *Courier Forschungsinstitut Senckenberg*, **223**, 523–541.

LEBEDEV, O. 2003. New early Famennian tetrapods from the Oryol region (Russia). *In*: SCHULTZE, H.-P., LUKSEVICS, E. & UNWIN, D. (eds) The Gross Symposium 2: Advances in Palaeoichthyology & IGCP 491 meeting (Riga, Latvia, Sept. 8–14, 2003). *Ichthyolith Issues, Special Publications*, **7**, 35–36.

LEBEDEV, O. A. 2004. A new tetrapod *Jakubsonia livnensis* from the Early Famennian (Devonian) of Russia and palaeoecological remarks on the Late Devonian tetrapod habitats. *In*: LUKSEVICS, E. & STINKULIS, G. (eds) The Second Gross Symposium 'Advances of palaeoichthyology' (Riga, 2003). *Acta Universitatis Latviensis*, **679**, 79–98.

LEBEDEV, O. A. & CLACK, J. A. 1993. Upper Devonian tetrapods from Andreyevka, Tula region, Russia. *Palaeontology*, **36**(3), 721–734.

LEBEDEV, O. A. & COATES, M. I. 1995. The postcranial skeleton of the Devonian tetrapod *Tulerpeton curtum* Lebedev. *Zoological Journal of the Linnean Society*, **114**, 307–348.

LELIÈVRE, H. 2002. Phylogénie des Brachythoraci (Vertebrata, Placodermi) et ajustement de la phylogénie à la stratigraphie. Les sites du Dévonien terminal, la caractérisation de leur milieu de dépôt par analyse de similitude de leur ichthyofaune. H. D. R. Sciences Naturelles, U.S.T.L., Villeneuve d'Ascq (12 décembre 2002).

LEWIS, P. C., GLEN, R. A., PRATT, G. W. & CLARKE, I. 1994. Explanatory notes. Bega—Mallacoota 1:250 000 Geological Sheet. SJ/55-4, SJ55-8. *Geological Survey of New South Wales*.

LOHEST, M. 1888. Recherches sur les poissons des terrains paléozoïques de Belgique. Poissons des Psammites du Condroz, Famennien supérieur. *Annales de la Société Géologique de Belgique, XV [1887–1888], Mémoire*, 112–203.

LONG, J. A. & GORDON, M. S. 2004. The greatest step in vertebrate history: a paleobiological review of the fish–tetrapod transition. *Physiological and Biochemical Zoology*, **77**, 700–719.

LUKSEVICS, E. 2001. Bothriolepid antiarchs (Vertebrata, Placodermi) from the Devonian of the north-western part of the East European Platform. *Geodiversitas*, **23**, 489–609.

LUKSEVICS, E. & ZUPINS, I. 2003. Taphonomic studies of the Devonian fish and tetrapod fossils from the Pavari site (Latvia). *In*: SCHULTZE, H.-P., LUKSEVICS, E. & UNWIN, D. (eds) The Gross Symposium 2: Advances in Palaeoichthyology & IGCP 491 meeting (Riga, Latvia, 8–14 Sept. 2003). *Ichthyolith Issues, Special Publications*, **7**, 37–38.

LUKSEVICS, E. & ZUPINS, I. 2004. Sedimentology, fauna, and taphonomy of the Pavari site, Late Devonian of Latvia. *In*: LUKSEVICS, E. & STINKULIS, G. (eds) The Second Gross Symposium: Advances of palaeoichthyology (Riga, 2003). *Acta Universitatis Latviensis*, **679**, 99–119.

MARSHALL, J. E. A., ASTIN, T. R. & CLACK, J. A. 1999. East Greenland tetrapods are Devonian in age. *Geology*, **27**, 637–640.

MAWSON, R. & TALENT, J. A. 2003. Conodont faunas from sequences on or marginal to the Anakie Inlier (central

Queensland, Australia) in relation to Devonian transgressions. *Bulletin of Geosciences (Czech Geological Survey)*, **87**, 335–358.

MAZIANE, N., HIGGS, K. T. & STREEL, M. 1999. Revision of the Late Famennian zonation scheme in eastern Belgium. *Journal of Micropalaeontology*, **18**, 117–125.

MEYER-BERTHAUD, B., WENDT, J. & GALTIER, J. 1997. First record of a large *Callixylon* trunk from the Late Devonian of Gondwana. *Geological Magazine*, **134**, 847–853.

MILES, R. S. 1968. The Old Red Sandstone antiarchs of Scotland: family Bothriolepididae. Palaeontographical Society Monographs, London, **122**, 1–130.

MYKURA, W. 1991. Old Red Sandstone. *In*: CRAIG, G. Y. (ed.) *Geology of Scotland* (3rd edn). The Geological Society, London, 297–346.

NEWMAN, M. J. 2005. A systematic review of the placoderm genus *Cosmacanthus* and a description of acanthodian remains from the Upper Devonian of Scotland. *Palaeontology*, **48**, 1111–1116.

ODIN, G. S. 1994. Geological Time Scale (1994). *Comptes Rendus de l'Académie des Sciences, Paris*, **318**, II, 59–71.

PAN, J., HUO, F., CAO, J., GU, Q., LIU, S., WANG, J., GAO, L. & LIU, C. 1987. [*Continental Devonian System of Ningxia and its biotas.*] Geological Publishing House, Beijing [In Chinese, with English abstract].

PASKEVICIUS, J. 1997. *The Geology of the Baltic Republics*. Vilnius University & Geological Survey of Lithuania, Vilnius.

RAYMOND, O. 1998. Dulladerry Volcanics. *In*: POGSON, D. J. & WATKINS, J. J. (compilers) Explanatory notes. Bathurst 1:250 000 Geological Sheet. SI/55-8. *Geological Survey of New South Wales*, 214–222.

RITCHIE, A., WANG, S., YOUNG, G. C. & ZHANG, G. 1992. The Sinolepidae, a family of antiarchs (placoderm fishes) from the Devonian of South China and Eastern Australia. *Records of the Australian Museum*, **44**, 319–370.

RUTA, M. & COATES, M. I. 2003. Bones, molecules, and crown-tetrapod origins. *In*: DONOGHUE, P. C. J. & SMITH, M. P. (eds) *Telling the Evolutionary Time—Molecular Clocks and the Fossil Record*. CRC Press, Boca Raton/Systematic Association, Special Volume Series/Palaeontological Association, 224–262.

RUTA, M., COATES, M. I. & QUICKE, D. L. J. 2003. Early tetrapod relationships revisited. *Biological Reviews*, **78**, 251–345.

SÄVE-SÖDERBERGH, G. 1932. Preliminary note on Devonian stegocephalians from East Greenland. *Meddelelser om Grønland*, **98**, 1–211.

SCHULTZE, H.-P. 1997. Umweltbedingungen beim Übergang von Fisch zu Tetrapode [Paleoenvironment at the transition from fish to tetrapod.] *Sitzungsberichte der Gesellschaft Naturforschender Freunde zu Berlin, N. F.*, **36**, 59–77.

SCHULTZE, H.-P. 1999. The fossil record of the intertidal zone. *In*: HORN, M. H., MARTIN, K. L. M. & CHOTKOWSKI, M. A. (eds) *Intertidal Fishes: Life in Two Worlds*. Academic Press, San Diego & London, 373–392.

SCHULTZE, H.-P. & CLOUTIER, R. 1996. Comparison of the Escuminac Formation ichthyofauna with other late Givetian/early Frasnian ichthyofaunas. *In*: SCHULTZE, H.-P. & CLOUTIER, R. (eds) *Devonian Fishes and Plants of Miguasha, Quebec, Canada*. Verlag Dr. Friedrich Pfeil, München, 348–368.

SCHULTZE, H.-P. & MAPLES, C. G. 1992. Comparison of the Late Pennsylvanian faunal assemblage of Kinney Brick Company Quarry, New Mexico, with other Late Pennsylvanian Lagerstätten. *In*: ZIDEK, J. (ed.) Geology and paleontology of the Kinney Brick Quarry, Late Pennsylvanian, central New Mexico. *New Mexico Bureau of Mines and Mineral Resources, Bulletin*, **138**, 231–242.

SCHULTZE, H.-P., MAPLES, C. G. & CUNNINGHAM, C. R. 1994. The Hamilton Konservat-Lagerstätte: Stephanian terrestrial biota in a marginal-marine setting. *In*: ROLFE, W. D. I., CLARKSON, E. N. K. & PANCHEN, A. L. (eds) Volcanism and early terrestrial biotas. *Transactions of the Royal Society of Edinburgh: Earth Sciences*, **84**(1993), 443–451.

SCHULTZE, H.-P., LUKSEVICS, E. & UNWIN, D. (eds) 2003. The Gross Symposium 2: Advances in Palaeoichthyology & IGCP 491 meeting (Riga, Latvia, Sept. 8–14, 2003). *Ichthyolith Issues, Special Publication*, **7**, University of Latvia [abstracts].

SCOTESE, C. R. 2002. PALEOMAP Project: Plate tectonic maps and continental drift animations. World Wide Web address: Arlington, Texas, http://www.scotese.com

SCOTESE, C. R. & MCKERROW, W. S. 1990. Revised World maps and introduction. *In*: MCKERROW, W. S. & SCOTESE, C. R. (eds) *Palaeozoic Palaeogeography and Biogeography*. Geological Society, London, Memoirs, **12**, 1–21.

SHUBIN, N. H., DAESCHLER, E. B. & COATES, M. I. 2004. The early evolution of the tetrapod humerus. *Science*, **304**, 90–93.

SOROKIN, V. S. 1978. Etapy razvitya severo-zapada Russkoy platformy vo Franskom veke [Stages of development of the north-western part of the Russian platform in the Frasnian]. Zinatne, Riga. [In Russian].

STEMMERIK, L. & BENDIX-ALMGREEN, S. E. 1998. Carboniferous age for the East Greenland 'Devonian' basin: Paleomagnetic and isotopic constraints on age, stratigraphy, and plate reconstructions: Comment. *Geology*, **26**, 284–285.

STREEL, M. 2005. Subdivision of the Famennian Stage into four Substages and correlation with the neritic and continental miospore zonation. *In*: 32nd International Geological Congress: Subcommission on Devonian Stratigraphy business meeting (Florence, 23 Aug. 2004). Extended abstract. *Subcommission on Devonian Stratigraphy Newsletter*, **21**, 15–17.

STREEL, M. & LOBOZIAK, S. 1996. 18B: Middle and Upper Devonian miospores. *In*: JANSONIUS, J. & MCGREGOR, D. C. (eds) *Palynology: Principles and Applications*. Vol. 2: Applications. Ch. 18: Paleozoic spores and pollen. American Association Stratigraphic Palynologists Foundation, College Station, Texas, 579–587.

STREEL, M. & LOBOZIAK, S. 2000. Correlation of the proposed conodont based Upper Devonian substage boundary levels into the neritic and terrestrial miospore zonation. *Subcommission on Devonian Stratigraphy Newsletter*, **17**, 12–14.

STREEL, M. & MARSHALL, J. E. A. 2006. Devonian–Carboniferous boundary global correlations and their paleogeographic implications for the Assembly of Pangaea. *In*: WONG, TH. (ed.) *Proceedings of the XVth International Congress on Carboniferous and Permian Stratigraphy* (Utrecht, 2003). Royal Netherlands Academy of Arts and Sciences, 481–496.

STREEL, M., FAIRON-DEMARET, M. & LOBOZIAK, S. 1990. Givetian-Frasnian phytogeography of Euramerica and western Gondwana based on miospore distribution. *In*: MCKERROW, W. S. & SCOTESE, C. R. (eds) *Palaeozoic Palaeogeography and Biogeography*. Geological Society, London, Memoirs, **12**, 291–296.

STREEL, M., CAPUTO, M. V., LOBOZIAK, S. & MELO, J. H. G. 2000. Late Frasnian–Famennian climates based on palynomorph analyses and the question of the Late Devonian glaciations. *Earth-Science Reviews*, **52**, 121–173.

TALENT, J. A., MAWSON, R., *ET AL*. 2000. Devonian palaeobiogeography of Australia and adjoining regions. *In*: WRIGHT, A. J., YOUNG, G. C., TALENT, J. A. & LAURIE, J. R. (eds) Palaeobiogeography of Australasian faunas and floras. *Association of Australasian Palaeontologists, Memoirs*, **23**, 167–257.

THOREZ, J., STREEL, M., BOUCKAERT, J. & BLESS, M. J. M. 1977. Stratigraphie et paléogéographie de la partie orientale du synclinorium de Dinant (Belgique) au Famennien supérieur: un modèle de bassin sédimentaire reconstitué par analyse pluridisciplinaire sédimentologique et micropaléontologique. *Mededelingen van de Rijks Geologische Dienst, N.S.*, **28**, 17–28.

TRAVERSE, A. 2003. Dating the earliest tetrapods: A Catskill palynological problem in Pennsylvania. *In*: WILDE, V. (ed.) Studies on fossil and extant plants and floras. Dedicated to Friedemann Schaarschmidt on the occasion of his 65th birthday. *Courier Forschungsinstitut Senckenberg*, **241**, 19–49.

TREWIN, N. H. (ed.) 2002. *The Geology of Scotland* (4th edn). The Geological Society, London.

VALIUKEVICIUS, J. & OVNATANOVA, N. 2005. The Early Famennian conodonts and fishes of Lithuania. *Geologija*, **49**, 21–28.

VOROBYEVA, E. I. 1977. Morfologija i osobennosti evolyutsii kisteperykh ryb [Morphology and peculiarities of the evolution of the crossopterygian fishes.] *Akademia Nauk SSSR, Trudy Paleontologischeskogo Instituta*, **163**, 1–239; Nauka, Moskva [In Russian].

WARREN, J. W. & WAKEFIELD, N. A. 1972. Trackways of tetrapod vertebrates from the Upper Devonian of Victoria, Australia. *Nature*, **228**, 469–470.

WARREN, A., JUPP, R. & BOLTON, B. 1986. Earliest tetrapod trackway. *Alcheringa*, **10**, 183–186.

WEBBY, B. D. 1972. Devonian geology of the Lachlan Geosyncline. *Journal of the Geological Society of Australia*, **19**, 99–123.

WILLIAMS, E. A., FRIEND, P. F. & WILLIAMS, B. P. J. 2000. A review of Devonian time scales: databases, construction and new data. *In*: FRIEND, P. F. & WILLIAMS, B. P. J. (eds) *New Perspectives on the Old Red Sandstone*. Geological Society, London, Special Publications, **180**, 1–21.

YOUNG, G. C. 1981. Biogeography of Devonian vertebrates. *Alcheringa*, **5**, 225–243.

YOUNG, G. C. 1993. Middle Palaeozoic macrovertebrate biostratigraphy of eastern Gondwana. *In*: LONG, J. A. (ed.) *Palaeozoic Vertebrate Biostratigraphy and Biogeography*. Belhaven Press, London, 208–251.

YOUNG, G. C. 1996. Devonian (Chart 4). *In*: YOUNG, G. C. & LAURIE, J. R. (eds) *An Australian Phanerozoic Timescale*. AGSO/Oxford University Press, Melbourne, 96–109.

YOUNG, G. C. 1999. Preliminary report on the biostratigraphy of new placoderm discoveries in the Hervey Group (Upper Devonian) of central New South Wales. *In*: BAYNES, A. & LONG, J. A. (eds) Papers in vertebrate palaeontology. *Records of the Western Australian Museum, Supplement*, **57**, 139–150.

YOUNG, G. C. 2003. North Gondwanan mid-Palaeozoic connections with Euramerica and Asia; Devonian vertebrate evidence. *Courier Forschungsinstitut Senckenberg*, **242**, 169–185.

YOUNG, G. C. 2005a. An articulated phyllolepid fish (Placodermi) from the Devonian of central Australia: implications for non-marine connections with the Old Red Sandstone continent. *Geological Magazine*, **142**, 173–186.

YOUNG, G. C. 2005b. New phyllolepids (placoderm fishes) from the Middle–Late Devonian of southeastern Australia. *Journal of Vertebrate Paleontology*, **25**, 261–273.

YOUNG, G. C. & TURNER, S. 2000. Devonian microvertebrates and marine-nonmarine correlation in East Gondwana: Overview. *In*: BLIECK, A. & TURNER, S. (eds) Palaeozoic Vertebrate Biochronology and Global Marine/Non-Marine Correlation—Final Report of IGCP 328 (1991–1996). *Courier Forschungsinstitut Senckenberg*, **223**, 453–470.

YOUNG, G. C., SHERWIN, L. & RAYMOND, O. L. 2000a. Late Devonian: Hervey Group. *In*: LYONS, P., RAYMOND, O. L. & DUGGAN, M. B. (eds) Explanatory Note—Forbes 1:250,000 Geological Sheet S155-7, 2nd edn. *AGSO Record*, 2000/20, 125–149.

YOUNG, G. C., LONG, J. & BURROW, C. 2000b. Vertebrata. *In*: TALENT, J. A., MAWSON, R., *ET AL*. Devonian palaeobiogeography of Australia and adjoining regions. *In*: WRIGHT, A. J., YOUNG, G. C., TALENT, J. A. & LAURIE, J. R. (eds), *Palaeobiogeography of Australasian Faunas and Floras*. Association of Australasian Palaeontologists, Memoirs, **23**, 209–219 and 250.

ZEIBA, S. & VALIUKEVICIUS, J. 1972. Novye dannye o famenskikh konodontakh yuzhnoy Pribaltiki [New data on the Famennian conodont fauna of the southern Peribaltic]. *Geografiya i Geologiya*, **IX**, 167–171 [In Russian, with Lithuanian and German abstracts].

ZHU, M. 2000. Catalogue of Devonian vertebrates in China, with notes on bio-events. *In*: BLIECK, A. & TURNER, S. (eds) Palaeozoic Vertebrate Biochronology and Global Marine/Non-Marine Correlation—Final Report of IGCP 328 (1991–1996). *Courier Forschungsinstitut Senckenberg*, **223**, 373–390.

ZHU, M., WANG, N.-Z. & WANG, J.-Q. 2000. Devonian macro- and microvertebrate assemblages of China.

In: BLIECK, A. & TURNER, S. (eds) Palaeozoic Vertebrate Biochronology and Global Marine/Non-Marine Correlation—Final Report of IGCP 328 (1991–1996). *Courier Forschungsinstitut Senckenberg*, **223**, 361–372.

ZHU, M., AHLBERG, P. E., ZHAO, W. & JIA, L. 2002. First Devonian tetrapod from Asia. *Nature*, **420**, 760–761.

ZIEGLER, W. & SANDBERG, C. A. 1990. The Late Devonian standard conodont zonation. *Courier Forschungsinstitut Senckenberg*, **121**, 1–115.

Middle to Late Famennian successions at Ain Jemaa (Moroccan Meseta)—implications for regional correlation, event stratigraphy and synsedimentary tectonics of NW Gondwana

S. I. KAISER[1], R. T. BECKER[2] & A. EL HASSANI[3]

[1]*State Museum of Natural History Stuttgart, Rosenstein 1, D-70191 Stuttgart, Germany (e-mail: kaiser.smns@naturkundemuseum-bw.de)*

[2]*Geologisch-Paläontologisches Institut, Westfälische Wilhelms-Universität, Corrensstr. 24, D-48149 Münster, Germany (e-mail: rbecker@uni-muenster.de)*

[3]*Institut Scientifique, Université Mohammed V Agdal, Avenue Ibn Batuta, B.P. 703, 10196 Rabat, Morocco (e-mail: elhassani@israbat.ac.ma)*

Abstract: Two pelagic successions near Ain Jemaa (Oulmès region, Moroccan Meseta) are dated by conodonts and ammonoids and provide new data on the discontinuous Middle to Upper Famennian faunal and facies evolution in the region. Upper Devonian shales and nodular limestones are assigned to the new Bou Gzem Formation, which is subdivided into three members. The Upper Member consists of black shales that are correlated with the globally widespread, transgressive black shale interval of the Hangenberg Event. The overlying quartzites are interpreted as prodeltaic deposits and assigned to the new Ta'arraft Formation that probably correlate with the major regressive phase of the Hangenberg Event. Contemporaneous ('Strunian') coarse siliciclastics have a wide distribution in different structural units of the Meseta. Both studied sections display a long sedimentary gap but of different extent at the base of the black shales. Comparison with other regions of Hercynian Morocco suggest an influence of Eohercynian tectonics on sedimentation, leading to extreme condensation and/or non-deposition, whilst other Meseta areas show evidence of contemporaneous reworking on uplifted structural highs and massive shedding of mass flows, conglomerates and turbidites into adjacent pelagic basins. Data from Oulmès and other Meseta regions suggest a timing of tectophases as early Middle Famennian (starting within the *marginifera* Zone) and Upper Famennian (starting within the Middle *expansa* Zone), interrupted by transgressive pulses of the global *Annulata* and Dasberg Events.

The Moroccan Meseta consists of folded Palaeozoic sediments, which are disconformably overlain by flat-lying Mesozoic and Cenozoic successions, and widespread granites related to Cenozoic volcanism (Michard 1976). As part of the Southern Variscides, the Meseta was influenced by Devonian/Carboniferous syn- and postsedimentary tectonic movements (e.g. Pique & Michard 1989, El Kamel *et al.* 1992; El Hassani & Tahiri 1994, 2000), which resulted in a complex framework of structural units (e.g. Pique & Michard 1981; Pique 1994), each with a rather different sedimentary, faunal and synsedimentary tectonic history. The Meseta area includes autochthonous areas as well as nappe and olistolistrom units that experienced considerable displacement and resedimentation/embedding during the middle Carboniferous peak of Variscan orogeny. The reconstruction of the Upper Devonian faunal, facies and structural evolution of the Meseta is important for a range of reasons. The area forms the southernmost part of the large-scale Variscan orogenic belt and, therefore, is a key segment for the understanding of the timing, patterns and processes of the Gondwana–Laurussia collision that evoked this major Phanerozoic mountain-building episode. Comparisons with the weakly deformed stable cratonic parts of NW Gondwana, the Anti-Atlas region (e.g. Wendt 1985), are especially important. Pique (1975) coined the term 'revolution famennienne' for Hercynian Morocco since Eovariscan crustal deformation, up to metamorphism in the Midelt area (dated as 367 ± 7 Ma, Pique & Michard 1981), seems to have started within the Famennian. At this time and continuing into the Lower Carboniferous, large parts of the Meseta, however, were still part of an extensional depositional regime (e.g. Pique & Kharbouch 1983; Beauchamp & Izart 1987). A better understanding of Eovariscan tectonics needs to be based on more-detailed stratigraphical data that allow a precise correlation of individual structural units, a refined reconstruction of facies and palaeogeographic changes, and, thus, a precise timing of block faulting, subsidence and uplift.

From: BECKER, R. T. & KIRCHGASSER, W. T. (eds) *Devonian Events and Correlations*.
Geological Society, London, Special Publications, **278**, 237–260. DOI: 10.1144/SP278.11
0305-8719/07/$15 © The Geological Society of London 2007.

Currently there is a huge discrepancy between palaeomagnetic (e.g. Kent & Van der Voo 1990; Bachtadse et al. 1995) and biogeographical data (e.g. Meyer-Berthaud et al. 1997; Becker & House 2000a) concerning the spatial distance between Gondwana and Laurussia in Upper Devonian time. The palaeomagnetic model suggests a several thousand kilometre wide Prototethys whereas faunas and floras strongly support a single biogeographical province stretching from the Anti-Atlas to the Rhenohercynian Zone and a rather narrow ocean, as proposed by Heckel & Witzke (1979), Becker (2001), El Hassani et al. (2001, 2003), and Carls (2001, 2003). This unresolved contradiction underlines the need for more detailed faunal and sedimentary data from Hercynian Morocco. Geochemical data also hold valuable clues: the neodymium isotopic composition of conodonts (Dopieralska et al. 2001; Dopieralska 2003) from the Montagne Noire (southern France) and the Meseta are very similar, indicating a common sediment source from exposed Precambrian crust.

Finally, the Upper Devonian was characterized by a complex sequence of global sedimentary and evolutionary events (House 1985; Becker 1993b; Walliser 1996), including the mass extinctions at the Frasnian–Famennian (Upper Kellwasser Event) and just before the Devonian–Carboniferous boundary (Hangenberg Event). A correct understanding of these major environmental perturbations requires knowledge of event patterns and stratigraphy in as many regions as possible but data from the Moroccan Meseta are still rather restricted or preliminary (e.g. Tahiri et al. 1992; Walliser et al. 1995; Lazreq 1999; Becker 1993a, c; Becker & House 2000a). In the frame of high resolution biostratigraphical, facies and geochemical investigations of the global Hangenberg Event (Kaiser 2005), Middle to Upper Famennian successions of the Oulmès region have been studied for the first time in detail and provide new insights into our understanding of the regional facies history, of event patterns and, by comparison with other Meseta regions, of Eohercynian tectonic movements in a crucial segment of NW Gondwana.

The chronostratigraphical subdivision of the Famennian into four formal substages follows recent proposals to the International Subcommission on Devonian Stratigraphy, with the Middle Famennian starting with the Lower *marginifera* Zone, the Upper Famennian with the *postera* or *styriacus* Zone, and with the Uppermost Famennian starting at the base of the Upper *expansa* Zone. The ammonoid zonation of the Upper Devonian (UD) follows Becker & House (2000b). Ammonoids and trilobites are kept in the museum of the Institute of Geology and Palaeontology, WWU Münster (GPIM numbers), conodonts are kept in the collection of the Staatliches Museum für Naturkunde, Stuttgart (SMNS numbers).

Sedimentary and faunal successions at Ain Jemaa

Previous investigations

The two studied Devonian sequences close to the Ain Jemaa village belong tectonically to the SW–NE-running Khouribgá–Oulmès Zone (or anticline), which is the eastern part of the Central Meseta (Pique & Michard 1981) or of the North–Central Domain (Pique 1994). Outcrops lie approximately 13 km north of Oulmès (Fig. 1), close to the track coming from Oulmès. The palaeontological significance of the area was first highlighted by Termier (1938), who reported the presence of a rather diverse Lower Famennian ammonoid fauna along strike of a single band of outcrops: Tabourit (entrance to the Tiliouine Gorge), Bou Gzem (=Jebel Bougouezzame, ca. = Ain Jemaa 2), Sidi bou Sif, and Ain Djema (or Ain Dram, ca.=Ain Jemaa 1). Cogney (1967) provided a very brief overview of Middle and Upper Devonian sedimentation at Bou Gzem (=Bou Keziam), with trilobite-bearing, dark Eifelian limestones (Ain Jemaa Member of the Slimane Formation, Zahraoui 1994), followed by reefal limestone roughly of Givetian age (Bou Sif Member of the Slimane Formation), by greenish-grey Frasnian shales, and by Lower Famennian nodular limestones that yielded pyritic or secondarily haematized tornoceratids and cheiloceratids (Termier & Termier 1950; Becker 1993a). Higher up, 3 m of sandy shale, 1 m of dark limestone with Middle Famennian trilobites (*Franconicabole*), 50 m of grey shales, and finally a thick quartzite unit forming the topographical peaks were noted. Alberti (1970) added biostratigraphical precision for the early Famennian by conodont sampling. His data suggest the presence in succession of the Upper/Uppermost *crepida* (level with *Palmatolepis glabra glabra* and *Pa. glabra pectinata*), Lower/Upper *rhomboidea* (level with *Pa. rhomboidea*), and Lower *marginifera* Zones (level with *Pa. quadrantinodosa inflexoidea*) in shaly, marly or solid, black to grey limestones.

The Upper Devonian near Ain Jemaa was subsequently briefly described by Tahiri (1991), Tahiri et al. (1992), Becker (1993a), and by Walliser et al. (1995). Similarities of the Frasnian shales with banded basinal shales (Bänderschiefer) of the Rhenish Massif were noted by Tahiri et al. (1992). A higher package of black shales was correlated with the black Kellwasser facies of Germany. Re-sampling of the goniatite fauna first (Becker

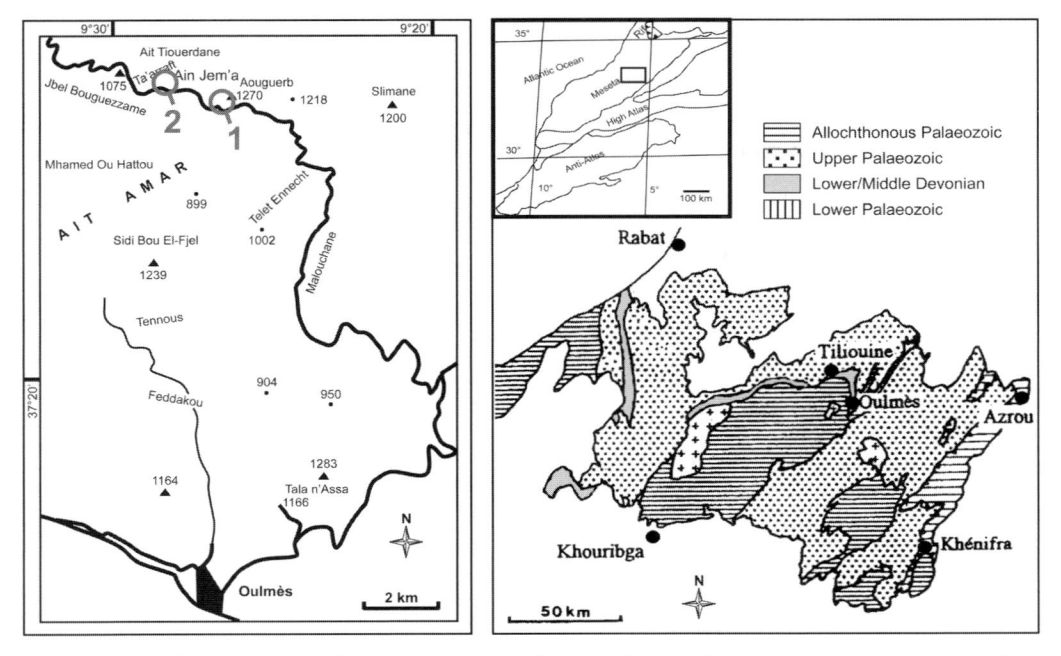

Fig. 1. Geological map (modified after Zahraoui *et al.* 2000) and studied locations in the Moroccan Meseta: 1=Ain Jemaa 1, 2=Ain Jemaa 2.

1993*a*) proved to be difficult but later attempts (2003/2004) were more successful and confirmed the presence of hitherto undescribed and new taxa, in association with pelecypods (*Loxopteria*) and small-sized brachiopods that are typical for a poorly oxygenated, subphotic outer shelf pelagic environment. The goniatite-bearing Lower Famennian succession was observed (Tahiri *et al.* 1992; Walliser *et al.* 1995) to pass upwards into yellowish, multicoloured shales with intercalated, silty-calcareous nodules with clymenids, among them *Platyclymenia annulata*, indicating UD IV-A (Becker & House 2000*b*) and the Upper *trachytera* conodont Zone. This suggests that the *Annulata* Event (House 1985; Becker 1993*b*; Becker *et al.* 2004) may be recognizable in the Oulmès area. A further indication is the report by Termier (1938) of *Erfoudites ungeri rotundolobatus* (taxonomy here revised) from Tabourit since the genus and species first enters in the *Annulata* Event beds (e.g. Becker *et al.* 2002, 2004). The sequence terminates with morphologically prominent quartzites, which yielded miospores indicating a 'Strunian' age (Tahiri 1991). This roughly comprises an interval from the Upper *expansa* to Middle *praesulcata* conodont Zone, and including the level of the German Hangenberg Sandstone (Bless *et al.* 1993). Becker (1993*a*) drew attention to the presence of an unfossiliferous black shale unit with some intercalated thin siltstone bands directly below the terminal Famennian quartzites. The post-sedimentary tectonics of the Oulmès region were summarized by Tahiri (1994).

New results

Our first detailed investigation of Middle and Upper Famennian conodont and event stratigraphy of the Oulmès area is based on two sections, Ain Jemaa 1 and 2, separated on strike by slightly less than 2 km distance (Carte du Maroc 1 : 100 000, sheet NI-29-XII-4, Khemisset, Figs 2–3). Section 1 lies close to the village and, coming from the East and passing the school, north of the winding, overall east–west-trending track. The reefal limestones of the Bou Sif Member form a cliff just north of the road, with some houses interspersed. On the back side, there is good outcrop of overlying Frasnian shales. Dark grey, brownish weathering shales with abundant styliolinids, some homoctenids, pelecypods and entomozoid ostracods (*Richterina, Nehdentomis*) yielded '*Phacops*' cf. *cryphoides* (GPIM B7A.7-1), a very poorly known small-eyed phacopid reported so far only by Richter & Richter (1926) from the early Upper Frasnian of Germany. The previous recognition of Kellwasser facies is supported by the discovery of poorly preserved, pyritic archoceratids (GPIM B6C.33-1) and of

Aulatornoceras sp. indet. (B6C.33-2) in black shales. The goniatite assemblage, as well as the presence of *Buchiola*, is typical for the Kellwasser facies but it is currently unknown whether both or only one of the Kellwasser units is present. Termier & Termier (1950, pl. CXLII, Figs 26–28) illustrated from the Bou Gzem a *Manticoceras* as a Lower Famennian *Lobotornoceras*.

Lower Famennian nodular limestones, which include a crinoidal marker bed, follow on the northern slope of the low hill. The sampled conodont beds lie further downslope in a N-trending valley in a small natural outcrop along a small brook (see Fig. 4). The Upper Devonian pelagic shales and nodular limestones above the reef limestones are here assigned to the new Bou Gzem Formation (type locality at Ain Jemaa 1), which can be subdivided into Lower (Frasnian grey and black shales), Middle (Famennian nodular limestone and shale), and Upper (latest Famennian black shale) Members. The beds dip rather steeply northwards. The ca. 50 m thick measured succession begins above a covered part, separating the Lower Famennian nodular beds, in the upper part of the Middle Member with silty shales containing orthoconic nautiloids and poorly preserved brachiopods (Bed 1a, Fig. 2; =Unit 6 of Cogney 1967). These silty shales are overlain by nodular limestones (Bed 1) with ostracodes, fish teeth, clymenids and abundant, well-preserved conodonts (Table 1). Collected macrofauna includes poorly preserved Kosmoclymeniidae indet. (evolute forms with weakly biconvex growth lines, GPIM B6C.33-3) and subevolute and compressed Cyrtoclymeniidae, probably *Protactoclymenia* sp. (growth lines concavo-convex, GPIM B6C.33-4). Thin sections of Bed 1 reveal a bioclastic wacke- to packstone (Fig. 5 part 1), following the limestone classification of Dunham (1962), or a floatstone after Embry & Klovan (1972). The limestone nodules contain less than 20% components (<1–2 mm), which consist of shell fragments of clymenids, pelecypods, ostracodes, trilobites and brachiopods. The conodont association of Bed 1 can be referred to the polygnathid–palmatolepid biofacies due to the high abundance of *Polygnathus communis communis*, *Palmatolepis gracilis gracilis* and *Pa. gracilis sigmoidalis*. Many specimens of *Branmehla inornata* and *Mehlina strigosa* were also found, whereas bispathodids (*Bispathodus aculeatus aculeatus*, *Bi. stabilis* Morphotype 2) and pseudopolygnathids (*Pseudopolygnathus* cf. *dentilineatus*) are less abundant. Only few specimens of *Br. gediki* and one specimen of *Pa. gracilis* cf. *expansa* were recovered.

Bed 2 is a bioclastic wacke- to packstone/floatstone (Fig. 5 part 2) with ca. 30% components (<1–2 mm), among them many shell fragments (orthoconic cephalopods, ammonoids, ostracodes, pelecypods, trilobites, brachiopods), fish teeth and conodonts (Table 1). The conodont biofacies is difficult to establish due to the low conodont abundance. The faunal assemblage is more or less the same as in Bed 1, but *Bi. aculeatus* and *Br. gediki* are missing, and the palmatolepids and polygnathids decrease in abundance, while *Br. inornata* increases. The overlying unfossiliferous, olive-green–grey silty shales (Bed 2b) are overlain by a 7 m thick, unfossiliferous pyritic black shale unit (Bed 2c, Upper Member of the Bou Gzem Formation), which is followed by an olive-green to grey shale unit (Bed 2d) with pyritic and haematite-rich nodules and concretions, indicating persisting poor oxygenation. The succession terminates with thick (>20 m), grey quartzitic sandstones (Bed 2e), that are here assigned to the new Ta'arraft Formation (after the Jebel Ta'arraft on the topographic map that forms the hill bordering the type locality and valley W of Ain Jemaa).

Section Ain Jemaa 2 is situated slightly less than 2 km W of Ain Jemaa 1, just S of a left curve of the winding track in a northwards-running small valley within the Bou Gzem (=Bouguezzam on the topographic map); it was previously briefly studied by Becker (1993a). As at Ain Jemaa 1, there is a partly overgrown stretch that separates the higher part of the Famennian from the folded, rather solid, yellowish Lower Famennian nodular limestones (lower part of Middle Member of the Bou Gzem Formation). The 29 m thick studied succession starts with grey silty shales (Bed 3a, Fig. 2) and nodular limestones (Bed 3), which can be regarded as a bioclastic mud- to wackestone with less than 5% components (~80 μm grain size; Fig. 5 part 3). Bed 3 yielded pelecypods, ostracodes, poorly preserved ammonoids, lithoclasts and abundant, well-preserved conodonts (Table 1), such as *Polygnathus communis communis*, *Mehlina strigosa*, *Icriodus* cf. *alternatus*, subspecies of *Palmatolepis glabra*, and subspecies of *Pa. quadrantinodosa*. The abundant specimens of *Pa. gracilis gracilis* indicate the palmatolepid biofacies. Similar to Ain Jemaa 1, a 4 m-thick, pyritic black shale horizon (Bed 4b, Upper Member of Bou Gzem Formation) with some thin (max. 3 cm) siltstone layers is intercalated within pyritic, grey and unfossiliferous silty shales (Beds 4a, 4c). Above, a series of thick, fine-grained quartzitic sandstones (Bed 4, new Ta'arraft Formation) dip with 38° to the NE. They lack graded bedding, show few ripples, and form locally the hill summits. These clastics are thought to represent a subtidal pro-deltaic environment.

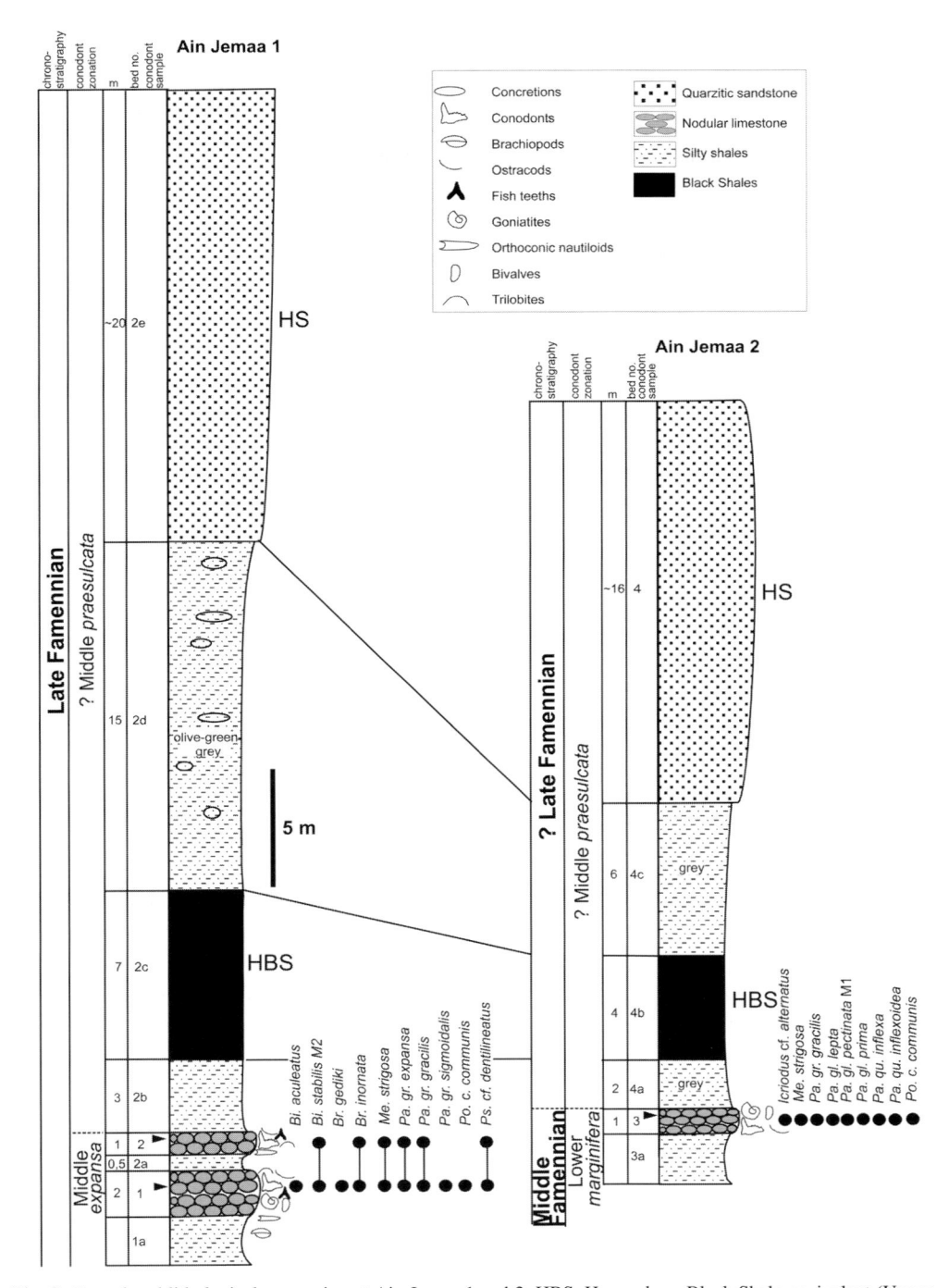

Fig. 2. Faunal and lithological succession at Ain Jemaa 1 and 2. HBS, Hangenberg Black Shale equivalent (Upper Member of Bou Gzem Formation); HS, Hangenberg Sandstone equivalent (Ta'arraft Formation).

Table 1. *Conodont faunas from sections Ain Jemaa 1 and 2.*

Ain Jemaa 1		
	Middle *expansa*	
	Bed 1	Bed 2
Bispathodus aculeatus aculeatus	3	
Bispathodus stabilis M2	5	2
Branmehla gediki	3	
Branmehla inornata	21	10
Mehlina strigosa	16	5
Palmatolepis gracilis expansa	1 cf.	1
Palmatolepis gracilis gracilis	24	3
Palmatolepis gracilis sigmoidalis	12	2
Polygnathus communis communis	43	9
Pseudopolygnathus cf. *dentilineatus*	2	3
fish teeth	xx	x
ostracodes	x	x

Ain Jemaa 2	
	Lower *marginifera*
	Bed 3
Icriodus cf. *alternatus*	9
Mehlina strigosa	6
Palmatolepis glabra lepta	3
Palmatolepis glabra pectinata M1	2
Palmatolepis glabra prima	13
Palmatolepis gracilis gracilis	42
Palmatolepis quadrantinodosa inflexa	1
Palmatolepis quadrantinodosa inflexoidea	2
Polygnathus communis communis	8

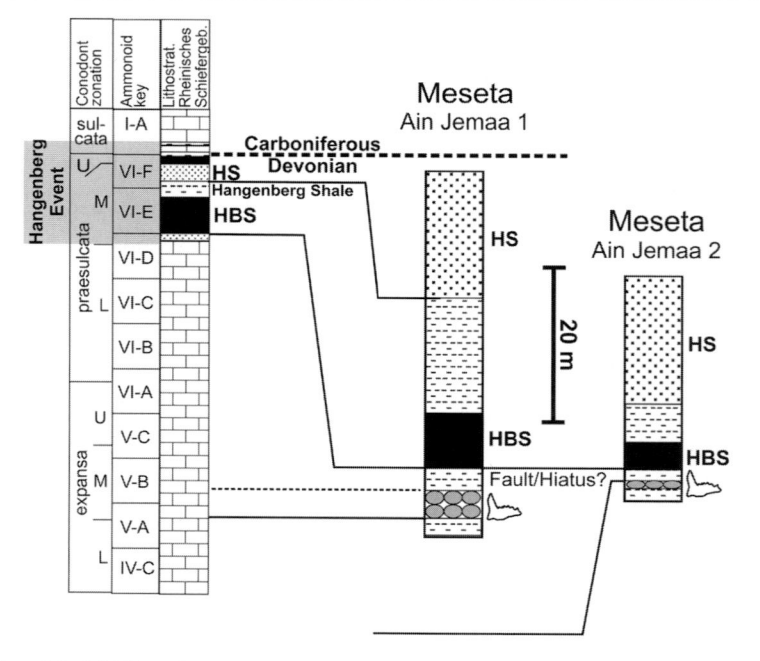

Fig. 3. Correlation of the Middle to Uppermost Famennian at Ain Jemaa with the bio- and lithostratigraphic succession of the Rhenish Massif. Abbreviations as in Figure 2.

Fig. 4. Outcrop photo of Ain Jemaa 1, showing the outcrop position of sampled nodular limestone, of Hangenberg Black Shale equivalents (Upper Member of Bou Gzem Formation), and of the Ta'arraft Formation (Hangenberg Sandstone equivalent).

Stratigraphical dating

Ain Jemaa 1

The occurrence of *Bispathodus aculeatus aculeatus*, the defining subspecies for the Middle *expansa* Zone, and the occurrence of *Branmehla gediki*, known from the Lower *expansa* to the lower part of the Middle *expansa* Zone (Capkinoglu 2000), indicate the basal Middle *expansa* Zone for Bed 1 at Ain Jemaa 1. This is supported to some extent by the polygnathid–palmatolepid biofacies of Bed 1, because the same conodont biofacies is widely known from European successions in the basal Middle *expansa* Zone, e.g. in the Carnic Alps (Kaiser 2005; Perri & Spalletta 1998; Spalletta *et al.* 1998). The presence of typical Upper Devonian V (Dasbergian) kosmoclymenids in Bed 1 is in good agreement with the conodont data; ammonoid–conodont correlation in the Tafilalt (Kaiser 2005) and in the Bergisch Gladbach area of Germany (Piecha 2005) showed that the base of the Dasbergian (UD V-A1) more or less coincides with the base of the Middle *expansa* Zone. The available faunal data suggest that Bed 1 correlates with the global Dasberg Event level (Becker 1993*b*) at the base of UD V.

The poorer conodont fauna of Bed 2 does not provide a precise age but the defining species of the *expansa* Zone was found. Similar to European successions, the bispathodid–branmehlid group increases in abundance above the basal Middle *expansa* Zone. There is no evidence for any marker species of the Upper *expansa* or *praesulcata* Zones that would suggest the presence of Upper Devonian VI (Wocklumian). The lack of *Bi. aculeatus aculeatus* shows that the Middle *expansa* Zone may be difficult to recognize in small faunas.

Conodont biofacies interpretations focus on the relative distribution of genera and interpretations are based on Sandberg & Ziegler (1979), Sandberg & Dreesen (1984), and Ziegler & Weddige (1999). Accordingly, the palmatolepid–bispathodid biofacies indicates deposition on the continental rise and lower slope, whereas the palmatolepid–polygnathid biofacies indicates deposition on the middle and upper slope of the outer shelf. Thus, the palmatolepid–polygnathid biofacies in the basal Middle *expansa* Zone at Ain Jemaa 1 (Bed 1) represents a moderately deep, pelagic, depositional environment, which is supported by the relatively common occurrence of clymenids.

In Europe, the palmatolepid–bispathodid biofacies that is thought to reflect deeper settings (when compared to a *Pa–Po*-biofacies, see above) began within the Middle *expansa* Zone (Kaiser 2005). The palmatolepid–bispathodid–(branmehlid) biofacies characterizes the interval between the late Upper *expansa* and Lower/Middle *praesulcata* Zone, prior to the Hangenberg Event (*Branmehla* is supposed to have lived in the same setting as *Bispathodus*, see Ziegler & Sandberg 1984). Compared to the conodont biofacies in Bed 2 at Ain Jemaa 1, the

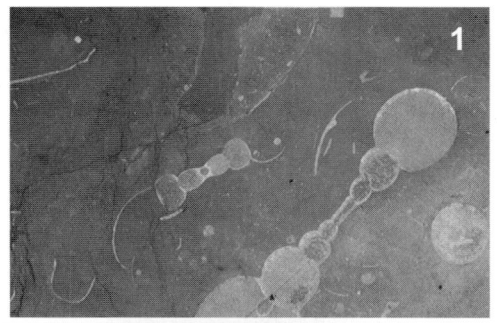

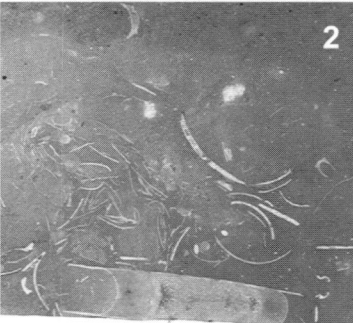

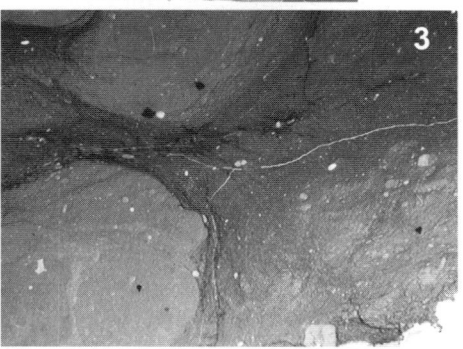

Fig. 5. Thin sections illustrating the microfacies of sampled nodular limestones near the top of the Middle Member of the Bou Gzem Formation (image widths 27 mm). **1**, Bioclastic floatstone/wacke- to packstone with kosmoclymeniid cross-sections, Ain Jemaa 1, sample AJ 1 (SMNS 66154-1), Bed 1, Middle *expansa* Zone. **2**, Bioclastic floatstone/wacke- to packstone with fragmented mollusc shells and orthoconic cephalopod, Ain Jemaa 1, sample AJ 2 (SMNS 66154-2), Bed 2, Middle *expansa* Zone. **3**, Bioclastic mud- to wackestone, Ain Jemaa 2, sample AJ 3 (SMNS 66154-3), Bed 3, Lower *marginifera* Zone.

similarities are obvious. *Branmehla* increases in abundance and this probably reflects a slight deepening of the depositional environment. However, *Polygnathus* continues to be frequent, suggesting that moderately deep conditions were maintained. The microfacies of Beds 1 and 2 indicate moderately high depositional energy (wacke- to packstone with large, partly fragmented fossil debris embedded in a fine matrix), in accordance with the conodont biofacies.

Although biostratigraphical evidence for the Upper *expansa* to Lower/Middle *praesulcata* Zone is lacking, the black shale horizon (Bed 2c) at Ain Jemaa 1 is interpreted as an equivalent of the Rhenish Hangenberg Black Shale. This assumption is mostly based on global sequence stratigraphy (Bless *et al.*, 1993; Becker 1996) and on the knowledge that the Hangenberg Black Shale has been recognized both to the North (Montagne Noire) and to the South (Anti-Atlas) of the Meseta (e.g. Kaiser 2005). As a consequence, the pre-event part of the Uppermost Famennian (UD VI) is either missing or extremely condensed (as on the Tafilalt Platform of the Anti-Atlas) and only represented by the argillaceous Bed 2b. The black shales represent a deepening phase and transgressive system tract (TST) above a regional sequence boundary. The subsequent grey shales (Bed 2d) can be correlated with the main, grey-green Hangenberg Shale of Germany, a highstand system tract (HST). The overlying quartzites of the Ta'arraft Formation (Bed 2e) can be regarded as an equivalent of the German Hangenberg Sandstone, which was deposited near the top of the Lower/Middle *praesulcata* Zone. This is in accordance with miospore results (Tahiri 1991), indicating a 'Strunian', still Uppermost Famennian age. The Hangenberg Sandstone and its equivalents have been interpreted (Bless *et al.* 1993; Van Steenwinkel 1993) to represent a slope fan of a lowstand system tract (LST) but a sequence boundary is only locally developed at its base (e.g. at the base of incised valleys). The gradual transition from shale to sandstone shows that there is no sequence boundary (unconformity) in the Oulmès area.

As an alternative interpretation, one would have to assume a significant black shale and subsequent regressive episode that is older and completely out of phase with the well-established global sea-level curve (Johnson *et al.* 1985; Bless *et al.* 1993; Becker 1996). Assuming extremely reduced but continuous late Famennian sedimentation in both studied sections, the grey and black shales above the last nodular limestone would have to be highly diachronous over a short distance. Such interpretations are unlikely but cannot be ruled out completely.

Ain Jemaa 2

Although the defining species *Palmatolepis marginifera marginifera* is missing at Ain Jemaa 2, a typical conodont assemblage of the Lower *marginifera* Zone (basal Middle Famennian) is evident in Bed 3, with *Pa. quadrantinodosa inflexoidea*, *Pa. quadrantinodosa inflexa* and *Mehlina strigosa* as recorded marker species (Table 1). The first subspecies became extinct at the top of the Lower *marginifera* Zone (Ziegler & Sandberg 1984). *Palmatolepis glabra prima* is abundant at Ain Jemaa 2, which is in accordance with observations in the same zone of European sections (e.g. Sardinia, Italy, Corradini 2003). A rather similar assemblage was recorded by Alberti (1970) from the youngest exposed solid limestone at Bou Keziam but our new sampling proved that there are locally no younger limestones in section 2. This may explain why Alberti was unable to reproduce Cogney' collection of Middle Famennian trilobites, such as *Franconicabole*. The palmatolepid biofacies of Bed 3 indicates deposition in a relatively deep pelagic environment, which is supported by the microfacies characteristics, reflecting low-energy deposition. However, it is remarkable that *Icriodus* is rather frequent (12% of the fauna) since the genus is normally common in more shallow settings of the middle shelf and not a typical element of palmatolepid faunas. Becker (1993a), however, has previously documented from the Montagne Noire Famennian icriodid blooms of more than 80% in black shale facies with rich pelagic, outer shelf goniatite faunas. A palmatolepid–icriodid biofacies observed in Sardinian sections was also regarded as pelagic (Corradini 2003, p. 76). It is therefore clear that the current conodont biofacies model needs further elaboration.

The grey shales of Bed 4a probably correlate with Bed 2b of Ain Jemaa 1. Therefore, there is a considerable hiatus above Bed 3, including at least the Upper *marginifera* to Middle *expansa* Zones, possibly also, as at Ain Jemaa 1, the Upper *expansa* and Lower *praesulcata* Zones. Bed 4b correlates, with reduced thickness that indicates stronger condensation and a more distal position with respect to the siliciclastic source, with the supposed Hangenberg Black Shale (Bed 2c) of section 1. It is followed by equivalents of the main Hangenberg Shale (Bed 4c) and of the Hangenberg Sandstone (Bed 4).

The laterally highly variable, very thin and incomplete (several zones are missing) sedimentary record of the Middle to Uppermost Famennian at Ain Jemaa is most likely a consequence of episodic non-deposition and extreme condensation over millions of years. There is no evidence for deposition and subsequent removal of beds; reworked conodonts have not been observed. The enduring lack of deposition is best explained by lasting relative sea-level fall, resulting in increased bottom turbulence, possibly including contourites and thermohaline currents (Hüneke 2006). Brief transgressive episodes, such as the *Annulata* Event at the base of UD IV, and the Dasberg Event at the base of UD V, seem to have allowed the episodic deposition of thin (condensed) pelagic limestone beds.

Comparisons with other Meseta regions allow to distinguish whether the recognized Middle and Upper Famennian relative sea-level fall at Ain Jemaa was triggered by regional tectonic movements or eustatic changes. The latter would have affected the whole sedimentary basin at the same time.

Comparison with the Famennian of other Meseta structural units (Figs 6–7)

El Hammam Zone

The Upper Devonian facies history changes at rather short distance from the Oulmès/Ain Jemaa region towards the North and East. East of the Oulmès Fault, the famous but still poorly studied Moulay Hassani section (Cogney 1967; Tahiri 1991, preliminary new data; MH in Fig. 6b) belongs to a more basinal setting, lacking reefal limestones. Therefore, it is assigned to a distinctive El Hammam structural unit (e.g. Tahiri & Lazraq 1988; Tahiri 1994). The Moulay Hassane Formation includes black shales and limestones with diverse, still unstudied Upper Frasnian (Lower Member, Tahiri 1991; Fadli 1994) and some Lower Famennian goniatites and brachiopods (Middle Member); the Kellwasser beds are obviously developed. Higher up, there is a thick sandstone unit (Upper Member; better re-assigned in future to a separate formation) that probably correlates with the (new) Ta'arraft Formation of Ain Jemaa. Currently, there is no record of Middle or Upper Famennian faunas and it is not yet known how much of this interval is present in the shales below the sandstones. It is likely that, as at Ain Jemaa 2, the Middle to (pre-Hangenberg Event) Uppermost Famennian is lacking. Further biostratigraphical results, including more miospore data, are needed.

Tiliouine area and Rabat–Tiflet Zone

Just N of the Ain Jemaa region, at Tiliouine (Tahiri & Hoepffner 1988, T in Fig. 6b), the (new) Bou Gzem Formation is completely missing above reefal limestones, or it is reduced to a very thin, probably incomplete unit of nodular limestone. This suggests that the reef complex was not drowned as much in the Frasnian as the reef of the Ain Jemaa area. Such regionally variable subsidence is an indication of Frasnian block tectonics. Above, there is a conglomerate with reworked Middle Devonian reefal and Upper Devonian nodular limestone, overlain by shales and sandstones that may partly correlate with the (new) Ta'arraft Formation. The sudden reworking and redeposition of already buried and lithified shallow (reef limestone)

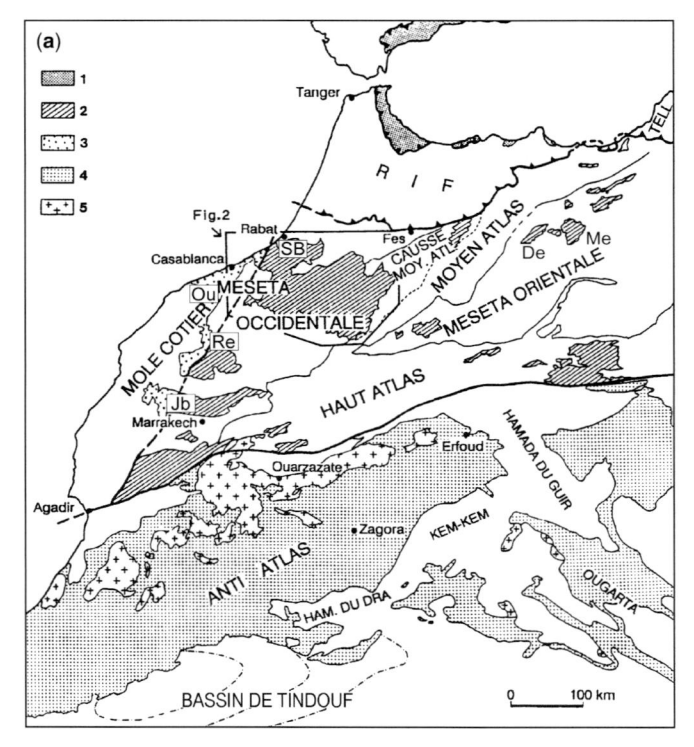

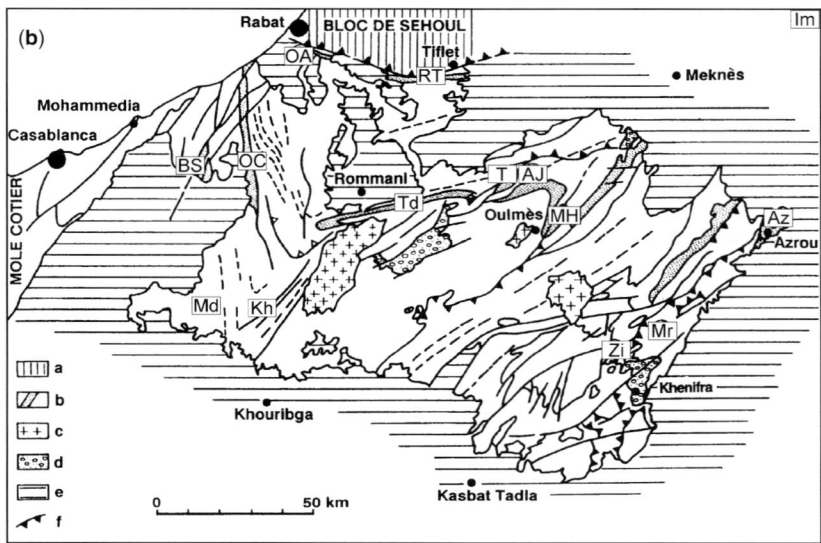

Fig. 6. Location of important Famennian successions of the Moroccan Meseta (updated from Walliser *et al.* 1995). Locality abbreviations in **a**: SB, Sidi Bettache Basin; Re, Mechra Ben Abbou (Rehamna); Jb, western Jebilet; Ou, Oulad-Abbou (Coastal Block, De, Debdou; Me, Mekam); in **b**: AJ, Ain Jemma (Oulmès Zone); MH, Moulay Hassne (El Hammam Zone); Td, Tiddas; T, Tiliouine; RT, Rabat-Tiflet Zone; OA, Oued Akrech S of Rabat; Kh, Khatoaut; Md, Mdakra Massif; OC, Oued Cherrat Zone; BS, Ben Slimane Zone; Mr, Mrirt; Az, Azrou; Zi, Ziar. Geological units in **a**: 1=Rif Palaeozoic, 2=Meseta Palaeozoic, 3=Palaeozoic of the Coastal Block, 4=Anti-Atlas Palaeozoic, 5=Precambrian granites. Units in **b**: a=Ordovician of the Caledonian Sehoul Block, b=Palaeozoic with dotted Devonian, c=Hercynian granites, d=Stephanian to Permian (Post-Variscan Palaeozoic), e=post-Palaeozoic cover, f=major faults.

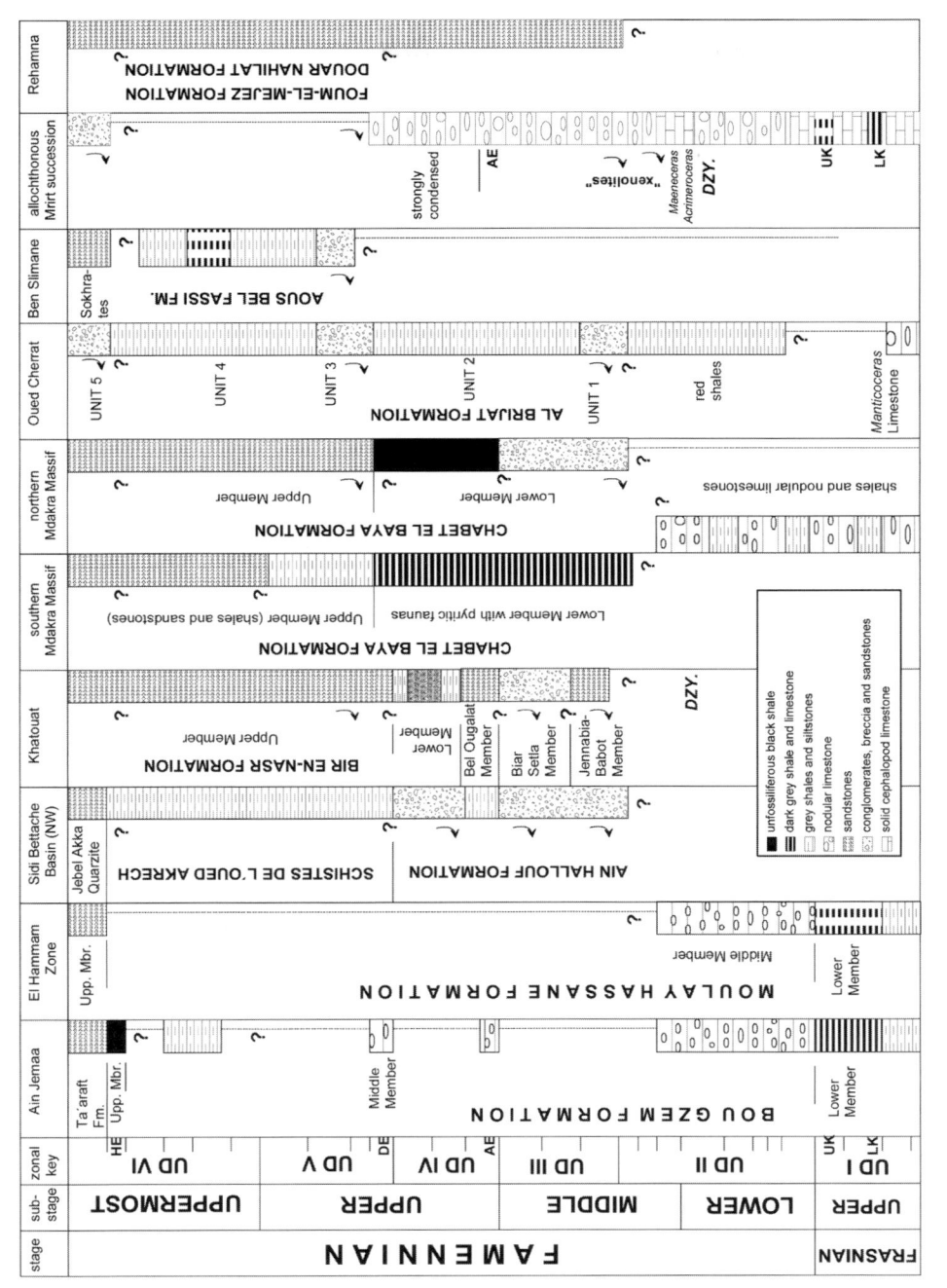

Fig. 7. Correlation of the lithostratigraphy of Ain Jemaa with other Meseta regions and with the chronostratigraphy. Subdivisions of substages (zonal keys, UD I to VI and subunits) follow the global ammonoid zonation and its correlation with the conodont succession as summarized by Becker & House (2000). LK, Lower Kellwasser Event; UK, Upper Kellwasser Event; AE, Annulata Event; DE, Dasberg Event; HE, Hangenberg Ebent; DZY, level of *Dzyduszyckia* limestones. Curved arrows indicate levels of local erosion and redeposition.

carbonates, and partly of overlying deeper marine (pelagic) sediments, indicate a significant Famennian rapid local uplift of the source area, probably caused by increasing block faulting. A basin-wide (eustatic) regression could have resulted in the erosion on topographic highs (seamounts) but the Givetian reef complexes were already drowned and partly covered by Frasnian deeper-water (subphotic) sediments prior to their Famennian exhumation. The eustatic model would thus require a very significant sea-level fall that even exposed pelagic sediments. It would not explain why some drowned reef areas suddenly became the subject of deep erosion and why neighbouring regions remained below the photic zone and were only affected by increased turbulence and conglomerate shedding.

The postulated (Tahiri & Hoepffner 1988) synsedimentary tectonic episode at Tiliouine seems to correlate with the Middle to Uppermost Famennian phase of non-deposition/extreme condensation at Ain Jemaa. Detailed biostratigraphical data, however, are not yet available. The Oulmès area occupied a distal position to the region of deep erosion. Similarities with the incomplete succession of the Rabat-Tiflet Zone to the northwest (El Hassani 1991, 2000; RT in Fig. 6b), where supposed late Famennian reworking and deposition of calcareous and siliceous conglomerates is well documented, suggest that the Tiliouine Upper Devonian belongs to a different structural zone than Ain Jemaa (Tahiri 1994) and is an eastward extension of the Sidi Bettache Basin of Pique (1983).

Sidi Bettache Basin

A similar situation as at Tiliouine and in the Rabat Tiflet Zone was described by Chakiri & Tahiri (2000) for the Tiddas area (Td in Fig. 6b) in the NE part (Zaer Zone) of the Sidi Bettache Basin (SB in Fig. 6a). Here, reworked Silurian and Devonian blocks, including Middle Devonian reefal limestones, are embedded between supposed Famennian shales and sandstones. Again, biostratigraphical data that could give more precise time constraints are not available. At the top of the Devonian, a 'Strunian' quartzite unit suggests the presence of a Hangenberg Sandstone equivalent.

In the NW part of the Sidi Bettache Basin, S of Rabat (Oued Akrech, OA in Fig. 6b), Famennian conglomerates as further evidence of reworking do occur in the Ain Hallaouf Formation, within a clastic, 'flyschoid' basinal facies with thick mass flow deposits indicating high subsidence (Pique & Kharbouch 1983; Pique 1984). As discussed above, a basinwide (eustatic) sea-level fall might lead to some erosion on seamounts but not to the strong facies differentiation of a shelf. The sudden appearance of neighbouring highs with major sediment exhumation and of continuous, thick, clastic deeper-water wedges underlines the significance of Eohercynian block movements. Intercalated shales with marker clymenids of the *annulata* Zone (UD IV-A, Lecointre 1926; Choubert & Faure-Muret 1961) prove that erosion and clastic shedding occurred within the central Meseta before and after this level, in the Middle and Upper Famennian. Following a more shaly unit (Schistes de l'Oued Akrech, probably Uppermost Famennian), the lenticular and shallow-marine sandbars of the Jebel Akala Quartzites most likely represent another equivalent of the Hangenberg Sandstone. This is supported by the presence of 'Strunian' miospores, including *Retispora lepidophyta* (Fadli 1994), which elsewhere disappear just before the D/C boundary. Overlying black shales yielded already Middle Tournaisian spores (identified by M. Streel).

Khatouat

In the southern prolongation of the Sidi Bettache Basin, Famennian successions occur in the Khatouat region and in the Mdakra Massif, separated by the Cherrat Ridge (Pique 1994, Fig. 18; see Fig. 6b). In the first region (KH in Fig. 6b), coquinas with the rhynchonellid *Dzieduszyckia* have been reported as allochthonous blocks in the vicinity of conglomerates of the Biar Setla Member, which is the middle member of the Fouizir Formation (Rolleau 1956; Hollard & Morin 1973; Fadli 1994). In the Mrirt area (see below), *Dzieduszyckia* occurs in the upper Lower (Uppermost *crepida* Zone) to basal Middle Famennian (Lower *marginifera* Zone). This suggests a poorly and incompletely (by reworking) preserved pelagic sedimentation in the Khatouat region until the basal Middle Famennian, perhaps followed by the undated sandstones of the Jennabia–Babot Member, which is the lower member of the Fouizir Formation (Fadli 1994; El Hassani & Benfrika 2000). Subsequently, the reworking of *Dzyduszyckia* limestones and the shedding of mass and debris flows, conglomerates and sandy turbidites formed the Biar Setla Member. In the Sibara region, similar conglomerates assigned locally to the Sequiet Abbes Formation (Zahraoui 1994; Fadli 1994) yielded conodonts, including *Pa. perlobata helmsi* and *Pa.* cf. *perlobata sigmoidea*. These give an overlapping age ranging from the Lower *trachytera* Zone (upper Middle Famennian, UD III-B/C) to the Upper *styriacus* Zone (basal Upper Famennian, UD IV-B), clearly post-dating the *Dzieduszyckia* coquinas. Acritarchs from shale levels of the Biar Setla Member and between greywackes of the overlying Bel Ougalat Member (the upper member of the Fouizir Formation), e.g. *Unellium piriforme* (Fadli

1994) unfortunately give only a broad Upper Devonian age or (e.g. *Winwaloeusia distracta*) range throughout the Devonian (see review of acritarch biostratigraphy in Le Hérisse *et al.* 2000).

The Lower Member of the subsequent Bir En-Nasr formation, consisting of shales and sandstones, yielded a Famennian miospore flora that seems to pre-date the entry of the globally widespread *Retispora lepidophyta* in the LL Zone. This assemblage of the VCo Zone is older than the base of the Middle *expansa* Zone (older than UD V, Dasbergian, Hartkopf-Fröder 2004). Therefore, all members of the Fouzir Formation and the lower part of the Bir En-Nasr Formation fall in the Middle to lower Upper Famennian. The coarse clastics and the major reworking events of the Fouizir Formation probably pre-date the *Annulata* Event.

The sandstones of the Upper Member of the Bir En-Nasr Formation, indicating a second peak of erosion and detrital discharge, accumulated during UD V/VI, which is supported by 'Strunian' spores (Fadli 1994). Currently, an equivalent of the Hangenberg Sandstone cannot yet be identified with certainty but the Upper Member, as the (new) Ta'arraft Formation of Ain Jemaa, is a prodeltaic unit that accumulated during a lowstand. Spores (*Retispora lepidophyta* and *Vallatisporites hystricosus*) from the Souk Jemaa Formation, a northern lateral equivalent of the Bir En-Nasr Formation, fall at least in the LE Zone, giving an UD VI (Uppermost Famennian, Wocklumian) age (Streel & Loboziak 1996), not necessarily excluding the Hangenberg Sandstone interval (LN Zone).

Mdakra area

The Famennian of the Mdakra Zone (Md in Fig. 6b) was deposited in a more distal position with respect to siliciclastic sources. The Chabet El Baya Formation (Fadli 1994) consists in the Oued Aricha area (southern part of the Mdakra Zone) of shales that include in its Lower Member important, poorly studied pyritic ammonoid faunas. Based on published data and illustrations (Termier 1936; Termier & Termier 1950, 1951; Roch 1950), the faunal evidence can be revised as follows. In the upper Oued Aricha valley, there are *Erfoudites ungeri rotundolobatus*, other *Erfoudites* species (det. '*Sporadoceras biferum*' and '*Sporadoceras* cf. *Ungeri*'), *Gundolficeras* aff. *bicaniculatum* (det. '*Lobotornoceras Sandbergeri*'), *Sporadoceras* cf. *orbiculare* (det. '*Sporadoceras* sp.'), *Prionoceras divisum sulcatum* (det. '*Imitoceras intermedium*'), ?*Protactoclymenia* sp. (det. '*Cyrto. angustiseptata*'), and platyclymenids (det. '*Clymenia* sp.'). All taxa are known from the Upper Devonian IV (lower Upper Famennian). The same age is indicated for an assemblage from Daichet, with *Erf. ungeri rotundolobatus*, *Sp. muensteri*, platyclymenids (det. '*Clymenia laevigata*' and '*Platyclymenia barrandei*'), and *Protoxyclymenia dubia*. An alleged '*Oxyclymenia undulata*' (Termier & Termier 1951), indicating younger, UD V, strata has not been illustrated and remains doubtful. However, lower Dasbergian (UD V) ammonoids are present at two other sections, N Bas Touil (*Kosmoclymenia* sp.) and Taicha (*Posttornoceras posthumum*, det. '*Discoclymenia cucullata*'). Hollard & Morin (1973) mentioned a 'cf. *Protornoceras planidorsatum*' from shales and sandstones of the Oued Zemrine, which might represent an *Armatites* (UD II-D to II-I, Becker 1993a) or a *Planitornoceras* (UD III, Becker *et al.* 2002), giving in any case an older age than all other faunas. The lower part of the Chabet El Baya Formation, therefore, spans approximately the Middle to lower Upper Famennian. Sandstones increase above the fossiliferous shales, in the upper part of the Upper Famennian.

In the northern part of the Mdakra Zone, near Beni Sekten, Termier & Termier (1951) report a sequence of upper Frasnian to lower Famennian shales and nodular limestones below the clastic Chabat El Baya Formation. This unit is lacking in other sections (Benfrika & Bultynck 2003), probably due to local non-deposition on a topographic high. The Chabet El Baya Formation begins with an important, probably Middle Famennian conglomeratic unit (Lower Member) that locally (Fadli 1994: section of Fig. 9; Benfrika & Bultynck 2003) cuts down into Middle Givetian (*rhenanus/varcus* Zone) reef limestone (Sidi Mohamed Smaine Formation) or even deeper, down to Eifelian shales and limestones (upper Sifi Ahmed Lemdoun Formation). This interval of truncation, reworking and redeposition is followed by undated 40 m of fossiliferous black shales. The Upper Member consists of sandy and conglomeratic calcarenites (Fadli 1994) with a 'Strunian' (Termier & Termier 1951) macrofauna. Records of *Mesoplica praelonga*, *Whidbornella caperata*, *Centrorhynchus letiensis* and *Prospira strunianus* from around Lalla Regraga indeed suggest an Uppermost Famennian age (see ranges given in Brice *et al.* 2005). In other sections, 'Strunian' quartzites are developed that either pre-date or correlate with the Hangenberg Sandstone. There are still no biostratigraphical data that give an age for the subsequent M'Garto Formation, consisting of shales and sandstones, although it has been correlated (Fadli 1994; Pique 1994) with the Bir En-Nasr Formation of the Khatouat Zone. This interpretation is contradicted by the miospore and ammonoid record. It is the Lower Member of the underlying Chabat El Baya Formation (Mdakra, UD IV/basal V) that mostly correlates with the Lower Member of the Bir En-Nasr Formation (Khatouat, UD IV), leaving probably a correlation of the Upper Member of the Chabat El Baya Formation (with 'Strunian' spores) with the Upper Member of the Bir En-Nasr Formation (with 'Strunian' brachiopods).

Oued Cherrat Zone

The narrow band of the Oued Cherrat Zone forms the western border of the Sidi Bettache Basin (OC in Fig. 6b). There, the Famennian Al Brijat Formation is characterized by siliciclastic sedimentation with evidence of tectonically controlled erosive events in a still-unknown source area and coarse clastic shedding into an adjacent, subsiding basin. Frasnian pelagic limestones are extremely condensed (of very low thickness) and incomplete (only some conodont zones recognizable) or locally completely missing. There is no evidence of conodont reworking, which suggests episodic and long-lasting Frasnian non-deposition. Synsedimentary reworking, however, is evidenced in the Famennian by three conglomeratic levels (Chalouan 1981: units 1, 3 and 5) that include exhumed Givetian reefal limestones regionally assigned to the Cakhrat-ach-Chleh Formation: one in the 'lower' Famennian (no biostratigraphical data), one in the Upper Famennian, and one near the top of the Famennian. The second level has been correlated with a conglomerate of the left bank of the Oued Cherrat that, based on *Bispathodus costatus* (with trends towards *Bi. spinulicostatus*), can be dated as higher part of the Middle *expansa* Zone (upper UD V). Deposition of this conglomerate, therefore, coincided with the onset of Upper Famennian non-deposition at Ain Jemaa. There are no reported black shales or clear Hangenberg Sandstone equivalents but the third conglomeratic level seems to form the top of the Devonian (Zahraoui *et al.* 2000) and may correlate with the Hangenberg lowstand just below the D/C boundary. The next-dated calcareous level of the region (Sidi-Jilali Limestone) falls already in the Middle Tournaisian (conodonts reported in Izart & Vieslet 1988).

Ben Slimane Zone

West of the Oued Cherrat, the Ben Slimane Zone (BS in Fig. 6b) represents again a more-basinal depositional complex but the Givetian to Lower Famennian may have been lost during a significant erosional event. Above a basal conglomerate, there are thick multicoloured shales and siltstones of the Aous Bel Fassi Formation that include in their upper part an important, small-sized, pyritic and diverse ammonoid fauna. Based on the presence of the goniatite *Effenbergia*, it can be precisely dated as UD VI-B (Becker & House 2000*b*), providing the only evidence in the Meseta of an Uppermost Famennian (Wocklumian), pre-Hangenberg pelagic assemblage. Overlying quartzites forming 'Sokhrates' (rock cliffs) can be correlated with the regressive Hangenberg Sandstone since they seem to include 'Strunian' brachiopods (Lecointre 1926; Fadli 1994).

Coastal Block

The Upper Devonian occurs in the Oulad–Abbou Syncline SW of Casablanca (Hollard 1967; Ou in Fig. 6a) and is said to contain Frasnian to basal Famennian shales and siltstones with rhynchonellids. Sediments of the higher Famennian are not known in the area. It is intriguing that reverse parts of the Upper Devonian are preserved in the neighbouring Ben Slimane and Coastal Blocks. This suggests an influence of strongly different local block movements.

Rehamna and Jebilet

In the Rehamna to the SW of the Ben Slimane Zone and Sidi Bettache Basin (Re in Fig. 6a), the Famennian is preserved in two areas, in the Foum El Mejez Graben and in an area east of Mechra Ben Abbou (Hollard *et al.* 1982; El Hassani & Benfrika 2000). The shales and sandstones of the Foum–el-Mejez Formation may represent large parts of the Famennian but there are no reliable biostratigraphical data for the lower succession. The Frasnian is lacking, probably due to post-sedimentary erosion. Brachiopod faunas from the upper part of the Foum–el-Mejez Formation are more or less the same as in the Douar Nahilat Formation from the area near Mechra Ben Abou (Hollard *et al.* 1982). The latter has been assigned to a molasse-type basin that was gradually filled by clastics. The brachiopods *Sphenospira julii* and *Mesoplica praelonga*, if correctly identified (modern revision of the material is needed, oral comm. D. Brice, Lille), are typical for the Uppermost Famennian ('Strunian', UD VI) but the first species may range upwards from UD V (Brice *et al.* 2005). The quartzite bars in the upper part of the succession at Foum–el-Mejez and the reported shallowing upwards at Douar Nahilat may well correlate with the Hangenberg Regression but new miospore data are needed.

Farther south of the Rehamna, the Devonian of the western Jebilet (Jb in Fig. 6a) seems to continue the Rehamna facies (Hollard 1967) but the Famennian of the Jbel Ardouz, the main outcrop near Mzoudia, is not well studied (very brief reviews in Huvelin 1977; Tahiri 1983; and El Hassani & Benfrika 2000). Sandstones with Famennian brachiopods are described to lie transgressive and with a basal conglomerate (Roch 1950) on (?Givetian) reefal limestone.

Ziar-Azrou Zone

The East–Central part of the Meseta (Pique & Michard 1981) includes autochthonous and allochthonous Upper Devonian successions, with the latter having been transported as nappe units from the East. The facies dominated by pelagic limestones differs strongly from the clastic deposits of Famennian basins to the west but there are strong similarities with the Montagne Noire (Becker 1993c; Becker & House 2000a). This suggests the existence of relatively uniform oceanic shelf areas in Upper Devonian time, stretching from the eastern border of the Central Meseta to the southern part of the French Massif Central and beyond, to the northern Variscides. The sedimentation of this vast region was determined contemporaneously by identical palaeoceanographical and palaeoecological changes, a fact that is not consistent with regionally different plate tectonic regimes of a large ocean with oceanic spreading.

The allochthonous Famennian successions are best exposed around Mrirt (Termier 1936; Agard et al. 1958; Hollard et al. 1970; Hollard & Morin 1973; Becker 1993c, 1994; Lazreq 1999; Walliser et al. 1995, 2000; Mr in Fig. 6b). Coquinas of the recently (Balinski & Biernat 2003) revised rhynchonellid *Dzieduszyckia* are intercalated in the pelagic succession or occur as exotic, reworked clasts in chaotic olistolite complexes (Becker & House 2000a) and may represent vent communities (Dubé 1989). The oldest coquinas have been dated as Uppermost *crepida* Zone (based on *Pa. glabra lepta*, Hollard et al. 1970), the youngest occur together with early sporadoceratids, giving an UD II-G age (Becker & House 2000b), which is supported by unpublished conodont sampling at Bou Ounebdou. It is very intriguing that the same interval is characterized by small to 1 m large 'xenolites' (Walliser et al. 2000) that document an important reworking event culminating low in the Middle Famennian. It is unlikely that the strong correlation of the Mrirt redeposition phase with the onset of condensed, discontinuous sedimentation at Ain Jemaa to the west was just a coincidence.

The subsequent Middle Famennian near Mrirt is composed of solid to nodular, pelagic limestones with rare macrofauna. An as yet unreported level with platyclymenids in limestone preservation indicates basal Upper Famennian (UD IV) equivalents of the *Annulata* Event. The youngest-recorded conodonts (Lazreq 1999) fall in the upper part of the Middle *expansa* Zone, which is supported by old (Termier 1936; Termier & Termier 1950) records of UD V clymenids, such as *Cymaclymenia* and *Gonioclymenia*. The highly condensed nature of the Upper Famennian and the age of the youngest pelagic faunas resemble, again, the situation at Ain Jemaa. Agard et al. (1955) report a conglomerate that lies with angular unconformity on pelagic limestones. Already, Termier (1936) described shallow-water brachiopods from around the D/C boundary of Mrirt but their relation to the Upper Famennian nodular limestones is not clear. However, all aspects together suggest an Uppermost Famennian, significant relative sea-level fall.

Farther to the SW, in the Ziar area (Zi in Fig. 6b), the Upper Devonian is similarly developed as near Mrirt (Termier 1936; Lazreq 1999; Walliser et al. 1995). There are reports of *Dzieduszyckia*-bearing olistolites (Hollard & Morin 1973; Mullin et al. 1976; Ager et al. 1976) of the *Annulata* Event, and of a condensed subsequent Upper Famennian with *Protoxyclymenia* (upper UD IV) and *Kosmoclymenia* (UD V, Termier 1936; Termier & Termier 1950). Latest Famennian beds have not yet been recorded and most outcrops have not been studied in detail.

The Upper Devonian of autochthonous outcrops between Mrirt and Azrou is much less known but seems to be incomplete. Middle Famennian 'griotte' limestones with conodonts and sporadoceratids lie disconformably on the Middle Devonian (Termier 1936; El Hassani & Benfrika 2000). At Dechra Ait Abdallah, an allochthonous breccia in the Viséan yielded Termier (1936) an Uppermost Famennian brachiopod fauna. Agard et al. (1955) report from the adjacent Jebel Tikanit conglomerates and yellow calcareous sandstones with Uppermost Famennian brachiopods, including *Mesoplica praelonga* and *Prospira struniana*. The similarities between the auto- and allochthonous Famennian sequences in the Ziar–Azrou Zone provide evidence that both originally were deposited laterally to one another, with more complete deposition towards the East.

Eastern Meseta

Unfortunately, there are no Famennian data for the Eastern Meseta. The preserved Givetian/Frasnian of Debdou and Mekam (Marhoumi et al. 1983; De and Me in Fig. 6a) is siliciclastic and shows no affinities with the Ziar–Azrou Zone or the Montagne Noire.

Fes area

Local occurrences of Upper Devonian sediments crop out in the tectonic window at Immouzer du Kandar S of Fes and to the N of Azrou (Cygan et al. 1990, Fig. 6a). Above basal Frasnian reef breccia, there are mostly fine-grained to marly neritic Frasnian and Famennian limestones with an important Upper Famennian (UD IV/V) shallow-water brachiopod fauna (Brice et al. 1983). The Fes area Famennian is very distinctive from the other structural units and is not a continuation of the Ziar–Azrou Zone. The age of an overlying flyschoid succession is not known.

Conclusions

The biostratigraphy and lithology of the two studied sections at Ain Jemaa must be interpreted with respect to the complex tectonic structure of the Meseta. Hints of faults were not observed at Ain Jemaa in the field. Grey and black shales overlie nodular limestones of the Lower *marginifera* Zone (basal Middle Famennian, section 2), or of the Middle *expansa* Zone (Upper Famennian, section 1). The black shales represent a major deepening of the depositional environment and are interpreted as equivalents of the anoxic Hangenberg Black Shale and of widespread Uppermost Famennian black shales elsewhere. The onset of black shale deposition caused the main extinction episode of the Hangenberg Event (Becker 1996). Consequently, the overlying grey, silty shales in both studied sections are interpreted to represent an equivalent of the Rhenish Hangenberg Shales. The subsequent thick sandstone deposits at Ain Jemaa (new Ta'arraft Formation) probably reflect the main lowstand episode of the Hangenberg Event in the upper part of the Middle *praesulcata* Zone and are the time-equivalent of the Rhenish Hangenberg Sandstone. Based on miospore data, the distinctive Hangenberg eustatic lowstand has been correlated with a brief but important glacial episode of southern Gondwana (South America, South Africa, e.g. Streel *et al.* 2000; Almond *et al.* 2002).

The similarities of the conodont biofacies evolution observed in southern Europe and the Moroccan Meseta support the model of a narrow oceanic shelf without significant spreading or oceanic crust separating Laurussia and Gondwana in the Famennian. Interestingly, these biofacies observations are in contrast to recent studies in the Anti-Atlas to the South by Kaiser (2005), where the basal Middle *expansa* Zone is dominated by bispathodids and palmatolepids, and higher parts of the same zone are dominated by palmatolepids and polygnathids. Following Becker & House (2000*a*), the biofacies and tectonic history of the Anti-Atlas and the Meseta differ strongly across the South Atlas Fault, which is considered as a major tectonic and palaeogeographical suture (e.g. Michard *et al.* 1982; Fig. 6a).

The new biostratigraphical data show that the Middle to Uppermost (pre-Hangenberg) Famennian succession of the Oulmès area is extremely condensed and incomplete, interrupted briefly by the deposition of thin nodular limestones during the *Annulata* and Dasberg Events. Since there is no evidence of reworking, the discontinuous sedimentation was probably caused by relative sea-level fall that led to increased bottom turbulence. The latter effect is known to have effectively suppressed the settling of fine siliciclastic and calcareous muds for long periods on many German pelagic seamounts (e.g. Hüneke 2006).

The Ain Jemaa data in comparison with the still limited, often rather poor biostratigraphical record of other Meseta units, can contribute to a more precise dating of Eovariscan tectonics. There are at least two periods of non-deposition, one starting early in the Middle Famennian (within the Lower *marginifera* Zone, Ain Jemaa 2), one late in the Upper Famennian (within the Middle *expansa* Zone), with both lasting until the Hangenberg Event. Regional comparisons show some evidence for a Middle Famennian phase of strong condensation and/or non-deposition from Ain Jemaa eastwards, to the El Hammam Zone, and to the autochthonous units of the Ziar–Azrou Zone. Significant erosion in uplifted source areas and redeposition in adjacent, fast-subsiding troughs seems to have occurred at this time around the Sidi Bettache Basin (Tiliouine area, perhaps in the Tiddas area, south of Rabat, in the Khatouat Zone with the Biar Setla conglomerates, in the northern Mdakra Massif, and perhaps also in the Oued Cherrat Zone) and in the allochthonous Mrirt succession ('xenolite' generation). Upper to Uppermost Famennian condensation and/or non-deposition also continued southeastwards from Ain Jemaa to the El Hammam Zone and to the Mrirt allochthonous units. Conglomerates, sandstone wedges or mass flow deposits of this age are, again, characteristic for the wider Sidi Bettache Basin (Rabat–Tiflet Zone, Oued Akrech S of Rabat, Khatouat Zone, Oued Cherrat Zone, perhaps Ben Slimane Zone). In these areas, the pre-Hangenberg Uppermost Famennian is either missing, developed in neritic facies with 'Strunian' brachiopods, or, very restricted (Ben Slimane Zone), in pelagic ammonoid facies. Finally, sea-level-controlled pro-deltaic wedges equivalent to the Hangenberg Sandstone may be present at Ain Jemaa, in the El Hammam Zone, S of Rabat (Jebel Akala Quartzite), in the Khatouat and Mdakra Zones, in the Rehamna and Jebilet, and in the Ben Slimane Zone. Towards the E, in the Ziar–Azrou Zone, as well as in the northern Mdakra Massif, thin conglomerates, calcarenites and breccias with neritic fauna may have formed due to their distal position from the clastic sources. It is clear that more modern high-resolution stratigraphical and biogeographical data are needed for all discussed areas, especially revised and new records of ammonoids, conodonts, miospores and brachiopods. These will eventually provide a refined reconstruction of the Famennian crustal differentiation at the NW margin of Gondwana.

Systematic descriptions

The 16 taxa retrieved from the two sections at Ain Jemaqa are based on Pa-elements. They are listed below alphabetically; some are illustrated in Fig. 8. First authors, revisions, as well as selected references are given. The conodont biostratigraphy of the Upper Devonian follows Sandberg (1979) and Ziegler & Sandberg (1984, 1990).

Bispathodus aculeatus aculeatus (Branson & Mehl 1934)

1934a *Spathodus aculeatus* Branson & Mehl, 186, pl. 17, figs 11, 14.
1974 *Bispathodus aculeatus aculeatus* Ziegler, Sandberg & Austin, 101, pl. 1, fig. 5; pl. 2, figs 1–8.
Range: From the basal Middle *expansa* to the Lower Carboniferous basal *crenulata* Zone (Ziegler *et al.* 1974; Ziegler & Sandberg 1984; Clausen *et al.* 1989).

Bispathodus stabilis (Branson & Mehl 1934) (Morphotype 2)

1934a *Spathodus stabilis* Branson & Mehl, 188–189, pl. 17, fig. 20.
1974 *Bispathodus stabilis* Ziegler, Sandberg & Austin, 103, 104, pl. 1, fig. 10; pl. 3, figs 1–3.
1991 *Bispathodus stabilis* Morphotype 2 Johnston & Chatterton, 180, pl. 2, figs 1, 2.
Range: From the base of the Lower *expansa* to the Lower Carboniferous *anchoralis* Zone (Ziegler & Sandberg 1984; Clausen *et al.* 1989).

Branmehla gediki (Capkinoglu 2000)

2000 *Branmehla gediki* Capkinoglu, 101, pl. 4, figs 1–6.
Range: From the Lower *expansa* to the basal Middle *expansa* Zone (Capkinoglu 2000).

Branmehla inornata (Branson & Mehl 1934)

1934a *Spathodus inornatus* Branson & Mehl, 185, pl. 17, fig. 23.
1957 *Spathognathodus inornatus* Bischoff & Ziegler, 166, pl. 13, figs 4–6.
1959 *Branmehla inornata* Hass, 381–382, pl. 50, fig. 3 (further synonymy).
Range: From the Uppermost (?Upper) *marginifera* Zone (Ziegler & Sandberg 1984) to the Lower Carboniferous (Kaiser 2005).

Icriodus cf. *alternatus* (Branson & Mehl 1934) (Figs 8.12–8.16)

1934a *Icriodus alternatus* Branson & Mehl, p. 225, pl. 13, figs 4–6.
1984 *Icriodus alternatus alternatus* Sandberg & Dreesen, p. 158, pl. 2, figs 5, 11.
2003 *Icriodus alternatus alternatus* Corradini, p. 92, pl. 2, figs 9–12.
Range: Known from the Upper *rhenana* Zone to the Uppermost *crepida* Zone (Schülke 1999). At Ain Jemaa, it ranges much higher than previously reported.

Mehlina strigosa (Branson & Mehl 1934)

1934a *Spathodus strigosus* Branson & Mehl, 187, pl. 17, fig. 17.
1989 *Mehlina strigosa* Metzger, 517, figs 14.13, 14.15.
Range: From the *rhomboidea* (Ziegler & Sandberg 1984) to the Lower Carboniferous Lower *duplicata* Zone (Luppold *et al.* 1994, Müssenberg).

Palmatolepis glabra lepta (Ziegler & Huddle 1968)

1968 *Palmatolepis glabra lepta* Ziegler & Huddle, 380–381.
1973 *Palmatolepis glabra lepta* Sandberg & Ziegler, 103, pl. 2, figs 3, 16.
1994 *Palmatolepis falcata* Metzger, 626, 629, figs 10.21–10.29.
Remarks: In multi-element taxonomy (e.g. Metzger 1994) *Pa. glabra lepta* falls in synonymy with *Nothognathella(?) falcata* Helms 1959.
Range: From the Uppermost *crepida* (Ziegler & Sandberg 1990, early morphotype) to the middle part of the Upper *trachytera* Zone (typical morphotype, Ziegler & Sandberg 1984).

Palmatolepis glabra pectinata Ziegler 1962

1962 *Palmatolepis glabra pectinata* Ziegler, 8, pl. 2, figs 3–5.
1973 *Palmatolepis glabra pectinata* Morphotype 1. Sandberg & Ziegler, 104, pl. 2, figs 4, 12–15; pl. 5, fig. 14.
Remarks: Morphotype 1 (Fig. 7.2) has a high, short parapet. It resembles a transitional form between *Pa. glabra prima* and *Pa. glabra lepta* (Sandberg & Ziegler 1973).
Range: Morphotype 1 ranges from the Uppermost *crepida* (Ziegler & Sandberg 1990) to the top of the Upper *rhomboidea* Zone (Sandberg & Ziegler 1973); Morphotype 2 ranges to the top of the Upper *marginifera* Zone (Ziegler & Sandberg 1984).

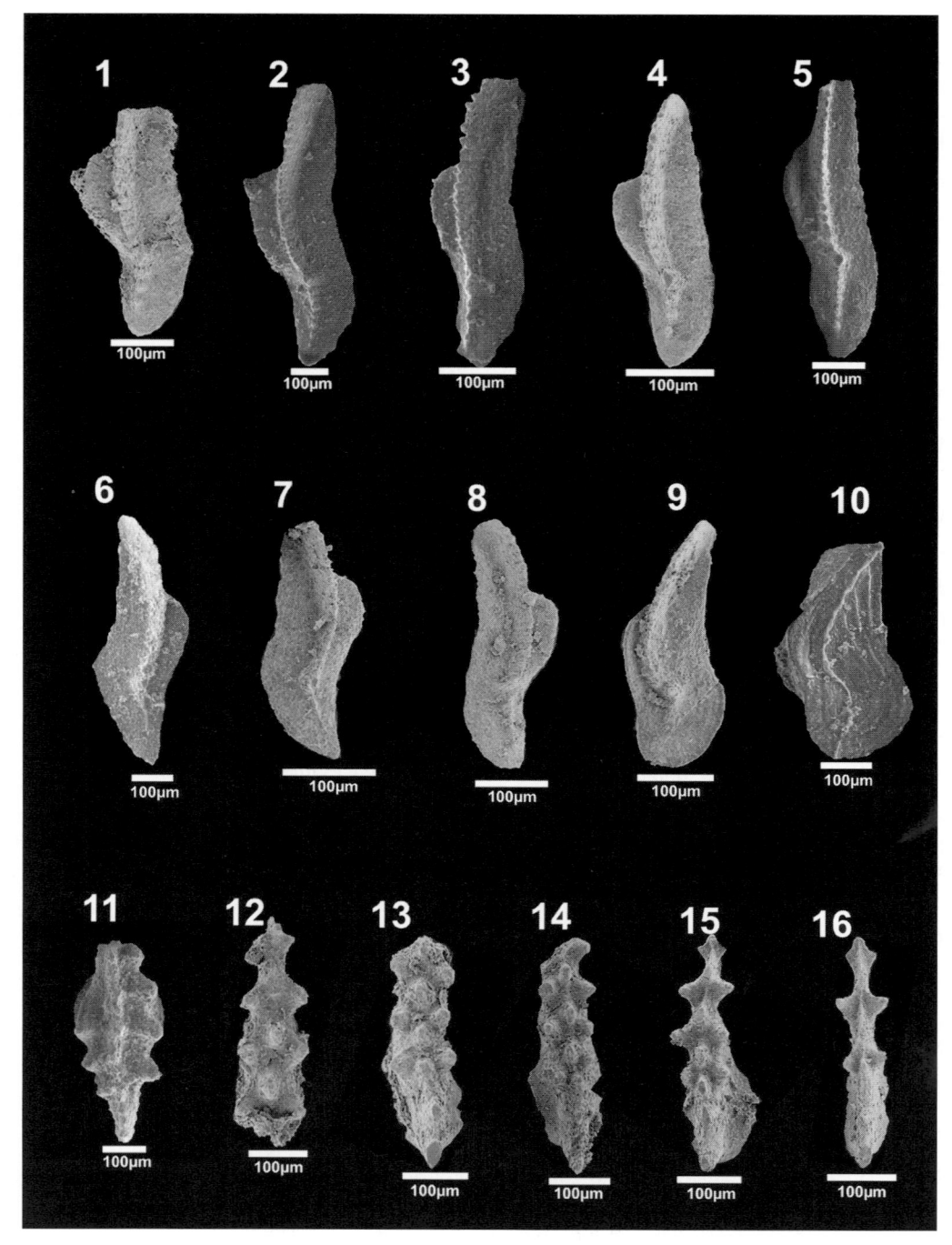

Fig. 8. Condonts from Ain Jemaa, all, apart from **11**, from Ain Jemaa 2, sample AJ3, Bed 3. **1–2** and **4–8**, *Palmatolepis glabra prima* Ziegler & Huddle 1968 (SMNS 66170-2...7). **3**, *Pa. glabra prima* Ziegler & Huddle 1968, transitional form to *Palmatolepis glabra pectinata* Ziegler 1962, Morphotype 1 (SMNS 66170-8). **9**, *Pa. quadrantinodosa inflexa* Müller 1956, upper view (SMNS 66170-9). **10**, *Pa. quadrantinodosa inflexa* Müller 1956, lower view (SMNS 66170-10). **11**, *Pseudopolygnathus* cf. *dentilineatus* Branson 1934 (SMNS 66169-2), Ain Jemaa 1, sample AJ1, Bed 1. 2. **12–16**, *Icriodus* cf. *alternatus* Branson & Mehl 1934 (SMNS 66170-11...15), specimen **16**, is a juvenile form.

Palmatolepis glabra prima Ziegler & Huddle 1968 (Figs 8.1–2, 8.4–8.8)

1968   *Palmatolepis glabra prima* Ziegler & Huddle, 379–380.
1973   *Palmatolepis glabra prima* Sandberg & Ziegler, 103, pl. 2, figs 1, 7.
1999   *Palmatolepis glabra unca* Schülke, 37–38, pl. 4, figs. 1–11.
Remarks: In multi-element taxonomy (Schülke 1999), *Pa. glabra prima* is a synonym of *Palmatodella unca* Sannemann 1955.
Range: From the base of the Upper *crepida* to the top of the Upper *marginifera* Zone (Ziegler & Sandberg 1984).

Palmatolepis gracilis expansa Sandberg & Ziegler 1979

1979   *Palmatolepis gracilis expansa* Sandberg & Ziegler, 178, pl. 1, figs 6–8 (further synonymy).
Range: From the base of the Lower *expansa* Zone to the base of the Hangenberg Event (Ziegler & Sandberg 1984; Clausen *et al.* 1989; Luppold *et al.* 1994).

Palmatolepis gracilis gracilis Branson & Mehl 1934

1934a   *Palmatolepis gracilis* Branson & Mehl, 238, pl. 18, figs 2, 8.
1965   *Palmatolepis gracilis gracilis* Spassov, pl. 1, figs 1, 2.
Range: From the Upper *rhomboidea* to the basal Carboniferous *sulcata* Zone (Ziegler & Sandberg 1984; Clausen *et al.* 1989); it is possible that rare basal Carboniferous occurrences are based on reworked specimens (see Corradini 2003).

Palmatolepis gracilis sigmoidalis Ziegler 1962

1962   *Palmatolepis deflectens sigmoidalis* Ziegler, 56–57, pl. 3, figs 24–28.
1966   *Palmatolepis gracilis sigmoidalis* Klapper, 31, pl. 6, fig. 8.
Range: From the upper part of the Upper *trachytera* Zone to the top of the Upper *praesulcata* Zone (Ziegler & Sandberg 1984; Clausen *et al.* 1989).

Palmatolepis quadrantinodosa inflexa Müller 1956 (Fig. 8.9–8.10)

1956   *Palmatolepis quadrantinodosa inflexa* Müller, 30, pl. 10, fig. 5.
1973   *Palmatolepis quadrantinodosa inflexa* Sandberg & Ziegler, 105, pl. 4, figs 7–13.
Range: From the upper part of the Upper *rhomboidea* to the top of the Lower *marginifera* Zone (Sandberg & Ziegler 1973).

Palmatolepis quadrantinodosa inflexoidea Ziegler 1962

1962   *Palmatolepis quadrantinodsa inflexoidea* Ziegler, 74, pl. 5, figs 14–18.
1973   *Palmatolepis quadrantinodsa inflexoidea* Sandberg & Ziegler, 105, pl. 4, figs 1–3.
Range: From the base to the top of the Lower *marginifera* Zone (Sandberg & Ziegler 1973).

Polygnathus communis communis Branson & Mehl 1934

1934b   *Polygnathus communis* Branson & Mehl, 293, pl. 24, figs. 1–4.
1969   *Polygnathus communis communis* Rhodes, Austin & Druce, pl. 12, figs. 2a–5c.
Remarks: In a restricted generic taxonomy (Vorontzova in Barskov *et al.* 1991), *Po. communis* is the type-species of *Neopolygnathus*.
Range: From the Middle *crepida* to the Lower Carboniferous *anchoralis* Zone (Vorontzova 1996).

Pseudopolygnathus cf. *dentilineatus* sensu Sandberg & Ziegler 1979 (Fig. 8.11)

1979   *Pseudopolygnathus* cf. *dentilineatus* Sandberg & Ziegler, 183–184, pl. 3, figs 18–21.
Remarks: After Sandberg *et al.* (1972), *Ps. dentilineatus* first occurs at the base of the *sulcata* Zone in the Western United States. *Ps.* cf. *dentilineatus* first occurs in the Upper Famennian, as observed by Capkinoglu (2000), Flajs & Feist (1988) and Clausen *et al.* 1989 (Middle *praesulcata* Zone). At Trolp (Graz Palaeozoic), it occurs in the Upper *praesulcata* Zone (Nössing 1975), and in the Carnic Alps (Elferspitz) in the Lower *praesulcata* Zone (Ebner 1973).

Field work was conducted under DFG grant (Ste 670/4-3, a joint project of T. Steuber, Bochum, and R. T. Becker) and was also supported by the Institut Scientifique, Rabat. The field company and support of T. Steuber is appreciated. M. Streel (Liége) commented on some miospore assemblages. This paper is a contribution to IGCP 499 on 'Devonian terrestrial and marine environments: from continent to shelf'. The manuscript benefited from the kind reviews of M. Hüneke (Greifswald, Germany) and W. T. Kirchgasser (Potsdam, New York).

References

AGARD, J., MORIN, P., TERMIER, G. & TERMIER, H. 1955. Esquisse dune histoire géologique de la région de Mrirt (Maroc central). *Notes et Mémoires du Service géologique du Maroc*, **12**, 15–28.

AGARD, J., BALCON, J.-M. & MORIN, P. 1958. Etude géologique et metallogenique de la region mineralisee du Jebel Aouam (Maroc central). *Notes et Mémoires du Service géologique*, **132**, 1–127.

AGER, D. V., COSSEY, S. P. J., MULLIN, P. R. & WALLEY, C. D. 1976. Brachiopod ecology in mid-Palaeozoic sediments near Khenifra, Morocco. *Palaeogeography, Palaeoclimatology, Palaeoecology*, **20**, 171–185.

ALBERTI, H. 1970. Neue Trilobiten-Faunen aus dem Ober-Devon Marokkos. *Göttinger Arbeiten zur Geologie und Paläontologie*, **5**, H. Martin-Festschrift, 15–29.

ALMOND, J., MARSHALL, J. & EVANS, F. 2002. Latest Devonian and earliest Carboniferous glacial events in South Africa. *16th International Sedimentological Congress, Abstract Volume*, 11–12.

BACHTADSE, V., TORSVIK, T. H. & SOFFEL, H. C. 1995. Paleomagnetic constraints on the paleogeographic evolution of Europe during the Paleozoic. *In*: DALLMEYER, ET AL, (eds) *Pre-Permian geology of central and eastern Europe*, 567–578.

BALINSKI, A. & BIERNAT, G. 2003. New observations on rhynchonelloid brachiopod *Dzieduszyckia* from the Famennian of Morocco. *Acta Palaeontologica Polonica*, **48**, 463–474.

BARSKOV, I. S., VORONTZOVA, T. N., KONONOVA, L. I. & KUZMIN, A. V. 1991. *Index conodonts of the Devonian and Lower Carboniferous*. Moskovskiy gosudarstvenniy universitet, 183pp.

BEAUCHAMP, J. & IZART, A. 1987. Early Carboniferous basins of the Atlas-Meseta domain (Morocco): Sedimentary model and geodynamic evolution. *Geology*, **15**, 797–800.

BECKER, R. T. 1993a. Stratigraphische Gliederung und Ammonoideen-Faunen im Nehdenium (Oberdevon II) von Europa und Nord-Afrika. *Courier Forschungsinstitut Senckenberg*, **155**, 1–405.

BECKER, R. T. 1993b. Anoxia, eustatic changes, and Upper Devonian to lowermost Carboniferous global ammonoid diversity. *In*: HOUSE, M. R. (ed.) *The Ammonoidea: Environment, Ecology, and Evolutionary Change*. Systematics Association Special Volume, 47, 115–163.

BECKER, R. T. 1993c. Kellwasser Events (Upper Frasnian Upper Devonian) in the Middle Atlas (Morocco)— Implications for plate tectonics anoxic event generation. *In*: *Global Boundary Events, An Interdisciplinary Conference, Kielce–Poland, Sep. 27–29, 1993, Abstracts*, 9.

BECKER, R. T. 1994. Faunal and sedimentary succession around the Frasnian–Famennian boundary in the eastern Moroccan Meseta. *64. Jahrestagung der Paläontologischen Gesellschaft, 26–30. Sept. 1994, Budapest, Ungarn, Vortrags- und Posterkurzfassungen*, 44.

BECKER, R. T. 1996. New faunal records and holostratigraphic correlation of the Hasselbachtal D/C-Boundary Auxiliary Stratotype (Germany). *Annales de la Société géologique de Belgique*, **117**(1), 19–45.

BECKER, R. T. 2001. Palaeobiogeographic relationships and diversity of Upper Devonian ammonoids from Western Australia. *Records of the Western Australian Museum, Supplement*, **58**, 385–401.

BECKER, R. T. & HOUSE, M. R. 2000a. Sedimentary and faunal succession of the allochthonous Upper Devonian at Gara d'Mrirt (Eastern Moroccan Meseta). *Notes et Mémoires du Service géologique*, **399**, 109–114.

BECKER, R. T. & HOUSE, M. R. 2000b. Devonian ammonoid zones and their correlation with established series and stage boundaries. *Courier Forschungsinstitut Senckenberg*, **220**, 113–151.

BECKER, R. T., HOUSE, M. R., BOCKWINKEL, J., EBBIGHAUSEN, V. & ABOUSSALAM, Z. S. 2002. Famennian ammonoid zones of the eastern Anti-Atlas (southern Morocco). *Münstersche Forschungen zur Geologie und Paläontologie*, **93**, 159–205.

BECKER, R. T., ASHOURI, A. R. & YAZDI, M. 2004. The Upper Devonian *Annulata* Event in the Shotori Range (eastern Iran). *Neues Jahrbuch für Geologie und Paläontologie, Abhandlungen*, **231**(1), 119–143.

BENFRIKA, E. M. & BULTYNCK, P. 2003. Lower to Middle Devonian conodonts from the Oued Cherrat area and its southern extension (North-Western Meseta, Morocco). *Courier Forschungsinstitut Senckenberg*, **242**, 209–215.

BISCHOFF, G. & ZIEGLER, W. 1957. Die Conodontenchronologie des Mitteldevons und des tiefsten Oberdevons. *Abhandlungen des hessischen Landesamtes für Bodenforschung*, **22**, 1–136.

BLESS, J. M., BECKER, R. T., HIGGS, K., PAPROTH, E. & STREEL, M. 1993. Eustatic cycles around the Devonian–Carboniferous boundary and the sedimentary and fossil record in Sauerland (Federal Republic of Germany). *Annales de la Société géologique de Belgique*, **115**, 689–702.

BRANSON, E. R. & MEHL, M. G. 1934a. Conodonts from the Grassy Creek Shale of Missouri. *University of Missouri Studies*, **8**, 171–259.

BRANSON, E. R. & MEHL, M. G. 1934b. Conodonts from the Bushberg sandstone and equivalent formations of Missouri. *Missouri University Studies*, **8**, 265–300.

BRICE, D., CHARRIERE, A., DROT, S. & REGNAULT, J. 1983. Mise en évidence, par des faunes de brachiopodes, de l'extension des formations dévoniennes dans la boutonnière d'Immouzer du Kandar. *Annales de la Societe géologique du Nord*, **53**, 445–458.

BRICE, D., LEGRAND-BLAIN, M. & NICOLLIN, J.-P. 2005. New data on Late Devonian and Early Carboniferous brachiopods from NW Sahara: Morocco, Algeria. *Annales de la Société Géologique du Nord*, **12**, (2ème série), 1–45.

CAPKINOGLU, S. 2000. Late Devonian (Famennian) conodonts from Denizliköyü, Gebze, Kocaeli, north-western Turkey. *Turkish Journal of Earth Sciences*, **9**, 91–112.

CARLS, P. 2001. Kritik der Plattentektonik um das Rhenohercynicum bis zum frühen Devon. *Braunschweiger geowissenschaftliche Arbeiten, Horst Wachendorf-Festschrift*, **24**, 27–108.

CARLS, P. 2003. Tornquist's Sea and the Rheic Ocean are illusive. *Courier Forschungsinstitut Senckenberg*, **242**, 89–109.

CHAKIRI, S. & TAHIRI, A. 2000. Genèse et démantelement de la plate-forme carbonatée dévonienne du Maroc central occidental. *Travaux de l'Institut*

Scientifique, Rabat, Série Géologie & Géographie Physique, **20**, 31–35.
CHALOUAN, A. 1981. Stratigraphie et structure du Paléozoique de l'Oued Cherrat: un segment du couloir de cisaillement de Meseta occidentale (Maroc). *Notes du Service géologique du Maroc*, **42**, 33–100.
CHOUBERT, G. & FAURE-MURET, A. 1961. Introduction stratigraphique. *Notes du Service géoloqique du Maroc*, **20**, 81–89.
CLAUSEN, C.-D., LEUTERITZ, K. & ZIEGLER, W. 1989. Ausgewählte Profile an der Devon/Karbon-Grenze im Sauerland (Rheinisches Schiefergebirge). *Fortschritte in der Geologie von Rheinland und Westfalen*, **35**, 161–226.
COGNEY, G. 1967. Sur le Dévonien de la région d'Oulmès (Maroc Central). *Compte Rendu Somnaires et Séances, Sociéte Géologique du France*, 7é série (9), 283–284.
CORRADINI, C. 2003. Late Devonian (Famennian) conodonts from the Corona Miziu Sections near Villasalto (Sardinia, Italy). *Palaeontographica Italica*, **89**, 65–116.
CYGAN, C., CHARRIERE, A. & REGNAULT, S. 1990. Datation par Conodontes du Dévonien (Emsien–Frasnian basal) dans le substratum du Moyen-Atlas au Sud de Fes (Maroc); implications paléogéographiques. *Géologie Méditeranée*, **17**, 321–330.
DOPIERALSKA, J. 2003. Neodymium isotopic composition of conodonts as a palaeoceanographic proxy in the Variscan oceanic system. http://geb.uni-giessen.de/geb/volltexte/2003/1168/.
DOPIERALSKA, J., BELKA, E. & HEGNER, Z. 2001. Neodymium isotopic variations in conodonts as paleoceanographic proxies in the Variscan Sea during the Late Devonian. *15th International Senckenberg Conference, Joint meeting IGCP 421/SDS, May 2001, Abstract Volume*, 31.
DUBÉ, T. E. 1989. Tectonic significance of Upper Devonian igneous rocks and bedded barite, Roberts Mountains Allochthon, Nevada, U.S.A. *Canadian Society of Petroleum Geologists, Memoirs*, **14**, 235–249.
DUNHAM, R. J. 1962. Classification of carbonate rocks according to depositional texture. *American Association of Petroleum Geology, Memoir*, **1**, 108–121.
EBNER, F. 1973. Die Conodontenfauna des Devon/Karbon-Grenzbereiches am Elferspitz (Karnische Alpen, Österrreich). *Mitteilungen und Abhandlungen des Geologisch-Paläontologischen Bergbau-Landesmuseums Joanneum*, **33**, 12–78.
EL HASSANI, A. 1991. La Zone de Rabat–Tiflet: Bordure nord le la chaîne calédono-hercyniènne du Maroc. *Bulletin de l'Institut Scientifique*, **15**, 1–134.
EL HASSANI, A. 2000. The Rabat–Tiflet zone: Late Devonian–Early Carboniferous of Tiflet and Lower and Middle Devonian of Rabat. *Notes et Mémoires du Service Géologique*, **399**, 115–122.
EL HASSANI, A. & BENFRIKA, E. M. 2000. The Devonian of the Moroccan Meseta: biostratigraphy and correlations. *Courier Forschungsinstitut Senckenberg*, **225**, 195–209.
EL HASSANI, A. & TAHIRI, A. (eds) 1994. Géologie du Maroc central et de la Meseta orientale. *Bulletin de l'Institut Scientifique*, **18**, 1–214.

EL HASSANI, A. & TAHIRI, A. 2000. The eastern part of central Morocco (western Meseta). *Notes et Mémoires du Service géologique du Maroc*, **399**, 89–92.
EL HASSANI, A., TAHIRI, A. & WALLISER, O. H. 2001. The Variscan crust between Gondwana and Laurasia. *15th International Senckenberg Conference, Joint meeting IGCP 421/SDS, May 2001, Abstracts*, 35.
EL HASSANI, A., TAHIRI, A. & WALLISER, O. H. (2003). The Variscan Crust between Gondwana and Baltica. *Courier Forschungsinstitut Senckenberg*, **242**, 81–87.
EL KAMEL, F., EL HASSANI, A. & DAFIR, J. E. 1992. Présence d'une tectonique synsédimentaire dans le Dévonien inférieur des Rehamna septentrionaux (Meseta marocaine occidentale). *Bulletin de l'Institut Scientifique*, **16**, 37–43.
EMBRY, A. F. & KLOVAN, J. E. 1972. Absolute water depth limits of Late Devonian paleoecological zones. *Geologische Rundschau*, **61**, 672–686.
FADLI, D. 1994. Le Famenno-Tournaisien. *Bulletin de l'Institut Scientifique*, **18**, 57–70.
FLAJS, G. & FEIST, R. 1988. Index conodonts, trilobites and environment of the Devonian-Carboniferous boundary beds at La Serre (Montagne Noire, France). *Courier Forschungsinstitut Senckenberg*, **100**, 53–107.
HARTKOPF-FRÖDER, C. 2004. Palynostratigraphy of upper Famennian sediments from the Refrath 1 Borehole (Bergisch Gladbach-Paffrath Syncline; Ardennes-Rhenish Massif, Germany). *Courier Forschungsinstitut Senckenberg*, **251**, 77–88.
HASS, W. H. 1959. Conodonts from the Chappel Limestone of Texas. *United States Geological Survey, Professional Papers*, **294**, 365–399.
HECKEL, P. H. & WITZKE, B. J. 1979. Devonian world palaeogeography determined from distribution of carbonates and related lithic paleoclimatic indicators. *Special Papers in Palaeontology*, **23**, 99–123.
HELMS, J. 1959. Conodonten aus dem Saalfelder Oberdevon. *Geologie*, **8**, 634–677.
HOLLARD, H. 1967. Le Dévonien du Maroc et du Sahara Nord-Occidental. *In:* OSWALD, D. H. (ed.) *International Symposium on the Devonian System*, **1**, 203–244, Calgary.
HOLLARD, H. & MORIN, P. 1973. Les gisements de *Dzieduszyckia* (Rhynchonellida) du Famennien inférieur du Massif Hercynien central du Maroc. *Notes et Mémoires du Service géologique du Maroc*, **33**, 7–14.
HOLLARD, H., LYS, M., MAURIER, P. & MORIN, A. 1970. Precision sur l'age Famennien inférieur de *Dzieduszyckia* (Rhynchonellida) du Massif hercynien central du Maroc. *Compte Rendu, Academie des Sciences, Série D*, **270**, 3177–3180.
HOLLARD, H., MICHARD, A., JENNY, P., HOEPFFNER, C. & WILLEFERT, S. 1982 Stratigraphie du Primaire de Mechra-Ben-Abbou, Rehamna. *Notes et Mémoires du Service géologique*, **303**, 13–34.
HOUSE, M. R. 1985. Correlation of mid-Palaeozoic ammonoid evolutionary events with global sedimentary perturbations. *Nature*, **313**, 17–22.
HÜNEKE, H. (2006). Erosion and deposition from bottom currents during the Givetian and Frasnian: Response to intensified oceanic circulation between Gondwana and

Laurussia. *Palaeogeography, Palaeoclimatology, Palaeoecology*, **234**, 146–167.

HUVELIN, P. 1977. Etude géologique et gitologique du Massif Hercynien des Jebilet (Maroc Occidental). *Notes et Mémoires du Service géologique*, **232 bis**, 1–307.

IZART, A. & VIESLET, J.-L. 1988. Stratigraphie, sédimentologie et micropaléontologie du Famennian, Tournaisian et Viséen du bassin de Sidi-Bettache et de ses bordures (Meseta marocaine nord-occidentale). *Notes du Service géologique du Maroc*, **44**(334), 7–41.

JOHNSON, J. G., KLAPPER, G. & SANDBERG, C. A. 1985. Devonian eustatic fluctuations in Euramerica. *Geological Society of America, Bulletin*, **96**, 567–587.

JOHNSTON, D. I. & CHATTERTON, B. D. E. 1991. Famennian conodont biostratigraphy of the Palliser Formation, Rocky Mountains, Alberta and British Columbia, Canada. *Geological Survey of Canada, Bulletin*, **417**, 163–183.

KAISER, S. I. 2005. Mass extinctions, climatic and oceanographic changes at the Devonian/Carboniferous boundary. Ph.D. Thesis, Ruhr University Bochum, 1–156.

KENT, D. V. & VAN DER VOO, R. 1990. Palaeozoic palaeogeography from palaeomagnetism of the Atlantic-bordering continents. *In*: MCKERROW, W. S. & SCOTESE, C. R. (eds) *Palaeozoic Palaeogeography and Biogeography*, Geological Society, London, Memoir, **12**, 49–56.

KLAPPER, G. 1966. Upper Devonian and Lower Mississippian conodont zones in Montana, Wyoming, and South Dakota. *University of Kansas, Paleontological Contributions*, **3**, 1–43.

LAZREQ, N. 1999. Biostratigraphie des conodontes du Givétien au Famennien du Maroc central – Biofaciès et événement Kellwasser. *Courier Forschungsinstitut Senckenberg*, **214**, 1–111.

LECOINTRE, G. 1926. Recherches géologiques dans la Meseta marocaine. *Mémoires de l'Société des Sciences Naturelles de Maroc*, **14**, 1–158.

LE HÉRISSE, A., SERVAIS, T. & WICANDER, R. 2000. Devonian acritarchs and related forms. *Courier Forschungsinstitut Senckenberg*, **220**, 195–205.

LUPPOLD, F. W., CLAUSEN, C.-D., KORN, D. & STOPPEL, D. 1994. Devon/Karbon-Grenzprofile im Bereich von Remscheid-Altenaer Sattel, Warsteiner Sattel, Briloner Sattel und Attendotn-Elsper Doppelmulde (Rheinisches Schiefergebirge). *Geologie und Paläontologie in Westfalen*, **29**, 7–69.

MARHOUMI, R., HOEPFFNER, C., DOUBINGER, J. & RAUSCHER, R. 1983. Données nouvelles sur l'histoire hercynienne de la Meseta orientale au Maroc: l'age dévonien des schistes de Debdou et du Mekam. *Compte rendu de l'Academie des Sciences, Paris*, **297** (Série II), 69–72.

METZGER, R. A. 1989. Upper Devonian (Frasnian-Famennian) conodont biostratigraphy in the subsurface of north-central Iowa and southeastern Nebraska. *Journal of Paleontology*, **63**, 503–524.

METZGER, R. A. 1994. Multielement reconstructions of *Palmatolepis* and *Polygnathus* (Upper Devonian, Famennian) from the Canning Basin, Australia, and Bactrian Mountain, Nevada. *Journal of Paleontology*, **68**, 617–647.

MEYER-BERTHAUD, B., WENDT, J. & GALTIER, J. 1997. First record of a large *Callixylon* trunk from the late Devonian of Gondwana. *Geological Magazine*, **134**, 847–853.

MICHARD, A. 1976. Éléments de géologie marocaine. *Notes et Mémoires du Service géologique*, **252**, 1–408.

MICHARD, A., YAZIDI, A., BENZIANE, F., HOLLARD, H. & WILLEFERT, S. 1982. Foreland thrusts and olistromes on the pre-Sahara margin of the Variscan orogen, Morocco. *Geology*, **10**, 253–256.

MÜLLER, K. J. 1956. Zur Kenntnis der Conodonten-Fauna des europäischen Devons, 1. Die Gattung Palmatolepis. *Abhandlungen der Sencken-bergischen naturforschenden Gesellschaft*, **494**, 1–70, pl. 1–11.

MULLIN, P., BENSAID, G. & KELLING, M. 1976. Les nappes hercyniennes au sud-est du Maroc central; une nouvelle interprétation. *Compte Rendu de l'Academie des Sciences, Paris, Série D*, **282**, 827–830.

NÖSSING, L. 1975. Die Sanzenkogel-Schichten (Unterkarbon), eine biostratigraphische Einheit des Grazer Paläozoikums. *Mitteilungen des Naturwisse-nschaftlichen Vereins der Steiermark*, **105**, 79–92.

PERRI, M. C. & SPALLETTA, C. 1998. Late Famennian conodonts from the Malpasso section (Carnic Alps, Italy). *Giornale di Geologia, seria 3^a, Special Issue, ECOS VII Southern Alps Field Trip Guidebook*, **60**, 220–227.

PIECHA, M. 2005. Late Famennian conodonts from the Refrath 1 Borehole (Bergisch Gladbach-Paffrath Syn-cline; Ardennes-Rhenish Massif, Germany). *Courier Forschungsinstitut Senckenberg*, **251**, 253–265.

PIQUE, A. 1975. Différenciation des aires de sédimentation au Nord-Ouest de la Meseta marocaine: la distension dévono-dinantienne. *Comptes rendus de l' Academie des Sciences Paris*, **281**, 767–770.

PIQUE, A. 1984. Faciès sédimentaire et évolution d'un bassin: le Bassin dévono-dinantien de Sidi-Bettache (Maroc nord-occidental). *Bulletin de l'Société géologique de France*, **26**(6), 1015–1023.

PIQUE, A. 1994. Géologie du Maroc. *Les domaines régionaux et leur évolution structurale*. El Maarif Al Jadida, Rabat.

PIQUE, A. & KHARBOUCH, F. 1983. Distension intracontinentale et volcanisme associé la Meseta Marocaine Nord-Occidentale au Dévonien-Dinantien. *Bulletin des Centres de Recherches, Exploration-Production de Elf-Aquitaine*, **7**, 377–387.

PIQUE, A. & MICHARD, A. 1981. Les zones structurales du Maroc Hercynien. *Sciences Géologiques, Bulletin*, **34**, 135–146.

PIQUE, A. & MICHARD, A. 1989. Moroccan Hercynides: A Synopsis. The Paleozoic sedimentary and tectonic evolution at the northern margin of West Africa. *American Journal of Science*, **289**, 286–330.

RICHTER, R. & RICHTER, E. 1926. Die Trilobiten des Oberdevons. *Abhandlungen der Preußischen Geologischen Landesanstalt, Neue Folge*, **99**, 1–314 + 12 pls.

ROLLEAU, R. 1956. Découverte d'une *Halorella* (Brachiopode fossile) dans la région du Khatouat. *Compte Rendu, Societe des Sciences naturelles physique du Maroc*, **22**(1), 26.

RHODES, F. H. T., AUSTIN, R. L. & DRUCE, E. C. 1969. British Avonian (Carboniferous) conodonts and their value in local and intercontinental correlation. *British Museum (Natural History) Bulletin, Geology Supplement*, **5**, 1–313.

ROCH, E. 1950. Histoire stratigraphique du Maroc. *Notes et Mémoires du Service géologique du Maroc*, **80**, 1–422.

SANDBERG, C. A. 1979. Devonian and Lower Mississippian Conodont Zonation of the Great Basin and Rocky Mountains. *Brigham Young University Geology Studies*, **26**, 87–106.

SANDBERG, C. A. & DREESEN, R. 1984. Late Devonian icriodontid biofacies models and alternate shallow-water conodont zonation. *Geological Society of America, Special Paper*, **196**, 143–175.

SANDBERG, C. A. & ZIEGLER, W. 1973. Refinement of standard Upper Devonian conodont zonation based on sections in Nevada and West Germany. *Geologica et Palaeontologica*, **7**, 97–122.

SANDBERG, C. A. & ZIEGLER, W. 1979. Taxonomy and biofacies of important conodonts of Late Devonian *styriacus*-Zone, United States and Germany. *Geologica et Palaeontologica*, **13**, 173–212.

SANDBERG, C. A., STREEL, M. & SCOTT, R. A. 1972. Comparison between conodont zonation and spore assemblages at the Devonian–Carboniferous boundary in the western and central United States and in Europe. *Compte Rendu, 7e Congress de Stratigraphie et Géologie du Carbonifere*, **1**, 179–203.

SANNEMANN, D. 1955. Oberdevonische Conodonten (to IIα). *Senckenbergiana lethaea*, **36**, 123–156.

SCHÜLKE, I. 1999. Conodont multielement reconstructions from the early Famennian (Late Devonian) of the Montagne Noire (Southern France). *Geologica et Palaeontologica, Sonderband*, **3**, 1–124.

SPALLETTA, C., PERRI, M. & PONDRELLI, M. C. (1998). Late Famennian conodonts from the Rio Boreado section (Carnic Alps, Italy). *Giornale di Geologia, seria 3ª, Special Issue, ECOS VII Southern Alps Field Trip Guidebook*, **60**, 214–219.

SPASSOV, C. 1965. Unterkarbon in Bulgarien. *Review of the Bulgarien Geological Society*, **26**(2), 157–166.

STREEL, M. & LOBOZIAK, S. 1996. Chapter 18B—Middle and Upper Devonian miospores. *In*: JANSONIUS, J. & MCGREGOR, D. C. (eds), Palynology: principles and application, *American Association of Stratigraphic Palynologists Foundation*, **2**, 575–587.

STREEL, M., CAPUTO, M. V., LOBOZIAK, S. & MELO, J. H. G. 2000. Late Frasnian–Famennian climates based on palynomorph analyses and the question of the Late Devonian glaciations. *Earth Science Reviews*, **52**, 121–173.

TAHIRI, A. 1983. Lithostratigraphie et structure du Jbel Ardouz—Maroc hercynien. *Bulletin de l'Institut Scientifique, Rabat*, **7**, 1–16.

TAHIRI, A. 1991. Le Maroc central septentrional: stratigraphie, sédimentologie et tectonique du Paléozoïque; un exemple de passage des zones internes aux zones externes de la chaîne hercynienne du Maroc. *Thèse doctorale, ès-Sciences, Université de Bretagne occidentale*, Brest, 311pp.

TAHIRI, A. 1994. Tectonique hercynienne de l'anticlinorium de Khouribga-Oulmès et du synclinorium de Fourhal. *Bulletin de l'Institute Scientifique*, **18**, 125–144.

TAHIRI, A. & HOEPFFNER, C. 1988. Importance des mouvements distensifs au Dévonien supérieur en Meseta nord-occidentale (Maroc); les calcaires démanteles de Tilouine et la ride d'Oulmès, prolongement oriental de la ride des Zaèr. *Compte Rendu, Academie des Sciences Paris, Série II*, **306**, 223–226.

TAHIRI, A. & LAZRAQ, N. 1988. Précisions stratigraphiques sur le Dévonien de la ride d'El Hammam au N d'Oulmès.–*Bulletin de l'Institut Scientifique*, **12**, 47–51

TAHIRI, A., EL HASSANI, A. & WALLISER, O. H. 1992. Etude comparée du Dévonien de la Meseta marocaine occidentale et des massifs hercyniens allemands. *Bulletin de l'Institut Scientifique*, **16**, 78.

TERMIER, H. 1936. Etudes géologiques sur le Maroc central et le Moyen-Atlas septentrional, Tome I-IV. *Protectorat de la Republique Francaise au Maroc, Service des Mines et de la Carte géologique, Notes et Mémoires*, **33**, 1–1566.

TERMIER, H. 1938. Nouveaux affleurements de Famennien dans le Maroc central. *Compte Rendu Sociéte Géologique de France*, **5é série**, 8, 40–42.

TERMIER, H. & TERMIER, G. 1950. Paléontologie Marocaine. II. Invertebres de l'Ere primaire. Fascicule III, Mollusques. *Protectorat de la Republique Francaise au Maroc, Service géologique, Notes et Mémoires*, **78**, 1–246.

TERMIER, H. & TERMIER, G. 1951. Stratigraphie et paléobiologie des terrains primaires de Benhamed. *Notes du Service géologique du Maroc*, 5(85), 48–105.

VAN STEENWINKEL, M. 1993. The Devonian–Carboniferous boundary: Comparison between the Dinant Synclinorium and the northern border of the Rhenish Slate Mountains. A sequence stratigraphic view. *Annales de la Société géologique de Belgique*, **115**(2), 665–681.

VORONTZOVA, T. N. 1996. The Genus *Neopolygnathus* (Conodonta): Phylogeny and Some Questions of Systematics. *Palaeontologicheskiy Zhurnal*, 1996(2), 82–84.

WALLISER, O. H. 1996. Global Events in the Devonian and Carboniferous. *In*: WALLISER, O. H. (ed.), *Global Events and Event Stratigraphy in the Phanerozoic*, 225–250, Springer, Berlin.

WALLISER, O. H., EL HASSANI, A. & TAHIRI, A. 1995. Sur le Dévonien de la Meseta marocaine occidentale. *Courier Forschungsinstitut Senckenberg*, **188**, 21–30.

WALLISER, O. H., EL HASSANI, A. & TAHIRI, A. 2000. Mrirt: A key area for the Variscan Meseta of Morocco. *Notes et Mémoires du Service géologique*, **399**, 93–108.

WENDT, J. 1985. Disintegration of the continental margin of northwestern Gondwana: Late Devonian of the eastern Anti-Atlas (Morocco). *Geology*, **13**, 815–818.

ZAHRAOUI, M. 1994. Le Dévonien inférieur et moyen. *Bulletin de l'Institut Scientifique*, **18**, 43–56.

ZAHRAOUI, M., EL HASSANI, A. & TAHIRI, A. 2000. Devonian outcrops of the Oued Cherrat shear zone. *Notes et Mémoires du Service géologique*, **399**, 123–128.

ZIEGLER, W. 1962. Taxonomie und Phylogenie oberdevonischer Conodonten und ihre stratigraphische Bedeutung. *Abhandlungen des Hessischen Landesamtes für Bodenforschung*, **38**, 1–166.

ZIEGLER, W. & HUDDLE, J. W. 1968. Die *Palmatolepis glabra*-Gruppe (Conodonta) nach der Revision der Typen von Ulrich & Bassler durch J. W. Huddle. *Fortschritte in der Geologie von Rheinland und Westfalen*, **16**, 377–368.

ZIEGLER, W. & SANDBERG, C. A. 1984. Palmatolepis-based revision of upper part of standard Late Devonian conodont zonation. *Geological Society of America, Special Papers*, **196**, 179–194.

ZIEGLER, W. & SANDBERG, C. A. 1990. The Late Devonian Standard Conodont Zonation. *Courier Forschungsinstitut Senckenberg*, **121**, 1–115.

ZIEGLER, W., SANDBERG, C. A. & AUSTIN, R. 1974. Revision of *Bispathodus* group (Conodonta) in the Upper Devonian and Lower Carboniferous. *Geologica et Palaeontologica*, **8**, 97–112.

ZIEGLER, W. & WEDDIGE, K. 1999. Zur Biologie, Taxonomie und Chronologie der Conodonten. *Paläontologische Zeitschrift*, **73**, 1–38.

Brachiopod faunal changes across the Devonian–Carboniferous boundary in NW Sahara (Morocco, Algeria)

D. BRICE[1,2], M. LEGRAND-BLAIN[3] & J.-P. NICOLLIN[1,2]

[1]*Laboratoire de Paléontologie stratigraphique, Faculté Libre des Sciences et Technologies, 41 rue du Port, 59046 Lille Cedex, France*

[2]*Institut Supérieur d'Agriculture, 48 boulevard Vauban, 59046 Lille Cedex, France (e-mail: d.brice@isa-lille.fr, jp.nicollin@isa-lille.fr)*

[3]*'Tauzia', Cours Général de Gaulle, 33170 Gradignan, France (e-mail: legrandblain@wanadoo.fr)*

Abstract: Based on previous systematic studies of productid, rhynchonellid and spiriferid brachiopods from NW Sahara (Morocco and Algeria), we recognize three successive faunas near the Devonian–Carboniferous boundary. A 'Lower Fauna', late Famennian in age [IV(?)-V and lower VI(?) Zones], and an 'Upper Fauna', early Tournaisian in age, are present in southern Morocco (Assa, Akka, Zemoul areas) and in Algeria, Timimoun area. A third 'Intermediate Fauna', with few taxa, and differing according to the areas, is identified in southern Morocco. Northwards, in Tafilalt–Ma'der basins, rare brachiopods, found above a 'Hangenberg Black Shale' equivalent, are in spite of taxonomic differences related to the 'Upper Fauna'. This important renewal of faunas could be in relation to the main lithological variations.

The Devonian–Carboniferous boundary beds of NW Sahara are characterized by prolific brachiopod assemblages in large shelf areas. Ammonoids and conodonts are rare in these regions, except in the more basinal Tafilalt-Ma'der, so the benthic faunas have considerable utility for biostratigraphy.

'Neritic-pelagic correlations and events' was the theme of the Subcommission on Devonian Stratigraphy meeting held in Rabat, Morocco, in 2004. Participants in a field trip to the western Dra valley, southern Morocco, visited a Devonian–Carboniferous section (Kheneg Lakahal, south of Assa) in the neritic facies, which provided numerous brachiopods and some conodonts (Brice *et al.* 2004; Kaiser *et al.* 2004*a*).

We have examined, and comment in this paper, on Devonian–Carboniferous brachiopods now available from South Morocco and neighbouring NW Algerian Sahara (Fig. 1). These faunas are from a series of sections (Fig. 1), from west to east: (1) Kheneg Lakahal, collected in 2003–2004 by R. T. Becker, S. Z. Aboussalam, H. Nübel, S. Kaiser, D. Brice and J.-P. Nicollin (Becker *et al.* 2004; Kaiser *et al.* 2004*a*); (2–6) sections in the Dra–Zemoul area, collected in 1950–1960 by H. Hollard and collaborators, partly cited by Hollard (1971, 1981) and described by Brousmiche (1975), Legrand-Blain (1995*a, b*), Brice & Nicollin (2000) and Nicollin & Brice (2001); (7–9) Ma'der-Tafilalt, collected recently by R. T. Becker and collaborators; (10) the Gourara-Timimoun region, NW Algerian Sahara, collected in 1960–1970 by J. Conrad and M. Legrand-Blain, and studied by Conrad & Termier (1970), Sartenaer (1975), Conrad *et al.* (1986) and Legrand-Blain (1995*a, b*). The systematic descriptions, affinities, locations and ranges of the entire brachiopod faunas are published in a separate paper (Brice *et al.* 2005).

Geological setting of the main investigated sections

Morocco

(1) The most complete stratigraphical sections, unfortunately inaccessible now, are exposed in the central Anti-Atlas Mountains of the Oued Zemoul area (Choubert *et al.* 1971); cephalopods cited from that region by Hollard (1971) need revision (Kaiser *et al.* 2004*a*, p. 71). In two sections, at Rich el Bergat and Jfeirat (Figs 1, 2c–d), the sediments consist of an alternation of clymenid shales and brachiopod coquinas in the lower part. In the 'Tazout Group' (Hollard 1981*b*, table 5), consisting of Tazout 1, Tazout 2 and Tazout 3 Sandstone Formations, these lower beds correspond partly to the Tazout 1 Formation, poorly dated, Famennian V (Hollard 1971), or VI (Hollard 1981*a*), and partly to the overlying Tazout 2 Formation, 'Etroeungt Zone' (Hollard 1971, table 1). The top

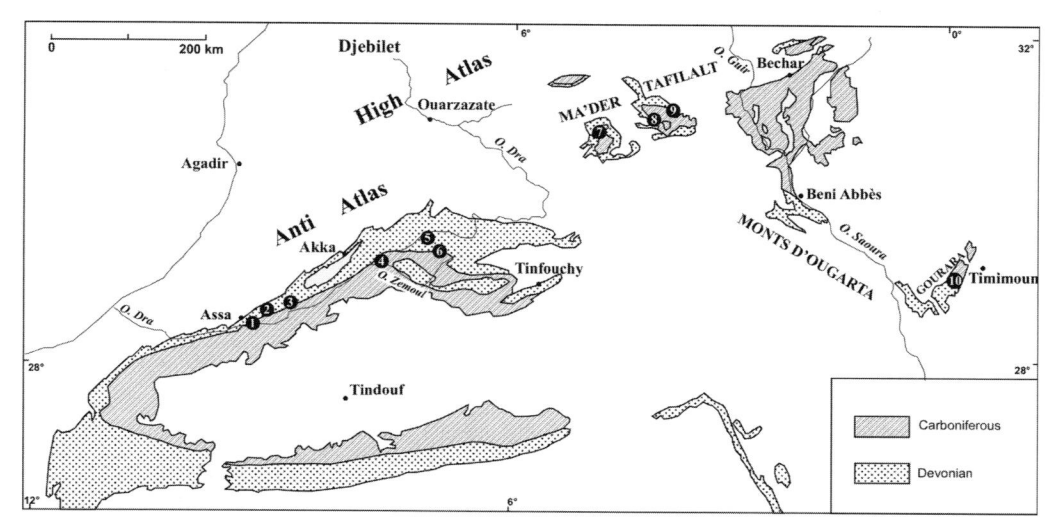

Fig. 1. Location of areas studied in southern Morocco and western Algeria. 1 Kheneg Lakahal; 2 Hassi Rharouar; 3 Kheneg Aftes, 50 km E of Assa; 4 Rich el Bergat; 5 Dfeif; 6 Jfeirat; 7 Bou Tlidat; 8 El Atrous; 9 M'Karig; 10 Gara el Kahla. Locations of sections: (sections 3–6) according to Hollard 1968, 1971. Geological maps of the outcrops are available after Choubert *et al.* 1969 (sections 1–3), Choubert *et al.* 1971 (sections 4–6), Destombes *et al.* 1988 (section 7), Destombes *et al.* 1986 (sections 8, 9), Kaiser 2005, pp. 48 and 68 (sections 7–9, 1), Ebbighausen *et al.* 2004 (section 10).

of the latter is locally eroded, and the lithologies between Tazout 2 and 3, variable conglomerates and red sandstones are interpreted as coastal deposits. In the lower part of Tazout 3, a *Gattendorfia* cf. *crassa* horizon present in three sections: Rich el Bergat, Dfeif, Jfeirat constitutes an important regional marker bed. Two related ammonoid horizons and assemblages, described from Tafilalt and Gourara (Ebbighausen *et al.* 2004; Bockwinkel & Ebbighausen 2006; Korn *et al.* 2006) are successively: *Gattendorfia-Eocanites–Gattendorfia-Kahlacanites*, probably middle Early Tournaisian in age. At Rich el Bergat, the thickness of sediments up to the *Gattendorfia* bed is 380 m. The upper Tazout 3 levels are middle–late Tournaisian in age (Legrand-Blain 2002, table 3).

(2) In the western Anti-Atlas Mountains, in the Assa area of the Dra Valley (Figs 1, 2a–b), the sediments become more sandy, with the three successive sandstone ridges of Tazout 1, 2 and 3, outcropping mainly at Kheneg Aftes, 50 km east of Assa (Hollard 1971). In spite of the absence of *Gattendorfia* in the western Anti-Atlas, field correlations with the Zemoul area are well established. Westwards, the thickness of Tazout Sandstones quickly decreases (Cavaroc *et al.* 1976), as shown on the geological map (Choubert *et al.* 1969). The Kheneg Lakahal section, 18–20 km SE of Assa, was recently visited by R. T. Becker and collaborators. The Tazout Group consists of: (1) 130 m thick sandstones and siltstones, with brachiopod coquinas, 'Maader Talmout Member' in local lithostratigraphy (Becker *et al.* 2004, p. 9; Kaiser *et al.* 2004a); (2) unfossiliferous siltstones, 'Kheneg Lakahal Member'; (3) an unnamed upper ridge. The fossiliferous Maader Talmout beds are equivalent to Hollard's Tazout 1–2 (Becker, written comm. 2006).

(3) The Tafilalt–Ma'der area in the eastern Anti-Atlas Mountains is famous for Devonian–Carboniferous cephalopod faunas, in basin palaeotopography (Hollard 1968; Korn 1999). Sediment, facies and thickness differ considerably in different localities. From west to east, at the base of regressive siliciclastics, the 'Grès de l'Aguelmous-n-ou-Fezzou' in Ma'der, and 'Grès d'Ouaoufilal' in Tafilalt (Hollard 1971; geological maps of Todhra-Ma'der by Destombes *et al.* 1988, and Tafilalt–Taouz by Destombes *et al.* 1986): 'Hangenberg Black Shale' equivalent has been described and considered as reflecting a major event, a major mass extinction in the basal Middle *praesulcata* Zone = top of the *Epiwocklumeria applanata* VI-D2 Subzone, just before the Devonian–Carboniferous boundary (Belka *et al.* 1999; Kaiser 2005, p. 74).

Brachiopods, though rare in the Tafilalt-Ma'der Devonian/Carboniferous basins, are important for biostratigraphy.

(a) In the Bou Tlidat section of Ma'der, NE of Fezzou, rare spiriferids are associated with the ammonoid *Acutimitoceras* (*Stockumites*), preserved in sideritic nodules (Becker written comm. 2004).

The spiriferids occur in condensed pelagic shales and sandstones, reflecting 'the basal Tournaisian transgression' (Kaiser et al. 2004b, p. 72), immediately overlying the Aguelmous Sandstones. The first appearance of *Syringothyris* is contemporaneous in this Ma'der locality and in the 'Upper Fauna' at Gara el Kahla, Algeria.

(b) In the El Atrous section of the southern Tafilalt, 20 km NW of Taouz, just above nodular shales with *Parawoklumeria*, the lower Ouaoufilal Formation consists of a 'Hangenberg Black Shale' equivalent, that is followed by a thick succession of azoic shales and turbidites; the upper Ouaoufilal Formation is a 55 m thick, rippled, mostly deltaic, brachiopod-bearing sandstone (Kaiser & Becker written comm. 2004). It yields a poor rhynchonellid fauna of *Centrorhynchus* (?) sp., a Famennian genus, near its base; higher in the sandstones, *Hemiplethorhynchus* (?) sp. of Tournaisian age, occurs. The overlying shales, the basal Oued Znaigui Formation, are unfossiliferous; middle Tournaisian ammonoids occur slightly higher in the formation (Korn et al. 2002).

(c) In the M'Karig section of the NE Tafilalt, NE of Ouidane Chebbi, a *Acutimitoceras* (*Stockumites*) bed is recognized above a 'Hangenberg Black Shale' equivalent, followed by about a 10 m siliciclastic succession (Kaiser 2005). According to this author, the Devonian–Carboniferous boundary transition can be fixed with the occurrence of the *Acutimitoceras* (*Stockumites*) fauna. This level has yielded a spiriferid *Tylothyris* (?) sp., a Tournaisian genus associated with *Acutimitoceras* (*Stockumites*) *intermedius*.

Algeria

In the Gara el Kahla section of the Gourara–Timimoun area, 30 km SW of Timimoun, the stratigraphical succession (Conrad et al. 1986) consists of (I) Kahla Mudstone with *Gonioclymenia* (Famennian, zone V). (II) The approximately 280 m thick Lower Kahla Sandstone contains brachiopod coquinas, succeeded by azoic deltaic sandstones and shales. (III) The 130 m thick Upper Kahla Sandstone contains successively: (i) transgressive sandstones yielding abundant brachiopod faunas, Early Tournaisian in age (ii) claystones including a thin ammonoid horizon, recently described (Ebbighausen et al. 2004) with *Kahlacanites* nov. gen., *Gattendorfia* cf. *crassa*, *Acutimitoceras*, of probable late Early Tournaisian age; (iii) siltstones and sandstones; and (iv) *Acrocanites* shales of Upper Tournaisian age.

Brachiopod faunas

We recognize and name two quite different associations, a Lower Fauna and an Upper Fauna and, incidentally, in rare cases, an Intermediate Fauna. The Lower Fauna occurs in all sections from west to east, except in the Tafilalt-Ma'der.

Lower Fauna

In the Assa area, southwestern Morocco: the Lower Fauna occurs at Hassi Rharouar and Kheneg Aftes sections, in the Tazout 1 Sandstone Formation (Fig. 2); at Kheneg Lakahal, in the lower part of the Maader Talmout Member (Fig. 3). It is characterized by the strophomenids *Leptagonia* cf. *analoga* (Phillips 1836), *Schuchertella* sp., the productids *Hamlingella talmouti* Legrand-Blain in Brice et al. 2005, *Whidbornella* cf. *pauli radiata* (Paeckelmann 1931), *Semiproductus* sp. 1, *Kahlella* sp., *Acanthatia* (?) sp. 1, *Mesoplica praelonga* (Sowerby 1840), *Spinocarinifera* aff. *lotzi* (Paeckelmann 1931), *Spinocarinifera* aff. *inflata* (Sokolskaya 1948) and *Ericiata* cf. *chonetiformis* (Krestovnikov & Karpychev 1948); by the rhynchonellids *Centrorhynchus* sp. cf. *lucida* (Veevers 1959), *Paurogastroderhynchus lakahalensis* Brice et al. 2005 and *Megalopterorhynchus* sp.; and by the spiriferids *Cyrtospirifer pseudorigauxia* Nicollin 2005 in Brice et al. 2005, *Cyrtospirifer* sp. 1 aff. *leboeufensis* Greiner 1957, *Cyrtospirifer* sp. 2 aff. *oleanensis* Greiner 1957, *Cyrtospirifer* sp. 3 aff. *warrenensis* Greiner 1957, *Cyrtospirifer* sp. 4, *Sphenospira* cf. *julii* (Dehée 1929), *Dichospirifer zemoulensis* Nicollin & Brice 2001, *Prospira struniana* (Gosselet 1879), *Prospira* cf. *struniana* and *Eobrachythyris hollardi* Brice & Nicollin 2000.

In the Zemoul area, central southern Morocco, in sections at Rich el Bergat, Dfeif and Jfeirat, the Lower Fauna in Tazout 1 correlatives is characterized by the productids *Whidbornella* cf. *pauli radiata* (Paeckelmann 1931), *Kahlella* sp., *Acanthatia* (?) sp. 1, *Steinhagella* cf. *membranacea* (Phillips 1841), *Mesoplica praelonga*, *Semiproductus* sp. 1, *Spinocarinifera* aff. *lotzi*, and *Ericiata* cf. *chonetiformis;* by the rhynchonellids *Centrorhynchus* sp. cf *lucida*, *Paurogastroderhynchus lakahalensis*; and by the spiriferids *Cyrtospirifer pseudorigauxia*, *Cyrtospirifer* sp. 1 aff. *leboeufensis*, *Sphenospira* cf. *julii*, *Dichospirifer zemoulensis*, *Prospira struniana*, *Prospira* sp. 2, and *Hollardospirifer draensis* Nicollin 2005 in Brice et al. 2005.

In the Gourara, Timimoun area of Algeria, in section at Gara el Kahla, the Lower Kahla Sandstone, about 100 m above a *Gonioclymenia* and *Prionoceras* horizon, is characterized by the Lower Fauna (Fig. 4). It contains the productids *Hamlingella* sp. 1, *Whidbornella* cf. *pauli radiata*, *Kahlella meyendorffi* Legrand-Blain 1995, *Kahlella* sp., *Steinhagella* cf. *membranacea*, *Mesoplica praelonga* and *Mesoplica* (*s.l.*) *nigeraeformis* (Martynova 1961); the rhynchonellids *Centrorhynchus* sp. cf. *lucida*,

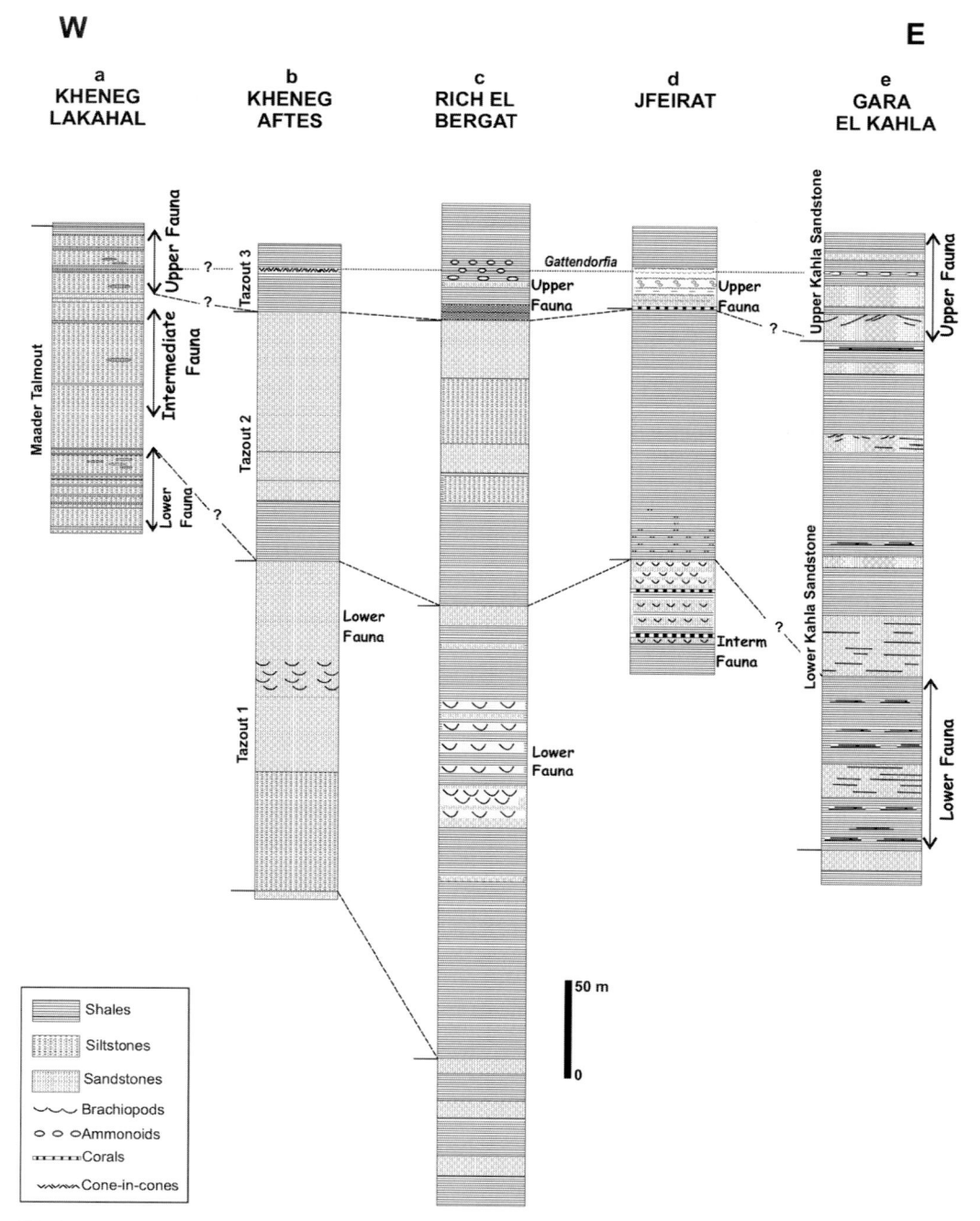

Fig. 2. Lithological correlations between southern Morocco (a–d) and western Algeria (e) sections, from west to east. (**a**) Kheneg Lakahal after Kaiser *et al.* 2004. (**b**) Kheneg Aftes (50 km E of Assa). (**c**) Rich el Bergat. (**d**) Jfeirat after Hollard 1971. (**e**) Gara el Kahla after Conrad & Termier 1985. *Notes*: The *Gattendorfia* gr. *crassa* level has been chosen as time reference line. The names of lithological units are indicated on the left part of the stratigraphic columns. The Lower, Intermediate and Upper Faunas are indicated on the right part of the stratigraphical columns when possible.

Fig. 3. Stratigraphical distribution of brachiopod faunas in the Kheneg Lakahal section, southern Morocco (after Kaiser et al. 2004). *Notes*: Lower, Intermediate and Upper Faunas have been distinguished. Conodont levels are indicated with * mark. Chonetids are present in the units +0 (SK 10, SK 14) and +1 (lower part BN +1). Ambocoeliids are present in the unit+1 (lower part BN +1).

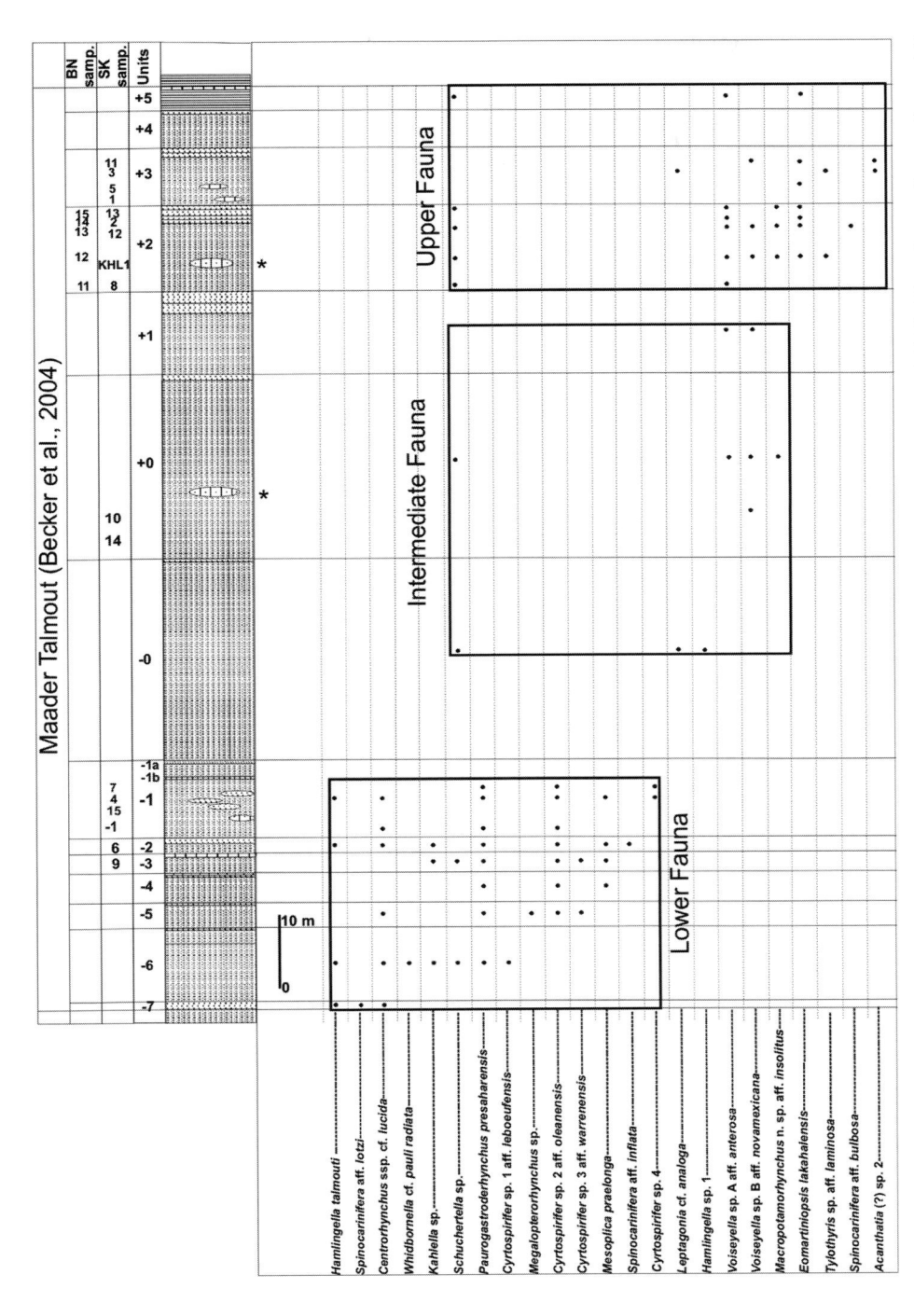

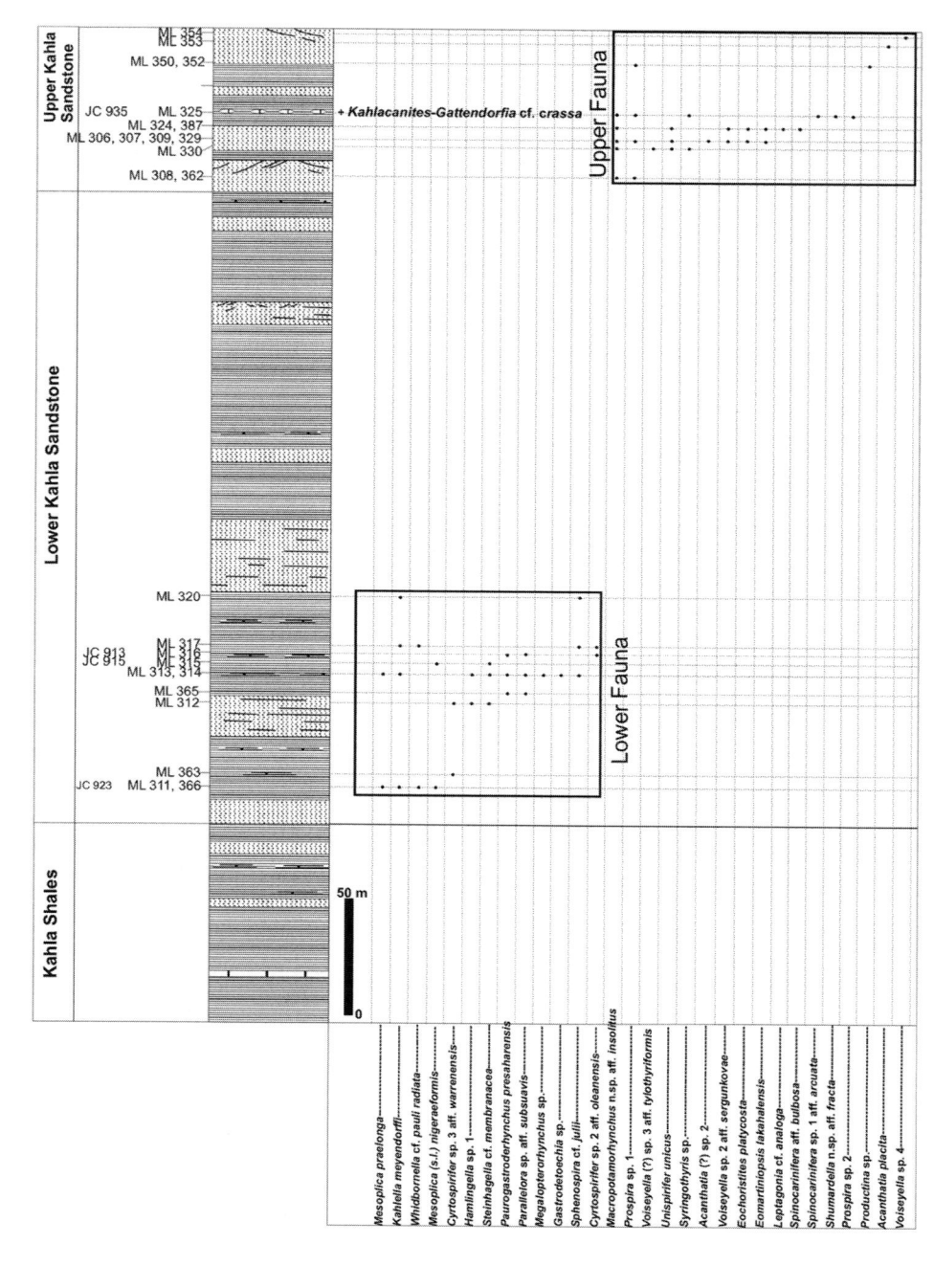

Fig. 4. Stratigraphical distribution of brachiopod faunas in the Gara el Kahla section, western Algeria. A Lower and Upper Faunas have been distinguished.

Paurogastroderhynchus lakahalensis, Megalopterorhynchus sp., and *Gastrodetoechia* sp.; and the spiriferids *Cyrtospirifer* sp. 2 aff. *oleanensis* Greiner 1957, *Sphenospira* cf. *julii*, and *Parallelora* aff. *subsuavis* (Plodowski 1968).

Comments on the Lower Fauna

In this fauna, there are index taxa, among them the productid *Mesoplica praelonga*, the types of which are from the Lower Pilton Formation of Devonshire (UK); the occurrence contains the LE to basal LN palynozones (O'Liathain 1993) (=*Wocklumeria* Zone, up to (?) basal *Acutimitoceras* (*Stockumites*) Zone). The genera *Mesoplica*, *Hamlingella* and *Steinhagella* are known only from the Upper Famennian (Brunton *et al.* 2000). The rhynchonellid genera *Paurogastroderhynchus*, *Megalopterorhynchus* and *Gastrodetoechia* are also only known from the Upper Famennian (Sartenaer 1975), however, the first genus characterizes a brachiopod zone in the *postera* and Lower *expansa* conodont Zones (Upper Famennian) in South Transcaucasus (Rzhonsnitskaya & Mamedov 2000); *Centrorhynchus* is known in the Upper/Uppermost Famennian of western Europa and Asia (Afghanistan and Iran). The Lower Fauna is also characterized by the last appearance of the spiriferid genus *Cyrtospirifer*.

Hollardospirifer draensis appears only in the Zemoul area. After Hollard (1981*a*), the upper part of the Dra Group where this species has been found mainly corresponds to open marine facies with goniatites and clymenides. This is the echo of the general transgression occurring at the end of the Famennian. Westwards, in the Assa area at the same period, the predominance of sandstones indicates much less open marine conditions.

Upper Fauna

The index taxa of the Lower Fauna are absent. New brachiopod faunas differ between western, central and eastern areas; a few brachiopods are also found in the Tafilalt-Ma'der northern (basinal) areas, above the 'Hangenberg Black Shale' equivalent.

In the Assa area, SW Morocco, the Upper Fauna occurs at Kheneg Lakahal section (Fig. 3) in the upper part of the Maader Talmout Member, probably equivalent to Tazout 2 Sandstone Formation (Becker, written comm. 2006). It is characterized by the strophomenids *Leptagonia* cf. *analoga* and *Schuchertella* sp.; by the productids, *Acanthatia* (?) sp. 2, *Spinocarinifera* aff. *bulbosa* (Havlicek 1984), by the rhynchonellid *Macropotamorhynchus* n. sp. aff. *insolitus* Carter 1987; and by the spiriferids *Voiseyella* sp. A aff. *anterosa* (Campbell 1957), *Voiseyella.* sp. B aff. *novamexicana* (Miller 1881), *Eomartiniopsis lakahalensis* Brice *et al.* 2005, *Unispirifer unicus* Havlicek 1984, and *Tylothyris* aff. *laminosa* (M'Coy 1844).

In the central area, in the Rich el Bergat, Dfeif and Jfeirat sections of the Zemoul area, the Upper Fauna is found in basal Tazout 3 beds, below the *Gattendorfia* horizon (Fig. 4). It is characterized by the productids *Semiproductus* (?) sp. 2 and *Spinocarinifera* sp. aff. *arcuata*; and by the spiriferids *Prospira* sp., *Eobrachythyris jacquemonti* Brice & Nicollin 2000, and *Eochoristites platycosta* (Havlicek 1984).

In Algeria, in the Gara el Kahla section of the Gourara Timimoun area, at the base of the Upper Kahla Sandstone but below and within the *Kahlacanites–Gattendorfia* horizon (Fig. 4), the Upper Fauna is characterized by the strophomenid *Leptagonia* cf. *analoga*; by the productids *Acanthatia* (?) sp. 2, *Productina* sp., *Spinocarinifera* aff. *bulbosa, Spinocarinifera* sp. 1 aff. *arcuata*; by the rhynchonellids *Macropotamorhynchus* n. sp. aff. *insolitus, Shumardella* n. sp. aff. *fracta* Carter 1988 (all of which occur in ferruginous sandstones above and below the *Kahlacanites–Gattendorfia* level); and by the spiriferids *Prospira* sp. 1, *Prospira* sp. 2, *Unispirifer unicus, Voiseyella* sp. 2 aff. *sergunkovae* (Bublichenko 1971), *Voiseyella* (?) sp. 3 aff. *tylothyriformis* (Krestovnikov & Karpychev 1948), *Eochoristites platycosta, Eomartiniopsis lakahalensis* and *Syringothyris* sp. indet.

In the north, in Tafilalt-Ma'der, only a few Upper Fauna brachiopods have been reported. At Bou Tlidat, Ma'der, spiriferids associated with the index ammonoid *Acutimitoceras* (*Stockumites*) belong to *Syringothyris* aff. *uralensis* Nalivkin 1975 and *Parallelora* (?) sp. At El Atrous, Tafilalt, the rhynchonellid *Centrorhynchus* (?) sp., a Devonian genus has been found in deltaic sandstones; it is followed by *Hemiplethorhynchus* (?) sp., of Tournaisian age.

Comments of the Upper Fauna

(1) In the Gara el Kahla section, Algeria, the basal Upper Kahla Sandstone, where the Upper Fauna appears in lumachellic sandy limestones, underlies the *Kahlacanites–Gattendorfia* horizon. That ammonoid association is a good marker of probable middle Early Tournaisian, above the main part of the Early Tournaisian (Ebbighausen *et al.* 2004); consequently, the appearance of the Upper Brachiopod Fauna indicates a transgressive episode, earliest Tournaisian in age.

(2) The Upper Fauna is more diverse compared to the Lower Fauna; index taxa are missing or differ between areas. For instance, the rhynchonellid genus *Macropotamorhynchus*, present in the Assa area, west southern Morocco and in the Gourara–Timimoun area in Algeria, is not recognized in

Zemoul (central southern Morocco) and Tafilalt areas of eastern Morocco. On the other hand, *Hemiplethorhynchus* (?) has only been identified from the Tafilalt area.

(3) Some productids (*Spinocarinifera* aff. *bulbosa*) and spiriferids (*Unispirifer unicus*, *Eochoristites platycosta*) are similar to taxa described from the Tournaisian of Libya (Havlicek 1984; Mergl & Massa 1992).

(4) The presence at Kheneg Lakahal section, of the rhynchonellid genus *Macropotamorhynchus* known from the lower Tournaisian in western and eastern Europe, Asia, North America and Australia, and of the spiriferid *Eomartiniopsis* known in Tournaisian or contemporary series in Eurasia and North America (Brice *et al.* 2005, p. 28), suggests a Tournaisian age, confirmed by the presence of diverse productids (Brice *et al.* 2005).

(5) *Syringothyris* is known in N. America from the upper Famennian (Carter 1988), but becomes ubiquitous during the Lower Carboniferous. In Russia, the genus is included in the list of earliest Carboniferous brachiopods appearing in the *Siphonodella sulcata* conodont Zone (Rzhonsnitskaya 1988, pp. 264, 268).

Intermediate Fauna

In the Assa area, SW Morocco, the Intermediate Fauna has been identified, mainly in units −0 to lower part of unit +2 of the Kheneg Lakahal section, as well as in the Akka area where the fauna is characterized by few taxa that are restricted to several levels. The taxa present at Kheneg Lakahal section are the strophomenids *Leptagonia* cf. *analoga* and *Schuchertella*, the productid *Hamlingella* sp. 1, the rhynchonellid *Macropotamorhynchus* n. sp. aff. *M. insolitus*, and the spiriferids *Voiseyella* sp. A aff. *anterosa* and *Voiseyella* sp. B aff. *novamexicana*. Sample +0.14 (unit + 0) of the Kheneg Lakahal section has yielded many poorly preserved chonetids, and units +1 and +2.10 has yielded abundant ambocoeliid spiriferids. The presence of *Hamlingella* in the lowermost level of the Intermediate Fauna at Kheneg Lakahal section suggests a latest Famennian age; however, rhynchonellid taxa of the lower fauna are missing (extinct ?) and the spiriferid genus *Voiseyella* appears for the first time. Moreover, the lower part of unit +0 (Fig. 3) yielded some conodonts giving a possible uppermost Famennian age (Kaiser *et al.* 2004, p. 97); the conodont fauna is diverse (Kaiser *et al.* 2004) but does not give a very precise age within the latest Famennian.

In the central area, in the northern Zemoul, the Intermediate Fauna is present at Dfeif in the lower part of upper Famennian VI Zone and at Sidi el Mouynir in Tazout sandstones. The few species present are mostly different from those of the Assa area, being *Semiproductus* sp. 1, *Kahlella* sp., *Steinhagella* cf. *membranacea*, *Ericiata* cf. *chonetiformis*, *Cyrtospirifer pseudorigauxia* and *Prospira struniana*.

In the eastern area, in the Timimoun area, the Intermediate Fauna has not been recognized in deltaic sandstones and shales of the Gara el Kahla section that contains only some ambocoeliid spiriferids.

Comments of the Intermediate Fauna

Among spiriferids of the Intermediate Fauna, we have pointed out:

- the occurrence of some species of *Voiseyella* in the Assa area; this genus is particularly known from the Carboniferous;
- the presence of *Prospira struniana* in the Zemoul area; this species is restricted to the latest Famennian (Strunian) in the type locality (Avesnois—France) (Brice *et al.* 2005), but in the Assa area it appears earlier (Nicollin & Brice 2001; Brice *et al.* 2005).

However, the Intermediate Fauna is much too different between the different areas to make comparisons possible.

Correlation between the three Faunas of southern Morocco and the Algerian Sahara

The correlation between the Lower Fauna from the Kheneg Lakahal (SW Morocco) and the Gara el Kahla (Algerian Sahara) sections is based on the presence of upper Famennian [IV(?)−V−lower VI(?) Zones] brachiopod taxa that they have in common, i.e. the strophomenid *Leptagonia* cf. *analoga*, the productids *Whidbornella* cf. *pauli radiata*, *Mesoplica praelonga* and *Steinhagella* cf. *membranacea*; the rhynchonellids *Paurogastroderhynchus*, *Megalopterorhynchus* and *Gastrodetoechia*; and the spiriferids *Cyrtospirifer* sp. 2 aff. *oleanensis*, and *Sphenospira* cf. *julii*. Among these, there are five index taxa: *Mesoplica praelonga*, *Paurogastroderhynchus*, *Megalopterorhynchus*, *Gastrodetoechia* and *Cyrtospirifer*.

Correlation is also possible between the Upper Fauna of the same sections, based on the presence of common taxa, such as the strophomenid *Leptagonia* cf. *analoga* (cf. Intermediate Fauna ?), the productids *Spinocarinifera* aff. *bulbosa*, *Acanthatia* (?) sp. 2, the rhynchonellid *Macropotamorhynchus* n. sp. aff. *fracta* and the spiriferids *Prospira* sp. 2, *Eochoristites platycosta* and *Eomartiniopsis lakahalensis*. The named productids, rhynchonellids and spiriferids are mostly known from the Lower Carboniferous in the Gara el Kahla section of the Algerian Sahara: occurring shortly before or associated with the *Kahlacanites–Gattendorfia* gr. *crassa* horizon, an important and precise marker above the main part of the lower Tournaisian.

The Intermediate Fauna is impoverished and not diverse in southern Morocco; it has no equivalent in the Gara el Kahla section of the Algeria Sahara where corresponding sediments are probably azoic.

Interpretation and conclusions

An important renewal of productid, rhynchonellid and spiriferid brachiopod faunas occurred in the southern Moroccan and Algerian Sahara around the Devonian–Carboniferous boundary. This renewal has been mainly documented in two sections, about 900 km apart, at Kheneg Lakahal, SE of Assa, and at Gara el Kahla, 30 km SW of Timimoun, where brachiopods were collected bed-by-bed.

We have recognized a Lower Fauna, upper Famennian in age [zones IV(?) – V – lower VI(?)], characterized by common markers (Figs 2 and 3) and present from west to east in the Assa, Akka, Zemoul areas of southern Morocco and in the Timimoun area, Algerian Sahara (but not in the NE part of southern Morocco, Ma'der-Tafilalt). We have also recognized an Upper Fauna, Tournaisian in age, that is well defined at Kheneg Lakahal (southern Morocco) and Gara el Kahla (Algerian Sahara) and characterized in both sections by common markers (Figs 2 and 3). In other areas, it is either absent or represented by often different taxa, as in the Ma'der-Tafilalt. Such differences can be partly explained by different environmental conditions. An Intermediate Fauna of latest Famennian (probable Strunian) age, impoverished in taxa, has been recognized only in southern Morocco in the Kheneg Lakahal section (Assa area) and in the Zemoul area but with distinct taxa. At Kheneg Lakahal, brachiopods are small-sized, restricted to thin intervals in siltstones facies and associated with ambocoeliid spiriferids. In the Gara el Kahla section, brachiopods are often absent or represented by a few ambocoeliids. All these features suggest unfavourable environmental conditions, such as might exist in a deltaic environment, for example.

These recognized brachiopod succession at the end of Devonian, whether caused by extinction or impoverishment, may be correlated with a global regression (cf. cycle 9, in Becker *et al.* 2004). This regression is more or less reflected in the lithologies, and may due to climatic variations, but tectonic movements can also be recognized in some sectors: they could explain the great variation of thickness between Kheneg Lakahal and Kheneg Aftes in the Assa area (cf. Fig. 2). The brachiopod succession may reflect the 'Hangenberg Event', characterized in pelagic facies by the sudden onset of black shales just before the Devonian–Carboniferous boundary, such as that locally present in Ma'der–Tafilalt, but still not recognized in SW Morocco.

In contrast, the great renewal of brachiopod faunas in the Lower Carboniferous is probably caused by or associated with a general transgression (the beginning of a new cycle, following cycle 9 in Becker *et al.* 2004).

In this paper, we tried to use available brachiopod faunas. However, faunas from southern Morocco were mainly sampled for cartographic purposes: they were not adapted or insufficient to illustrate the complete brachiopod evolution at the end of Devonian. This is much regrettable because brachiopods can be very useful for biostratigraphy in facies which are rarely favourable to conodonts or ammonoids. Moreover, important areas such as the Zemoul, where mixed neritic–pelatic sequences occur, are unfortunately not accessible today for security reasons.

We agree with Becker *et al.* (2004, p. 3) that 'research is still in an exploration stage' in this important area.

The authors are indebted to Henri Hollard†, R. Thomas Becker, Westfälische Wilhems-Universität, Münster, and Sandra Kaiser, Ruhr-Universität, Bochum, Germany, for the loan of their brachiopod collections, Agnès Rage, for providing access to Hollard's collection housed in the Museum National d'Histoire Naturelle (Paris). Many thanks to Peter von Bitter (Toronto) for reviewing and improving the English version of the manuscript before submission. The manuscript benefited from the reviews of Dr. Ma Xueping and R. Thomas Becker, as well as the editorial help of Bill Kirchgasser.

References

BECKER, R. T., JANSEN, U., PLODOWSKI, G., SCHINDLER, E., ABOUSSALAM, Z. S. & WEDDIGE, K. 2004. Devonian litho- and biostratigraphy of the Dra Valley area – an overview. *In*: EL HASSANI, A. (ed.) Devonian neritic–pelagic correlation and events in the Dra valley (western Anti-Atlas, Morocco). *Documents de l'Institut Scientifique*, Rabat, **19**, 3–18.

BELKA, Z., KLUG, C., KAUFMANN, B., KORN, D., DÖRING, S., FEIST, R. & WENDT, J. 1999. Devonian conodont and ammonoid succession of the eastern Tafilalt (Ouidane Chebbi section), Anti-Atlas, Morocco. *Acta Geologica Polonica*, **49**, 1–23.

BOCKWINCKEL, J. & EBBIGHAUSEN, V. 2006. A new ammonoid fauna from the *Gattendorfia-Eocanites* Genozone of the Anti-Atlas (Early Carboniferous; Morocco). *Fossil Record* **9**, 87–129.

BRICE, D., LEGRAND-BLAIN, M., NICOLLIN, J.-P., BECKER, R. T. & KAISER, S. 2004. Brachiopod biostratigraphy around the Devonian–Carboniferous boundary in Morocco and surrounding localities in Algerian Sahara. *In*: Devonian neritic–pelagic correlation and events. IUGS Subcommission on Devonian Stratigraphy, International Meeting on Stratigraphy, Rabat, Morocco, March 1–10, Abstracts, 15–18.

BRICE, D., LEGRAND-BLAIN, M. & NICOLLIN, J.-P. 2005. New data on Late Devonian and Lower Carboniferous Brachiopods from NW Sahara: Morocco, Algeria. *Annales Société Géologique du Nord*, **12** (2ème série), 1–45.

BRICE, D. & NICOLLIN, J.-P. 2000. *Eobrachythyris* Brice, 1971, an index genus (Spiriferid brachiopod) for the Late Devonian and Early Carboniferous in Southern Anti Atlas (N. Africa) North Gondwana. Travaux de l'Institut Scientifique, Rabat, *Série Géologie & Géographie Physique*, **20**, 57–68.

BROUSMICHE, C. 1975. Etude de quelques Productida (Brachiopoda) du Maroc présaharien. *Annales de Paléontologie*, **61**, 119–163.

BRUNTON, C. H. C., LAZAREV, S. S., GRANT, R. E. & JIN YU, Gan 2000. Productidina. *In*: KAESLER, R. L. (ed.) *Treatise on Invertebrate Palaeontology. Part H: Brachiopoda Revised*. The Geological Society of America & The University of Kansas, **3**, 424–609.

BUBLICHENKO, N. L. 1971. Brakhiopody nizhnego karbona Rudnogo Altaia (Tarkhanskaya svita). (Lower Carboniferous Brachiopoda of the Rudny Altai), **8**, 1–189. Alma-Ata.

CAMPBELL, K. S. W. 1957. A Lower Carboniferous Brachiopod-Coral Fauna from New South Wales. *Journal of Paleontology*, **31**, 34–98.

CARTER, J. L. 1988. Early Mississippian brachiopods from the Glen Park Formation of Illinois and Missouri. *Bulletin of Carnegie Museum of Natural History*, **27**, 1–82.

CAVAROC, V. V., PADGETT, G., STEPHENS, D. G., KANES, W. H., BOUDDA, A. & WOOLLEN, I. D. 1976. Late Paleozoic of the Tindouf Basin, North Africa. *Journal of Sedimentary Petrology*, **46**, 77–88.

CHOUBERT, G., DESTOMBES, J. & HOLLARD, H. 1971. Carte géologique des plaines du Dra au sud de l'Anti-Atlas central. Feuilles Agadir Tissint, Oued Zemoul au 1/200 000. *Notes et Mémoires du Service Géologique du Maroc*, **219**.

CHOUBERT, G., FAURE-MURET, A., DESTOMBES, J. & HOLLARD, H. 1969. Carte Géologique du Flanc Sud de l'Anti-Atlas occidental et des Plaines du Dra. Feuille Foum el Hassane-Assa au 1/200 000. *Notes et Mémoires du Service Géologique du Maroc*, **159**.

CONRAD, J., MASSA, D. & WEYANT, M. 1986. Late Devonian regression and Early Carboniferous transgression on the northern African platform. *Annales de la Société géologique de Belgique*, **109**, 113–122.

CONRAD, J. & TERMIER, G. 1970. Trilobites tournaisiens du Sahara nord-occidental et central. *Bulletin de la Société d'Histoire Naturelle de l'Afrique du Nord, Alger*, **60**, 67–79.

DEHEE, R. 1929. Description de la faune d'Etroeungt. Faune du passage du Dévonien au Carbonifère. *Mémoires de la Société géologique de France*, n. s., **5**, 1–6.

DESTOMBES, J. & HOLLARD, H., *ET AL*. 1986. Carte géologique du Maroc. Tafilalt–Taouz au 1/200 000. *Notes et Mémoires du Service Géologique du Maroc*, **244**.

DESTOMBES, J. & HOLLARD, H., *ET AL*. 1988. Carte géologique du Maroc. Todrha–Ma'der (Anti-Atlas oriental, zones axiale et périphérique Nord et Sud). au 1/200 000. *Notes et Mémoires du Service Géologique du Maroc*, **243**.

EBBIGHAUSEN, V., BOCKWINKEL, J., KORN, D. & WEYER, D. 2004. Early Tournaisian ammonoids from Timimoun (Gourara, Algeria). Mitteilungen aus dem Museum für Naturkunde, Berlin, *Geowissenschaftliche Reihe*, **7**, 133–152.

GOSSELET, J. 1879. Nouveaux documents pour l'étude du Famennien. Tranchées de chemin de fer entre Féron et Sémeries. Schistes de Sains. *Annales de la Société géologique du Nord*, **6**, 389–399.

GREINER, H. 1957. "*Spirifer disjunctus*": its evolution and paleoecology in the Catskill Delta. *Peabody Museum of Natural History Yale University, Bulletin*, **11**, 1–75.

HAVLICEK, V. 1984. Diagnoses of new brachiopod genera and species. Part 2. *In*: SEIDL, K. & RÖHLICH, P. (eds) *Explanatory booklet. Geological map of Libya, 1/250 000*, (NG 33-2), sheet Sabha, Industrial Research Centre, Tripoli, 63–67.

HOLLARD, H. 1968. Le Dévonien du Maroc et du Sahara nord-occidental. *In*: OSWALD, D. H. (ed.) *International Symposium on the Devonian System, 1967*, Alberta Society of Petroleum Geologists, Calgary, Alberta, **1**, 203–244.

HOLLARD, H. 1971. Sur la transgression dinantienne au Maroc présaharien. *Compte rendu 6ème Congrès International Stratigraphie Géologie Carbonifère*, Sheffield 1967, **3**, 923–936.

HOLLARD, H. 1981*a*. Principaux caractères des formations dévoniennes de l'Anti-Atlas. *Notes Service Géologique, Maroc*, **42**, 15–22.

HOLLARD, H. 1981*b*. Tableaux de corrélation du Silurien et du Dévonien de l'Anti-Atlas. *Notes Service Géologique, Maroc*, **42**, 23, 5 tabl.

KAISER, S. 2005. Mass extinctions, climatic and oceanographic changes at the Devonian/Carboniferous boundary. *Dissertation an der Fakultät für Geowissenschaften Bochum*, 1–156.

KAISER, S., BECKER, R. T., BRICE, D., NICOLLIN, J. P., LEGRAND-BLAIN, M., ABOUSSALAM, S. Z., EL HASSANI, A. & NÜBEL, H. 2004*a*. Sedimentary succession and neritic faunas around the Devonian–Carboniferous boundary at Kheneg Lakahal south of Assa (Dra Valley, SW Morocco). *In*: EL HASSANI, A. (ed.) *Devonian Neritic–Pelagic Correlation and Events in the Dra Valley (Western Anti-Atlas, Morocco)*. Documents de l'Institut Scientifique, Rabat, **19**, 69–74.

KAISER, S., BECKER, R. T. & STEUBER, T. 2004*b*. Sedimentology and sea-level change around the Devonian–Carboniferous boundary in southern Morocco. *In*: EL HASSANI (ed.) *Devonian Neritic–Pelagic Correlation and Events*; Subcommission on Devonian Stratigraphy International meeting, March 1–10, 2004. Abstracts, 71–72.

KORN, D. 1999. Famennian Ammonoid stratigraphy of the Ma'der and Tafilalt (Eastern Anti-Atlas, Morocco). *Abhandlungen Geologische Bundesanstalt*, **54**, 147–179.

KORN, D., KLUG, C., EBBIGHAUSEN, V. & BOCKWINKEL, J. 2002. Palaeogeographical meaning of a Middle Tournaisian ammonoid fauna from Morocco. *Geologica et Palaeontologica*, Marburg, **36**, 79–86.

KORN, D., BOCKWINKEL, J. & EBBIGHAUSEN, V. 2006. The Tournaisian and Viséan ammonoid stratigraphy in North Africa. *Kölner Forum für Geologie und Paläontologie* **15**, 70–71.

KRESTOVNIKOV, V. N. & KARPYSHEV, V. S. 1948. Fauna i stratigrafiia sloev Etroeungt reki Zigan (Iuzhnyi Ural). *Akademiia Nauk SSSR Geologicheskii Institut, Trudy*, **66**, 29–66.

LEGRAND-BLAIN, M. 1995a. Les Brachiopodes Productida au passage Dévonien-Carbonifère sur le craton nord-saharien. *118ème Congrès national des Sociétés historiques et scientifiques, Pau 1993, 4ème Colloque de Géologie africaine*, 425–444.

LEGRAND-BLAIN, M. 1995b. Relations entre les domaines d'Europe occidentale, d'Europe méridionale (Montagne Noire) et d'Afrique du Nord à la limite Dévonien-Carbonifère: les données des brachiopodes. *Bulletin Société belge de Géologie*, **103** (1–2 pour 1994), 77–97.

LEGRAND-BLAIN, M. 2002. Le Strunien et le Tournaisien au Sahara algérien: limites, échelles lithostratigraphiques et biostratigraphiques régionales. *Mémoires Service Géologique de l'Algérie*, **11**, 61–85.

MARTYNOVA, M. V. 1961. Stratigrafiia i brakhipody famenskogo iarusa zapadnoi chasti Tsentral'nogo Kazakhstana. (Stratigraphy and Brachiopods of the Famennian Stage of Western Central Kazakhstan). *Moskovsky Gosudarstviennyi Universitiet, Materialy po geologii centralnogo Kazakhstana*, **2**, 1–151.

MERGL, M. & MASSA, D. 1992. Devonian and lower carboniferous brachiopods and bivalves from Western Libya. *Biostratigraphie du Paléozoïque Université Claude Bernard—Lyon*, **1**, 1–117.

MILLER, S. A. 1881. Subcarboniferous fossils from the Lake-Valley Mining District of New Mexico, with descriptions of new species. *Cincinnati Society of Natural History, Journal*, **4**, 306–315.

NALIVKIN, D. V. 1975. Brakhiopody. *In*: STEPANOV, D. L. & GARANJ, I. M. (eds) *Paleontologicheskii Atlas Kamenno-ugol'nykh Otlozhenii Urala. Vsesoiuznyi Neftianoi Nauchno-Issledovatel'skii Geologo-Razvedo-chnyi Institut (VNIGRI)*, Trudy (Moscow), **383**, 154–203.

NICOLLIN, J.-P. & BRICE, D. 2001. Systematics, biostratigraphy and biogeography of four Famennian Spiriferid brachiopods from Morocco. *Geologica Belgica*, **3**, 173–189.

O'LIATHÀIN, M. 1993. Stratigraphic palynology of the Upper Devonian–Lower Carboniferous succession in North Devon, Southwest England. *Annales de la Société Géologique de Belgique*, **115**, 649–659.

PAECKELMANN, W. 1931. Die fauna des deutschen Unterkarbons. 2. Teil: Die Productinae und Productusaehnlichen Chonetinae. *Abhandlungen der Preussichen Geologischen Landesanstalt*, N. F. **136**, 1–442.

PHILLIPS, J. 1836. *Illustrations of the geology of Yorkshire. Part 2, The Mountain Limestone District*. John Murray, London, 1–253.

PHILLIPS, J. 1841. *Figures and Descriptions of the Palaeozoic Fossils of Cornwall, Devon and West Somerset*. Longman Brown, Green & Longmans, London, 1–231.

PLODOWSKI, G. 1968. Neue Spiriferen aus Afghanistan. *Senckenbergiana lethaea*, **49**, 251–257.

RZHONSNISTSKAYA, M. A. 1988. The Brachiopoda of the Devonian/Carboniferous boundary deposits on the USSR territory. *In*: GOLUBSOV, V. K. (ed.) *The Devonian Carboniferous Boundary at the Territory of the USSR*. Joint Stratigraphic committee of the USSR, Byelorussian Geological Research Institute, Minsk. *Nauka I Technica*, p. 262–271 [in Russian].

RZHONSNISTSKAYA, M. A. & MAMEDOV, A. B. 2000. Devonian stage boundaries in the southern Transcaucasus. *In*: BULTYNCK, P. (ed.) *Subcommission on Devonian Stratigraphy. Recognition of Devonian Series and Stages Boundaries in Geological Areas. Courier Forschungsinstitut Senckenberg*, **225**, 197–213.

SARTENAER, P. 1975. Rhynchonellides du Famennien supérieur du Sahara occidental (Algérie). *Bulletin de l'Institut Royal des Sciences Naturelles de Belgique, Bruxelles, Sciences de la Terre*, **51**, 1–12.

SOKOLSKAYA, A. N. 1948. Evolutsiya roda *Productella* Hall i smeknykh s nim form v Paleozoe podmoskovnoy kotloviny. (Evolution of the Genus *Productella* Hall and Related Forms in the Palaeozoic of the Moscow Region). *Trudy Paleontologicheskogo Instituta*, **14**, 1–167.

SOWERBY, J. DE, C. 1840. On the physical structure and older stratified deposits of Devonshire. *In*: SEDGWICK, A. & MURCHISON, R. I. (eds), *Transactions of the Geological Society of London*, (2nd series), **5**, 633–703.

VEEVERS, J. J. 1959. Devonian and Carboniferous Brachiopods from North-Western Australia. Bureau of Mineral Resources. *Bulletin of Australian Geology and Geophysics*, **55**, 34.

Index

Page numbers in italics, e.g. *153*, refer to figures. Page numbers in bold, e.g. **121**, signify entries in tables.

Acadian Orogeny 39, 43, 71, 72, 75, 76
Acadian Tectophase 70, 71
Achanarras Fish Bed level of Caithness 135
Achanarras horizon 133, *134*, 135
    external drainage *144*
    fish fauna 140–1
    Kernavė correlation 146–7
Achanarras lake 138–40, *141*
    calcium supply 142
    climate 139–40
    depth 138–9
    evaporation rates 142
    maximum extent 139, 142, 149
    outflow 143
    water budget 141–3, *143*
    water chemistry 142–3
Acritarchs 178, *180*, 248–9
    distribution 187–8
Agoniatites zone 90
Aguelmous Sandstones 263
Ahrerouch Formation 159, *160*
Ain Hallaouf Formation 248
Ain Jemaa, Morocco 237–60
    beds 240
    conodonts 238, 239, *242*, 253–5, *254*
    correlation with Rhenish *242*
    depositional environment 243–4, 252
    faunal succession 238–42, *241*
    goniatite assemblage 238–9, 240
    lithological succession *241*
    microfacies *244*
    outcrop photo *243*
    sections 239–42, 243–5
    sedimentary succession 237, 238–42
    stratigraphical dating 243–5
    taxa descriptions 253–5, *254*
    tectonic structure 252
    trilobites 245
Al Brijat Formation 250
Al-normalised elemental ratios 206, 208–9, *210*
Alden Pyrite Bed 110
Algeria 263
    brachiopod faunal changes 261–71
Alternating Gradient Force Magnetometer (AGFM 2900) 201
ammonoids 1, 2, 101, 158, 173, 237, 238, 263, 267
    bioseismometer 5
    biostratigraphy 87–90, 110–11
    extinctions 83
    pyritic 249
    zones 84
Anajdam, Morocco 197, 199, 202, *207*
    carbonate content *204*
    Crystallinity Index Standard (CIS) *203*
    magnetic hysteresis *205*, *206*
    magnetic susceptibility *204*
    stratigraphical column *200*

anetoceratids 27
Annulata Event *237, 239, 245, 249, 251*
anoxia 133, 173, 174, 186, 188, 211
    photic zone 191, 193
ansatus zone 109, *124*
Anti-Atlas, Morocco 157–72
    Anticlinorium 9, *10*
Aous Bel Fassi Formation 250
Appalachian Basin 39–79, 105, 106, *107*, 128
    biostratigraphy 39, 53, 75
    correlation of strata 39, 43–53, 75
    depositional cycles 65, 73–4, *74*, 76
    Eifelian deposition 71–3
    Eifelian palaeogeography *85*
    Emsian correlations 43–5, *44*
    Emsian deposition 70–1
    geography *42–3*
    geological setting 39–41
    Pragian deposition 70
    sea-level 73
    sequence stratigraphy 65–73, *69*
    stratigraphic nomenclature 53–8
    stratigraphic synthesis 53–65, *66–7*
    study localities *40*
Appohimchi Faunal subprovince 128
Aquetuck Member 45, *48–9*, 71
Ardennes, Belgium 9
Arkona Formation 109, 110, 111, 118
Arkona Shale *117*
aryl isoprenoids 173, 184, *185*, 193
    ratio 191, *191*
Assa Formation 16, 17–21, 31
    brachiopod faunas 17
    conodont faunas 17
    distribution 17
    tentaculitids 19
    trilobites 19
Australia, tetrapods 222, 226
australis zone 85–7, 99, 101
Azrou–Kenifra Basin 198

Baggy Sandstone 3
Bakoven Member *54–5, 56–8, 60–1*, 73, 85, 87, 90–1, 93, 98–9, 110
Baltic region 145
barite 209, 212
Beaverdam Member 45, *47*, 70
Beechwood Member 109, 111, 121, *124*
Belgium 9, 149, 186, 187
    tetrapods 220, 224–5, 228
Bell Shale 87, 93, 94, 98, 111, 114, 118
Bellepoint Member 50, *58*, 59, 65
bentonite *54, 56–7, 59, 60–1, 63–4*, 72, 83, 96, 176, 177
    K-bentonite event stratigraphy 90–1
    marker bed 50
    Sprout Brook 45, *46*
    see also Tioga K-bentonites

Bergisch Gladbach–Paffrath Syncline 157–72, *161, 163*, 186, 187
  brachiopod localities 161–2
  chronostratigraphy *164*
  faunal correlations 162–5
  goniatites 160–2, *168*
  lithostratigraphy *162*
Berkley Member 119, 121
Biar Setla Member 248
Bidwell Bed 123
biofacies
  depth range 126
  studies 212
biogenic productivity 197
biogeographical data 228
biomarker maturity indicator 190–1
biostratigraphy 5, 39, 41, 53, 75, 84–90, *86*, 108–12, 222–7, *224*, 265–6
bioturbation 97, 98, 112
Bir En-Nasr Formation 249
bispathodids 243, 252
bivalve biostratigraphy 111–12
Blasdell Bed 123
Blue Beds 116, 119, 121
Bobs Ridge Member *57*, 71
Bohemian subdivision 9
bone bed 94, 96
Bou Gzem Formation 237, 238, 240, *244*, 245
Bou Tlidat, Morocco 262
Bou-Ounebdou 197, 198–9, 202, *207*
  carbonate content *204*
  Crystallinity Index Standard (CIS) *203*
  magnetic hysteresis *205, 206*
  magnetic susceptibility *204*
  stratigraphical column *200*
brachiopods 1, 9, 17, *19, 20*, 21, **22–3**, 26–7, 29–30, 40, 93, 111, 114, 116, 119, 158–60, 161–2, 239, 250, 251
  biostratigraphy 111, *265–6*
  common taxa 268
  environmental conditions 269
  evolutionary lineages 11
  faunal changes 261–71
  Intermediate Fauna 261, 268, 269
  Lower Fauna 261, 263–7, 269
  marker 157
  Upper fauna 261, 267–8, 269
  zonation 41
Brint Road Member 119
Büchel Formation 157, 158, 160, 161
  Tornoceras 169
Büdesheimer Bach borehole 176, *177*
  geochemistry 180–4, **183**
  materials and methods 177–8
  palynofacies 178–80, *179, 180*
Büdesheimer Goniatitenschiefer Formation 176–7, 187
Butternut Member 111, 116, 118, 119, 121

C. cuspidatus 146
Cabrieroceras bed 51, 87
Caithness Flagstone Group 135
Cakhrat-ach-Chleh Formation 250
calcite 139
  diamagnetic contribution 206

Caledonian Orogen 135
Campine Basin, Belgium 186
Canowindra fish fauna 226–7
carbon
  isotopes 148, 176, 188, 192, 212
  total organic (TOC) 173, 180, *181*, 182, 192
carbonate 83, 125, 139, 180
  content 202, *204*
  flux 142–3
  production 142, 209–11, 212
Cardiff Member 92, 98, 114
Carlisle Centre Member *48*
Catskill Formation 223
Centerfield Member 109, 110, 111, 118, 121, 123–4
cephalopods 92, 261
Chabat El Baya Formation 249
Chenango Member 116, 118, 121
Cherry Valley Limestone 110, 112, 114
Cherry Valley Member 51, 53, *54–5, 60–3*, 73, 87, 91–2, 93, 97–9, 101
Chestnut Street Beds 51, 53, *60–2*, 87, 91, 93, 97–9, 101, 102, 110
China, tetrapods 222, 226
chitinozoans 180, 188–9
Chittenango Member 92, 98, 110, 114
Chotec bio event 75
chromatography
  gas (GC) 178
  medium-pressure liquid (MPLC) 178
chronostratigraphy 11, 31, 41, *164*
clastic input 206, *207*, 210
  synorogenic 39
clay minerals 200–2, 212
  diagenesis 202
  identification 200
Cole Hill submember 116, 118, 119
colour alteration index (CAI) 202
Columbus Formation 45, 50, *58*, 65, 72, 75, 84, 85, 90–1, 93, 97, 99
  and Onondaga Formation 58–65
  zones 94–6
condensed horizons 41, 91, 96, 112, 118, 252
Conodont Alteration Index (CAI) 174–6, 185
conodonts 1, 9, 11–17, *14–15, 20*, **22–3**, 25–7, 41, 84–7, 94, 101, 109–10, 147–8, 176, 237–9, *242*, 248, 251–5, *254*, 268
  apatite 197
  biostratigraphy 53, 109–10
  dating 144–5
  neodymium isotope composition 238
  zones 29, 53, *66–7*, 75, 83, *108*, 223, 224, 225, 226
coral 1, 41, 59, 72, 124
  see also reefs
Coral Marl Member 159, 160
costatus zone 84–5, 101
crepida zone 251
crinoidal marker bed 240
Crystallinity Index Standard (CIS) 200, 202, *203*, 209

dacryoconarid, nowakiid zonation 90
Dacryoconarid Events 90
Dalejian 27, 28, 29, 30
Dartmoor, UK, ammonoids 2
Dasberg Event 237, 243, 245

INDEX

Delaware Fauna 99–101
Delaware Formation 51, 72, 73, 84, 87, 90–2, 99, 111, 114–16
  zones 94–6
Delphi Station Member 116, 118, 119
depositional processes, allocyclic 128
desiccation polygons 137–8, *138*, 140
detrital input
  carbonate production 209–11, *211*
  and climate 212–13
  indicators 206
Devonian
  Carboniferous boundary 261–71
  global cooling 230
  land-sea interactions 133–55
  Late 219–35
  Lower 9–37, 39–79
  Middle 9–37, 39–79, 83–104, 105–30, 157–72
  T-R cycles 73, *74*, 98, 101
  tectonics 230
Devonian Correlation Table (DCT) 11, *20*
Devonian System, The 4
Dfeif, Morocco 262, 263, 267, 268
diagenesis 206, 209
Dinant Basin, Belgium 187
dinoflagellates 187
diparilis zone 124
dolomite 139
Douar Nahilat Formation 250
Dra Group 267
Dra Valley, Morocco 2, 9–37, 157–72, *163*
  brachiopods *20*, **22–3**, 158–60
  chronostratigraphy *164*
  conodonts *20*, **22–3**
  faunal correlations 165–6
  geological map *159*
  goniatites *168*
  map *158*
  section correlation *12–13*
  Uncites 168
drill cores 108
Dublin Member 97, 98, 99
Dublin Shale 91, 94, 96, 99
Dundee Formation 84, 85, 87, 92–3, 114
dysoxia 133, 186, 188, 189, 211

East Berne Member 53, *60–3*, 73, 75, 90, 92, 93, 98, 101, 114
ecological restructuring 83
ecological–evolutionary sub-units (EESUs) 83
economic deposits 41
Edgecliff Member *48*, 50, *54–5, 56–7*, 59, 65, 71–2, 74
Eifel Mountains 173–96
  faunal correlations 162
  lithostratigraphy *162*
Eifelian 9, 29, 31, 39–79
  bioevents 83–104
  depositional sequences 83–104
  nomenclature **68**
  terrestrial and marine 149–50
eiflius zone 87, 147
El Atrous, Morocco, section 263

Emsian 4, 9, 26, 27, 28, 29, 30, 31, 39–79
  cycles 45
  nomenclature **68**
  subdivision of 45
ensensis zone 84, 87, 116, 118, 147, 148
erosion 106, 197
Esopus Formation 39, 41, 45, *46–7*, 55–8, 65, 70, 74, 75
Estonia 133, 145
Eumetabolotoechia 111, 116
eutrophication 188, 192
Eversole Member *58*, 59
excavatus zone 25, 27–8
expansa zone 243, 245, 251
extinction 5, 99, 133, 148, 191, 269
  episode 252
  mass 1, 173–96, 197, 238, 262
  phases 174
  widespread 83

facies
  change 9, 114, 127
  control 109
  cross-correlation 1, 157
  lateral variation 31, 94
  vertical changes 31
falling stage systems tract 43, 65, 70, 71, 72–3, 112
falsiovalis zone 199
Famennian 4, 197–217, 237–60, 246
fauna
  carryover 99, *100*
  changes 39
  holdover 99, *100*
  migration 125
  provinciality 105–31
  turnover 84
Fimbrispirifer 111
Findlay Arch 106, 108, 125, 127, 128
  Plum Brook Shale 119
  Silica Shale 119–21
fish
  bones 72, 96
  fauna 146
  teeth 72, 91
flooding surface 70, 72, 73, 112
  maximum 43, 121, 123, 138
fluorescent microscopy 177
forebulge 105, 128
Forschungsinstitut Senckenberg 11
fossils 94, 96
  guide fossils 29, 109
  index species 165, 245, 267
  marker 1, 245
  micro marine 146
  teeth 72
  trace 71
Fouizir Member 248, 249
Foum-el-Mejez Formation 250
Frasnian 197–217
  mass extinction 173–96
  reef 3

gamma-ray logging 148
Gara el Kahla, Algeria 263, *264, 266*, 267, 268, 269

gastropod biostratigraphy 112
Gedinnian 11, 16
Geminospora lemurata 144
geochemical analysis 197–217
Givetian 4, 90
    brachiopod-goniatite correlation 157–72
Glenerie Formation 70
Global Stratotype Sections and Points (GSSP) 1, 9–11, 84
Gondwana 1, 237–8
goniatites 2, 9, 27, 30–1, 41, 116, 158, 160–2, *168*, 238–9, 240, 250
    zones 53, *66–7*, *108*, 157
Gosseletia 111
graben 135, 136, 139, 141
Greenland 135, 136
    tetrapods 220, 223
Gumaer Island Member 45, *48–9*, 71

Hadrophyllum Bed 96, 97, 99, 101
Halihan Hill Bed 53, 83, 90, 91, 92, 93, 97, 98, 99, 101, 109, 110, 114
Hamilton Fauna 83, 101, 114, 116
Hamilton Group 39, *60–1*, 83, 91, 98, *107*, 112, *113*, 114, 128, 150
    correlations 51–3, 105–31
Hangenberg Black Shale 244, 245, 252, 261, 262, 267
Hangenberg Event 1, 5, 237, 238, 250, 252, 269
Hangenberg Sandstone 245, 248, 249, 252
Hardangerfjord Shear Zone 149
Hassi Rharouar, Morocco 263
hassi zone 198
hemiansatus zone 87, 109, 116, 118
Hercynian facies 9
Hervey Group 226–7
Highland Boundary Fault 135, 136, 149
highstand system tract 43, 71, 72, 98, 102, 112, 116, 121, 244
Holy Cross Mountains, Poland 186, 187, 188, 189, 192, 193
hopane/moretane ratio 184, *184*, 185, 190, 192
Hornstein Member 157
House, Michael Robert 1–8, *2, 3*
Housean Pits 2
Hungry Hollow Formation 109, 111, 123–4
Hungry Hollow Limestone 118, 121
Hunsrück-Schiefer 27
Huntersville Chert 70
Huntersville Formation 45, *47, 49*, 50, *57*, 71, 75
Hurley Member 51, 53, *54–5, 60–4*, 73, 75, 87, 91, 98, 110, 114
hydrocarbons, aromatic 184
hydrogen index (HI) 173, 180, *181*, 189, 190, 192
hypoxia 174
hystricospores 186

IGCP projects *see* UNESCO-IGCP projects
Indiana
    Beechwood Member 124
    Silver Creek Member 116
    Swanville Member 121
inorganic geochemistry 201–2
    analysis 206–9

insolation 149, 150
    maximum 133
    seasonal 141
Institut Scientifique University Mohammed V, Rabat 1–2
International Subcommission on Devonian Stratigraphy (SDS) 1, 11, 157, 198, 222, 238
International Union of Geological Sciences (IUGS) 1
inversus zone 28, 29
Ipperwash Formation 111, 125
Ipperwash Limestone 121, 123
isotope record 1, 148
    see also carbon isotopes; oxygen isotopes
Ivy Point Member 121

Jaycox Member 121, 123
Jennabia–Babot Member 248
Jfeirat, Morocco 261, 262, 263, *264*, 267

Kačák Event 75, 83–104, 133–55, *150*
    -*otomari* Event 83, 90, 148
    correlation 101–2
    definitions of 148–9
Kahla Mudstone 263
Kahla Sandstone 263, 267
Kashong Member 111, 125
Kaskaskia Supersequence 65, 70, 76
Kellwasser Event 1, 5, 174
Kellwasser Horizons 173, 174, 176, 188, 191–3, 197, 198, 212
Kernavė Member 133, 145–6, 149
    age 147–8
    Archanarras correlation 146–7
    conodont assemblage 147–8
    palaeogeography 146
    palynomorphs *147*
Ketleri Formation 225
Kettle Point Formation 125
Khebchia Formation 29–30, 32
    age 30
    brachiopod fauna 29–30
    conodont fauna 29
    distribution 29
    facies 29
    goniatites 30
    ostracodes 30
    tentaculids 29
    trilobites 30
Kheneg Aftes, Morocco 262, 263, *264*
Kheneg Lakahal, Morocco 262, 263, *265*, 267, 268, 269
Khouribgá–Oulmès Zone 238
Khovanshchina Formation 225
kindlei zone 17, 21
Klondike Member *58*, 59, 65
kockelianus zone 87, 99, 101, 147, 148lacustrine cycles 136, 137, *137*, 145

lagoonal facies 158
Lahnstein substage 27
Latvia, tetrapods 220–2, 225
Laurussia, collision 237–8
Ledyard Member 110, 121, 123, 124
Leicester Pyrite Bed 111
Levanna Member 116, 118

Lewis Centre Bed 96, 97, 99
Light Hydrocarbon Preference Index (LHPI) 182
Lincoln Park Shale 110
Lincoln Park submember *60-1*, 91, 98
linguiformis zone 198
lithostratigraphy 5, *162, 241*
Lmhaïfid Formation 11-16, 31
  age 16
  brachiopod faunas 16
  conodont faunas 11-16, *14-15*
  distribution 11
  facies 11
  graptolites 16
  trilobites 16
Lochkovian 15, 17
Lomme Formation 149
Lower Eday Sandstone Formation 141
lowstand system tract 41, 70, 72, 244
Ludlowville Formation 105, *107*, 109, 110, 111, 112, 118, 121, *122*

Maader Talmout Member 263, 267
Maenioceras Zone 90, 167-9
magnetic hysteresis *210*, 212
  loops 201, *201*
  properties 205-6, *205, 206*
magnetic susceptibility 1, 148-9, 201, 202, *204*, 210, *210*, 212
magnetism, rock 197-217, 238
magnetite 201, 209
  low-coercivity 197, 205
Mahantango Formation *64*
Marblehead Member 97
Marcellus Formation 51, *57*, 72, 93
Marcellus Shale 58, 76, 83
Marcellus subgroup 39, 83, 91, 94, 98, *107*, 112
marginifera zone 244-5
Marietta Bed 118
marker beds 39, 50, 51, 91, 114, 126, 192, 240, 262
marker species 111, 165, 245, 267
Martenberg fauna 165
mass spectrometry 178
Mdâouer-el-Kbîr Formation 25-8, 31
  age 27-8
  brachiopod faunas 26-7
  conodont faunas 25-6, 27
  distribution 25
  facies 25
  goniatites 27
  tentaculititds 26, *28*
Menteth Member 125
Merrimbula Group 226
Merzâ-Akhsaï Formation 17, 21-5, 31
  age 25
  brachiopod faunas 21-5, **22-3**
  brachiopods *18-19*
  conodont faunas 21
  distribution 21
  facies 21
  trilobites 25
M'Garto Formation 249
Michigan Basin 106
Michigan Basin Faunal subprovince 128
Midland Valley 136

Milankovitch 5
  cyclicity 114, 135, 145, 149
  insolation 141
miospore 178, 191, 249
  composition and preservation 186-7
  data 229, 250, 252
  dinocyst ratio 187
  diversity 197
  marine phytoplankton ratio 186-7
  source proximity 186-7
  tetrads 186
  zones 223, 225
M'Karig, Morocco 263
Moonshine Falls Bed 121, 123
Moorehouse Member 50, *54-5, 56-8*, 59, 65, 72, 75, 90, 98
Moroccan Institutions 11
Moroccan Meseta 237-60
  Ben Slimane Zone 250
  Coastal-Block 250
  comparison of units 245-52, *246*
  Eastern Meseta 251
  El Hammam Zone 245
  Famennian successions *246*
  Fes area 251-2
  geological map *239*
  Jebilet 250
  Khatouat 248-9
  Mdakra area 249-50
  Oued Cherrat Zone 250
  Oulmés area 252
  Rabat-Tiflet Zone 245-8
  Rhemna 250
  Sidi Bettache Basin 248
  Tiliouine area 245-8
  Ziar-Azrou Zone 256
Morocco 2
  brachiopod faunal changes 261-71
  Frasnian-Famennian boundary 197-217
  geological setting 198-9, 261-3
  structural subdivision *199*
Moscow Formation 105, 109, 110-11, 112, 124-5
Mottville Member 98, 110, 111, 116, 118, 119, 121
Moulay Hassane Formation 245
Murder Creek Bed 123

Nadsnezha Beds 225
Nedrow Member 50, *54-5, 56-8*, 65, 71-2, 85
  black beds 59, 72
Needmore Formation 45, *47, 49*, 50, *55, 56*, 70, 71, 72
neritic-pelagic correlation 9-37, 157, 261
Neu-Oos Cypridinenschiefer Formation 177, 193
New Albany Shale 111, 124
New York
  Ludlowville Formation 121, *122*
  Moscow Formation 124-5
  Oatka Creek Formation 114
  Skaneateles Formation 116-18
  stratigraphy *88*, 91-2
North America
  ammonoid biostratigraphy 87-90, 110-11
  basal datum 108-9
  biogeography 128
  biostratigraphy *86*, 108

North America (*Continued*)
  bivalve biostratigraphy 111–12
  brachiopod biostratigraphy 111
  conodont biostratigraphy 84–7, 109–10
  dacryoconarid biostratigraphy 90
  Eifelian sequences 83–104
  faunal comparisons 99–101, *100*, 128
  faunal migration 105
  gastropod biostratigraphy 112
  geological setting 106–8
  Middle Devonian 105–31
  regional stratigraphy 91–8
  sequence stratigraphy 98–9
  time-rock chart *126*
North Vernon Limestone 98
Norway 136, 149

Oatka Creek-Mount Marion Formation 39, 41, 51, *54–5, 60–4*, 65, 73, 76, 83–4, 87, 90–2, 94, 98, 105, 109–12, 114–16
  units thickness *115*
Ohio
  Bloomville 96–7
  Delaware Formation 114–16
  Prout Formation 123–4
  Sandusky 97–8
  stratigraphy 94–8, *95*
Old Red Sandstone 134, 144
Olentangy Formation 110, 111
Olentangy Shale 96, 108, 119
Onondaga Fauna 99, 101
Onondaga Formation 39, 41, 45–50, *52–8*, 58–65, 71, 72, 75, 83, 84, 90, 93, 98, 99
Ontario, Canada 128
  Arkona Formation 118
  Bell Shale 114
  Ipperwash Beds 125
  London–Sarnia area 93
  Niagra Peninsula-Selkirk area 93
  stratigraphy *89*, 92–4
  Widder Formation 121–3, *122*
Oos Plattenkalk Formation 176, 187, 188, 190, 191, 193
Orcadia Formation 135
Orcadian Basin 133, 134–6, 149, *150*
  carbonate 139
  development 141
  external drainage 149–50
  map of *134*
  palaeogeography *144*
organic matter 209
  amorphous (AOM) 173, *179*, 185–6, 191
  preservation 173–96
  soluble (SOM) 182
  total (TOM) 190
Oriskany Formation 55, 65, 70, 74, 76
Oriskany Sandstone 39
Orkney succession *145*
Orthoceras Limestone 15, 17
ostracodes 1, 30–1, 173
Otisco Member 121
otomari Event *see* Kačák Event
Otsego Member 53
Oued Mzerreb Member 157, 159

Oued-el-Mdâouer Formation 16–17, 31
  age 17
  conodont faunas 16–17
  facies 16
  thickness 16
  trilobites 17
Oui-n'-Mesdoûr Formation 28–9, 31–2
  age 28
  brachiopods 28
  conodont faunas 28
  distribution 28
  facies 28
  goniatites 28
  tentaculitids 28
Ourthe valley, Belgium 186
oxygen isotopes 197, 230

palaeo-oxygenation indexes 212
palaeomagnetic data 238
palaeoredox proxies 181, *182*, **183**, 189, 206–8, *207, 208*
palmatolepids 243, 245, 252
palynofacies analysis 185–9
palynological samples 177, 223
palynomorphs 1, 192
Paper Mill Bed 116, 118, 119
Paracyclas 111, 116
Paraspirifer 111
Parastringocephalus, locality and description 166–7
Pavari, Latvia 225
Pennsylvania, tetrapods 220, 223
Peppermill Gulf Bed 118, 119
Petrolia Member 124
Pilton Formation 267
pireneae zone 21
Plattenkalk Formation 157, 158, 160, 162
playa lake 135
Plum Brook Formation 87, 109, 110, 111
Plum Brook Shale 98, 108, 119, *120*, 121
Pole Bridge Bed 116, 118, 119
Polygnathus eiflius zone 108
Pompey Member 110, 116, 118, 119
Portland Point Member 125
Portwood Member 111
Pragian 9, 11, 16–17, 25, 39–79
  nomenclature **68**
prasinophytes 173, 178–80, *180*, 188, 191, 192
precipitation/evaporation ratio 140
pristane/phytane ratio 181, 189–90, *190, 191*
productivity
  palaeoproductivity proxies 208–9
  primary productivity evolution 211–12
Prototethys 1, 238
Prout Formation 109–10, 111, 123–4, 125
Prout Limestone 118, 119
Prüm Syncline 173, 174–6, 188, 193
  geological setting 176–7
Ptchopteria 111
pumilo Beds 165
Purcell Limestone 90
Purcell Member *64*, 73
pyrite *54*, 72, 110–11, 118, 121, 123, 125, 186
  fauna 118, 119, 249
  paramagnetic 209

Quarry Hill Member 45, 46–7, 70

redox conditions 208, 211–12
    see also palaeoredox proxies
reefs *3*, 160, 161, 245–8
    cyanobacterial calcimicrobe algae 173
regressive surface, maximum 43
regressive system tract 73–5, 112, 116, 118, 269
reworking 109–10, 118, 124, 248, 251
    syn-sedimentary 250
Rhamphodopsis threiplandi 146
Rheic Ocean 150, *150*
Rheinisches Schiefergebirge, Germany 9
rhenanus zone 109, 110, 124, 198
Rhenish Massif 157–72, *242*
    faunal correlations 165
    outcrop pattern *175*
Rhenish Slate Mountains *161*
Rhenish subdivision 9, 11, 17
Rich el Bergat, Morocco 261, 262, 263, *264*, 267
Rock Glen Member 123
Rockport Quarry Formation 118
Rogers City Formation 93, 99
Russia, tetrapods 220–2, 225–6

Sahara
    brachiopod faunas 261–71
    lithological correlations *264*
    study locations 261, *262*
Sandusky Limestone 94
Sandwick Fish Bed, Orkney 136, 137–8, *137, 138*
    age 143–6
    laminations 139, *140*
Saugerties Member 45, *48–9*, 71
Sbskro Formation 83
Scaat Craig Beds 223–4
Schoharie Formation 45, *48–9*, 58, 59, 65, 70, 71, 74, 75
scolecodonts 180
    fossilization potential 189
    post mortem transport 189
Scotland, tetrapods 220, 223–4
sea-level
    changes 1, 5, 210, *211*
    curve 39, *113*
    Euramerican 39, *74*
    eustatic 1, 5, 39, 71, 73, 105–31, 133, 192
    fall 192, 245, 248, 251, 252
    palaeogeographical implications 127–8
    relative 125–6
    rise 101, 197
    sedimentological implications 126–7
    trends 39, 73, 157
    unconformities 41
sediments
    cyclic 5, 39, 41, 50, 59, 105, 125–6, 136–7, 145
    fluvial 138
    geochemistry 189–91
    maturity 185, 191
    organic rich 197
Selinsgrove Member *55, 56*, 72
Sellersburg Formation 108, 109, 111, 112, 121, 124

Seneca Member 50, *54–5, 56–8, 60–1*, 72, 75, 90, 98
sequence stratigraphy 1, 5, 39–79, *69, 98–9*, 105
    3rd & 4th order 112
    boundaries 112
    foreland basin correlation 112–14
    parasequences 41, 72, 114
Sequiet Abbes Formation 248
serotinus zone 26
Shriver Formation 70
Sidi el Mouynir 268
Siegenian 11, 18, 21, 25, 27
Silica Formation 109, 110, 111, 119–21
    Blue Beds 116, 119, 121
Silica Shale 119–21, *120*
Silver Creek Member 111, 112, 116, 121
Skaneateles Formation 87, 91, 92, 94, 105, *107*, 109–12, 116–21, *120*
    thinning of *117*
Slate Run 91, 94, 96
Solsville Member 98
Souk Jemaa Formation 249
Spafford Member 121, 123
Spawn Hollow Member 45, *46–7*, 70
Spinocyrtia 111, **112**
spores 144, *145*
    zone assemblage 146, 223
    see also miospore
Stafford-Mottville Limestone 91
Stafford-Mottville Member 94, 116
starvation surface 112, 124, 127
    maximum 43, 73, 98
sterane proportion *184*, 185, 190, 193
sterane/hopane ratio 173, 182–4, *184*, 190, 191, 192
Stony Hollow Fauna 90, 91, 99–101, 102, 116
Stony Hollow Member 53, *60–4*, 73, 83, 98, 99
Stony Point Beds 123
Stratford Member 94, 97, 98
stratigraphy *89*, 92–4
    event 237–60
    high resolution 39, 75
    markers 45
    nomenclature 53–8, **68**, 76
    regional correlation 237–60
    synthesis 39–79
    use of subdivisions 9–11
    see also biostratigraphy; chronostratigraphy; lithostratigraphy; sequence stratigraphy
Stringocephalus 157–9, 160, *163*
    locality and description 166
stromatolites 137
Stromness Group 135, 137, *137*, 145, 148
Stroud Bed *55, 56, 58*, 65, 72
styliolinids 173
Subsinucephalus, locality and description 166
sulcatus zone 17
sulphur, total (TS) 180
Swanville Member 98, 109, 121

Ta'arraft Formation 237, 245, 248
Tafilalt-Ma'der basins 261
Taghanic Unconformity 110
taxa 145
taxonomy 145, 161, 166–70
Tazout Group 261–2

Tazout Sandstone Formation 263, 267, 268
tectonics 105–31, 245–8, 269
    erosion 250
    synsedimentary 237–60
    timing of 237
Ten Mile Creek Dolostone 125
Ten Mile Creek Formation 109, 111, 123–4
tentaculitids 9, 26, *28*, 28, 30, 173
tetrapods 220, 223, 224–5, 228
    ages 227
    biostratigraphical subdivisions 222–7, *224*
    cluster analysis 228
    distribution of 222–7, *224*
    diversification of 219–35
    environment 219
    localities 219, 220, **221**, 222, *229*
    localities/taxa matrix 228
    migration route 229–30
    morphologies 219
    origin of 219
    palaeobiogeographical context 228–30, *229*
    palaeoenvironment 227–8
    phylogenetic relationships 229
Thedford Member 123
Thermal Alteration Index 174, 185
Tichenor Limestone 124–5
Tichenor Member 111
timorensis zone 98, 109, 119, 123, 124
Timrhanrhart Formation 30–1, 32
    age 31
    brachiopods 30–1
    conodont facies 30
    distribution 30
    facies 30
    goniatites 30–1
    ostracodes 30–1
    tentaculids 30
    trilobites 30–1
Tioga K-bentonites 50, 51, *52–3, 57*, 59, 65, 90, 91, 96
Tippecanoe Supersequence 65
Tornoceras 161
    locality and description 169, *170*
Tornquist Zone 149
trace element ratios 189, 202
transgressive system tract 41, 59, 65, 71, 72–5, 98, 102, 112, 114, 116, 119, 125, 126–7, 192, 244
Traverse Group 108
Trevoneites paffrathensis 169–70

triangularis zone 198, 199
trilobites 1, 16, 30–1, 173, 238, 245
trinorneohopane ratio 184, *184*, 190–1
Tully Formation 110, 111
Turkey Ridge Member *60–1, 64*, 73
Turkey Ridge Sandstone 51

Uncites 157–8, 159, 160, *163, 164*
    locality and description 167
unconformity 41–3, *66–7*, 70, 72, 73, 127
    erosive 71
    origin 41
    recognition 41
    subaerial 43
UNESCO-IGCP projects 11, 198
Union Springs Formation 39, 41, 50, 51, *54–8*, 58, *60–4*, 65, 72, 75, 76, 83, 84, 90–2, 98, 99, 112
    units thickness *115*
Ussher Society 4, 5

varcus zone 123
Variscan Orogeny 198, 237, 252
Venice Member 93, 94, 97, 98, 99
vent communities 251
vitrinite reflectance 174–6, *175*, 185

Walden Cliffs Bed 123
Wallbridge Unconformity 39, 45, 65–70, 71, 76
Wanakah Member 110, 121, 123, 124
water chemistry 142
weathering 209, 212
    chemical 188
    intensity 205, 206
Widder Formation 110, 111, 121–3, *122*, 125
Wildcat Valley Formation 45
Wiltwyck Member 45, *46–7*, 70
Windom Member 110–11, 125

X-ray diffraction (XRD) 200, 202
xenolites 251

Zhongning Formation 226
zircon U-Pb dating 174, 227
Zlíchov Formation 27, 28
Zlichovian stage 25–9